Lecture Notes in Computer Science 8421

Commenced Publication in 1973
Founding and Former Series Editors:
Gerhard Goos, Juris Hartmanis, and Jan van Leeuwen

Sourav S. Bhowmick   Curtis E. Dyreson
Christian S. Jensen   Mong Li Lee
Agus Muliantara   Bernhard Thalheim (Eds.)

Database Systems for Advanced Applications

19th International Conference, DASFAA 2014
Bali, Indonesia, April 21-24, 2014
Proceedings, Part I

 Springer

Volume Editors

Sourav S. Bhowmick
Nanyang Technological University, Singapore
E-mail: assourav@ntu.edu.sg

Curtis E. Dyreson
Utah State University, Logan, UT, USA
E-mail: curtis.dyreson@usu.edu

Christian S. Jensen
Aalborg University, Denmark
E-mail: csj@cs.aau.dk

Mong Li Lee
National University of Singapore, Singapore
E-mail: leeml@comp.nus.edu.sg

Agus Muliantara
Udayana University, Badung, Indonesia
E-mail: muliantara@cs.unud.ac.id

Bernhard Thalheim
Christian-Albrechts-Universität zu Kiel, Germany
E-mail: thalheim@is.informatik.uni-kiel.de

ISSN 0302-9743   e-ISSN 1611-3349
ISBN 978-3-319-05809-2   e-ISBN 978-3-319-05810-8
DOI 10.1007/978-3-319-05810-8
Springer Cham Heidelberg New York Dordrecht London

Library of Congress Control Number: 2014934170

LNCS Sublibrary: SL 3 – Information Systems and Application, incl. Internet/Web
and HCI

Typesetting: Camera-ready by author, data conversion by Scientific Publishing Services, Chennai, India

Printed on acid-free paper

Springer is part of Springer Science+Business Media (www.springer.com)

Preface

It is our great pleasure to present to you the proceedings of the 19th International Conference on Database Systems for Advanced Applications, DASFAA 2014, which was held in Bali, Indonesia. DASFAA is a well-established international conference series that provides a forum for technical presentations and discussions among researchers, developers, and users from academia, business, and industry in the general areas of database systems, web information systems, and their applications.

The call for papers attracted 257 research paper submissions with authors from 29 countries. After a comprehensive review process, where each paper received at least three reviews, the Program Committee accepted 62 of these, yielding a 24% acceptance rate. The reviewers were as geographically diverse as the authors, working in industry and academia in 27 countries. Measures aimed at ensuring the integrity of the review process were put in place. Both the authors and the reviewers were asked to identify potential conflicts of interest, and papers for which a conflict was discovered during the review process were rejected. In addition, care was taken to ensure diversity in the assignment of reviewers to papers. This year's technical program featured two new aspects: an audience voting scheme for selecting the best paper, and poster presentation of all accepted papers.

The conference program includes the presentations of four industrial papers selected from thirteen submissions by the Industrial Program Committee chaired by Yoshiharu Ishikawa (Nagoya University, Japan) and Ming Hua (Facebook Inc., USA), and it includes six demo presentations selected from twelve submissions by the Demo Program Committee chaired by Feida Zhu (Singapore Management University, Singapore) and Ada Fu (Chinese University of Hong Kong, China).

The proceedings also includes an extended abstract of the invited keynote lecture by the internationally known researcher David Maier (Portland State University, USA). The tutorial chairs, Byron Choi (Hong Kong Baptist University, China) and Sanjay Madria (Missouri University of Science and Technology, USA), organized three exciting tutorials: "Similarity-based analytics for trajectory data: theory, algorithms and applications" by Kai Zheng (University of Queensland, Australia), "Graph Mining Approaches: From Main memory to Map/reduce" by Sharma Chakravarthy (The University of Texas at Arlington, USA), and "Crowdsourced Algorithms in Data Management" by Dongwon Lee (Penn State University, USA). The panel chairs, Seung-won Hwang (Pohang University of Science and Technology, South Korea) and Xiaofang Zhou (University of Queensland, Australia), organized a stimulating panel on database systems for new hardware platforms chaired by Aoying Zhou (East China Normal University, China). This rich and attractive conference program of DASFAA 2014 is accompanied by two volumes of Springer's *Lecture Notes in Computer Science* series.

Beyond the main conference, Shuigeng Zhou (Fudan University, China), Wook-Shin Han (Pohang University of Science and Technology, South Korea), and Ngurah Agus Sanjaya, (Universitas Udayana, Indonesia), who chaired the Workshop Committee, accepted five exciting workshops: the Second International DAS-FAA Workshop on Big Data Management and Analytics, BDMA; the Third International Workshop on Data Management for Emerging Network Infrastructure, DaMEN; the Third International Workshop on Spatial Information Modeling, Management and Mining, SIM³; the Second International Workshop on Social Media Mining, Retrieval and Recommendation Technologies, SMR; and the DASFAA Workshop on Uncertain and Crowdsourced Data, UnCrowd. The workshop papers are included in a separate proceedings volume also published by Springer in its *Lecture Notes in Computer Science* series.

The conference would not have been possible without the support and hard work of many colleagues. We would like to express our gratitude to the honorary conference chairs, Tok Wang Ling (National University of Singapore, Singapore) and Zainal Hasibuan (University of Indonesia, Indonesia), for their valuable advice on many aspects of organizing the conference. Our special thanks also go to the DASFAA Steering Committee for its leadership and encouragement. We are also grateful to the following individuals for their contributions to making the conference a success:

- General Chairs - Stéphane Bressan (National University of Singapore, Singapore) and Mirna Adriani (University of Indonesia, Indonesia)
- Publicity Chairs - Toshiyuki Amagasa (University of Tsukuba, Japan), Feifei Li (University of Utah, USA), and Ruli Manurung (University of Indonesia, Indonesia)
- Local Chairs - Made Agus Setiawan and I. Made Widiartha (Universitas Udayana, Indonesia)
- Web Chair - Thomas Kister (National University of Singapore, Singapore)
- Registration Chair - Indra Budi (University of Indonesia, Indonesia)
- Best Paper Committee Chairs - Weiyi Meng (Binghamton University, USA), Divy Agrawal (University of California at Santa Barbara, USA), and Jayant Haritsa (Indian Institute of Science, India)
- Finance Chairs - Mong Li Lee (National University of Singapore, Singapore) and Muhammad Hilman (University of Indonesia, Indonesia)
- Publication Chairs - Bernhard Thalheim (Christian-Albrechts-University, Germany), Mong Li Lee (National University of Singapore, Singapore) and Agus Muliantara (Universitas Udayana, Indonesia)
- Steering Committee Liaison - Rao Kotagiri (University of Melbourne, Australia)

Our heartfelt thanks go to the Program Committee members and external reviewers. We know that they are all highly skilled scientists with many demands on their time, and we greatly appreciate their efforts devoted to the timely and careful reviewing of all submitted manuscripts. We also thank all authors for submitting their papers to the conference. Finally, we thank all other individuals who helped make the conference program attractive and the conference successful.

April 2014

Sourav S. Bhowmick
Curtis E. Dyreson
Christian S. Jensen

Organization

Honorary Conference Co-Chairs

Tok Wang Ling   National University of Singapore, Singapore
Zainal Hasibuan   University of Indonesia, Indonesia

Conference General Co-Chairs

Stéphane Bressan   National University of Singapore, Singapore
Mirna Adriani   University of Indonesia, Indonesia

Program Committee Co-Chairs

Sourav S. Bhowmick   Nanyang Technological University, Singapore
Curtis E. Dyreson   Utah State University, USA
Christian S. Jensen   Aalborg University, Denmark

Workshop Co-Chairs

Shuigeng Zhou   Fudan University, China
Wook-Shin Han   POSTECH, South Korea
Ngurah Agus Sanjaya   University Udayana, Indonesia

Tutorial Co-Chairs

Byron Choi   Hong Kong Baptist University, Hong Kong
Sanjay Madria   Missouri University of Science & Technology,
     USA

Panel Co-Chairs

Seung-won Hwang   POSTECH, South Korea
Xiaofang Zhou   University of Queensland, Australia

Demo Co-Chairs

Feida Zhu   Singapore Management University, Singapore
Ada Fu   Chinese University of Hong Kong, Hong Kong

Industrial Co-Chairs

Yoshiharu Ishikawa   Nagoya University, Japan
Ming Hua   Facebook Inc., USA

Best Paper Committee Co-Chairs

Weiyi Meng   Binghamton University, USA
Divy Agrawal   University of California Santa Barbara, USA
Jayant Haritsa   IISc, India

Steering Committee Liaison

R. Kotagiri   University of Melbourne, Australia

Publicity Co-Chairs

Toshiyuki Amagasa   University of Tsukuba, Japan
Feifei Li   University of Utah, USA
Ruli Manurung   University of Indonesia, Indonesia

Publication Co-Chairs

Bernhard Thalheim   Christian-Albrechts-University, Kiel, Germany
Mong Li Lee   National University of Singapore, Singapore
Agus Muliantara   University Udayana, Indonesia

Finance Co-Chairs

Mong Li Lee   National University of Singapore, Singapore
Muhammad Hilman   University of Indonesia, Indonesia

Registration Chairs

Indra Budi   University of Indonesia, Indonesia

Local Co-Chairs

Made Agus Setiawan   University Udayana, Indonesia
I. Made Widiartha   University Udayana, Indonesia

Web Chair

Thomas Kister   National University of Singapore, Singapore

Research Track Program Committee

Nikolaus Augsten	University of Salzburg, Austria
Srikanta Bedathur	IIIT Delhi, India
Ladjel Bellatreche	Poitiers University, France
Boualem Benatallah	University of New South Wales, Australia
Bishwaranjan Bhattacharjee	IBM Research Lab, USA
Cui Bin	Peking University, China
Athman Bouguettaya	CSIRO, Australia
Seluk Candan	Arizona State University, USA
Marco A. Casanova	Pontifícia Universidade Católica do Rio de Janeiro, Brazil
Sharma Chakravarthy	University of Texas Arlington, USA
Chee Yong Chan	National University of Singapore, Singapore
Jae Woo Chang	Chonbuk National University, South Korea
Sanjay Chawla	University of Sydney, Australia
Lei Chen	Hong Kong University of Science & Technology, Hong Kong
James Cheng	Chinese University of Hong Kong, Hong Kong
Reynold Cheng	University of Hong Kong, Hong Kong
Gao Cong	Nanyang Technological University, Singapore
Sudipto Das	Microsoft Research, USA
Khuzaima Daudjee	University of Waterloo, Canada
Prasad Deshpande	IBM India, India
Gill Dobbie	University of Auckland, New Zealand
Eduard C. Dragut	Purdue University, USA
Cristina Dutra de Aguiar Ciferri	Universidade de São Paulo, Brazil
Sameh Elnikety	Microsoft, USA
Johann Gamper	Free University of Bozen-Bolzano, Italy
Shahram Ghandeharizadeh	University of Southern California, USA
Gabriel Ghinita	University of Massachusetts Boston, USA
Le Gruenwald	University of Oklahoma, USA
Chenjuan Guo	Aarhus University, Denmark
Li Guoliang	Tsinghua University, China
Ralf Hartmut Gting	University of Hagen, Germany
Takahiro Hara	Osaka University, Japan
Haibo Hu	Hong Kong Baptist University, Hong Kong
Mizuho Iwaihara	Waseda University, Japan
Adam Jatowt	Kyoto University, Japan
Panos Kalnis	King Abdullah University of Science & Technology, Saudi Arabia
Kamal Karlapalem	IIIT Hyderabad, India
Panos Karras	Rutgers University, USA
Norio Katayama	National Institute of Informatics, Japan
Sangwook Kim	Hanyang University, South Korea

Hiroyuki Kitagawa	University of Tsukuba, Japan
Jae-Gil Lee	KAIST, South Korea
Sang-Goo Lee	Seoul National University, South Korea
Wang-Chien Lee	University of Pennsylvania, USA
Hou U. Leong	University of Macau, China
Ulf Leser	Humboldt University Berlin, Germany
Hui Li	Xidian University, China
Lipyeow Lim	University of Hawai, USA
Xuemin Lin	University of New South Wales, Australia
Sebastian Link	University of Auckland, New Zealand
Bin Liu	NEC Lab, USA
Changbin Liu	AT & T, USA
Boon Thau Loo	University of Pennsylvania, USA
Jiaheng Lu	Renmin University, China
Qiong Luo	Hong Kong University of Science & Technology, Hong Kong
Matteo Magnani	Uppsala University, Sweden
Nikos Mamoulis	University of Hong Kong, Hong Kong
Sharad Mehrotra	University of California Irvine, USA
Marco Mesiti	University of Milan, Italy
Prasenjit Mitra	Penn State University, USA
Yasuhiko Morimoto	Hiroshima University, Japan
Miyuki Nakano	University of Tokyo, Japan
Wolfgang Nejdl	University of Hannover, Germany
Wilfred Ng	Hong Kong University of Science & Technology, Hong Kong
Makoto Onizuka	NTT Cyber Space Laboratories, Japan
Stavros Papadoupoulos	Hong Kong University of Science & Technology, Hong Kong
Stefano Paraboschi	Università degli Studi di Bergamo, Italy
Sanghyun Park	Yonsei University, South Korea
Dhaval Patel	IIT Rourkee, India
Torben Bach Pedersen	Aalborg University, Denmark
Jian Pei	Simon Fraser University, Canada
Jeff Phillips	University of Utah, USA
Evaggelia Pitoura	University of Ioannina, Greece
Pascal Poncelet	Université Montpellier 2, France
Maya Ramanath	IIT New Delhi, India
Uwe Röhm	University of Sydney, Australia
Sherif Sakr	University of New South Wales, Australia
Kai-Uwe Sattler	Ilmenau University of Technology, Germany
Markus Scheider	University of Florida, USA
Thomas Seidl	Aachen University, Germany
Atsuhiro Takasu	National Institute of Informatics, Japan
Kian-Lee Tan	National University of Singapore, Singapore

Nan Tang	Qatar Computing Research Institute, Qatar
Dimitri Theodoratos	New Jersey Institute of Technology, USA
Wolf Tilo-Balke	University of Hannover, Germany
Hanghang Tong	CUNY, USA
Kristian Torp	Aalborg University, Denmark
Vincent Tseng	National Cheng Kung University, Taiwan
Vasilis Vassalos	Athens University of Economics and Business, Greece
Stratis Viglas	University of Edinburgh, UK
Wei Wang	University of New South Wales, Australia
Raymond Wong	Hong Kong University of Science & Technology, Hong Kong
Huayu Wu	Institute for Infocomm Research, Singapore
Yinghui Wu	University of California at Santa Barbara, USA
Xiaokui Xiao	Nanyang Technological University, Singapore
Jianliang Xu	Hong Kong Baptist University, Hong Kong
Bin Yang	Aarhus University, Denmark
Man-Lung Yiu	Hong Kong Polytechnic University, Hong Kong
Haruo Yokota	Tokyo Institute of Technology, Japan
Xike Xie	Aalborg University, Denmark
Jeffrey Xu Yu	Chinese University of Hong Kong, Hong Kong
Aoying Zhou	East China Normal University, China
Wenchao Zhou	Georgetown University, USA
Roger Zimmermann	National University of Singapore, Singapore

Industrial Track Program Committee

Alfredo Cuzzocrea	ICAR-CNR and Unversity of Calabria, Italy
Yi Han	National University of Defense Technology, China
Kaname Harumoto	Osaka University, Japan
Jun Miyazaki	Tokyo Institute of Technology, Japan
Yang-Sae Moon	Kangwon National University, South Korea
Chiemi Watanabe	University of Tsukuba, Japan
Kyoung-Gu Woo	Samsung Advanced Institute of Technology, South Korea
Chuan Xiao	Nagoya University, Japan
Ying Yan	Microsoft Research, Asia, China
Bin Yao	Shanghai Jiaotong University, China

Demonstration Program Committee

Palakorn Achananuparp	Singapore Management University, Singapore
Jing Gao	University at Buffalo, USA

Yunjun Gao   Zhejiang University, China
Manish Gupta   Microsoft Bing Research, India
Hady Lauw   Singapore Management University, Singapore
Victor Lee   John Carroll University, USA
Zhenhui Li   Penn State University, USA
Siyuan Liu   Carnige Mellon University, USA
Weining Qian   East China Normal University, China
Victor Sheng   University of Central Arkansas, USA
Aixin Sun   Nanyang Technological University, Singapore
Yizhou Sun   Northeastern University, USA
Jianshu Weng   Accenture Analytics Innovation Center,
                           Singapore
Tim Weninger   University of Notre Dame, USA
Yinghui Wu   University of California at Santa Barbara, USA
Peixiang Zhao   Florida State University, USA

External Reviewers

Ibrahim Abdelaziz   Soumyava Das
Ehab Abdelhamid   Ananya Dass
Yeonchan Ahn   Jiang Di
Cem Aksoy   Aggeliki Dimitriou
Amin Allam   Lars Döhling
Yoshitaka Arahori   Philip Driessen
Nikolaos Armenatzoglou   Ines Faerber
Sumita Barahmand   Zoé Faget
Christian Beecks   Qiong Fang
Brigitte Boden   Xing Feng
Selma Bouarar   Sergey Fries
Ahcene Boukorca   Chuancong Gao
Sebastian Breß   Ming Gao
Yilun Cai   Azadeh Ghari-Neat
Yuanzhe Cai   Gihyun Gong
Jose Calvo-Villagran   Koki Hamada
Mustafa Canim   Marwan Hassani
Brice Chardin   Sven Helmer
Wei Chen   Silu Huang
Sean Chester   Fuad Jamour
Ricardo Rodrigues Ciferri Min-Hee Jang
Xu Cui   Stéphane Jean

Minhao Jiang
Salil Joshi
Akshar Kaul
Georgios Kellaris
Selma Khouri
Jaemyung Kim
Henning Koehler
Hardy Kremer
Longbin Lai
Thuy Ngoc Le
Sang-Chul Lee
Hui Li
John Liagouris
Wenxin Liang
Xumin Liu
Cheng Long
Yi Lu
Yu Ma
Zaki Malik
Xiangbo Mao
Joseph Mate
Jun Miyazaki
Basilisa Mvungi
Adrian Nicoara
Sungchan Park
Youngki Park
Paolo Perlasca
Peng Peng
Jianbin Qin
Lizhen Qu
Astrid Rheinländer
Avishek Saha
Shuo Shang
Jieming Shi
Juwei Shi
Masumi Shirakawa
Md. Anisuzzaman Siddique

Thiago Luís Lopes Siqueira
Guanting Tang
Yu Tang
Aditya Telang
Seran Uysal
Stefano Valtolina
Jan Vosecky
Sebastian Wandelt
Hao Wang
Shenlu Wang
Xiang Wang
Xiaoyang Wang
Yousuke Watanabe
Huanhuan Wu
Jianmin Wu
Xiaoying Wu
Fan Xia
Chen Xu
Yanyan Xu
Zhiqiang Xu
Mingqiang Xue
Da Yan
Shiyu Yang
Yu Yang
Zhen Ye
Jongheum Yeon
Adams Wei Yu
Kui Yu
Qi Yu
Chengyuan Zhang
Zhao Zhang
Zhou Zhao
Jingbo Zhou
Xiangmin Zhou
Linhong Zhu
Anca Zimmer
Andreas Zuefle

Table of Contents – Part I

Graph Data Management

Spatio-temporal Data Management

Database for Emerging Hardware

Data Mining

Probabilistic and Uncertain Data Management

Web and Social Data Management

Table of Contents – Part II

Data Mining

Spatio-temporal Data Management

Graph Data Management

Security, Privacy and Trust

Web and Social Data Management

Keyword Search

Data Stream Management

Data Quality

Industrial Papers

Demo Papers

Tutorials

Challenges for Dataset Search

David Maier, V.M. Megler, and Kristin Tufte

Computer Science Department, Portland State University
{maier,vmegler,tufte}@cs.pdx.edu

Abstract. Ranked search of datasets has emerged as a need as shared scientific archives grow in size and variety. Our own investigations have shown that IR-style, feature-based relevance scoring can be an effective tool for data discovery in scientific archives. However, maintaining interactive response times as archives scale will be a challenge. We report here on our exploration of performance techniques for Data Near Here, a dataset search service. We present a sample of results evaluating *filter-restart* techniques in our system, including two variations, *adaptive relaxation* and *contraction*. We then outline further directions for research in this domain.

Keywords: data discovery, querying scientific data, ranked search.

1 Introduction

In the past, most scientific data was "one touch": it was collected for a specific experiment or project, analyzed by one or a small number of investigators, then used no further. The advent of shared scientific repositories – where data is collected prospectively and used by multiple scientists – is bringing an era of "high-touch" data. Each observation is now used tens or hundreds of times. A premier example of high-touch data is the Sloan Digital Sky Survey (SDSS), where a systematic imaging of the night sky has produced more than 5000 publications.[1] Such shared repositories not only provide higher returns on the investment in data collection, they enable new scientific discoveries, because of the broader spatial and temporal coverage. However, simply providing a shared data does not guarantee high "touchability" on its own. In a repository such as SDSS, with a handful of kinds of data coming from a common instrument, it is relatively easy for a scientist to determine what data is available that might match his or her needs. But in other domains, such as environmental observation, the number and variety of datasets grows over time, as new sensors and platforms are developed and deployed more widely. The situation can reach the point where no one person knows the full extent of the holdings of the repository. Scientists have a harder time finding relevant datasets for their work, and there is a true danger that increased growth in a repository actually decreases the touchability of data.

While there is a wide array of software tools for analysis and visualization of scientific data, there are fewer systems supporting data discovery. We heard this complaint

---

[1] http://www.sdss3.org/science/publications.php

S.S. Bhowmick et al. (Eds.): DASFAA 2014, Part I, LNCS 8421, pp. 1–15, 2014.

from colleagues and visitors at the NSF-sponsored Center for Coastal Margin Observation and Prediction (CMOP). While CMOP provides interfaces, similar to those in other archives, to navigate its data holdings, their effective use requires prior knowledge of the kinds of data and their organization. Furthermore, those interfaces do not support finding datasets with particularly properties, such as "high salinity and low pH." Inspired by their needs, we have been investigating search tools to help locate datasets relevant to their information needs. We have adopted an approach to ranked search of datasets based on traditional Information Retrieval architectures, such as are used in web search. We prototyped our ideas in a service called Data Near Here (DNH), which we have deployed over the CMOP observation archive. Most searches execute quickly, and initial reaction is good, based on user studies with scientists, but we are concerned with scaling our approach as the archive expands.

We report here on some of the performance techniques we have examined, including a sampling of our evaluation results. We find that the *filter-restart* technique from top-k query processing can be effective in many cases. We have obtained further improvements by adjusting the cutoff threshold based on partial search results. The adjustment can be both downward between stages (*adaptive relaxation*) and upward within a stage (*contraction*). This work suggests many other avenues of investigation for scaling dataset search, some of which we describe in our final section.

2   Description of Data Near Here

At the high level, our architecture follows the search architecture used by most Internet search engines today. The adaptation to dataset search is shown in Fig. 1. The architecture consists of two sections, Asynchronous Indexing and Interactive Search, each of which communicates with a shared Metadata Catalog. Asynchronous Indexing consists of the *Crawl, Read* and *Extract Features* processes. The *Crawl* process identifies datasets to summarize. The *Read* process takes each dataset identified by the crawl process and attempts to read it. The *Extract Features* process asynchronously summarizes each read dataset into a set of features and stores the summary in a metadata catalog. The feature extraction process encapsulates dataset partitioning, hierarchy creation and feature extraction functions.

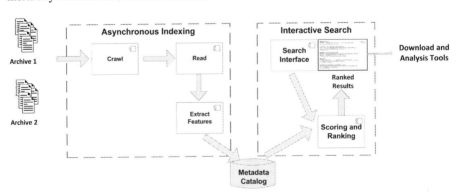

Fig. 1. High-level dataset search architecture

The primary processes in Interactive Search are the *Search Interface* and the *Scoring-and-Ranking* task. Searchers use the *Search Interface* to specify their search criteria. The Search Interface passes the search criteria to the Scoring-and-Ranking component, and displays the results returned. *Scoring-and-Ranking* accesses the metadata catalog, identifies the highest-ranked summaries for the given search criteria, and returns them to the Search Interface.

The *Metadata Catalog* stores the features and dataset summaries resulting from the asynchronous indexing and makes them available for use by other processes, such as the Scoring-and-Ranking process.

Our prototype implementation, Data Near Here [14, 15], has now been deployed for use by our scientists for over six months. The dataset summaries in DNH generally consist of the bounds of each variable, plus some file-level metadata. The one exception currently is latitude and longitude, which are jointly summarized as a geospatial footprint that is a geometry such as point, polyline or polygon. Search queries are similar to dataset summaries. DNH scores datasets against queries using a similarity function that considers distance and overlap between a dataset and the query, scaling different terms by the query extent. For example, if the query specifies a range of one day, then the previous and following day are considered a close match, whereas if the query specifies a month, the previous and following month are considered close.

Fig. 2 shows the primary user interface, which includes on a single webpage the search interface and ranked search results. The search interface is akin to the "advanced search" used by some text search engines, but adapted to scientific data.

Fig. 2. Search interface for "Data Near Here", showing a sample search for a geographic region (shown as a rectangle on the map) and date range, with temperature data in the range 6:11C. Result datasets (or subsets) are shown by markers and lines in the output pane, together with their relationship to the search region. In the ranked list of answers, no full matches for the search conditions were found; two partial matches to a search with time, space and a variable with limits are shown in the text panel, and the top 25 are shown on the map.

The search results are displayed as brief summaries of the datasets, akin to the "snippets" of a document shown in web search. Details of a dataset's metadata and contents can be viewed by clicking on an individual snippet. Because geospatial location is so important to our user base, the spatial extent of the search is shown on the map interface (the square white box), and the spatial extent of the datasets in the search results list are also shown on the map (here, diagonal lines and markers). However, our work is not limited to this kind of geospatial-temporal search.

One issue for scientists in finding relevant data is the mismatch between the scales of the data they seek, the scales of observation, and the partitioning of data for processing and storage. Multiple scientists might use the same datasets, but have very different scales of data of interest. Fig. 3 shows a dataset containing from a science cruise; the overall dataset is split into smaller segments representing specific sections of the cruise. Lynda, a microbiologist, wants data for a fairly short time period, since a different time in the tidal cycle might not reflect her sampling conditions. In Fig. 3, the most interesting portion of data for her is the cruise segment from July 28, 10-12 a.m., since it is closest in time and space to her time and location of interest. Another part of the cruise track passes her sampling location, but it is not close enough in time to be very relevant. Joel, an oceanographer, is looking for simultaneous low oxygen and lower pH; for him, the most relevant data is for the whole day of August 1 – a much larger portion of the dataset. Our metadata catalog supports such dataset hierarchies, and our search interface is aware of them. It considers datasets at all hierarchy levels for the results list, and may sometimes return both parent and child datasets. For example, it might return a single leg of a cruise as highly relevant to a query, while the full cruise might appear later in the list with lower overall relevance.

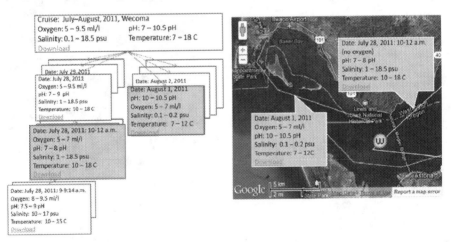

Fig. 3. Example of a dataset hierarchy: a dataset containing several million environmental observations (taken at 3 millisecond intervals) during a specific 2-month science cruise, segmented into a hierarchy. The white line on the map shows the cruise track, and the marker "w" shows the location of Lynda's water sample. The most detailed level is a single simplified segment or leg of a cruise, often covering part of an hour or a few hours; these segments are aggregated into successively longer and more complex cruise segments, and lastly into an entire cruise dataset. The most relevant portion to Lynda is shown shaded on the left in the hierarchy, while the most relevant portion to Joel is shown shaded on the right. From [13].

3   Performance Study

3.1   Filter-Restart

While most searches of the current CMOP archive take a few seconds, we are interested in understanding the effects of further growth on interactive response times. While replicating the catalog can handle larger numbers of users, the number of catalog entries could increase by a factor of 100 in the coming years, which could increase response times for searches. In our prototype, the search engine's tasks of performing the search and identifying the top-k results account for the majority of the elapsed time seen by the user, and thus are our focus. This section briefly presents some of techniques we have explored to improve search performance, based on pre-filtering and exploiting the dataset hierarchy. (We have also investigated what affect the choice of hierarchy has on search efficiency, but do not report those results here [13].)

We extend the basic algorithm we use to identify the top-k results for a search to add a cutoff filter. We further refine that method with relaxation – a technique that reduces the filter's cutoff score if too few results are returned and re-issues the search. This combination of a filter and relaxation is called filter-restart.

3.2   Background

Our goal is to maintain search response in an "interactive response" range, while we grow the size or complexity of the underlying metadata collections. One area we investigated is query evaluation techniques to improve top-k query performance [8]. We wondered if these techniques could be applied to our metadata catalogs containing hierarchical metadata.

Metadata Catalog versus Index. Classic Information Retrieval systems primarily achieve good performance by using inverted indexes, where each word in a document is treated as a key, and the documents in which the word occurs are stored as values. In contrast, our datasets may contain real numbers and searches are generally for ranges rather that for a specific value, so an inverted index is not an appropriate indexing technology. Our catalog is index-like; however, the hierarchical structure is not driven primarily by performance considerations; instead, the hierarchy structure carries semantic meaning. However, these semantics can help search. For example, we have defined a parent dataset as containing the bounds of all its children. If the closest edge of a parent dataset is "far" from our search, we know that no child dataset can be close; if a parent is completely inside our search area, all children will be, too. Where practical and effective, we would like to use indexing concepts with our metadata to improve performance.

We implemented our catalog in PostgreSQL with B-tree indexes for non-spatial data and R-trees from PostGIS. Our search queries use the database indexes to navigate our catalog entries. We expect the majority of catalog accesses to come from identifying the top-k similar entries to a search; but our similarity function does not easily translate into a query against an index. Rather than a request for a specific set or range of identified rows, we need the top-k entries subject to a function that considers the center and the end-points of the data range, relative to another range given query time. Thus, while index algorithms inform our work, search performance over our catalog is not directly predictable from index performance statistics. There are similarities between our approach and, for example, Hellerstein and Pfeffer's RD-trees ("Russian Doll" trees) [7]. An implementation of our system using RD-tree indexes might provide faster performance than the current combination of B-trees trees and R-trees, though perhaps not by the orders of magnitude we desire.

Several other scientific fields have developed specialty indexes for scientific data, with the most advanced work in the field of astronomy [10–12, 16, 17]. Unlike our case, these systems generally assume they are dealing with homogeneous data.

Top-k Evaluation Techniques. Our search engine latency is primarily driven the execution time of (possibly multiple) SQL queries to identify the top-k results. We could process the entire metadata collection, computing the score of each entry, and then sort the resulting list. This approach does not scale well, and the sort is a blocking operation. These issues have led to development of improved top-k techniques and algorithms. We wished to explore the use of such top-k techniques and adapt them to our setting. We examined the classification of top-k approaches by Ilyas et al. [8] and used their taxonomy dimensions to identify those that apply to our setting. In particular, for the query-model dimension, we use a top-k selection query; we have data and query certainty; our infrastructure provides both sorted and random data access methods; we desire an application-level technique; and we use only monotone ranking functions. However, most the techniques identified there had limitations relative to our requirements [13]. The most appropriate seemed to be Filter-Restart techniques such as that of Bruno et al. [3], which operate by applying range-selection queries to limit the number of entries retrieved. A selection query is formulated that only returns entries above an estimated cutoff threshold (the *filter*), and the retrieved entries are ranked. Note that if the threshold is underestimated, too many entries are retrieved and performance suffers. Conversely, it may be overestimated, in which case too few entries are retrieved, and the query must be re-formulated and re-issued with a lower threshold (the *restart*). Filter-restart techniques rely for their performance improvement on the filter reducing the number of entries returned without compromising the quality of the result.

Filter-Restart and Cutoff Scores. It has long been known that limiting the results returned from a query by using a filter can dramatically improve response times [4]. To reduce the number of rows returned from the database, we add a filter to the search to return a subset of the rows guaranteed to contain the top-k results, by applying a formula derived from our scoring formula and adding a cutoff threshold. To be useful, the filter must be selective, and the time to evaluate the filter and process the results should be less than just evaluating all the rows without the filter.

A challenge for the filter-restart class of techniques is determining how to initially set the cutoff threshold for a given top-k search. As Chaudhuri et al. noted [5], there is as yet no satisfactory answer to estimating result cardinalities and value distributions in order to identify an optimal cutoff value, leaving this estimation as an open area for research. Bruno et al. translate a top-k query into a range query for use with a traditional relational RDBMS [3]; they perform this translation for three monotonic distance functions, SUM, EUCLIDEAN DISTANCE and MAX. They use database catalog statistics to estimate the initial starting score, using certain distribution functions. Their experiments show that assuming uniformity and independence of distributions was computationally expensive and leads to many restarts [3]. They also note that "no strategy is consistently the best across data distributions. Moreover, even over the same data sets, which strategy works best for a query q sometimes depends on the specifics of q." For our primary use case (observational data), it is more common for attributes to be positively or negatively correlated than for them to be independent. Our data may also be strongly clustered, temporally, spatially, or on some observational variable. Thus, we expect to see issues similar to those found by other researchers in identifying the best cutoff value to use.

When an initial cutoff score fails to return sufficient matches (defined by a range such as $10 \le k \le 50$) , the search must be restarted with a revised cutoff score. This technique is often called *relaxation*, which is defined by Gaasterland as "generalizing a query to capture neighboring information" [6]. As with the initial cutoff score, we must identify a revised cutoff score to use for the revised search.

3.3 Basic Algorithm with Filter

Our basic algorithm first processes the given search terms to build an SQL query to retrieve entry information from the metadata tables. We then retrieve all roots from the database and score them. Any entry we deem likely to have "interesting" children we add to a list of parents; we then produce SQL to retrieve the children of parents in the list, and repeat the process at each next level of the hierarchy. When we get no rows returned from the database or we find no entries with likely interesting children at some level of the hierarchy, we terminate. Thus, rather than a single top-k query we issue an SQL query for (potentially) each level of any hierarchy within our forest of hierarchies. The scored entries are sorted and the top-k returned.

At first glance it may appear that we will score every entry in the catalog. If all entries in the catalog are roots (and therefore have no children), then a table scan is required. In contrast, if the ranges of the relevant variables in a parent entry are completely within the search term's range, we deem the children not interesting, and only return the parent. If the closest and furthest edges of the variable's range have the same score, we also deem the children not interesting, as the children's scores cannot be higher than the score of the closest edge of the variable. Again, only the parent entry is returned. In these cases, only a subset of the entire catalog's entries is scored.

The description so far is without the filter. The filter must operate over interval summaries and avoid false negatives. Therefore, for a particular cutoff score (*lowestScore*), our filter must return (at least) all entries that would achieve a higher score using our similarity measure. If insufficient matches are found from an application of the filter, we must restart the query with a "looser" filter. To implement the filter, we add terms to the SQL query to limit entries returned to:

- Those we estimate will have a final score for the entry that is higher than the cutoff score; that is, entries where the center of the data range has a score of *lowestScore* or higher. We may increase a summary's score if the data range overlaps the search term; however, we never decrease a score below that of the center of the data range. This filter term excludes entries that will not be adjusted upwards and that cannot have a score higher than *lowestScore*.

- Those whose children may have higher scores than the cutoff score; that is, entries that have children and where the closest edge to the search term has a score of *lowestScore* or higher. These parents may have a child with scores higher than the parent (if, for example, a child's bounds are near the parent's closest edge). If the closest edge has a score lower than *lowestScore*, not only will the parent not have a possible score higher than *lowestScore*, but none of its children will either; thus, "far" parents and their children are not retrieved.
- Entries that span at least one search term; or, where the closest edge has a score higher than *lowestScore*. These entries will have their scores adjusted upwards based on the amount of overlap with the search term. They also may have children whose bounds are completely within the search term, and thus are perfect matches for that term.

For geospatial search terms, we developed a slightly different filter, in order to reduce the number of heavyweight spatial calculations. In addition to storing the entry's shape, we add two columns to our data model. While building the metadata entry, one column is updated with the shape's centroid, and the other with the maximum radius of the shape, allowing us to quickly bound the minimum distance to the shape.

Since performance (in terms of elapsed time and resources used) is driven largely by the number of rows returned from the SQL query, applying the filter should improve performance. The first SQL query in any search only evaluates the root entries in the forest of trees. We access successive levels of the hierarchy (i.e., children) by parent_id, and the filter acts to reduce the number of child entries returned from each level. Our initial cutoff score is a default value that empirically works will for our archive. However, if the cutoff score is set too high to find k entries with equal or higher scores, insufficient (or no) rows will be returned from the initial walk through the hierarchy, and a restart is required.

3.4   Relaxation

If fewer than k entries are found with sufficiently high scores, we restart the search with a lower starting score, that is, we *relax* our filter. We developed and experimented with three relaxation techniques:

- Naïve: In our simplest technique, whenever we do not find sufficient results at our starting filter score, we reissue the query after having reduced the filter score by a fixed (arbitrary) number.
- Adaptive: Our second technique uses information calculated while processing the entries returned from a query to adjust (in our case, lower) the starting score for subsequent queries.
- Contraction: This technique modifies adaptive relaxation by reversing relaxation. That is, if we discover that we are retrieving more than the desired number of results using a specific filter score, we increase the score for subsequently issued SQL queries during the same search.

3.5   Performance Tests

Methods and Data. We tested filtering and the three types of relaxation: Naïve (N), Adaptive (AD), and Contraction (CN). We focus on the performance of a single server: if single-server performance is sufficient for the expected workload, no additional scaling approaches are needed, while if more servers are needed, the number will be defined largely by the single-server performance. To understand the performance characteristics of our approach apart from the (potentially confounding) impact of underlying disk hardware, we restrict ourselves to data sizes where the working set of the metadata catalog can fit in memory (2GB of buffer pools). We created two collections over which to run our tests. Both summarize real-world data from a non-CMOP archive of interest to our scientists. The smaller is approximately 5 times the size of CMOP's current collection, while the larger is 25 times the size. (We have also experimented with a dataset 200 times larger.) We implemented the search evaluation techniques described earlier, while retaining the current design of Data Near Here.

To test our techniques over different hierarchy "shapes", we developed 8 hierarchies, each with different aggregation characteristics. In addition, we keep a ninth "no hierarchy", with every leaf treated as a root. We developed three test suites of searches, using search terms that we know from experience to be frequently used by our scientist and with very different performance characteristics: time-only, space-only, and time-and-space. We report here some of our results with the time-only and space-only suites. Each query in a test suite was run 5 times against each target hierarchy. For each search, the maximum and minimum response times were discarded and the response times for the remaining three queries were averaged, as is common in database performance measurement. Whenever a different table or hierarchy was accessed, the first search was discarded, to force the relevant indexes (and, possibly, data) to be loaded into memory, simulating a "warm cache".

All tests were run on a 2 quad-core 2.13 GHz Intel Xeon system with 64 GB main memory, running Ubuntu 3.2.0 with Apache 2.2.22, PHP 5, PostgreSQL 9.1 and PostGIS 2.0. The PHP space limit per request was increased to 2 Gb, and the time limit to 1,000 seconds. PostgreSQL 9.1 only uses a single core per user request and the flow through the application is sequential; as a result, a search uses only one processor.

Test Collection. We scanned two sets of data from NOAA for use in the performance tests.[2] The first two are chlorophyll-a concentration data from NASA's Aqua Spacecraft[3], in two different formats: Science Quality (herein abbreviated to Collection s) and Experimental (herein referred to as Collection e). The data is gathered by the Moderate Resolution Imaging Spectroradiometer (MODIS). The data scanned differs in density (0.05° for Science Quality versus 0.0125° for Experimental), and has differences in the algorithms used to produce the data.

We created a configurable scanner that reads the subset of data for our geographic area of interest (the region between latitude 40 to 50, longitude -122 to -127), and creates a leaf dataset record for each (configurable) block. The blocks were configured

---

[2]   To reduce confusion, we will refer to the results of scanning each of these sets of data as a "catalog".

[3]   http://coastwatch.pfeg.noaa.gov/coastwatch/CWBrowser.jsp

to be 0.25° latitude by 0.25° longitude; each block is treated as a separate "dataset" from the perspective of the metadata collection. Where there are missing values for cloud cover and other reasons, the actual area of each block represented in the data may be substantially smaller than the nominal area. The scanner checks the physical locations of the data points reported, and represents the physical extent of the valid data by a convex hull of the valid data points. The leaf data counts for Collection s are 192,554 leaf records, with 962,770 entries in the variables table (5.5 times the size of our current catalog). Collection e has 930,261 leaf records, with 4,653,105 entries in the variables table, or 4.8 times the size of s.

Test Hierarchies. We developed 8 different hierarchies, and a configurable hierarchy creation program. The hierarchies represent different patterns of aggregation of space and time, with the number of levels varying from one (all entries are leaves) to 5 (root, three intermediate levels, and leaf). Each hierarchy was built over the s and the e collections. In the performance tables below, a designation "5e" means the e data with Hierarchy 5; similarly, "5s" means the s data with Hierarchy 5.

Search Suites. Since almost all searches include time or space or both search terms, we developed three search suites: one with queries containing only a time search term; one with queries with a space search term only; and one with queries containing both a time term and a space term. We present some results for the first two here.

Time search suite. This search suite consists of 77 searches. The searches range in duration from 1 day to a search spanning 44 years (1/1/1970 to 1/1/2014), which is longer than our entire temporal coverage. Some searches were defined that match exactly the various levels of the hierarchies; others are offset from and overlapping hierarchy groupings, but covering the same durations; and some are both longer and shorter than various hierarchy grouping levels.

Space search suite. This suite consists of 20 searches. The smallest search's spatial extent is 4.4 sq.Km, while the largest search spans 2 degrees latitude by 2 degrees longitude for an area of approximately 52,000 sq.Km., which is a bit more than 8% of our entire geographic coverage area. As with the time searches, spatial searches were created that aligned with hierarchy groupings, were offset from or overlapping hierarchy groupings, and were both larger and smaller than hierarchy grouping levels.

In all cases, we requested $10 \leq k \leq 50$, that is, a minimum number of 10 and a maximum of 50 entries to be returned from a search.

Time Search Suite Results. We first ran tests for the time suite to assess the benefit of the filter alone (without restart) [13]. Adding a filter without restart capabilities means that a search may not result in the minimum requested entries. For most of the hierarchies, adding the filter reduces the response time to one-fifth or less of the response time without the filter. For "None" and no filter, every search failed to complete (either exhausting PHP memory or time limits) for the e collection; when the filter was added, nearly four-fifths of the searches completed, although the mean response time for those queries was very high compared to the other hierarchies. For the other hierarchies, almost all searches for the e collection completed.

Space Search Suite Results. For consistency, we tested the space search suite against the same collections and hierarchies as used for the time search suite. Many of the "no filter" searches did not complete, so we do not include those results. We report results for the *e* collections while including one *s* collection (2s) for comparison. Of the 20 searches in the space suite, 7 resulted in more than one iteration using naïve relaxation; four searches had two iterations only. Three searches had more than 2 iterations; one search iterated 8 times before exiting and returning fewer than the minimum desired results.

For the searches with two iterations (that is, the searches that might be affected by AD), AD used the same final cutoff score as Naïve relaxation; this means that the revised cutoff calculated by AD after the first iteration was either higher than the "minimum cutoff adjustment", or the same as Naïve's cutoff score.

Table 1 summarizes the total number of iterations, average rows returned from SQL, the average number of matches returned from the searches, and response times for the three searches with more than two iterations. All response times are significantly longer than for the time-search suite, as the spatial comparisons used in the distance measure take more time to calculate than the numeric comparisons used for one-dimensional variables such as time. In every case, AD reduced the number of iterations and reduced the response time by more than 50%, even when the number of rows returned from the database (Avg. SQL Rows) was almost identical to those returned for N. This is because the *lowestScore* was adjusted down by more than naïve_reduction for all affected searches, and in most cases the search found sufficient matches during the first restart.

Table 1. Space Search Suite, N vs. AD: Summary of (Three) Searches with More Than Two Iterations. "# Iterations" is the total number of iterations across all searches in the suite.

Hierarchy	Naïve (N)				Adaptive (AD)			
	# Iterations	Avg. SQL Rows	Avg. Matches Found	Mean Response (s)	# of Iterations	Avg. SQL Rows	Avg. Matches Found	Mean Response (s)
1e	17	20,419	36	33.8±6.2	7	20,229	24	14.4±2.2
2s	19	7,688	33	14.2±3.5	6	7,444	15	4.5±0.2
2e	17	33,302	36	61.6±7.4	8	33,074	31	29.9±2.4
3e	17	22,505	35	31.5±5.3	7	21,754	24	13.2±2.9
5e	17	22,484	35	32.9±4.3	7	21,732	24	13.8±3.2
6e	17	22,984	35	32.9±4.3	7	22,232	24	13.8±3.2

Table 2 reports the average rows retrieved by SQL and response time for the subset of searches affected by CN for the space-search suite, when run against different hierarchies. As shown in Table 2, CN increased the final cutoff scores for 13 of the 20 space queries for hierarchy 1e, but only for between 4 and 7 searches over the other hierarchies. We see that CN results in lower mean response times and lower variability (lower standard deviations) for affected searches in every hierarchy tested. CN reduces the number of rows processed by approximately 20% to nearly 50%, while reducing response time by approximately 50% and reducing variability in all cases.

We conclude that for this search suite, the combination of the filter with adaptive relaxation and contraction provides the best results across all hierarchies (as it did for the other search suites [13]). While these techniques do not help all searches, they also do not negatively impact the searches that they do not improve. As multiple iterations tend to be associated with long response times, the searches with the longest response times tend to be the ones most improved.

Discussion. The overall goal in our performance work is to understand, improve and predict the performance characteristics of our approach. We showed that as the meta-data collection size increases, our filter allows us to complete searches that would otherwise time-out or exhaust memory. We have experimented with an additional, modified filter for space searches that may provide additional benefits by allowing use of spatial indexes (which are not used by our current spatial measure); however, the improvement achieved of approximately 1/3 is less than the order-of-magnitude needed to make a substantial difference to the scalability.

Over all search suites (including time-space), less than one third of the searches required restart. For the time search suite, restarts only occurred in one hierarchy. For searches that required restart, Adaptive (which estimates a revised cutoff score based on the matches found so far) reduced the number of restarts required by half or more over Naïve (which uses a fixed revision to the cutoff score). For affected searches, AD reduced response time by around half, and reduced the variability of response times across the searches within the search suite by around half as well. Thus, Adaptive relaxation achieves its goal of reducing latency over Naïve. For all suites, adding Contraction (successively increasing the cutoff threshold as the number of desired matches is reached) reduced the response time and variability; the improvement was around 20-25% for the affected searches. However, the improvement gained by adding Contraction was lower than the improvement of Adaptive over Naïve.

Based on these and other tests, we see that hierarchies can, depending on their organization and on the specific search, further improve performance compared to no hierarchy. For most hierarchies compared, the increase in response times between the *s* and *e* collections was, for most searches, smaller than the increase in the

Table 2. Space Search Suite, AD vs. CN, for searches affected by CN only. "# Affected" is the number of searches with increased low score.

Hie rar chy	# Aff ect ed	Adaptive (AD)			Contraction (CN)		
		Avg. SQL Rows (1000's)	Avg. Matches Found (1000's)	Mean Response (s)	Avg. SQL Rows (1000's)	Avg. Matches Found (1000's)	Mean Response (s)
1e	13	115.2	39.2	35.0±58.4	82.1	3.5	29.6±24.8
2s	5	65.4	33.3	21.3±6.3	44.1	9.0	14.7±3.6
2e	7	239.5	114.5	78.2±69.3	159.0	30.4	54.3±34.3
3e	5	286.7	147.6	93.9±44.2	212.8	91.2	66.4±25.4
5e	4	333.3	181.8	110.6±48.4	272.1	128.7	87.0±23.8
6e	4	333.7	181.8	110.6±48.4	272.5	128.7	87.0±23.8

number of entries between the collection sizes. This reduction supports the added value of taking advantage of hierarchies to improve response times for a given collection size.

In the currently deployed version of the search engine, we use the filter with Adaptive relaxation and Contraction, and the centroid-and-max-radius based spatial filter.

All search types perform well for our current catalog size, and for the expected organic growth of our catalog over the next 5 years. To support larger archives, we may need to modify how we store metadata on variable ranges, and learn to distinguish and replace hierarchies that lead to long response times.

4   Further Challenges

We cover here additional issues in dataset searching.

- Selecting a good initial cutoff score can avoid restarts, and it might be possible to use metadata catalog statistics to guide the choice. Bruno et al. [3] attempt to adapt to the underlying data by taking a sample of the data, and running a training workload on it to identify a cutoff score. Other queries from a similar workload can then use the same cutoff score. To use this technique, we would need a "search similarity function" that can compare the current search to other searches, then use the cutoff score from the most similar search. There might be other techniques for selecting a good initial cutoff.
- Further refining the filters we construct could cut down on the number of dataset summaries for which we have to compute summaries. Currently, the filter is constructed of "per term" conditions that operate independently. There might be advantages to constructing filters that consider multiple terms jointly.
- An important question, given the widely different performance results from the search suites, is how to predict which hierarchies will provide the good performance for queries or query classes, if there is more than one choice. We explored a range of hypotheses for how to predict the best hierarchy to use, but without clear success [13]. The best-performing hierarchy varies according to the type of search terms and the density of data close to the combination of terms. If we cannot accurately determine a good hierarchy a priori, there are options to consider:

  (a) Start a search on each hierarchy and provide the response from the hierarchy that comes back first, canceling the others.
  (b) Choose a single "good enough" hierarchy approach as a default, and further develop performance methods to maintain performance in acceptable ranges.
  (c) Keep every set of catalog tables small enough to give fast response from each, and add an integrator to combine the top-k from across the various lists.
  (d) Provide partial results back early once a search with long response times is recognized (e.g., many results are returned from an SQL query, or a restart is required), and continue to refine them as more results become available.
  (e) Try additional methods to predict hierarchy performance.

- As noted, we wish to understand the performance of a single server; if single-server performance is sufficient for the expected workload, no additional scaling approaches are needed. When single-server performance is found insufficient even after optimization, throughput can be further scaled by adding servers. In our current archive, we use different hierarchies for different types of data, mixing all of them in a single metadata collection. Thus, a single search may be traversing many different hierarchy shapes simultaneously, which could be parallelized. It is a bit harder to partition a hierarchy for parallel search, as only a few branches of a hierarchy might apply to a specific search.

- An implicit assumption in our current similarity function is that data values are uniformly distributed in the regions and intervals captured in the summaries. For some features (such as time), that assumption is fairly accurate, but not for all. Initial tests show that our summary-based similarity function nevertheless closely tracks one that considers individual observations. However, incorporating more information about values distributions in a dataset, including joint distributions of correlated variables, might yield benefits.

- Another major performance aspect is the effect of varying the complexity of the queries themselves. We do not know the "usual" number of search terms, but that number will have a large effect on response times. User studies in Internet search have found 5 to be a common number for text retrieval [1, 2, 9]; but that number might not be typical in our setting. Further research might characterize dataset-search complexity, and develop techniques for accelerating complex queries.

- We know that retrieving data from the database is a large component of our search engine latency. Our initial data design was chosen for ease of implementation, but constrains our performance. We have a fairly general schema for variables that makes it easy to add a new variable, but requires joins when we have search terms for more than one variable. A materialized view that pivots the variables table save the join cost.

- One aspect of scaling we have not treated here is the diversity of variable naming and meaning as an archive grows. Especially as we try to support search over multiple archives, we will encounter different variable names that represent the same quantity, and the same name used for different quantities. We explore this problem and possible solutions in other work [14].

These issues mainly consider issues that arise in our IR-style approach to dataset search (though any approach will need to deal with the variable diversity issues mentioned in the last bullet). Obviously, there could well be other effective approaches. We are also interested in how our techniques can be adapted to other domains where temporal and geospatial aspects are not as dominant.

Given the growth in size and variety of scientific archives and the amount of time spent by scientists in locating relevant data, maintaining interactive response times will be a challenge. Our scientists have found significant value in the search capability we've already provided in Data Near Here. Scaling this initial work into a multi-repository search engine through IR-style data discovery can help increase the touchability and reuse of data that was previously "one touch". We thereby create higher returns on the investment in data collection and enable new scientific discoveries.

Acknowledgments. This work is supported by NSF award OCE-0424602. We thank the staff of CMOP for their support.

References

1. Ageev, M., et al.: Find it if you can: A game for modeling different types of web search success using interaction data. In: Proceedings of SIGIR (2011)
2. Aula, A., et al.: How does search behavior change as search becomes more difficult? In: Proc. of the 28th International Conference on Human Factors in Computing Systems, pp. 35–44 (2010)
3. Bruno, N., et al.: Top-k selection queries over relational databases: Mapping strategies and performance evaluation. ACM Trans. Database Syst. TODS 27(2), 153–187 (2002)
4. Carey, M.J., Kossmann, D.: On saying "enough already!" in SQL. ACM SIGMOD Rec. 26(2), 219–230 (1997)
5. Chaudhuri, S., et al.: Integrating DB and IR technologies: What is the sound of one hand clapping. In: CIDR 2005, pp. 1–12 (2005)
6. Gaasterland, T.: Cooperative answering through controlled query relaxation. IEEE Expert 12(5), 48–59 (1997)
7. Hellerstein, J.M., Pfeffer, A.: The RD-tree: An index structure for sets. University of Wisconsin-Madison (1994).
8. Ilyas, I.F., et al.: A survey of top-k query processing techniques in relational database systems. ACM Comput. Surv. CSUR. 40(4), 11 (2008)
9. Jansen, B.J., et al.: Real life, real users, and real needs: A study and analysis of user queries on the web. Inf. Process. Manag. 36(2), 207–227 (2000)
10. Koposov, S., Bartunov, O.: Q3C, Quad Tree Cube: The new sky-indexing concept for huge astronomical catalogues and its realization for main astronomical queries (cone search and Xmatch) in open source database PostgreSQL. In: Astronomical Data Analysis Software and Systems XV. pp. 735–738 (2006)
11. Kunszt, P., et al.: The indexing of the SDSS science archive. Astron. Data Anal. Softw. Syst. 216 (2000)
12. Lemson, G., et al.: Implementing a general spatial indexing library for relational databases of large numerical simulations. Scientific and Statistical Database Management, 509–526 (2011)
13. Megler, V.M.: Ranked Similarity Search of Scientific Datasets: An Information Retrieval Approach (PhD Dissertation in preparation) (2014)
14. Megler, V.M.: Taming the metadata mess. IEEE 29th International Conference on Data Engineering Workshops (ICDEW), pp. 286–289. IEEE Computer Society, Brisbane (2013)
15. Megler, V.M., Maier, D.: Finding haystacks with needles: Ranked search for data using geospatial and temporal characteristics. In: Bayard Cushing, J., French, J., Bowers, S. (eds.) SSDBM 2011. LNCS, vol. 6809, pp. 55–72. Springer, Heidelberg (2011)
16. Singh, G., et al.: A metadata catalog service for data intensive applications. In: Proceedings of the 2003 ACM/IEEE Conference on Supercomputing, p. 33 (2003)
17. Wang, X., et al.: Liferaft: Data-driven, batch processing for the exploration of scientific databases. In: CIDR (2009)

Secure Computation on Outsourced Data: A 10-year Retrospective

Hakan Hacıgümüş[1], Bala Iyer[2], and Sharad Mehrotra[3]

[1] NEC Labs America
hakanh@acm.org
[2] IBM Silicon Valley Lab.
balaiyer@us.ibm.com
[3] University of California, Irvine, USA
sharad@ics.uci.edu

Abstract. This paper outlines the "behind the scene" story of how we wrote the DASFAA 2004 paper on secure computation on outsourced data to support SQL aggregation queries. We then describe some of the research we have done following the paper.

1 Paper Background

This paper was a result of a successful collaboration between IBM Silicon Valley Laboratory and the University of California, Irvine that explored the idea of outsourcing database management to third-party service providers. At the time, service-based software architectures and software-as-a-service were emerging concepts. Such architectures, made possible by the advances in high-speed networking and ubiquitous connectivity, provided numerous advantages: it alleviated the need for organizations to purchase and deploy expensive software and hardware to run the software on, deal with software upgrades, hire professionals to help maintain the infrastructure etc. Instead, in the software as a service model (also known also, at the time, as the application service provider (ASP) model), the end-users could rent the software over the Internet and pay for what they used. The model was very attractive in the light of the fact that around that time, people costs were beginning to dominate the overall costs of IT solutions. The ASP model, helped bring the people costs down significantly since the task of administration and software maintenance was now outsourced to service providers bringing in economy of scales.

Our collaboration started on a fateful SIGMOD Conference meeting that brought authors of this paper together. The key question we asked was: Given the emerging software-as-a-service, what will it take to provide data management as a service (or the DAS model as we called it)? In the DAS model, the service provider would provide seamless mechanisms for organizations to create, store, access their databases. Moreover, the responsibility of data management, i.e. database backup, administration, restoration, database reorganization to reclaim space, migration to new versions without impacting availability, etc. will

S.S. Bhowmick et al. (Eds.): DASFAA 2014, Part I, LNCS 8421, pp. 16–27, 2014.
© Springer International Publishing Switzerland 2014

be taken over by service providers. Users wishing to access data will now access it using the hardware and software at the service provider instead of their own organizations computing infrastructure. The application will not be impacted due to outages in software, hardware, and networking changes or failures at the database service provider sites. The DAS model will alleviate the problem of purchasing, installing, maintaining and updating the software and administering the system. Instead, users will be able to use the ready system maintained by service provider for its database needs.

1.1   Technical Challenges

Intuitively, we felt that database-as-a-service (or DAS model as we called it) would inherit all the advantages of the software-as-a-service model – even more so, given that for organizations data management, given its complexity, is a significant portion of the overall IT expense. Furthermore, data management, at the time (as is still the case) has a significant people cost requiring professionals to create, implement, and tune data management solutions. Given, what we felt was a unique opportunity business opportunity; we explored new challenges that would arise if we were to adopt a service-oriented model for data management. Our first paper on the topic appeared in the ICDE 2002 conference that raised multiple challenges in supporting the DAS model. In particular, we identified three challenges: a) appropriate user interfaces through which database services and data is presented to the end-user - the interface must be easy to use, yet be sufficiently powerful to allow users to build complex application; b) hiding the additional latencies that arise due to remote data access that sits across the network at the service provider sites; and c) most importantly, the challenge of data confidentiality – it was (and still is) well recognized that organizations view data to be amongst their most valuable assets. In the DAS model, the users data sits on the premises of the third-party service providers where it could be used / misused in ways that can no longer be controlled by data owners. The confidentiality challenge consisted of two distinct security problems: First, if the DAS model was to succeed, it was imperative that mechanisms be developed to empower service providers to provide sufficient security guarantees to the end-user. Appropriate firewall technologies, of course, help prevent malicious outsider attacks. Given the value of data to organizations, and the danger of litigations given data thefts, our goal became using encryption mechanisms to protect users confidentiality even if data at rest (in disks) gets stolen. Data management when data is stored encrypted in disks was less understood at the time (and to date has not been fully addressed). It led to a variety of challenges such as what granularity should the data be encrypted at (e.g., page level, record level, field level), what type of encryption technique should be used to encrypt data, how should key-management be performed, how should the pages/files be organized, how should indices be built upon encrypted data, what are the performance implications of encryption, what are the impacts on query optimization and processing, etc. Many of these challenges were addressed by us in subsequent papers [15, 10–12].

The second confidentiality challenge came from the issue whether end-users (data owners) will indeed trust service providers with their data (irrespective of whether or not service providers follow best practices to prevent data from being stolen through malicious outsiders). There could be several reasons for distrust. The service provider model allows providers to be anywhere over the Internet, possibly outside international boundaries and hence outside the jurisdictional control of the legal systems accessible to data owners. Furthermore, as has been validated through numerous security surveys since then, insider attacks at the service providers are a major vulnerability in service-oriented models. Even if the service provider is a well-intentioned honest organization, a disgruntled or a malicious employee (of the service provider) could steal data for either personal gains or defaming a service provider.

This second challenge led us to explore a new approach to data management in the context of DAS model wherein the data is encrypted appropriately at the data owner site and is never visible in plaintext at the service provider. This way, even if there is the data stored at the service provider is made visible, data confidentiality is preserved since the data is encrypted. This was an ideal solution for the DAS model. However, the approach led us to the challenge of implementing SQL queries (i.e., relational operators such as selections, joins, etc.) over encrypted data representation. Cryptographic techniques that allow predicates to be evaluated over encrypted data representation (now known as searchable encryption) had not developed at the time [1, 2, 4, 21]. Even to date, such techniques are not efficient enough to be considered as a general solution that works well under most/all conditions. Instead, we devised an information theoretic approach which we called bucketization that allowed us to evaluate predicates though at the technique could admit false positives. This led us directly to partitioned computation of SQL queries collaboratively between the service provider and the client side wherein part of the predicate evaluation would be done at the service provider with some filtering at the client side to determine the true answer. This work on partitioned execution of SQL queries appeared in ACM SIGMOD 2002 conference wherein it was recently recognized through a SIGMOD Test-of-Time award in 2012.

1.2   Aggregations over Encrypted Data

While our SIGMOD paper addressed how predicate searches in SQL queries could be done collaboratively by appropriately partitioning the computation, it, nonetheless, had a gaping hole when it came to supporting even the simplest aggregation queries. To see this consider a simple aggregation query such as SELECT sum(salary) from EMPLOYEE that adds up the total salary of all the employees in an organization. While bucketization was devised to support predicate evaluations, and could hence be used to implement selections, joins, and through the algebra for partitioned computation, fairly complex SQL queries, it could not support aggregation efficiently. A query such as above would require the entire EMPLOYEE table (or at least the projection to the salary field to be transmitted across the network to the client side where it would be decrypted and then added. While bucketization provided us with a way to evaluate predicates, it did not provide us with a way to compute on the encrypted data directly. What we needed was to incorporate homomorphic encryption techniques into the data representation to enable computing over encrypted data into the DAS model.

Homomorphic encryption techniques are a form of encryption that allow specific form of computations to be performed on cipher text such that the resulting output is the encrypted form of the true result of the computation. Thus, decrypting the resulting cipher text would yield the true answer to the query. Homomorphic encryption would permit us to compute the above aggregation query efficiently – if, for instance, the salary field was encrypted using such an encryption, we could add the salaries over the encrypted representation on the server, and then transmit the result to the client where the result could be decrypted. This would completely eliminate the need to transmit the whole relation to the client side bringing tremendous savings.

Homomorphic encryption can be classified as partially homomorphic that allows only one operation (i.e., either multiplication, or addition) to be performed over the encrypted domain or fully homomorphic that allows both the operations to be evaluated in cipher text. Examples of partially homomorphic techniques are unpadded RSA that allows multiplication, and Paillier cryptosystem that supports addition over encrypted representation. Given that aggregation queries could involve both multiplication and addition, we required a fully homomorphic encryption. The need and the power of the fully homomorphic cryptosystem has been long recognized. A secure solution to such a encryption method, however, remained an open problem in cryptography for over 30 years and has only recently been solved by the work of Gentry [6, 27, 5], which has recently led to a large number of innovations in homomorphic cryptosystems. To date, however, an efficient implementation of fully homomorphic system to be of practical value in general remains illusive to the best of our knowledge.

2   Aggregation Queries over Encrypted Data

Our focus and goal at the time, was not as much on improving cryptography, but rather on exploring how efficient query processing could be implemented in the DAS model if such a fully homomorphic crypto system were to be made available. To make progress, we chose the fully homomorphic cryptosystem, PH (Privacy Homomorphism), proposed originally in [26] even though it was known not to be fully secure. We give a background on PH as follows:

2.1   Background on Privacy Homomorphisms

Definition of PH: Assume $\mathcal{A}$ is the domain of unencrypted values, $\mathcal{E}_k$ an encryption function using key k, and $\mathcal{D}_k$ the corresponding decryption function, i.e., $\forall a \in \mathcal{A}$, $\mathcal{D}_k(\mathcal{E}_k(a)) = a$. Let $\tilde{\alpha} = \{\alpha_1, \alpha_2, \ldots, \alpha_n\}$ and $\tilde{\beta} = \{\beta_1, \beta_2, \ldots, \beta_n\}$ be two (related) function families. The functions in $\tilde{\alpha}$ are defined on the domain $\mathcal{A}$ and the functions on $\tilde{\beta}$ are defined on the domain of encrypted values of $\mathcal{A}$. $(\mathcal{E}_k, \mathcal{D}_k, \tilde{\alpha}, \tilde{\beta})$ is defined as a privacy homomorphism if $\mathcal{D}_k(\beta_i(\mathcal{E}_k(a_1), \mathcal{E}_k(a_2), \ldots, \mathcal{E}_k(a_m))) = \alpha_i(a_1, a_2, \ldots, a_m) : 1 \leqslant i \leqslant n$. Informally, $(\mathcal{E}_k, \mathcal{D}_k, \tilde{\alpha}, \tilde{\beta})$ is a privacy homomorphism on domain $\mathcal{A}$, if the result of the application of function α_i on values may be obtained by decrypting the result of β_i applied to the encrypted form of the same values.

Given the above general definition of PH, we next describe a specific homomorphism proposed in [26] that we will use in the remainder of the paper. We illustrate how the PH can be used to compute basic arithmetic operators through an example.

- The key, $k = (p, q)$, where p and q are prime numbers, is chosen by the client who owns the data.
- $n = p \cdot q$, p and q are needed for encryption/decryption and are hidden from the server. n is revealed to the server. The difficulty of factorization forms the basis of encryption.
- $\mathcal{E}_k(a) = (a \bmod p, a \bmod q)$, where $a \in \mathbb{Z}_n$. We will refer to these two components as the p *component* and q *component*, respectively.
- $\mathcal{D}_k(d_1, d_2) = d_1 q q^{-1} + d_2 p p^{-1} \pmod{n}$, where $d_1 = a \pmod p$, $d_2 = a \pmod q$, and q^{-1} is such that $q q^{-1} = 1 \pmod p$ and p^{-1} is such that $p p^{-1} = 1 \pmod q$.
$$(1)$$
- $\tilde{\alpha} = \{+_n, -_n, \times_n\}$, that is addition, subtraction, and multiplication in mod n.
- $\tilde{\beta} = \{+, -, \times\}$, where operations are performed componentwise.

Example: Let $p = 5, q = 7$. Hence, $n = pq = 35, k = (5, 7)$. Assume that the client wants to **add** a_1 and a_2, where $a_1 = 5, a_2 = 6$. $\mathcal{E}(a_1) = (0, 5), \mathcal{E}(a_2) = (1, 6)$ (previously computed) are stored on the server. The server is instructed to compute $\mathcal{E}(a_1) + \mathcal{E}(a_2)$ componentwise (i.e., without decrypting the data). The computation $\mathcal{E}(a_1) + \mathcal{E}(a_2) = (0+1, 5+6) = (1, 11)$. The result, $(1, 11)$ is returned to the client. The client decrypts $(1, 11)$ using the function $(d_1 q q^{-1} + d_2 p p^{-1}) \pmod{n} = (1 \cdot 7 \cdot 3 + 11 \cdot 5 \cdot 3) \pmod{35} = 186 \pmod{35}$, which evaluates to 11, the sum of 5 and 6.[1] The scheme extends to **multiplication** and **subtraction**.

2.2   Extensions to PH

Notice that PH could directly be used to support aggregation in SQL queries since it is fully homomorphic. We were, however, faced with a few challenges listed below which became the topic of our technical contribution in this DAS-FAA 2004 paper. The full discussion can be found in [7].

Division Operation: The PH we use does not support division. SQL aggregation, however, requires the handling of division. As an example, consider the SQL clause SUM (expense / currency_ratio) on a table where expenses are listed in different currencies. Let us consider the first two terms of this summation; $\frac{a_1}{c_1} + \frac{a_2}{c_2}$. The numerator and denominator of the sum are $a_1 c_2 + a_2 c_1$ and $c_1 c_2$, respectively. Both only involve addition and multiplication, which are supported by the PH. The inability of the PH to handle division is compensated in query processing. The numerator and the denominator for the division are

---

[1] n is selected in such a way that results always fall in $[0, n)$; for this example $[0, 35)$.

computed separately and returned to the client for final computation. Then, the client decrypts both and performs the division.

Floating Point Number Representation: As defined, our PH is defined over the integer domain [26]. SQL has a float or real number data type. Our PH needs to be extended to handle real number arithmetic. Our suggestion is to treat real number as fractions. Arithmetic of fractions may be carried out using the technique used for division operation. Numerators and denominators are computed separately by the server on encrypted data and both sent to the client for final decryption and division. Consider the following example for floating point arithmetic.

Example 1. Let $p = 11, q = 13$. Hence, $n = pq = 143, k = (11, 13)$. Let us assume that the user wants to compute $x = a_1 + a_2$ for $a_1 = 1.8, a_2 = 0.5$. Hence, $\mathcal{E}(a_1) = ((7, 5), (10, 10)), \mathcal{E}(a_2) = ((5, 5), (10, 10))$.

$$\mathcal{E}(a_1) + \mathcal{E}(a_2) = \frac{\mathcal{E}(a_1^n)}{\mathcal{E}(a_1^d)} + \frac{\mathcal{E}(a_2^n)}{\mathcal{E}(a_2^d)} = \frac{\mathcal{E}(a_1^n) + \mathcal{E}(a_2^n)}{\mathcal{E}(a_1^d)} = \frac{(7,5) + (5,5)}{(10,10)} = \frac{(12,10)}{(10,10)} \pmod{143}$$

This result is sent to the client, and the client performs decryption as follows:
$a_1 + a_2 = \frac{12 \cdot 13 \cdot 6 + 66 \cdot 11 \cdot 6}{10 \cdot 13 \cdot 6 + 10 \cdot 11 \cdot 6} = \frac{23}{10} \pmod{143} = 2.3$

Handling Negative Numbers: Negative numbers can be dealt by offsetting the range of numbers. To see the need for this, recall that arithmetic is defined on modulo n in PH. For example, the numbers 33 and -2 are indistinguishable when we represent them in modulo 35. That is, 33 (mod 35) $\equiv$ -2 (mod 35).

Let the maximum value that can be represented in the computer system be $v_{max} > 0$ and the minimum value be $v_{min} = -v_{max}$. Then the possible value range would be $[v_{min}, v_{max}]$. We first map this range into another range, which is $[0, 2v_{max}]$. In this new range, any value $x < v_{max}$ is interpreted as a negative number relative to value of v_{max}. By defining that, an actual value x is mapped to a shifted

After these definitions, arithmetic operations should be mapped accordingly as well. For example, addition is mapped as $x + y = x' + y' - (v_{max} - v_{min})$ and similarly subtraction is mapped as $x - y = x' - y' + (v_{max} - v_{min})$. It is obvious that the server should have the encrypted value for $(v_{max} - v_{min})$, which will be computed with same encryption scheme and stored once. To illustrate the problem we stated above and the mapping scheme, consider the following example.

2.3   Security Extension to the PH

In this section we show how we addressed a security exposure, test for equality by the server, in the given PH scheme.

Preventing Test for Equality: Picking n such that $n > v_{max} - v_{min}$ (as we did above) enables the server to test for equality. Say x and y are two numbers encrypted as (x_p, x_q) and (y_p, y_q), respectively. Let $z = x * y$, which implies in the encrypted domain, $(z_p, z_q) = (x_p, x_q) * (y_p, y_q)$. The server could start adding (x_p, x_q) to itself every time checking the equality between the sum and

(z_p, z_q). When the equality is satisfied, the server learns the unencrypted value of y. Thus, $(z_p, z_q) = \underbrace{(x_p, x_q) + (x_p, x_q) + \ldots + (x_p, x_q)}_{y \ times}$.

We plug this exposure by adding random noise to the encrypted value. We encrypt an original value x as follows; $\mathcal{E}(x) = (x \pmod{p} + R(x) \cdot p, x \pmod{q} + R(x) \cdot q)$, where $R(x)$ is a pseudorandom number generator with seed x. $R(x)$ value is generated during every insertion/update by the client. This prevents equality testing for the server and the server cannot remove the noise without knowing p and q. The noise is automatically removed at the client upon decryption. In the presence of noise, the following decryption function should be used in place of equation (1): $\mathcal{D}_k(d_1, d_2) = (d_1 \bmod p)qq^{-1} + (d_2 \bmod q)pp^{-1} \pmod{n}$ This equation is true because noise had been added in multiples of p for the first and in multiples of q in the second term. The modulo of each (mod p) and (mod q) term removes the added noise.

Another benefit of introducing the noise is that p and q components are no longer stored in modulo p and q, respectively. It makes it additionally difficult for the server to guess their values.

2.4   Query Processing over Encrypted Data

While PH provided us with a solution to implement the simple aggregation query about adding salaries of employees we discussed above, we were next faced with a challenge of computing aggregation queries that contain predicates. Most aggregation SQL queries are not as simple as the one used above to motivate the need for a fully homomorphic encryption approach. Consider, for instance, now a slightly modified query such as SELECT sum(salary) FROM employee WHERE age > 45 and age < 50. The bucketization not only selected terms which were part of the answer, but also false positives (which were eliminated on the client side). But this causes a problem since we cannot compute the answer at the server for aggregation. The basic idea we came across, which led us to the DASFAA paper, was that in bucketization we could treat buckets as sure or not sure. If we distinguish between them, we can compute the aggregation over sure things on the client side. Since most aggregations can be computed in a progressive manner, we could compute part of the aggregation on the server and transfer the rest when we were unsure to client side, decrypt and compute the answer on the client side. The strategy would not just work, but would also tune well by controlling the maximum size of each buckets and ensuring that number of unsure buckets given queries is limited. We give an overview of the process as follows:

Given a query Q, our problem is to decompose the query to an appropriate query Q^S on the encrypted relations R^S such that results of Q^S can be filtered at the client in order to compute the results of Q. Ideally, we would like Q^S to perform bulk of the work of processing Q. The effectiveness of the decomposition depends upon the specifics of the conditions involved in Q and on the server side representation R^S of the relations involved. Consider, for example, a query to retrieve sum of salaries of employee in $did = 40$. If did is a field-level encrypted field, the server can exactly identify records that satisfy the condition by utilizing the equality between the client-supplied values and the encrypted values stored on the server. In such a case, aggregation can be fully performed on the *salary* attribute of the selected tuples exploiting the PH representation. If, on the other hand, the condition were more complex, (e.g., `did > 35 AND did < 40`), such a query will be mapped to the server side by mapping the *did* to the corresponding partitions associated with the *did* field that cover the range of values from 35 to 40. Since the tuples satisfying the server side query may be a superset of the actual answer, aggregation cannot be completely performed at the server. Our strategy is to separate the qualified records into those that certainly satisfy the query conditions, and those that may satisfy it - the former can be aggregated at the server, while the latter will need to be transmitted to the client, which on decrypting, can filter out those that do not, and aggregate the rest. The strategy suggests a natural partitioning of the server side query Q^S into two queries Q_c^S and Q_m^S as follows:

• **Certain Query** (Q_c^S): That selects tuples that *certainly* qualify the conditions associated with Q. Results of Q_c^S can be aggregated at the server.
• **Maybe Query** (Q_m^S): That selects *etuples* corresponding to records that *may* qualify the conditions of Q but it cannot be determined for sure without decrypting. The client decrypts these *etuples*, and then selects the ones that actually qualify and performs the rest of the query processing.

To finalize the computation, the client combines results from these queries to reach the actual answers. We next discuss how a client side query Q is translated into the two server side representations Q_c^S and Q_m^S.

3   Follow-up Work

The work described above on database-as-a-service was part of a successful collaboration between IBM and UCI that culminated in a Ph.D. dissertation of Hakan Hacıgümüş. Our follow up work, motivated and influenced in a large part by our initial papers on DAS has explored numerous challenges that arose when we started developing the DAS model. We discuss some of these additional challenges we explored subsequently below.

First, our approach to partitioned computation was based on supporting secure indexing tags by applying bucketization (a general form of data partitioning), which prevents the service provider from learning exact values but still allows it to check if the record satisfies the query predicate. Bucketization provided a natural sliding scale confidentiality that allowed us to explore a tradeoff between security and performance – e.g., if we map all data values to the same bucket, the approach provides perfect security (since adversary cannot distinguish between any data) but it incurs heavy overhead since now the service provider cannot prune records based on query predicates. Likewise, if we map each distinct value has a different bucket, the adversary gains significant knowledge about the data, but can also perform potentially perfect pruning in the context of query processing. Our subsequent work explored the natural tradeoff between security and performance that results in using bucketization as an underlying technique for supporting SQL queries [13, 9, 14]. Another angle we explored the DAS model was towards exploiting additional benefits that the DAS model intrinsically provides. First amongst these is that outsourcing naturally creates the ability to seamlessly share data with others. Data sharing has, and to a large degree still remains a significant challenge – organizations often develop mutual (pairwise) data sharing protocols that implement their sharing policies. Individuals often share data with each other using diverse mechanisms such as emails, posting on publicly accessible websites, or physically sharing sharing data using memory sticks etc. DAS can significantly reduce the burden of data sharing, if a user could, besides outsourcing its data / query processing, could also outsource the task of data sharing to the database service. Such a solution would not just alleviate the responsibility of organizations to share the data, but it will also result result in improved performance and availability since the data (to be shared) already resides on the service providers. To a large degree, the technology has already adopted such an approach at least at the individual levels with the emergence and proliferation of services such as dropbox and google drive. Our work, however, explored sharing in the context when service providers were untrusted (and hence data was appropriately encrypted). Another benefit that outsourcing provides, particularly, in the context of personal data management, is that it empowers users to access data from anywhere, anytime, and from any device. This is particularly useful for data such as Web history, bookmarks, account information, passwords, etc .that a user may wish to access remotely using a mobile device or using different machines from environments that are not necessarily trusted. We explored mechanisms wherein trusted computation can be performed using the outsourced data model through the help of a small-footprint trusted proxy [18, 17, 16, 20, 19].

Since the time we explored the DAS model, the computing field has witnessed a major shift in computing paradigm – fueled by the advances in virtualization and high-speed networking, cloud computing model is emerging that can roughly be summarized as X as a service , where X can be virtualized infrastructure (e.g., computing and/or storage), a platform (e.g., OS, programming language execution environment, databases, web servers, etc.), software applications (e.g., Google apps), a service or a test environment, etc. Much like what motivated the industrial trend towards software-as a service (and also our work on DAS), the primary motivator of cloud computing is the utility computing model (aka pay-as-you go model) where users get billed for the computers, storage, or any resources based on their usage with no up-front costs purchasing the hardware/software or of managing the IT infrastructure. The cloud provides the illusion of limitless resources that one can tap into in times of need, limited only the the amount one is willing to spend on renting the resources. Our work on database as a service naturally fits the cloud model but the cloud offers many new challenges and opportunities that we did not originally explore as part of our work on DAS.

First, unlike our assumption in DAS, where the resources were assumed to be very limited on the client side, in the cloud setting organizations may actually possess significant resources that meets majority of their storage and computational needs. For instance, in the cloud setting data may only be partially outsourced, e.g., only non-sensitive part of the data may be kept on the cloud. Also, it may be only at peak query loads that the computation needs to be offloaded to the cloud. This has implications from the security perspective since much of the processing involving sensitive data could be performed at the private side. In DAS, since the goal was to fully outsource the data and computation, the focus of the solutions was on devising mechanism to compute on the encrypted representation (even though such techniques may incur significant overhead). In contrast, in the cloud environments, since local machines may have significant computational capabilities, solutions that incur limited amount of data exposure of sensitive data (possibly at a significant performance gain) become attractive. Second, while our DAS work primarily dealt with database query workload, in a cloud setting, we may be interested in more general computation mechanisms (i.e. not only database workloads). For instance, map-reduce (MR) frameworks are used widely for large-scale data analysis in the cloud. We may, thus, be interested in secure execution of MR jobs in public clouds.

Another challenge arises from the potential autonomy of the cloud service providers. It is unlikely that autonomous providers will likely implement new security protocols and algorithms (specially given significant overheads associated with adding security and the restrictive nature of cryptographic security for a large number of practical purposes). For instance, it is difficult to imagine Google making changes to the underlying storage models, data access protocols and interfaces used by its application services (such as Google Drive, Picasa, etc.) such that users can store/search/process data in encrypted form. This calls for a new, robust and more flexible approach to implement privacy and confidentiality of data in cloud-based applications.

Our current research related to DAS is exploring the above challenges in the context of cloud computing[2]. In particular, we have explored a risk-aware data processing framework for cloud computing that, instead of focusing on techniques to eliminate possibility of attacks, provides mechanisms to limit / control the exposure risks. Different ways to steer data through the public and private machines and the way data is represented when it is on public machines exhibit different levels of risks and expose a tradeoff between exposure risks and system specific quality metrics that measure the effectiveness of a cloud based solution. Given such a tradeoff, the goal of the risk aware computing changes from purely attempting to maximize the application specific metrics to that of achieving a balance between performance and sensitive data disclosure risks. We have explored such a risk based approach for various contexts – HIVE and SparQL queries over relational data /RDF data as well as lookup queries over key value stores [25, 24, 22, 23, 3].

Another avenue of our work on cloud computing addresses the issue of autonomy of service providers by exploring an middleware-based approach for security where end-clients connect to cloud services through a security layer entitled

---

[2] This work is part of the Radicle Project (http://radicle.ics.uci.edu) at UCI which is funded through in part by the NSF Grant CNS 1118127.

CloudProtect. CloduProtect empowers users with control over their data that may store in existing cloud-based applications such as Box, Google drive, Picasa, Google Calendar, etc. It is implemented as a intermediary that intercepts http requests made by the clients, transforms the request to suitably encrypt/decrypt the data based on the users confidentiality policies before forwarding the request to the service provider. Encrypted representation of data may interfere with the users experience with the service - while requests such as CRUD operations (i.e., create, read, update, delete) as well as possibly search can be performed on encrypted representation, operations such as Google translate requires data to be in plaintext. CloudProtect keeps track of how data is stored on the server and based on the operation request may launch an exception protocol wherein encrypted data may be brought back to the client, decrypted and restored at the server before the service request is submitted to the server. Through the exception protocol, CloudProtect provides continued seamless availability of Web services. Furthermore, CloudProect supports mechanisms to adapt the data representation at the service providers to strike a balance between the risk of data exposure and the service usability (i.e., the number of times an exception protocol is invoked to meet users request).

Hakan Hacıgümüş followed up with the research work in the larger context of data management systems in the cloud. In that context, we worked on numerous challenges that need to be overcome before the database systems can be successfully integrated with the cloud delivery platforms. Some of the research areas include, elastic transactional database systems, large scale, flexible Big Data Analytics systems in the cloud, workload and resource management in cloud databases, data-rich mobile applications in the cloud. Many of the areas are explored in the CloudDB project at the NEC Laboratories of America [8].

References

1. Bellare, M., Boldyreva, A., O'Neill, A.: Deterministic and efficiently searchable encryption. In: Menezes, A. (ed.) CRYPTO 2007. LNCS, vol. 4622, pp. 535–552. Springer, Heidelberg (2007)
2. Boneh, D., Waters, B.: Conjunctive, subset, and range queries on encrypted data. In: Vadhan, S.P. (ed.) TCC 2007. LNCS, vol. 4392, pp. 535–554. Springer, Heidelberg (2007)
3. Canim, M., Kantarcioglu, M., Hore, B., Mehrotra, S.: Building disclosure risk aware query optimizers for relational databases. PVLDB 3(1) (2010)
4. Curtmola, R., Garay, J.A., Kamara, S., Ostrovsky, R.: Searchable symmetric encryption: improved definitions and efficient constructions. In: Proc. of ACM CCS (2006)
5. Gentry, C.: A Fully Homomorphic Encryption Scheme. Ph.D. Thesis, Stanford University (2009)
6. Gentry, C.: Fully homomorphic encryption using ideal lattices. In: Proc. of STOC (2009)
7. Hacıgümüş, H.: Privacy in Database-as-a-Service Model. Ph.D. Thesis, Department of Information and Computer Science, University of California, Irvine (2003)
8. Hacıgümüş, H.: NEC Labs Data Management Research. SIGMOD Record 40(3) (2011)

9. Hacıgümüş, H., Hore, B., Iyer, B.R., Mehrotra, S.: Search on encrypted data. In: Secure Data Management in Decentralized Systems, pp. 383–425 (2007)
10. Hacıgümüş, H., Iyer, B., Mehrotra, S.: Query Optimization in Encrypted Database Systems. In: Zhou, L.-z., Ooi, B.-C., Meng, X. (eds.) DASFAA 2005. LNCS, vol. 3453, pp. 43–55. Springer, Heidelberg (2005)
11. Hacıgümüş, H., Mehrotra, S.: Performance-Conscious Key Management in Encrypted Databases. In: Proc. of DBSec (2004)
12. Hacıgümüş, H., Mehrotra, S.: Efficient Key Updates in Encrypted Database Systems. In: Proc. of DBSec (2005)
13. Hore, B., Mehrotra, S., Hacıgümüş: Managing and querying encrypted data. In: Handbook of Database Security, pp. 163–190 (2008)
14. Hore, B., Mehrotra, S., Tsudik, G.: A privacy-preserving index for range queries. In: Proc. of VLDB (2004)
15. Iyer, B., Mehrotra, S., Mykletun, E., Tsudik, G., Wu, Y.: A Framework for Efficient Storage Security in RDBMS. In: Bertino, E., Christodoulakis, S., Plexousakis, D., Christophides, V., Koubarakis, M., Böhm, K. (eds.) EDBT 2004. LNCS, vol. 2992, pp. 147–164. Springer, Heidelberg (2004)
16. Jammalamadaka, R.C., Gamboni, R., Mehrotra, S., Seamons, K.E., Venkatasubramanian, N.: gvault: A gmail based cryptographic network file system. In: Proc. of DBSec (2007)
17. Jammalamadaka, R.C., Gamboni, R., Mehrotra, S., Seamons, K.E., Venkatasubramanian, N.: idataguard: middleware providing a secure network drive interface to untrusted internet data storage. In: Proc. of EDBT (2008)
18. Jammalamadaka, R.C., Gamboni, R., Mehrotra, S., Seamons, K.E., Venkatasubramanian, N.: A middleware approach for outsourcing data securely. Computers & Security 32 (2013)
19. Jammalamadaka, R.C., Mehrotra, S., Venkatasubramanian, N.: Pvault: a client server system providing mobile access to personal data. In: Proc. of StorageSS (2005)
20. Jammalamadaka, R.C., van der Horst, T.W., Mehrotra, S., Seamons, K.E., Venkatasubramanian, N.: Delegate: A proxy based architecture for secure website access from an untrusted machine. In: Proc. of ACSAC (2006)
21. Kamara, S., Papamanthou, C., Roeder, T.: Dynamic searchable symmetric encryption. In: ACM Conference on Computer and Communications Security (2012)
22. Khadilkar, V., Oktay, K.Y., Kantarcioglu, M., Mehrotra, S.: Secure data processing over hybrid clouds. IEEE Data Eng. Bull. 35(4) (2012)
23. Oktay, K.Y., Khadilkar, V., Hore, B., Kantarcioglu, M., Mehrotra, S., Thuraisingham, B.M.: Risk-aware workload distribution in hybrid clouds. In: IEEE CLOUD (2012)
24. Oktay, K.Y., Khadilkar, V., Kantarcioglu, M., Mehrotra, S.: Risk aware approach to data confidentiality in cloud computing. In: Bagchi, A., Ray, I. (eds.) ICISS 2013. LNCS, vol. 8303, pp. 27–42. Springer, Heidelberg (2013)
25. Pattuk, E., Kantarcioglu, M., Khadilkar, V., Ulusoy, H., Mehrotra, S.: Bigsecret: A secure data management framework for key-value stores. In: Proc. of IEEE CLOUD (2013)
26. Rivest, R.L., Adleman, L.M., Dertouzos, M.: On Data Banks and Privacy Homomorphisms. In: Foundations of Secure Computation (1978)
27. van Dijk, M., Gentry, C., Halevi, S., Vaikuntanathan, V.: Fully homomorphic encryption over the integers. IACR Cryptology ePrint Archive (2009)

Online Indexing and Distributed Querying Model-View Sensor Data in the Cloud

Tian Guo[1], Thanasis G. Papaioannou[2], Hao Zhuang[1], and Karl Aberer[1]

[1] School of Computer and Communication Sciences
Ecole Polytechnique Fédérale de Lausanne (EPFL)
{tian.guo,hao.zhuang,karl.aberer}@epfl.ch
[2] Information Technologies Institute
Center for Research & Technology Hellas (ITI-CERTH)
thanasis.papaioannou@iti.gr

Abstract. As various kinds of sensors penetrate our daily life (e.g., sensor networks for environmental monitoring), the efficient management of massive amount of sensor data becomes increasingly important at present. Traditional sensor data management systems based on relational database lack scalability to accommodate large-scale sensor data efficiently. Consequently, distributed key-value stores in the cloud is becoming the prime tool to manage sensor data. Meanwhile, model-view sensor data management stores the sensor data in the form of modelled segments. However, currently there is no index and query optimizations upon the modelled segments in the cloud, which results in full table scan in the worst case. In this paper, we propose an innovative model index for sensor data segments in key-value stores (KVM-index). KVM-index consists of two interval indices on the time and sensor value dimensions respectively, each of which has an in-memory search tree and a secondary list materialized in the key-value store. This composite structure enables to update new incoming sensor data segments with constant network I/O. Second, for time (or value)-range and point queries a MapReduce-based approach is designed to process the discrete predicate-related ranges of KVM-index, thereby eliminating computation and communication overheads incurred by accessing irrelevant parts of the index table in conventional MapReduce programs. Finally, we propose a cost based adaptive strategy for the KVM-index-MapReduce framework to process composite queries. As proved by extensive experiments, our approach outperforms in query response time both MapReduce-based processing of the raw sensor data and multiple alternative approaches of model-view sensor data.

1 Introduction

As various kinds of sensors penetrate our daily life, our planet is undertaking the vast deployment of sensors embedded in different devices that monitor various phenomena for different applications of interest, e.g., air/electrosmog pollution, radiation, early earthquake detection, soil moisture, permafrost melting, etc.

S.S. Bhowmick et al. (Eds.): DASFAA 2014, Part I, LNCS 8421, pp. 28–46, 2014.
© Springer International Publishing Switzerland 2014

The data generated by a large number of sensors are represented as streaming time series in which each data point is associated with a time-stamp and a sensor value. These raw discrete observations are taken as the first citizen in traditional relational sensor data management systems, which leads to numerous problems. For example, in order to discover the hidden knowledge in the raw sensor data, users usually adopt other third-party modeling tools(e.g., Matlab and Mathematica), which involve of tedious and time-consuming data extract, transform and load (ETL) processes. On the other hand, unbounded sensor data streams often have missing values and unknown errors, which also poses great challenges for traditional raw sensor data management.

To this end, various model-based sensor data management techniques [1,2,3,4] have been proposed. Models capture the inherent correlations (e.g., temporally and spatially) in the sensor data stream through splitting sensor data into disjoint segments and approximating each segment with different types of models (e.g., regression and probabilistic.). For instance, in mobile computing, trajectories data are often represented as a collection of disjoint segments with different polynomial models, which facilitates future user behaviour analysis and data mining. Moreover, models could help derive the missing values by an abstraction layer over the raw sensor data [3] and improve the query processing by accessing minimal amount of data [2,5]. However, current works on model-based sensor data management[3,2] are mostly based on materialized views or interval indices [6] of relational databases[6], which makes it difficult to meet the demands of the increasing amount of sensor data. Thus, it is imperative to design model-view sensor data management in the context of non-relational key-value stores and MapReduce framework, thereby improving the performance in terms of scalability and parallelism.

This paper is about how to manage modeled segments in the cloud rather. We design and implement the on-line indexing and distributed querying model-view sensor data based on distributed key-value stores and MapReduce computing in the cloud.

Our proposed model index characterizes each segment by its time and value intervals [3,2,4] in order to facilitate data processing as well as is able to efficiently update new arriving data segments.

As for query processing, due to continuously increasing size of sensor data, even less selective queries are related to huge number of modeled segments. Thus, we design a MapReduce based approach combining both model index searching and segment gridding to query the model-view sensor data, which also differentiates our work from traditional query methods over raw sensor data. Our contributions are summarized as follows:

- On-line model index in distributed key-value stores: we devise an innovative key-value represented model index that indexes both the time and value intervals of sensor data segments, referred to as *KVM-index*. The composite structure of KVM-index consisted of memory-resident search tree and index-model tables in the key-value store, can accommodate new sensor data segments efficiently in the on-line fashion.

- Index searching and distributed segment processing integrated approach: we propose a MapReduce based query processing approach that is able to process the separated parts of the index-model table found by the intersection and stabbing search of KVM-index.
- Cost model based adaptive approach for composite queries: for time and value composite queries, we design an adaptive strategy considering the number of waves and the data locality of MapReduce computing to decide which discrete splits, from the time or value dimension of KVM index to process.
- Experimental evaluation: our framework has been fully implemented, including on-line sensor data segmentation, modeling, KVM-index and related query processing algorithms. It has been thoroughly evaluated against a significant number of alternative approaches. As experimentally shown, our approach significantly outperforms in terms of query response time all other ones for time/value point, range and composite queries.

The remainder of this paper is as follows: Sec. 2 summarizes some related works on sensor data management, interval index and index based MapReduce optimization. In Sec. 3, we describe sensor data segmentation and modeling and querying model-view sensor data. The detailed designs of our KVM-index are presented in Sec. 4. In Sec. 5 and Sec. 6, we discuss the updating and query processing algorithms of *KVM-index* respectively. Then, in Sec. 7, we present thorough experimental results to compare our approach with traditional ones on both raw sensor data and modeled segments. Finally, in Sec. 8, we conclude our work and discuss our future work.

2   Related Work

Many researchers have proposed techniques for managing modeled segments of sensor data in relational databases. MauveDB [3] designed a model-based view to abstract underlying raw sensor data; it then used models to project the raw sensor readings onto grids for query processing. As opposed to MauveDB, FunctionDB [2] only materialized segment models of raw sensor data. Symbolic operators are designed to process queries using models rather than raw sensor data. However, both approaches in [3] and [2] do not take into account the role of an index in managing model-view sensor data.

Many index structures have been proposed to manage interval data [7,8,6]. However, they all are memory-oriented structure and cannot be directly applied to the distributed key-value cloud stores. Many efforts [8,7,9,10] have also been done to externalize these in-memory index data structures. The relational interval tree(RI-tree) [6] integrates interval tree into relational tables and transforms interval queries into SQL queries on relational tables. This method makes efficient use of built-in B+/-tree indices of RDBMS. The authors in [11] proposed an approach based on interval-spatial transformation, which leveraged multi-dimensional index to manage transformed interval data in spatial space. However, such approach needed to invoke three different range queries for one interval intersection query and therefore it is not applicable to our large-scale data in distributed environments. The latest effort to develop the interval index for distributed key-value stores [8] utilized MapReduce to construct a segment tree materialized in the key-value store. This approach outperforms the interval query processing provided by HBase (http://hbase.apache.org/). However, segment tree [8] is essentially a static interval index and requires data duplication to index segments, as is also the case with [12].

MapReduce parallel computing is an effective tool to access large scale of segment models of sensor data in cloud stores. Many researchers proposed index techniques to avoid data scan by MapReduce for low-selective queries [13]. The authors in [14] integrated indices into each file split, so that mappers can use index to only access predicate qualified data records. In [15], indices applicable to queries with predicates on multiple attributes via indexing different record attributes in different block replicas were designed. In [16], a split-level index is designed to decrease MapReduce framework overhead. However, above index based MapReduce optimizations are only employed in processing static data.

3 Preliminaries upon Model-View Sensor Data Management

In this section, we discuss some issues related to model-view sensor data management. First, we explain the modeled segments of sensor data and generalize the process of online segmenting sensor time series. Then the query types of our interest are presented. Finally, we describe some particular techniques for querying model-view sensor data.

3.1 Sensor Data Segmentation and Modelling

Various sensor data segmentation and modelling algorithms have been extensively researched, such as PCA,PLA DFT, etc.. [17,4,18]. The core idea is to fragment the time series from one sensor into *modeled segments*, and then approximates each segment by a mathematical function with certain parameters[17]. The mathematical model for each segment takes as dependent variable the sensor value and as independent variable the time-stamp. For simplicity, we will refer the modeled segment as segment in the rest of the paper. For example, in Fig. 1(a), the time series is divided into eight disjoint segments each of which

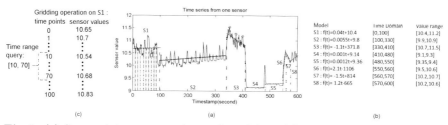

Fig. 1. (a) Sensor data segmentation and modeling; (b) linear regression function of each segment; (c) discrete time-stamp and sensor value pair set from segment gridding

is modeled by a linear regression function and has associated time domain and value range shown in Fig. 1(b). In model-view sensor data management, only the segment models are materialized and therefore the query processing is performed on the segments instead of the raw sensor data, as in [2].

3.2   Querying Model-View Sensor Data

First, we describe the categories of queries on model-view sensor data of our interest.

- Time point or range query: return the values of one sensor at a specific time point or during a time range.
- Value point or range query: return the timestamps or time intervals when the values of one sensor is equal to the query value or fall within the query value range. There may be multiple time points or intervals of which sensor values satisfy the query predicate.
- Composite query: the predicate of composite queries involves of both time and value dimension and the query processor returns a set of timestamp-sensor value pairs satisfying the time and value constraints.

In total, we attempt to support five kinds of queries over model-view sensor data. Different from traditional querying raw sensor data that directly returns the sensor data by index searching, the generic process to query model-view sensor data comprises the following two steps:

- Searching qualified segments: qualified modeled segments are defined as the ones of which time (resp. value) intervals intersect the query time (resp. value) range or cover the query point. This step makes use of index techniques to efficiently locate the qualified segments.
- Gridding qualified segments: From the perspective of end users, abstract functions of qualified segments are not user-friendly and difficult for data visualization[2]. Therefore, segment gridding is a necessary process[3,2], which converts the abstract functions into a finite set of data points. It applies three operations to each qualified segment: (i) It discretizes the time interval of one segment at a specific granularity to generate a set of time points. (ii) It calculates the sensor values at the discrete time points based on the models of segments. (iii) It filters out the sensor values that do not satisfy the query predicates. The qualified time or value points are returned as query results. Take the gridding operation on segment S1 in Fig.1(a) for an example. The gridding result is a set of discrete timestamp and value pairs shown in Fig.1(c), from which the predicate-satisfied pairs are extracted as results.

4   Key-Value Model Index

In this section, we describe the proposed key-value model index(KVM-index). Building the model index on top of distributed key-value stores enables high throughput, concurrency, scalability and distributed computing for querying sensor data. We first give a general structure and then discuss each component of the KVM-index.

4.1   Structure of KVM-Index

As each segment of sensor data has a specific time and value range, instead of indexing the mathematical functions of segments, our idea is to take the time and value interval as keys to index each segment, which allows the KVM-index to directly serve the queries proposed in Sec. 3.

Our KVM-index consists of the time and value dimension each of which is an interval index. For a sensor data segment in Fig. 2(a), the time and value intervals are respectively indexed by the KVM-index shown in Fig. 2(b) and (c). (As the structures of the time and value part of the KVM-index are similar, we just omit the details of the value part in Fig. 2(c).). The time (or value) interval index is a composite structure including the virtual searching tree (*vs-tree*) in memory and an index-model table in the distributed key-value store. The interval index takes effect identically on time and value intervals, therefore we will describe the generic design and more details can be found in [19].

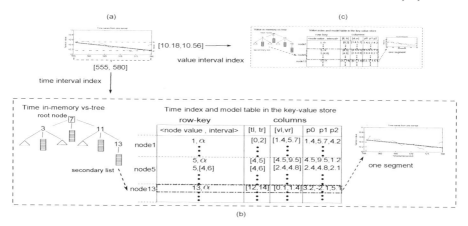

Fig. 2. Structures of KVM-index

4.2   In-memory *vs-tree*

In memory, we define a standard binary search tree as the virtual searching tree (*vs-tree*), as in the whole life of KVM-index, only the root value r of *vs-tree* is kept in memory and the values of other nodes can be online computed during the search process. The time (or value) interval of one segment to index is registered at the highest node overlapping the interval during the searching from the root node. This unique node, which is defined as the registration node τ for the time (or value) interval, is responsible for materializing the segment into the secondary list of τ in the key-value store. For negative sensor data values, we use simple shifting to have the data range start from 0 for convenience. Then, the domain of the *vs-tree* is defined as $[0, R]$. All the operations on the *vs-tree* are performed in memory.

4.3   Key-Value Index-Model Table

In the key-value store, we design a composite storage schema, namely index-model table, which enables one table not only to store the functions of segments, but also to materialize the structural information of the *vs-tree*, i.e., the secondary list (SL) of each tree node. A index-model table in the distributed key-value stores consists of a collection of splits, called regions. Each region will be hosted by a region server. All the region servers are coordinated by the master server for data distribution, fault-tolerance and query processing. As is shown in Fig.2(b), each row of the time (or value) index-model table corresponds to only one segment of sensor data, e.g., the segment in the black dotted rectangle in Fig.2(b). A row key consists of the node value and the interval registered at that node. The time and value domain and the coefficients of the function are all stored in different columns of the same row. This design of the row-key enables the SL of one node to form a consecutive range of rows, which facilitate the query processing discussed later. Fig.2(b) provides the examples of the SLs of node 1, 5 and 13, which are materialized both in the time and value index-model table.

In Fig.2(b), the α in some row-keys is a postfix to indicate that this row is the starting position of the SL of one node in the table. The selection of specific α should make sure that the binary representation of $< \tau, \alpha >$ is in front of any other $< \tau, l_t, r_t >$. This design is very helpful for accessing particular SL. For instance, if the query processor requires to access all the segments stored at node 5, then all the corresponding segments lie in the rows within the range $[< 5, \alpha >, < 6, \alpha >)$.

5   KVM-Index Updates

In this section, we present the algorithm of online updating new incoming segments into KVM-index. As is shown in Fig. 3(a), it includes two processes, namely searching the registration node in memory and materializing segments into the index-model table.

5.1   Searching Registration Node (rSearch)

The *rSearch* is responsible for locating the node τ at which the time (or value) interval $[l_t, r_t]$ should be registered. See the *rSearch* part in Fig.3(a). It is similar to binary search and the stop condition is that the $[l_t, r_t]$ overlaps the current node value.

```
Input: [l_t, r_t]. /* value and time intervals of one segment
M_i, r /* M_i denotes the coefficients of the segment
begin
    /* expand the domain if needed
    if (r_t > R) then
    |   r= 2^⌈log(r_t+2)⌉-1 - 1 /* expand to new root value
    end
    /* search registration node
    cur=r; /* current node on the search path h=log(r)-1;
    while (h ≥ 0) do
        if (l_t ≤ cur and r_t ≥ cur) then
        |   break; /* current node is the registration node
        end
        else
            if (l_t > cur) then
            |   cur= cur+ 2^h;
            end
            if (r_t < cur) then
            |   cur= cur- 2^h;
            end
            h=h-1;
        end
    end
    /* materialize the model of the segment.
    if the SL of node has been initialized then
    |   rowkey= <node|l_t|r_t>;
    end
    else
    |   rowkey= <node|α>;
    end
    insert [l_v, r_v]. [l_t, r_t], M_i into the time(or value) index-model table.
end
```

(a)

```
Input: time query range [l_t, r_t], root value r
Output: E /* node set for further MapReduce processing
begin
    cur=r; /* current node on the search path h=log(r)-1;
    while (h ≥ 0) do
        if (l_t ≤ cur and r_t ≥ cur) then
        |   break; /* cur is the registration node
        end
        else
            E = E ∪ cur
            if (l_t > cur) then
            |   cur= cur + 2^h;
            end
            if (r_t < cur) then
            |   cur= cur - 2^h;
            end
            h=h-1;
        end
    end
    /* search split D .
    Binary-search(cur, l_t.E);
    Binary-search(cur, r_t.E);
    E = E - E ∩ [l_t, r_t] /* intersection search result
end
```

(b)

Fig. 3. (a) Sensor data segment update algorithm of KVM-index; (b) Intersection search on the *vs-tree*

As sensors yield new data on the fly, the time (or value) domain of *vs-tree* should be able to catch up with the variation of the time(or value) domain of sensor data. Therefore, the update algorithm first invokes a *domain expansion process* to dynamically adjust the domain of current *vs-tree* if needed. Then, the registration node can be found on the validated *vs-tree*. Based on the following Lemma 1 and Lemma 2, KVM-index is able to decide when and how to adjust the domain of *vs-tree*. The detailed proofs can refer to [19].

Lemma 1. *For a segment M_i with time(value) interval $[l_t, r_t]$, its registration node lies in a tree rooted at $2^{\lceil \log(r_t+2) \rceil - 1} - 1$.*

Lemma 2. *For a segment M_i with time (value) interval (l_t, r_t), if the right endpoint r_t exceeds the current domain R, namely $r_t > R$, the domain of vs-tree needs to expand.*

For example, in Fig. 4(a) when model 1 is to be inserted, the *vs-tree* rooted at node 7 is still valid and model 1 is registered at node 1. However, when model 2 arrives, its right end-point, i.e., 21, exceeds the domain $[0, 14]$. The *vs-tree* is expanded to root at 15 and the new extended domain is the area enclosed by the dotted block. Then, model 2 can be updated successfully.

5.2 Materializing Segments

This process first constructs the row-key based on τ and then materializes the model of the segment into the corresponding row. See the materialization part

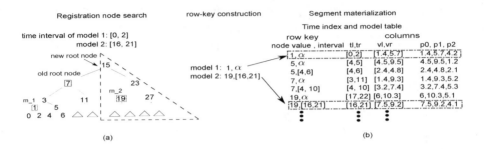

Fig. 4. Sensor data segment update in KVM-index

in Fig.3(a). The row key is chosen as $< \tau, \alpha >$ when no segment has been stored at the SL of τ. Otherwise, the time interval $[l_t, r_t]$ (or $[l_v, r_v]$ for value interval) is incorporated into the row key, i.e., $< \tau, l_t, r_t >$ (resp. $< \tau, l_v, r_v >$) to avoid different segments in the SL of τ overwrite each other. We illustrate the segment materialization process with the examples of $model1$ and $model2$ in Fig. 4. First, the KVM-index checks whether the starting rows with key $< 1, \alpha >$ and $< 19, \alpha >$ exist. Then, $< 1, \alpha >$ is constructed for model 1, whilst the KVM-index constructs the row key $< 19, [16, 21] >$ for model 2 as the SL of node 19 has been initialized.

6   Query Processing

In this section, we describe how to leverage KVM-index to query model-view sensor data. The novelties of our algorithm lie in three aspects. First, the in-memory time (or value) vs-$tree$ utilizes the intersection and stabbing search to efficiently locate the nodes which accommodate qualified segments. Second, different from the sequential scan in traditional disk-based clustered index, our proposed data access approach based on MapReduce is capable of processing the SLs discretely located in the index-model table for range and point queries. Finally, as for composite queries, we put forward an adaptive strategy to decide which dimension of KVM-index is more efficient for MapReduce to process.

6.1   Intersection and Stabbing Search on vs-$tree$

For time (or value) range and point queries, the corresponding operations on the time (or value) vs-$tree$ are the intersection and stabbing search [6,10]. Fig. 3(b) presents the intersection search algorithm. Given a time (resp. value) query range $[l_t, r_t]$, $rSearch$ is first invoked to find the registration node τ of $[l_t, r_t]$. Then, the searching path splits at τ and two binary searches are called to individually continue searching for the l_t and r_t. The stabbing search is similar to the standard binary search, so we omit the detailed description here and the dotted-linked nodes in Fig. 5(a) give an illustration. All the visited nodes during the intersection and stabbing search are recorded in a node set $\mathcal{E}$. Regarding the nodes in $[l_t, r_t]$, all the intervals registered at them must overlap $[l_t, r_t]$. The final results of the intersection search are the nodes in $\mathcal{E}$ outside of $[l_t, r_t]$. This design facilitates the split construction of MapReduce processing discussed later. For the stabbing search, there exists no such problem. For example, in Fig. 5(a), node 7 is the registration node of the query range $[6, 12]$. The links between nodes on the left sub-tree of the root represent the path of intersection search. The resulting $\mathcal{E}$ is extracted from the path, namely, $3, 5, 13, 15$. Then, we will analyse the property of the nodes in $\mathcal{E}$.

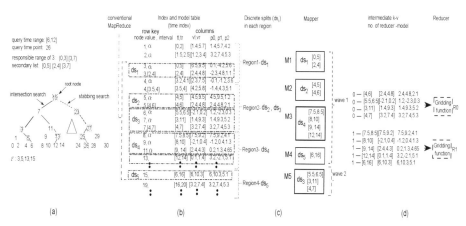

Fig. 5. Query processing with KVM-index and MapReduce

Theorem 1. *The SLs of the nodes in $\mathcal{E}$ have qualified segments while any node outside of the $\mathcal{E}$ and the query range does not have qualified segments.*

Proof. Define $[l_i^j, r_i^j]$ as the intervals registered at node j of vs-tree. The left and right child node of j are separately denoted by c_l^j and c_r^j. The current node that the search path passes by is denoted by v.

First, we focus on the nodes on the path from root to registration node. Suppose the searching path turns left from root, namely $r_t < r$. For root node, its left responsible range is $[0, r]$, and therefore there exists $l_i^r < r_t$ in the SL of root. As for the right sub-tree of root c_r^r, no intervals in the SL of c_r^r satisfies $l_i^{c_r^r} < r_t$, otherwise such intervals would be registered at r rather than c_r^r. Above analysis applies to the case when the searching path turn right and other nodes on the path.

Then let's see the path from the registration node down to l_t. If the search path turns left, namely, $l_t < v$, then node v is within the query range $[l_t, r_t]$ and all the intervals in the SL of v intersect $[l_t, r_t]$. For the case when the search path turns right, $l_t > v$, there exists intervals in the SL of v satisfying $r_i^v > l_t$. But for c_l^v, all the intervals in the SL of c_l^v address the $r_i^{c_l^v} < l_t$, so the left sub-tree of v is discarded. This part of analysis applies to the path from registration node down to r_t. $\square$

Based on above analysis, the *SLs* of nodes in $\mathcal{E}$ may also comprise some un-qualified segments. For instance, in Fig. 5(a) the segments with time intervals $[0, 5]$, $[2, 4]$ and $[3, 7]$ are all registered at node 3, whilst only the one with $[3, 7]$ satisfies the query range $[6, 12]$. Therefore, we need a segment filtering procedure for each *SL*. So far, after the in-memory search on the *vs-tree*, a distributed algorithm integrating both segment filtering and gridding is necessary to evaluate queries.

6.2   Distributed Segment Filtering and Gridding

We apply MapReduce to parallelize the segment filtering-gridding process in which each mapper takes a data split as its input. Due to the sequential data feeding mechanism, MapReduce constructs one split for each region of the key-value table before each mapper starts to scan the assigned split. However, the distribution of *SLs* in the index-model table poses challenges for conventional MapReduce.

Input: $\mathcal{E}$.
Output: Re_j, $(j = 1...n_r)$
/* node list mapped region j of the time(or value) table
begin
    for *each node i in* $\mathcal{E}$ do
        /*Consult the meta-data store to locate
        the region h_i that hosts the *SL* of node i
        $Re_{h_i} = Re_{h_i} \cup i$
        /* insert i into the node list of region Re_{h_i}
    end
end

(a)

Input: query range l_t, r_t (or l_v, r_v)
Re_j, $(j = 1...n_r)$./* n_r is the number of regions
$l_k, u_k, (k = 1...n_r)$ /* lower and upper boundary
of the rows in region k of the time(or value) table
Output: S_k, /*discrete splits set for each region
begin
    for *each region k* do
        if $l_t \leq r_k$ and $r_t \geq l_k$ then
            lb= $l_t \leq l_k$? $l_k : l_t$;
            rb= $r_t \geq r_k$? $r_k : r_t$;
            newSplit= the rows between
            $< lb >$ and $< rb >$;
            $S_k = S_k \cup$ newSplit;
        end
        for *each node j in* Re_k do
            newSplit= the rows between
            $< j, \alpha >$ and $< j + 1, \alpha >$;
            $S_k = S_k \cup$ newSplit;
        end
    end
end

(b)

Fig. 6. (a) SR-mapping algorithm; (b) disSplits algorithm

As the nodes in set $\mathcal{E}$ are discrete, the associated *SLs* correspond to discontinuous ranges of rows in the index-model table. For the time range query $[6, 12]$ in Fig.5(b), the blue dash-dotted blocks indicate the *SLs* of nodes in $\mathcal{E}$ and the red dash-dotted blocks represent the *SLs* of nodes in $[6, 12]$. One straightforward idea is to make MapReduce work on a sub-table covering all the *SLs* of nodes in $\mathcal{E}$ and query range $[l_t, r_t]$. This approach is introduced as a baseline one in Sec. 7. For example, in Fig. 5(b), the left red brace shows the minimum range of rows for MapReduce to process for range query $[6, 12]$. This approach is not intrusive for the kernel of MapReduce, but as shown in Fig. 5(b), MapReduce has to process additional *SLs* corresponding to the irrelevant nodes outside of $\{3, 5, 13, 15\}$ and $[6, 12]$, which consequentially causes a lot of extra computation and communication overheads.

Therefore, we propose a discrete data access method which includes two algorithms denoted by *SR-mapping* and *disSplits* shown in Fig. 6. The idea is to find the ownerships of the *SLs* in $\mathcal{E}$ among regions and afterwards to construct a split generation kernel which produces discrete splits only for the relevant *SLs* in each region. In Fig. 6(a), *SR-mapping* algorithm consults the meta-data store in the master of the distributed key-value store to build a mapping between the *SLs* of nodes in $\mathcal{E}$ and the regions hosting them.

The *SLs* of nodes in $[l_t, r_t]$ are continuous in the index-model table, thus the *disSplits* algorithm uses the range of row-keys of each region to locate consecutive *SLs* overlapping $[l_t, r_t]$ and produces one split for such *SLs* in the region. In this way, *disSplits* algorithm is able to overcome inefficient iterations over each *SL* of the node in $[l_t, r_t]$. Then, *disSplits* generates additional discrete splits for the *SLs* in $\mathcal{E}$. Moreover, *disSplits* algorithm can avoids generating the splits with overlapping *SLs*, since the nodes in the query range have been eliminated from $\mathcal{E}$ in the intersection search of *vs-tree*. For example, see the discrete splits constructed in each region in Fig. 5(c). With *SL-mapping* and *disSplits* algorithms, we completely eliminate the irrelevant *SLs* between the nodes in $\mathcal{E}$ and $[l_t, r_t]$ for MapReduce computing. Furthermore, the *SR-mapping* and *disSplits* are both conducted efficiently in-memory and incur no network I/O overhead.

As for the MapReduce based segment filtering and gridding, suppose that the number of reducers is P and each reducer is denoted by $0, \cdots, P$-1. For point queries, we use only one reducer for segment gridding. For range queries, the partition function f is responsible for assigning the qualified segments to different reducers. It is designed on the basis of query time (resp. value) range $[l_t, r_t]$ (resp. $[l_v, r_v]$) and the time (or value) interval $[l_i, r_i]$ of qualified segment M_i. The idea is that each reducer is in charge of one even sub-range $\frac{r_t - l_t}{P}$. Such a partition function f is given in Eq. 1.

$$f(r_i) = \begin{cases} l_t \leq r_i \leq r_t & \lfloor \frac{(r_i - l_t) * P}{r_t - l_t} \rfloor \\ r_i \geq r_t & P - 1 \end{cases} \qquad (1)$$

The functionalities of mappers and reducers are depicted in details below and also shown in Fig.5(d) for an illustration.

- **Mapper:** Each mapper checks the time (resp. value) interval $[l_i, r_i]$ of one segment i in its data split. For the qualified segments, the intermediate key is derived from the partition function $f(r_i)$. The model coefficients $< p_i^1, \cdots, p_i^n >$ form the value part of the intermediate key-value pair.
- **Reducer:** Each reducer receives a list of qualified segments $< p_0^1, \cdots, p_0^n >$, $< p_1^1, \cdots, p_1^n >, \cdots$ and invokes the segment gridding function for each segment.

So far, our MapReduce based segment filtering and gridding is able to process the *SLs* located by the intersection or stabbing search on *vs-trees* with the *SR-mapping* and *disSplits* algorithms.

6.3   Adaptive MapReduce Processing for Composite Queries

The composite query is a two-dimensional query on both time and value domain. Traditional methods to process composite queries is to build two-dimensional interval index (e.g., multi-dimensional segment tree, multi-dimensional interval tree, etc.), which results in heavy communication and computation cost for index maintenance in the distributed environment.

Our idea is to fit the composite query processing into the above KVM-index-MapReduce framework. KVM-index is responsible for filtering on one dimension and the discrete split generation. Then, the filtering on the other dimension is integrated into the mapping phase of the MapReduce processing. The overhead caused by adding the other predicate checking into the segment filtering in the mapping phase can be negligible. However, one emerging problem is which dimension (i.e., the time or value) of KVM-index to use for the initial filtering and discrete split construction, since the cost of processing the time or value index-model table vary greatly due to different *vs-trees* and discrete split generation. In our proposed cost model, we will take into account waves and data-transferring overhead of discrete splits which dominate the performance of MapReduce programs [14,15,16].

Waves are the number of rounds for one MapReduce program to invoke mappers to process all the splits n_s given the maximum mappers slots M available in a cluster. It is defined as

$$W = \frac{n_s}{M} \tag{2}$$

Take the mapping phase in Fig.5(c) for an example. Suppose the mapper slots for each region server and the cluster are respectively 1 and 4. Then for a query involving of 5 discrete splits, two waves of mappers are needed. The time and value dimensions of KVM-index produce different number of discrete splits n_s depending on the different selectivity of the composite query on the time and value dimension. As the maximum capacity of mapper slots is constant, waves are determined by n_s, which reflects different amount of data for MapReduce to process.

Data transferring affects the communication overheads of MapReduce programs and is related to the distribution of discrete splits among regions. The mapper task assignment mechanism in MapReduce and key-value stores takes advantage of data locality to avoid extra communication overhead, thus the mapper slots of one node are priorly allocated to the local discrete splits. For the nodes hosting less discrete splits than the number of waves, they have to fetch data from other ones after finishing all the local splits. For the nodes of which the number of local discrete splits is greater than that of waves, the redundant local discrete splits will be transmitted to other spare nodes in order to finish the mapping phase within the wave constraint. Therefore, we define the data transferring cost based on waves, the number of discrete splits $|S_i|$ in each region and the mapper slots m_i of each node. Here we assume that the regions of the index-model table are evenly distributed across the cluster and each node accommodates one region.

$$D_t = \sum_{i=1}^{n_r} g(\lfloor W \rfloor * m_i - |S_i|) \tag{3}$$

where $g(x) = \begin{cases} 0, x >= 0 \\ |x|, otherwise \end{cases}$ and n_r is the number of regions of the index-model table.

The first case of function $g(*)$ corresponds to un-saturated node, we set the cost as 0 since the data transferring cost will be added by the over-saturated one. Then, the total cost is defined as the weighted linear combination of waves and data transferring overhead as follows:

$$C = \alpha * W + (1 - \alpha)D_t, (0 \leq \alpha \leq 1) \qquad (4)$$

The weight parameter is user-specific and tunable. For a composite query, we first respectively invoke the *vs-tree* search, *SR-mapping* and *disSplits* of the time and value dimension of KVM-index. And then the query processor calculates the costs to decide which dimension to use for split construction and which dimension to be integrated into the segment filtering in mappers. So far, all the queries proposed in Sec. 3.2 can be evaluated under the KVM-index-MapReduce framework.

7   Experiments

In this section, we conduct extensive experiments to evaluate the update and query processing performance of our KVM-index and compare with other query processing approaches based on both raw and model-view sensor data.

7.1   Setup

We build our system on the Hbase and Hadoop in Cloudera CDH4.3.0. The experiments are performed on our private cloud that consists of 1 master and 8 slaves. The master node has 64GB RAM, 4TB disk space (4 x 1TB disks in RAID5) and 12 2.30 GHz (Intel Xeon E5-2630) cores. Each slave node has 6 2.30 GHz (Intel Xeon E5-2630) cores, 32GB RAM and 6TB disk space (3 x 2TB disks). All the nodes are connected via 1GB Ethernet.

The data set contains discrete accelerometer data from mobile phones each of which has one timestamp and three sensor values representing the coordinates. The size of the raw sensor data set is 22 GB including 200 millions data points. We simulate the sensor data emission in order to online segment and update sensor data into the KVM-index. We implement an online sensor data segmentation component [17] applying the PCA(piecewise constant approximation)[1], which approximates one segment with a constant value (e.g., the mean value of the segment). As how to segment and model sensor data is not the focus of this paper, other sensor data segmentation and modelling approaches can also be applied here. Provided that the segments are characterized by the time and value intervals, our KVM-index and related query processing techniques are able to manage them efficiently in the cloud. Finally, there are around 25 millions sensor data segments (nearly 15 GB) uploaded into the key-value store. Regarding the segment gridding, we choose 1 second as the time granularity which is application-specific.

7.2 Baseline Approaches

We develop four baseline approaches for querying sensor data to compare with, namely:

Raw data based MapReduce (RDM). We create two tables, which respectively take the timestamp and sensor value as the row-keys, such that the query range or point can be used as keys to locate the qualified data points. Then the query processor invokes MapReduce programs to access the data points of large size.

MapReduce (MR). This approach utilizes native MapReduce to process the whole index-model table for filtering the qualified segments in the map phase and performing the segment gridding in the reduce phase.

MapReduce+KVI (MRK). *MRK* is to leverage KVM-index to find a consecutive sub-index-model table which covers all the *SL*s of nodes visited by the search on the time(or value) *vs-tree* for MapReduce processing. See the red brace shown in Fig. 5(b) for an example.

Filter of the key-value store (FKV). HBase provides the filter functionality to support predicate-based row selection. The filter component sends the filtering predicate to each region server. Then all servers scan their local data in parallel and return the qualified segments. Then the query processor retrieves each returned qualified segment to conduct gridding locally.

7.3 Update

Fig. 7(a) exhibits the comparison of update performance between model-view based KVM-index and raw sensor data based approach in the simulation of sensor data on-line emission. The horizontal axis represents the data loading process. The vertical axis is the average time of update operations for the segments during the loading process of the previous 10% data. The KVM-index executes one update operation each time a new segment is produced while the raw sensor data approach updates a sensor data point each time. The average update time of KVM-index keeps relatively stable and is 1× greater than that of the raw sensor data approach, as it needs to materialize the segment information into the key-value store. But the amount of data uploaded by the raw data approach is much larger than model-view approach as we mentioned in Sec. 7.1.

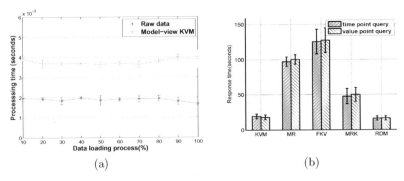

(a)   (b)

Fig. 7. (a) Comparison of sensor data update performances; (b) Point queries

7.4   Point Queries

The performance of time and value point queries are shown in Fig. 7(b). The response time of *KVM* is greater than *RDM*, because *RDM* is able to locate the qualified data points directly by the query point while *KVM* needs to invoke MapReduce for segment filtering and gridding. Our *KVM* only processes the *SLs* of nodes from the searching on *vs-tree*, and therefore outperforms the other model-view based approaches. *MRK* takes nearly 1× less time than *MR*, as *MRK* works on the sub-table derived from the searching path on the time (or value)*vs-tree*. *FKV* has the largest response time, since it needs to wait for server-side full table scan and to conduct the sequential segment gridding locally.

7.5   Range Queries

In Fig.8, we refer to query selectivity as the ratio of the number of qualified segments(or data points) over that of total segments (or data points) for model-view approaches (or *RDM*). As is depicted in Fig.8(a), *KVM* outperforms *MR* up to 2× for the low-selective time range queries. *KVM* starts mappers for relevant *SLs* and distributes the workload of segment filtering and gridding, thus even for queries with high selectivity 80% it still achieves 30% improvement over *MR*. The query time of *MRK* is nearly 1× greater than that of *KVM*, but 15% less than that of *MR*. Compared with full table scan of *MR*, *MRK* only processes the sub-table delimited by the KVM-index. Yet, as compared to *KVM*, *MRK* processes more redundant segments. Regarding the *RDM*, as the query selectivity

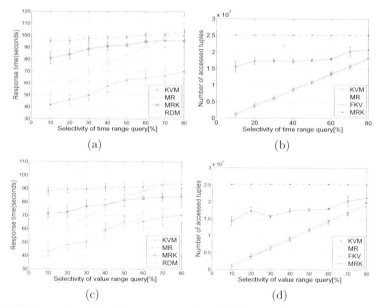

Fig. 8. (a) Query times for time range queries; (b) Accessed segments for time range queries; (c) Query times for value range queries; (d) Accessed segments for value range queries

increases, the amount of data to process for MapReduce becomes much larger than that of model-view approaches and therefore RDM consumes 10% more time at least than that of model-view approaches. As FKV needs to wait for each region server to finish the local data scanning, it takes 1000 to 5000 seconds for queries from selectivity from 10% to 80%. We omit to plot the results of FKV in Fig.8(a) for convenience.

We also analyse the numbers of accessed segments by each approach in Fig. 8(b), which exhibits how different access methods of the segments affect the query performance (as the size of raw sensor data set is much larger than that of segment set, we omit the RDM in Fig. 8(b)). Since MR scans the entire table, the number of accessed segments is the greatest. From an application's point of view, only qualified segments are returned for gridding in FKV, thus the amount of the processed segments matches the selectivity of queries. Our KVM processes a little more segments than FKV due to some unqualified segments in the SLs. MRK accesses more segments than FKV and KVM as the sub-index-model table processed by MRK comprises many irrelevant SLs. Regarding value range queries, the results present similar patterns as for the time range queries in Fig. 8 (c) and (d), hence we skip the detailed analysis.

7.6   Composite Queries

In this part, we focus on the composite queries with range constraints on the time and value dimension, since they are the most complex form of composite queries. Moreover, as the above experiments have verified the superiority of our KVM-index based query processing over the other model-view and raw sensor data based approaches, we only evaluate the cost adaptive approach itself. Our experimental methodology is to fix the selectivity of time dimension and observe the variation of query times with increasing selectivity on the value dimension. Limited by the space, we only show the results from three groups of experiments with different fixed selectivity on the time dimension (i.e., 10%, 40% and 70%) in Fig.9. The red line represents the results from the adaptive approach while the blue line indicates the response times of query processing on the dimension opposite to that of the adaptive approach. For instance, if the adaptive approach

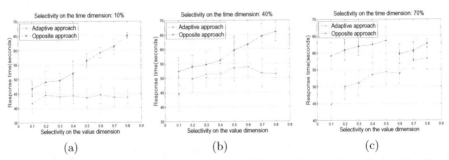

Fig. 9. Query times for composite queries of different fixed selectivity on the time dimension (a) 10%; (b) 40%; (c) 70%

Table 1. Selection probability of the time and value dimension of KVM-index

	$\beta = 10\%$	$\beta = 40\%$	$\beta = 70\%$
$P(t\mid s_v \leq \beta)$	0.58	0.32	0.25
$P(v\mid s_v \leq \beta)$	0.42	0.68	0.75
$P(t\mid s_v > \beta)$	0.9	0.94	0.92
$P(v\mid s_v > \beta)$	0.1	0.06	0.08

chooses the time dimension of KVM-index, the opposite one would work on the value part. Regarding the weight α in the cost model in Sec. 6.3, as the starting overhead of each wave of mappers is significant for CPU non-intensive jobs [14,15,16], we choose 0.65 to put more weight on the waves.

Fig. 9 shows that our adaptive approach can choose the appropriate dimension of KVM-index for efficient segment filtering and gridding. Moreover, as the difference of selectivity between the time and value dimension increases, our adaptive approach can outperform the opposite one up to 30%.

Next, we attempt to find out how our cost model in Eq.4 takes effects in the adaptive selection. Tab. 1 lists the probability that the time(or value) dimension of KVM-index has been chosen in different phases of each experiment group. For example, the first column $\beta = 10\%$ corresponds to the first group of experiment. s_v(or s_t) means the selectivity of composite queries on the value(or time) dimension. Then $P(t\mid s_v \leq \beta)$ represents the probability that the time dimension is selected when $s_v \leq \beta$. Intuitively, the dimension with low selectivity results into less splits and therefore only the value dimension should be selected when $s_v \leq \beta$, namely, $P(t\mid s_v \leq \beta)$ should be close to zero. However, since our cost model takes into account both the waves and the distribution of splits in the index-model table, $P(t\mid s_v \leq \beta)$ and $P(v\mid s_v \leq \beta)$ are comparable, which means that the distribution of splits affects the choice of the adaptive approach. When $s_v > \beta$, the amount of data to process becomes dominant in the cost model, thus $P(t\mid s_v > \beta)$ is high while $P(v\mid s_v > \beta)$ keeps low.

8 Conclusion

Model-based sensor data management is undergoing the transition from relational databases to the cloud environment. To the best of our knowledge, this is the first work to explore model index for managing model-view sensor data in the cloud. Different from conventional external-memory index structure with complex node merging and splitting mechanisms, our *KVM-index* consisted of two interval indices resident partially in memory and partially materialized in the key-value store, is easy and efficient to maintain in the dynamic sensor data yielding environment. Moreover, we designed a novel approach to process the discrete *SLs* in the index-model table. Based on this KVI-index-MapReduce framework, a cost model based adaptive query processing for composite queries is proposed. Extensive experiments show that our approach outperforms in terms of query response time and index updating efficiency not only traditional query processing methods over raw sensor data, but also all other approaches based on model-view sensor data.

Acknowledgement. This research is supported in part by the EU OpenIoT Project FP7-ICT-2011-7-287305 and OpenSense project under SNF Number 20NA21_128839.

References

1. Sathe, S., Papaioannou, T.G., Jeung, H., Aberer, K.: A survey of model-based sensor data acquisition and management. In: Managing and Mining Sensor Data, Springer (2013)
2. Thiagarajan, A., Madden, S.: Querying continuous functions in a database system. In: SIGMOD (2008)
3. Deshpande, A., Madden, S.: Mauvedb: supporting model-based user views in database systems. In: SIGMOD (2006)
4. Papaioannou, T.G., Riahi, M., Aberer, K.: Towards online multi-model approximation of time series. In: MDM (2011)
5. Bhattacharya, A., Meka, A., Singh, A.: Mist: Distributed indexing and querying in sensor networks using statistical models. In: VLDB (2007)
6. Kriegel, H.-P., Pötke, M., Seidl, T.: Managing intervals efficiently in object-relational databases. In: VLDB (2000)
7. Elmasri, R., Wuu, G.T.J., Kim, Y.-J.: The time index: An access structure for temporal data. In: VLDB (1990)
8. Sfakianakis, G., Patlakas, I., Ntarmos, N., Triantafillou, P.: Interval indexing and querying on key-value cloud stores. In: ICDE (2013)
9. Ang, C.H., Tan, K.P.: The interval b-tree. Inf. Process. Lett. 53 (1995)
10. Arge, L., Vitter, J.S.: Optimal dynamic interval management in external memory. In: 37th Annual Symposium on Foundations of Computer Science (1996)
11. Shen, H., Ooi, B.C., Lu, H.: The tp-index: A dynamic and efficient indexing mechanism for temporal databases. In: ICDE, pp. 274–281. IEEE (1994)
12. Kolovson, C.P., Stonebraker, M.: Segment indexes: dynamic indexing techniques for multi-dimensional interval data. SIGMOD Rec. 20 (1991)
13. Iu, M.-Y., Zwaenepoel, W.: Hadooptosql: a mapreduce query optimizer. In: EuroSys (2010)
14. Dittrich, J., Quiané-Ruiz, J.-A., Jindal, A., Kargin, Y., Setty, V., Schad, J.: Hadoop++: making a yellow elephant run like a cheetah (without it even noticing). VLDB Endow. 3 (2010)
15. Dittrich, J., Quiané-Ruiz, J.-A., Richter, S., Schuh, S., Jindal, A., Schad, J.: Only aggressive elephants are fast elephants. VLDB Endow. 5 (2012)
16. Eltabakh, M.Y., Özcan, F., Sismanis, Y., Haas, P.J., Pirahesh, H., Vondrak, J.: Eagle-eyed elephant: split-oriented indexing in hadoop. In: EDBT (2013)
17. Guo, T., Yan, Z., Aberer, K.: An adaptive approach for online segmentation of multi-dimensional mobile data. In: Proc. of MobiDE, SIGMOD Workshop (2012)
18. Ding, H., Trajcevski, G., Scheuermann, P., Wang, X., Keogh, E.: Querying and mining of time series data: experimental comparison of representations and distance measures. VLDB Endowment 1 (2008)
19. Tian Guo, T.G.P., Aberer, K.: Model-view sensor data management in the cloud. In: Proceedings of the 2013 IEEE International Conference on BigData. IEEE Computer Society (2013)

Discovery of Areas with Locally Maximal Confidence from Location Data

Hiroya Inakoshi, Hiroaki Morikawa, Tatsuya Asai,
Nobuhiro Yugami, and Seishi Okamoto

Fujitsu Laboratories Ltd., Kawasaki, Japan
{inakoshi.hiroya,morikawa.hiroaki,asai.tatsuya,
yugami,seishi}@jp.fujitsu.com

Abstract. A novel algorithm is presented for discovering areas having locally maximized confidence of an association rule on a collection of location data. Although location data obtained from GPS-equipped devices have promising applications, those GPS points are usually not uniformly distributed in two-dimensional space. As a result, substantial insights might be missed by using data mining algorithms that discover admissible or rectangular areas under the assumption that the GPS data points are distributed uniformly. The proposed algorithm composes transitively connected groups of irregular meshes that have locally maximized confidence. There is thus no need to assume the uniformity, which enables the discovery of areas not limited to a certain class of shapes. Iterative removal of the meshes in accordance with the local maximum property enables the algorithm to perform 50 times faster than state-of-the-art ones.

Keywords: Optimized association rule mining, Location data, GPS.

1 Introduction

The development of the Global Positioning System (GPS) led to the development of a number of low-cost GPS-equipped devices that can calculate their precise location [11]. GPS receivers are widely used in aircraft, ships, and motor vehicles for both business and personal use. They are extensively used for fleet management and asset tracking in the logistics and retail industries [2][8]. Outdoor sports fans enjoy riding bicycles equipped with GPS-loggers [13]. Even digital cameras are now being equipped with GPS [7]. The development of applications and services for collecting location information from GPS devices as well as from other sources such as global system for mobile communication and Wi-Fi networks enable people to share their experiences with friends through those location histories.

There are even advanced applications and services that collect location histories and associate them with ones from different data sources [6][12]. For example, traffic information for use in identifying congested roads and determining the best route to take is now automatically generated from different vendors of

S.S. Bhowmick et al. (Eds.): DASFAA 2014, Part I, LNCS 8421, pp. 47–61, 2014.

car navigation services [10]. Tourism and meteorology industries could associate geo-tagged photos sent from subscribers with their findings. Of course, such applications require less investment and operate at lower cost if location histories and/or geo-tagged contents are available.

Location data are a collection of records, or data points, each of which comprises a device identifier, time stamp, longitude, latitude, and other optional values:

$$(id, time, longitude, latitude, v_1, \cdots, v_m).$$

With the applications mentioned above, two kinds of geospatial information should be discoverable from location data: regions and trajectories. While both are important in advanced geospatial applications, here we focus on discovering regions because trajectories can be represented as sequences of regions.

There are many types of regions with various properties. Giannotti et al. [6] defined regions of interest (RoIs) and represented a trajectory as a sequence of RoI. An RoI is a popular, or dense, area of points. Their *PopularRegion* algorithm extracts dense and rectangular regions. Zheng et al. [12] defined *stay regions*, which are narrow clusters of stay points where a subsequence of a trajectory is within given distance and time thresholds. Their *ExtractStayRegion* algorithm extracts dense and rectangular regions of stay points with sizes within $d \times d$, where d is a given distance threshold. Convergence and divergence patterns represent aggregate and segregate motions, respectively, and could be used to extract circular or rectangular RoIs [14]. In these algorithms, regions are defined in an ad hoc manner, so they work well for their designed purpose, such as for recommending near-by activities or analyzing trip patterns.

More generalized algorithms for optimized association rule mining have been reported [3–5][9]. They are mathematically well formulated and applicable to records in two-dimensional space. It is easy to discover stay regions by using such algorithms, as the *ExtractStayRegion* algorithm does, once the location records have been processed to have an additional field representing the time difference from previous ones. Optimized association rule mining is therefore promising as a generic methodology for RoI discovery using geospatial records. Consequently, our goal is to develop a technique for discovering optimized association rules with a property favorable for geospatial data, i.e., a local maximum confidence.

After some preliminaries and a look at related work, we introduce the local maximum property over location data on a two-dimensional plane. We then describe our $O(n \log n)$ time algorithm for extracting closed regions of interest. It works by filtering irregular meshes in accordance with the local maximum property of confidence. Next, we present experimental results showing that the proposed *discoverMaximalClosure* algorithm is 50 times faster than Rastogi's [9], which discovers the top-k totally optimized confidence regions. We then present two applications and end with a summary of the key points.

2   Preliminaries

First we introduce some preliminaries related to optimized association rule mining of location data and then discuss related work as characterized on the basis of the preliminaries.

An association rule is defined over the attributes of a record and has the form $C_1 \rightarrow C_2$. Each association rule has associated support and confidence. Let the support for condition $C_i(i = 1, 2)$ be the number of records satisfying C_i. It is represented as $sup(C_i)$. The support for a rule in the form $C_1 \rightarrow C_2$ is then the same as the support for C_1 while its confidence is the ratio of the supports of conditions $C_1 \wedge C_2$ and C_1:

$$sup(C_1 \rightarrow C_2) = sup(C_1),$$
$$conf(C_1 \rightarrow C_2) = \frac{sup(C_1 \wedge C_2)}{sup(C_1)}.$$

A rule is *confident* if its confidence is not less than a given confidence threshold θ. A rule is *ample* if its support is not less than a given support threshold Z.

The hit and gain of a rule are defined as:

$$hit(C_1 \rightarrow C_2) = sup(C_1 \rightarrow C_2)conf(C_1 \rightarrow C_2),$$
$$gain(C_1 \rightarrow C_2) = hit(C_1 \rightarrow C_2) - \theta \ sup(C_1 \rightarrow C_2).$$

Let records be in the form $(x, y, a_1, \cdots, a_m)$, where x and y are longitude and latitude, respectively, and $a_j(j = 1, \cdots, m)$ are other attributes associated with position (x, y). A device identifier or time stamp could be one in a_j. Although many studies have considered various forms of the presumptive and objective conditions, here we consider the case in which both are conjunctions of atomic conditions, which are $a_j = v$ for nominal attributes, $a_j \in [l, u]$ for numeric attributes, and $(x, y) \in D \subset R^2$ for location attributes.

In the optimized association rule mining, the presumptive condition can have uninstantiated atomic conditions. Atomic conditions are uninstantiated (*resp.* instantiated) when one of v, $[l, u]$, and D is a variable (*resp.* a value). If the uninstantiated and instantiated conditions are separately written, the association rule $U \wedge C_1 \rightarrow C_2$ appears. If U_i denotes an instantiation of U, U_i can be obtained by replacing variables in U with values.

Optimized association rule mining is categorized into three types:

1. An *optimized support problem* is to discover confident instantiations that maximize the support.
2. An *optimized confidence problem* is to discover ample instantiations that maximize the confidence.
3. An *optimized gain problem* is to discover ample instantiations that maximize the gain.

The related works introduced in section 3 consider some or all of those types of problems. We concentrated on the optimized confidence problem because it was the most important problem for the geospatial applications in which we were interested.

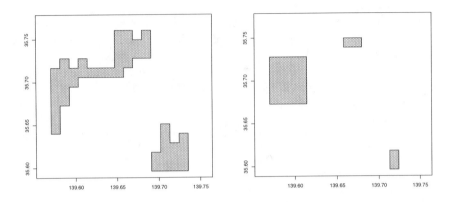

Fig. 1. Example of admissible, or x-monotone regions (left) and rectangular regions(right)

Example 1. Consider the database of a social messaging service. Let a record in the database be denoted by $(x, y, id, year, mon, mday, hour, text, photo)$, where *photo* is a Boolean attribute indicating whether a photo is attached to the message. Let the condition for position $(x, y) \in D$ be uninstantiated by letting D be a variable and the conditions $hour \in [l, u]$ and $photo = v$ be instantiated by letting $[l, u]$ and v be values. An uninstantiated association rule is thereby obtained.

$$((x, y) \in D) \wedge (hour \in [l, u]) \rightarrow (photo = v),$$
$$\text{where} \begin{cases} l = 5:00, u = 8:00, \\ v = True \end{cases}$$

To find an area where messages are likely attached to photos and posted in the early morning, we solve the optimized confidence problem by discovering ample instantiations of variable D such that the confidence of the uninstantiated rule above is maximized.

It is important to consider which class of region $D \subset R^2$ to choose. In related work, as described in section 3, admissible and rectangular regions were considered as illustrated in Fig. 1. An admissible region is a connected x-monotone region. A region is called x-monotone if its intersection with any vertical line is undivided. A rectangular region is obtained by making x and y independent numeric attributes. In two related studies, the entire region was separated into fixed grid-like meshes, or pixels, in the first stage, under the assumption that the points were uniformly distributed.

For cases in which the distribution of points is biased, the meshes and regions to be discovered should be defined in a general way.

Definition 1. *A set of meshes,* $\mathcal{M} = \{m_i \mid i = 1, \cdots, n\}$, *of the entire region* U *satisfies two conditions:*

$$U = \cup_{i=1,\cdots,n} m_i,$$
$$m_i \cap m_j = \phi \ \text{if} \ i \neq j.$$

A closed region comprising meshes is defined as a transitive closure of the relation $\mathcal{R} : \mathcal{M}^2 \mapsto \{True, False\}$, *such that*

$$\mathcal{R}(m_i, m_j) = True \ \text{iff} \ m_i \ \text{and} \ m_j \ \text{share a border, otherwise False.}$$

A grid-like mesh obviously follows the definition of mesh above, as either an admissible or rectangular region follows that of closed region above as well. A mesh built using a *quad-tree*, as shown in Fig. 2, is another example of mesh. It is useful when the distribution of points is biased because the number of points in each mesh should be nearly equal. Another example is a set of polygonal meshes like administrative districts. This is useful when meshes should be manually defined for some reasons. Our algorithm can handle these kinds of meshes and discover closed regions, i.e., transitive closures of the relation $\mathcal{R}$.

3   Related Work

Optimized association rule mining was first introduced by Fukuda et al. They formulated discovery of association rules for a single numeric attribute [5] and two-dimensional numeric attributes [4]. Their algorithms include an $O(n)$ time algorithm, where n is the number of meshes for an admissible region, that maximizes the gain achieved by using dynamic programming (DP) with fast matrix search, $O(n^{1.5})$ time algorithms for a single rectangular region that maximizes the gain, and approximation algorithms for a single rectangular region that maximizes the support or confidence.

These algorithms identify a single rectangle or admissible region because their goal is to keep the optimization rules simple enough for people to easily understand them. Note that they are based on the assumption that the uninstantiated numeric attributes are uniformly distributed. Therefore, they start by splitting relations into equal-sized buckets or grid-like meshes for one- or two-dimensional data, respectively.

Rastogi and Shim generalized the formulation of optimized association rule mining [9]. Their optimized association rules can have disjunctions over an arbitrary number of uninstantiated attributes, which can be either categorical or numeric. This means that the discovered instantiations can be multiple hyper rectangular regions.

They also showed that their problem is NP-hard and proposed algorithms that search through the space of instantiations in decreasing order of the weighted sum of the confidence and support, while using a branch and bound technique to prune the search space efficiently.

Table 1. Comparison of three algorithms for optimized rule mining of geospatial data

	Fukuda	Rastogi	Dobkin
assumes uniformity?	yes	yes	no
no. of regions	1	k	1
admissible regions	yes	no	no
rectangular regions	yes	yes	yes

Their algorithms require the enumeration of instantiations of numeric attributes, each of which has the form $a_i \in [l_i, u_i]$, before they start processing so that the identified instantiations do not overlap each other. The number of instantiations could be huge for numeric attributes since they take conbinations of values. The *pruneInstArray* reduces those instantiations that are never included in the final disjunctions of instantiations. Note that the numeric attributes are assumed to be uniformly distributed in Rastogi's algorithms as in Fukuda's algorithms.

Dobkin [3] presented an algorithm that solves optimized gain problems for a single rectangular region. It takes $O(N^2 \log N)$ time, where N is the number of points, and does not necessarily suppose that numeric attributes are uniformly distributed since it takes all distinct values of those numeric attributes in the records.

Table 1 compares the algorithms of Fukuda, Rastogi, and Dobkin.

4   Local Maximization of Confidence

Our goal was to develop an algorithm that does not require the numeric attributes to be uniformly distributed and that identifies multiple regions. Both features are important for discovering RoIs in geospatial data. We achieved both by introducing locally optimized association rule mining, which provides another mathematically sound formalization for discovering RoIs.

Definition 2. *A region $\tilde{D}$ is locally maximized with respect to confidence if*

$$sup(\tilde{D}) > Z \ and \ conf(D) < conf(\tilde{D}) \ for \ \forall D \supset \tilde{D}, \tag{1}$$

where $\tilde{D}, D \subset R^2$ or $\tilde{D}, D \subset 2^{\mathcal{M}}$.

Informally, this means that $\tilde{D}$ is locally maximal if there is no area D containing $\tilde{D}$ having a confidence greater than or equal to $\tilde{D}$.

The following lemma is trivially true:

Lemma 1. *Given a set of meshes $\mathcal{M} = \{m_i \mid i = 1, 2, \cdots, n\}$ and an uninstantiated association rule $U \wedge C_1 \to C_2$, let s_i and c_i be the support and confidence, respectively, of mesh m_i for the association rule instantiated by $U = ((x, y) \in m_i)$. For $\forall D \in 2^{\mathcal{M}}$,*

$$conf(D) = \frac{\sum_{m_i \in D} s_i c_i}{\sum_{m_i \in D} s_i} \leq \frac{1}{|D|} \sum_{m_i \in D} \frac{s_i c_i}{\min_{m_i \in D} s_i}, \tag{2}$$

where $|D|$ is the number of meshes in D.

For a naive algorithm that enumerates all possible transitive closures of $\mathcal{R}$, it could take exponential time to compute the locally maximized confidence regions, $\tilde{D} \in 2^{\mathcal{M}}$, since there are $\sum_{k=1,\cdots,n} {}_nC_k$ possible transitive closures.

The following theorem presents the condition that each mesh should satisfy when it is not included in a local maximum area. The theorem is proved by using Lemma 1.

Theorem 1. *For $\forall D, E \in 2^{\mathcal{M}}$ such that $D \subset E$, let $\bar{D} = E \setminus D$. Then $conf(E) < conf(D)$ holds if*

$$\frac{s_i c_i}{\min_{m_i \in \bar{D}} s_i} < conf(E) \text{ for } \forall m_i \in \bar{D}. \tag{3}$$

Proof. To keep the formulas simple, we introduce some invariants: $S_E = \sum_{m_i \in E} s_i$, $H_E = \sum_{m_i \in E} s_i c_i$, $S_{\bar{D}} = \sum_{m_i \in \bar{D}} s_i$ and $H_{\bar{D}} = \sum_{m_i \in \bar{D}} s_i c_i$. Since $E = D \cup \bar{D}$ and $D \cap \bar{D} = \phi$,

$$S_D = S_E - S_{\bar{D}},$$
$$H_D = H_E - H_{\bar{D}}.$$

Now $conf(D) - conf(E)$ is evaluated:

$$
\begin{aligned}
conf(D) - conf(E) &= \frac{H_D}{S_D} - \frac{H_E}{S_E} \\
&= \frac{S_E(H_E - H_{\bar{D}}) - (S_E - S_{\bar{D}})H_E}{S_D S_E} \\
&= \frac{S_{\bar{D}}}{S_D} \frac{S_{\bar{D}} H_E - S_E H_{\bar{D}}}{S_{\bar{D}} S_E} \\
&= \frac{S_{\bar{D}}}{S_D} \left(\frac{H_E}{S_E} - \frac{H_{\bar{D}}}{S_{\bar{D}}} \right) \\
&= \frac{S_{\bar{D}}}{S_D} \left(conf(E) - conf(\bar{D}) \right).
\end{aligned}
\tag{4}
$$

From lemma 1 and (3), we get

$$conf(\bar{D}) \leq \frac{1}{|\bar{D}|} \sum_{m_i \in \bar{D}} \frac{s_i c_i}{\min_{m_i \in \bar{D}} s_i} < conf(E). \tag{5}$$

From (4) and (5), $conf(D) > conf(E)$ is proved.

Theorem 1 implies that it is not necessary to enumerate all combinations of meshes and evaluate their confidence as a naive algorithm would do. Instead, it is sufficient to see if (3) holds for each individual mesh. However, (3) is unlikely to hold for most of the meshes if $\min_{m_i \in \bar{D}} s_i$ is too small. This is the case for car probe data and location data from other movable objects, where the position

---

Algorithm 1. $discoverMaximalClosure(\mathcal{D}, \mathcal{M}, C_1, C_2, Z)$

1: $\mathcal{M} \leftarrow buildQuadtree(\mathcal{D}, \mathcal{M}, C_1, Z)$
2: obtain $sup(m)$ and $hit(m)$ of $C_1 \rightarrow C_2$ for $\forall m \in \mathcal{M}$
3: $\Gamma \leftarrow \{\mathcal{M}\}$
4: **repeat**
5:      **for** $E \in \Gamma$ **do**
6:          $\bar{D} \leftarrow \{\}$
7:          $\sigma \leftarrow \min_{m \in E} sup(m)$
8:          **for all** $m \in E$ such that $hit(m)/\sigma < conf(E)$ **do**
9:              $\bar{D} \leftarrow \bar{D} \cup \{m\}$
10:         **end for**
11:         **if** $\bar{D} = \{\}$ or $sup(E \setminus \bar{D}) < Z$ **then**
12:             report E
13:             $\bar{D} = E$
14:         **end if**
15:         $\mathcal{M} \leftarrow \mathcal{M} \setminus \bar{D}$
16:     **end for**
17:     $\Gamma \leftarrow composeTransitiveClosure(\mathcal{M})$
18: **until** no report occurred at line 12

---

data are not distributed uniformly. In those cases, it is better to use meshes whose support counts are almost equal to each other.

With this in mind, we present an algorithm for discovering regions with locally maximal confidence in the following section. We also mention an algorithm for reducing the number of meshes to be checked to see if they share borders.

5   Algorithms

Algorithm 1 shows our algorithm for discovering locally maximal regions. The algorithm comprises two parts. The first part initializes the set of meshes and evaluates the support and hit count for each mesh(line 1–3). It does this by using a position database $\mathcal{D}$, an initial fixed-sized mesh set $\mathcal{M}$, the association rule to be optimized $C_1 \rightarrow C_2$, and the minimum support threshold Z. One such set of meshes is called a *quad-tree*. As illustrated in Fig. 2, it is obtained by recursively splitting a mesh evenly into four smaller meshes with horizontal and vertical borders. The algorithm continues to split the given mesh set $\mathcal{M}$ into finer meshes so that they have more (or less) than the minimum support threshold Z. The white and black points in the figure represent position data satisfying C_1 and $C_1 \wedge C_2$, respectively. The result is different size meshes with support counts less than $Z = 2$.

The second part of the algorithm identifies transitively connected groups of meshes with locally maximal confidence (line 4–18). The algorithm iteratively narrows a transitive closure by removing meshes that do not satisfy the local maximum property given by theorem 1 with σ determined for that transitive closure (line 8–10). Using $\min_{m \in E} sup(m)$ for σ is not any problem since

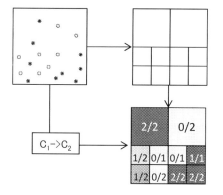

1. Split area in accordance with number of points in meshes

2. Count number of black and white points to fill hit and support counts of rule $C_1 \to C_2$ in meshes.

Fig. 2. Building of quad-tree as filling hit and support counts in meshes

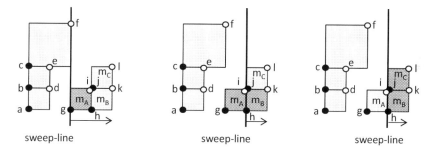

sweep-line   sweep-line   sweep-line

Fig. 3. Composing transitive closures. Polling a vertex in $Q = \{a, b, \cdots, l\}$ one another, the sweep-line moves from left to right. m_A is activated at the beginning (left). Polling h, activates m_B, which is found to share a border with m_A (middle). Polling i, then deactivates m_A. Next, polling j, activates m_C, which is found to share a border with m_B (right).

$\sigma \leq \min_{m \in \bar{D}} sup(m)$ always holds. By sorting $\mathcal{M}$ in ascending order of $conf(m)$, each mesh is visited only once by line 9. Therefore, the nested loop (line 5–16) iterates at most n times. If the transitive closure E is maximal and ample, then it is reported at line 12. The remaining meshes are processed by *composeTransitiveClosure* function, which composes transitive closures for the next iteration (line 17).

The computation time of *discoverMaximalClosure* is $O(n \log n)$ since *composeTransitiveClosure* takes $O(n \log n)$ time as shown below and the outer most loop iterates constant times proportional to Z.

It takes $O(n^2)$ time to naively check if the remaining meshes share borders with each other. Figure 3 shows an outline of the sweep-line algorithm used to reduce the number of meshes to be checked to see if they share borders. A sweep-line is a vertical line that makes regions *active* while it crosses over them. All regions are initially inactive. As the sweep-line moves from left to right, only the active regions are checked to see if they share borders with each other. If one shares a border with another, the pair is output. Such pairs are never missed

---

Algorithm 2. *composeTransitiveClosure*($\mathcal{M}$)

1: let Q be a list comprising left-bottom and right-top vertices of $\forall m \in \mathcal{M}$
2: sort Q in the order representing the time when regions become active or inactive
3: $A \leftarrow \{\}$
4: $E \leftarrow \{\}$
5: **while** Q is not empty **do**
6:     poll one from Q and call it v_1
7:     let m_1 be the mesh of v_1
8:     **if** v_1 is a left-bottom vertex, **then**
9:         $A \leftarrow A \cup \{m_1\}$
10:        $E \leftarrow E \cup \{(m_1, m_2) | \ m_2 \in A \ \text{s.t.} \ m_1.btm \leq m_2.top \ \textbf{and} \ m_1.top \geq m_2.btm\}$
11:    **else**
12:        $A \leftarrow A \backslash \{m_1\}$
13:    **end if**
14: **end while**
15: $V \leftarrow \cup_{(m_1,m_2) \in E} \{m_1, m_2\}$
16: report all connected graphs in $G(V, E)$ by traversing it in breadth first manner

---

because no two regions share a border unless both of them become active at the same time. A region becomes inactive again as the sweep-line moves away from the top of that region and thus is never examined with other regions. Once the sweep-line has traversed completely to the right, all pairs of adjacent regions have been identified and output. By traversing a graph comprising those pairs as its edges in a breadth first manner, the sweep-line algorithm composes all transitive closures.

Algorithm 2 implements this process. The sweep-line is emulated by using queue Q containing left-bottom and right-top vertices sorted in the order that represents the time when the regions become active and inactive, respectively (line 1–2). Due to space limitations, the algorithm in not explained in detail. A brief description of how it works is given in the caption of Fig 3. It computes in $O(n \log n)$ time even if the regions have arbitrary sizes and shapes as long as they can be represented as polygons. Interested readers can consult a text book on computational geometry, such as [1].

6   Experimental Results and Applications

This section presents experimental results for both artificial and realistic data sets to demonstrate the performance of our algorithm and the goodness of the areas discovered. Then, two example applications are presented. All the experiments were run on a Core i5-680 3.60GHz machine with 4GB of RAM running Windows.

6.1   Experiments

An experiment was done using artificial data to compute the performance of our *discoverMaximalClosure* algorithm with that of Rastogi's algorithm. Rastogi's algorithm was selected for comparison because it is the only one among the related ones that can discover multiple areas with mathematically sound definitions for the areas it discovers, as mentioned in section 3.

The artificial data comprised N uniformly distributed points, either enabled or disabled in accordance with a two-dimensional Gaussian mixture distribution of the K components:

$$p(\mathbf{x}) = \sum_{k=1}^{K} \pi_k \mathcal{N}(\mathbf{x}|\mu_k, \Sigma_k), \text{ where } \mathbf{x}, \mu_k \in R^2 \text{ for } k = 1, \cdots, K. \qquad (6)$$

Let $\pi_k = 1/K$ and $\Sigma_k = \mathrm{diag}(\sigma, \sigma)$ for simplicity, and let μ_k be determined randomly for $k = 1, \cdots, K$. To obtain an artificial data point, each point $\mathbf{x}$ was generated uniformly and then enabled if $p(\mathbf{x}) > 1/K$ or disabled otherwise by random processes.

Rastogi's and our algorithms are used to discover areas containing as much enabled points as possible while eliminating disabled points. This is achieved by discovering the areas that maximize the confidence of the rule:

$$C_1 \rightarrow C_2, \text{ where } \begin{cases} C_1 : \textit{True}, \\ C_2 : \textit{True} \text{ if the point is enabled, } \textit{False} \text{ otherwise} \end{cases}$$

Rastogi's algorithm requires three parameters, $n, M,$ and $minSup$, and that the entire rectangle be evenly separated into $\sqrt{n} \times \sqrt{n}$ grids. It can then discover a set of maximally confident M regions that have no less support than $minSup$. It is reasonable to set the parameters M to K.

Our $discoverMaximalClosure$ algorithm, on the other hand, requires one parameter, Z. Although it adaptively separates the entire rectangle in accordance with the given data by using the $buildQuadtree$ function, we used $\sqrt{n} \times \sqrt{n}$ grids to enable us to compare performance. This function discovers locally maximal transitive closures each of which has a support no less than Z.

Figure 4 shows the regions discovered by the two algorithms with $\sqrt{n} = 16$, as well as the black (enabled) or white (disabled) points generated with $K = 3$ and $N = 1000$. The entire rectangle was about $15.9km \times 19.6km$, and variance σ was set to $2.0km$.

We compared the processing times for $N = 10,000$, $K \in \{1, 2, \cdots, 10\}$, and $\sqrt{n} \in \{16, 32, 64\}$. The processing times were invariant with K for every $\sqrt{n}$, which are not shown due to space limitations. As shown in the left graph of Fig. 5, our $discoverMaximalClosure$ algorithm was about 50 times faster than Rastogi's for $\sqrt{n} = 32$. The right graph of Fig. 5 compares the probabilities of capturing enabled points in the discovered regions to evaluate the effectiveness of the two algorithms. The expected probability is defined as $\frac{\sum_{\mathbf{x} \in D} p(\mathbf{x})}{sup(D)}$, where D is a discovered region. The $discoverMaximalClosure$ was comparable or better in capturing the Gaussian mixture distribution than Rastogi's algorithm, although we do not understand why Rastogi's algorithm do not perform well in some cases.

Figure 6 shows the regions discovered by the two algorithms for realistic data that were not distributed uniformly. We could not disclose the detail of this

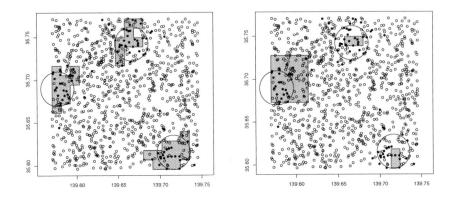

Fig. 4. Region discovered by *discoverMaximalClosure* algorithm (left) and by Rastogi's algorithm (right) for artificial data with $K = 3, N = 1000$ and $\sqrt{n} = 16$. Center and radius of each circle respectively represents mean and standard deviation of two-dimensional Gaussian mixture component.

data because of some contractual reasons, but it was from taxis as introduced in section 6.2. As described above, the *discoverMaximalClosure* algorithm adaptively separates the entire rectangle in accordance with the given data by using the *buildQuadtree* function while Rastogi's uses fixed grid-like meshes. Our algorithm is better for non-uniformly distributed data points because it discovers well-limited and high confidence regions. Rastogi's algorithm, on the other hand, would need to use finer grid-like meshes and take more time to discover such well-limited regions.

6.2 Application: Taxi Fleet Control

Given the results shown in Fig. 6, one application is taxi fleet control based on expected demand. The taxis would regularly report their location (every minute, for instance) and whether or not they were available. Let the data points be:

$$(id, time, longitude, latitude, pflag, cflag),$$

where *cflag* or *pflag* is *True* if the taxi identified by *id* is unavailable in the current time step or was unavailable in the previous one, respectively, and *False* otherwise.

Optimizing the confidence of the association rule given below would increase the probability of taxis picking up passengers in the discovered regions because there should be more demand for taxis and fewer available ones around:

$$C_1 : pflag = False,$$
$$C_2 : pflag = False \text{ and } cflag = True.$$

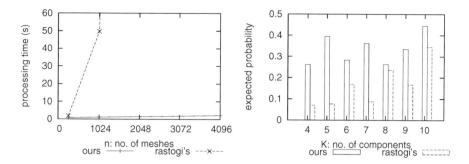

Fig. 5. Comparison between Rastogi's algorithm and our algorithm. Left graph plots processing times for variable grid size and fixed number of components ($K = 8$), right graph plots probabilities of capturing enabled points within regions.

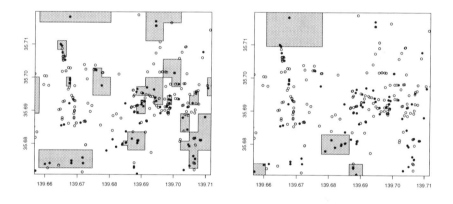

Fig. 6. Regions discovered by *discoverMaximalClosure* algorithm (left) and by Rastogi's algorithm (right) for floating car data that were not distributed uniformly

In other words, a point is enabled when a taxi picks up a passenger and disabled otherwise. Note that taxis unavailable in the previous time step are not taken into consideration. The quick processing speed of our algorithm enables taxis to obtain in realtime the locations of areas where they are more likely to pick up passengers without competing with other ones.

6.3   Application: Discovery of Sights from Social Messages

Given the popularity of social network services (SNSs), another application is discovering sights to see on the basis of location data and/or photos attached to text messages. People could identify attractive places to visit by discovering areas that maximize the confidence given by the following rule:

C_1 : *True* if message has location data, *False* otherwise,

C_2 : *True* if message has both location data and a photo, *False* otherwise.

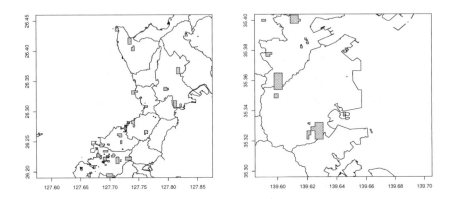

Fig. 7. Locations in Okinawa (left) and Yokosuka (right), Japan, where many messages with photos were posted

The gray areas in Fig. 7 represent attractive areas as determined from the confidence. Those in Okinawa (left diagram) include capes and beaches and several popular sights, like Shuri Castle, an aquarium, and memorial parks around Naha City. Those in Yokosuka (right diagram) include the aquarium on Hakkei Island, parks, and museums.

7   Conclusion

The locally optimized association rule mining has now been formalized. A theorem that holds for locally maximized two-dimensional areas was formulated, and from it an algorithm was derived that efficiently discovers confident areas with shapes that are not limited to being rectangular or x-monotone. It does not require the assumption that position data are uniformly distributed as is often the case in analyzing GPS data from vehicle and smartphones. Experimental results showed that our *discoverMaximalClosure* algorithm was 50 times faster than Rasitogi's, which discovers globally optimized, rectangular regions of interest. Applications include ones where it is essential to discover regions of interest. Its efficiency makes our algorithm well suited for helping taxis to determine where to go and pick up passengers. It can also be used to determine which sights to visit by using the location data and/or photos attached to text messages.

References

1. de Berg, M., Chong, O., van Kreveld, M., Overmars, M.: Computational Geometry - Algorithms and Applications, 3rd edn. Springer (2008)
2. Derekenaris, G., Garofalakis, J., Makris, C., Prentzas, J., Sioutas, S., Tsakalidis, A.: Integrating GIS, GPS and GSM Technologies for the Effective Management of Ambulances. Computers, Environment and Urban Systems 25(3), 267–278 (2001)

3. Dobkin, D.: Computing the Maximum Bichromatic Discrepancy, with Applications to Computer Graphics and Machine Learning. Journal of Computer and System Sciences 52(3), 453–470 (1996)
4. Fukuda, T., Morimoto, Y., Morishita, S., Tokuyama, T.: Data Mining with Optimized Two-Dimensional Association Rules. ACM Transactions on Database Systems 26(2), 179–213 (2001)
5. Fukuda, T., Morimoto, Y., Morishita, S., Tokuyama, T.: Mining Optimized Association Rules for Numeric Attributes. Journal of Computer and System Sciences 58(1), 1–12 (1999)
6. Giannotti, F., Nanni, M., Pinelli, F., Pedreschi, D.: Trajectory Pattern Mining. In: Proceedings of the 13th ACM SIGKDD International Conference on Knowledge Discovery and Data Mining, KDD 2007, pp. 330–339. ACM Press, New York (2007)
7. Mamei, M., Rosi, A., Zambonelli, F.: Automatic Analysis of Geotagged Photos for Intelligent Tourist Services. In: 2010 Sixth International Conference on Intelligent Environments, pp. 146–151. IEEE (2010)
8. Panahi, S., Delavar, M.R.: Dynamic Shortest Path in Ambulance Routing Based on GIS. International Journal of Geoinformatics 5(1), 13–19 (2009)
9. Rastogi, R., Shim, K.: Mining Optimized Association Rules with Categorical and Numeric Attributes. IEEE Transactions on Knowledge and Data Engineering 14(1), 29–50 (2002)
10. Teerayut, H., Witayangkurn, A., Shibasaki, R.: The Challenge of Geospatial Big Data Analysis. In: Open Source Geospatial Research & Education Symposium (2012)
11. Yoshida, D., Song, X., Raghavan, V.: Development of Track Log and Point of Interest Management System Using Free and Open Source Software. Applied Geomatics 2(3), 123–135 (2010)
12. Zheng, V.W., Zheng, Y., Xie, X., Yang, Q.: Collaborative Location and Activity Recommendations with GPS History Data. In: Proceedings of the 19th International Conference on World Wide Web - WWW 2010, pp. 1029–1038. ACM Press, New York (2010)
13. Zheng, Y., Zhang, L., Xie, X., Ma, W.Y.: Mining Interesting Locations and Travel Sequences from GPS Trajectories. In: Proceedings of the 18th International Conference on World Wide Web, WWW 2009, pp. 791–800. ACM Press, New York (2009)
14. Zheng, Y., Zhou, X. (eds.): Computing with Spatial Trajectories. Springer (2011)

Multi-way Theta-Join Based on CMD Storage Method

Lei Li, Hong Gao, Mingrui Zhu, and Zhaonian Zou

Massive Data Computing Research Lab
Harbin Institute of Technology, Heilongjiang, China
{thor.lilei,hit.mingrui}@gmail.com, {honggao,znzou}@hit.edu.cn

Abstract. In the era of the Big Data, how to analyze such a vast quantity of data is a challenging problem, and conducting a multi-way theta-join query is one of the most time consuming operations. MapReduce has been mentioned most in the massive data processing area and some join algorithms based on it have been raised in recent years. However, MapReduce paradigm itself may not be suitable to some scenarios and multi-way theta-join seems to be one of them. Many multi-way theta-join algorithms on traditional parallel database have been raised for many years, but no algorithm has been mentioned on the CMD (coordinate modulo distribution) storage method, although some algorithms on equal-join have been proposed. In this paper, we proposed a multi-way theta-join method based on CMD, which takes the advantage of the CMD storage method. Experiments suggest that it's a valid and efficient method which achieves significant improvement compared to those applied on the MapReduce.

Keywords: CMD, Multi-way Theta-Join.

1 Introduction

Big data is one of the most challenging and exciting problems we are facing in recent years. With the development of social network, e-commerce and the internet of things, large quantity of data is produced every day. Because of this tremendous volume of data, many old algorithms designed on traditional smaller data sets are not applicable any more, and new algorithms have to be proposed. Multi-way theta-join is one of these challenging problems, which demands join results from more than one join condition with operators include not just 'equal', but also 'bigger than' and 'smaller than'. With more operators and relations involved, multi-way theta-join query is more powerful in data analysis than ordinary equal-join query while the complexity and time consuming of it are soaring up at the same time. Assuming we have four relations each has 10 million records. The Cartesian space produced by the joins among these four relations is to the fourth power of 10 million, and that is quite a big number that none of the old join algorithms could handle efficiently. So it's really important and urgent to figure out a way to process multi-way theta join efficiently on the huge volume of data.

To solve these big data problems under current hardware and computation theory, many methods are proposed and MapReduce is one of the most popular solutions raised in recent years. MapReduce was first proposed by Google [6] [7] to demonstrate how they handled the big data problems in their business field. The storage

S.S. Bhowmick et al. (Eds.): DASFAA 2014, Part I, LNCS 8421, pp. 62–78, 2014.
© Springer International Publishing Switzerland 2014

and computation concepts were quite suitable to solve the problems that Google faced, and those methods became even more well-known and popular after Hadoop [9] was published, which is another implementation version of Google's big-table and MapReduce. MapReduce is famous for its simplicity, reliability in processing large volume of data on clusters, and it is adopted by many industrial companies to deal with their own business, like Google, Facebook, Alibaba and so on. Various solution providers like IBM, Oracle and SAS also use Hadoop to solve the problems their costumers confront. Somehow, MapReduce seems to become the skeleton key to solving all the big data problems. However, some argue that there are still some drawbacks in MapReduce that cannot be ignored [10]. First of all, they claim that MapReduce overlooks the importance of data schema undeniably. They believe that the separation of the schema from the application is a good idea while MapReduce just takes no advantage of it. And without a functional schema, it has no functionality to keep the garbage data out of the dataset. The second reason which inspires them is that MapReduce may neglect the importance of indexes which could accelerate the access to the data. In fact, as they criticized, it just uses the brute force as a processing option, which is so primitive that lots of needless works are involved. So they come up with the conclusion that MapReduce may not be a good choice for many usage scenarios. As far as our concern, MapReduce is still a powerful tool to cope with many problems in terms of easiness and steadiness, so we also conduct a series of experiments to see the performance of MapReduce on solving the multi-way theta-join problem.

Other solutions are based on traditional relational parallel databases which are provided by the database machine provider like Oracle and Teradata. Although Hadoop is popular nowadays, many companies still use these traditional databases to solve their problems. Baidu, the largest searching service provider in China, use MySQL to build its large database. Amazon provides part of their cloud service based on MySQL while Microsoft provides their SQL Azure on their SQL server. These solutions have nothing to do with Nosql at all, but they are still steady and fast enough to cope with the modern information explosion. It can be seen that if we adapt the traditional relation databases correctly, we can still take advantages of them to deal with modern big data problems.

When it comes to multi-way theta-join operation, which is a traditional problem especially when the data volume is large, deploying algorithms on parallel environment is inevitable. The solutions based on traditional database fall into three categories: hash-join [11] [12], sort-merge [11] [13] and nested-loop [10]. They are powerful tools to deal with join problems, but little attention has been paid on the storage paradigm which can affect efficiency dramatically. Other solutions presented lately mainly focus on MapReduce. Apler Okcan [2] and Xiaofei Zhang [3] provided their solutions in recent years. Although simple and efficient their algorithms may be, they still cannot overcome the internal drawbacks that MapReduce is born with. All they are trying to do is to split the data as delicately as possible and allocate the tasks as concurrently as possible, but the data transported through the network are still too large which, in fact, is multiple times bigger than the original dataset. Even after the data is sent to the reducers, what it is going to do is still a kind of calculation by force. Moreover, although the data have already been stored in HDFS, when the MapReduce program is started, those large volumes of data still have to be transported through the network between the step of the Map and the Reduce, and that is one of the most time consuming part of the big data calculation. As we mentioned above, despite the fact that MapReduce is becoming a mature solution in the field of large volume data processing, some problems have not been solved till now and many people are still working on them.

  So when we look back at what we have achieved in the field of parallel database, some shining ideas can still be applied to solve the problems brought by huge volume of data with some adjustments, and CMD [1] is surely one of them. CMD (Coordinate Modulo Distribution) is designed to achieve optimum parallelism for a very high percentage of range queries on multidimensional data. Its main thought is to decluster the data set based on its value ranges. Many declustering methods such as round-robin or hashed declustering, just concentrate on the values of a single attribute or a few key attributes so that operations carried on these attributes can achieve high performance. However, these methods will lose their efficiency when they are trying to deal with those non-partitioning attributes. So CMD is therefore proposed to support queries that involve those non-partitioning attributes. With the knowledge of every attribute's distribution, CMD storage can support the multi-way theta-join naturally. However, only a little attention has been paid on this area so far. So we bring this storage method back on stage again to exploit it on big data environment.

  In this paper, we propose techniques that enable efficient parallel execution of multi-way theta-join based on CMD. Our main contributions are as follows:

1. We propose a cost model to evaluate each operation of the multi-way theta-join query to find out the order of join sequence. Under our proposed cost model, we can figure out which part of the query should be processed first at every step, thus the optimum result can be achieved.
2. We take advantage of the CMD storage to determine which part of the data set should take part in the join operation and which part of the data should be sent to which node in the cluster. With such searching strategy, the control logic would be easier and the workload would be split into many little jobs. On the other hand, it is also a pruning skill, with which a great number of data would be abandoned and a much lower I/O cost can be achieved.

The rest of the paper is structured as follows. In Section 2, we briefly review the CMD storage and elaborate the application scenario for multi-way theta-joins. We formally define some terminology in Section 3 and present our multi-way theta-join algorithm in Section 4. Experiment result will be demonstrated at Section 5 and Section 6 concludes the paper.

2   Preliminary

2.1   CMD Storage Model

CMD storage model is a declustering method initially proposed to achieve high performance in range query on a parallel environment. Assume we have a relation R with d attributes: $R(A_1 A_2 \dots A_d) \subseteq D_1 \times D_2 \times \dots \times D_d$, where A_i is the ith attribute and D_i is the domain of A_i. We want to decluster this relation to store it on a cluster which has P storage nodes. The $D_1 \times D_2 \times \dots \times D_d$ is actually a hyperspace with every attribute acts as a dimension. By splitting every dimension into N intervals ($N = n_i P$, n_i is an integer factor), we can get $n_1 P \times \ldots \times n_d P$ hyper-rectangles. For example as shown in Figure 1, the horizontal dimension is A_1 and the vertical dimension is A_2. Assume we have a relation $R(A_1, A_2) \subseteq [0,800] \times [0,2000]$ and four storage nodes, $n1 = n2 = 2$. The inner tuples of numbers which are near the coordinate axes are the coordinates of the original hyper-space. Therefore, the hyperspace can be partitioned into 64 hyper-rectangles.

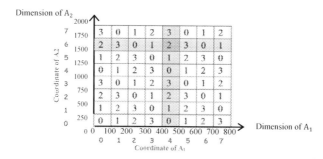

Fig. 1.

A CMD storage system consists of one master node which acts as a controlling node and several slave nodes which store the data and run the query. The P we introduced before is the number of slaves in CMD. CMD treats the positions of these intervals as coordinates on one dimension, as showed in the outer tuples of the numbers that are far from the coordinate axes in Figure 1. For example, the CMD coordinate of interval [0, 100) is 0 and the CMD coordinate of interval [600,700) is 6. A hyper-rectangle is the Cartesian product of d intervals $I_{1x_1} \times \ldots \times I_{dx_d}$($I_{ix_j}$ denotes the x_j^{th} interval of A_i), so $(x_1, \ldots, x_d)$ can be deemed as the coordinate of a hyper-rectangle. Then we can allocate these hyper-rectangles to different nodes by the function $CMD(x_1, \ldots, x_d) = (x_1 + \cdots + x_d) \bmod P$, which calculates the node number that stores the hyper-rectangle$(x_1, \ldots, x_d)$. Those hyper-rectangle from the same relation and stored in the same node is denoted as $R.CMD(x_1, \ldots, x_d)$.

Let's take a look at the example illustrated in Figure 1. Here we have a tuple (450, 1600) from R, the hyper-rectangle that contains it is [400, 500) × [1500, 1750), so the coordinate of this hyper-rectangle is (4, 6), and it is the second node that stores the tuple which can be calculated by (4+6) mod 4=2.

A relation stored as CMD method is called a CMD-relation. From now on, all the relations that we will mention will be CMD-relations.

There are a handful of advantages of the CMD method. First of all, since the data declustering is based on all dimensions, database operations that involve any of the partitioning dimensions can be performed very efficiently. Secondly, since the method is balanced in terms of the data volume, and neighboring regions are on separate nodes, maximum computing concurrency can be achieved. Moreover, since it can efficiently support range queries due to the benefits of the grid file structure and a join operation can be divided into several range queries, it can efficiently support join query naturally.

2.2   Multi-way Theta-Join

Multi-way theta-join is a query that involves more than two relations and in the WHERE clause there are a series of join operations whose join conditions are included in $\{<, \leq, =, >, \geq\}$. For example, the query *"Select R1.A from R1, R2, R3, R4 where R1.B=R2.F and R1.C<R4.O and R2.E>R4.N and R2.E<R3.I and R3.K=R4.O"* is a multi-way theta join. It is a powerful tool to deal with many intricate scenarios. However, when the data volume grows up, the memory demand will soar up by many times exponentially since the potential result space is the Cartesian product of all the related dimensions. It is unrealistic to fulfill this task on a standalone computer and should be run in a parallel environment. A facile method to conduct the join query is to deal with the conditions one by one, but a large amount of intermediate data would be produced if we take no actions of optimization. For example, two relations with 2G of data each may produce more than 10G of intermediate data which is a disaster to no matter what kind of algorithm we could think of. So we come up with a CMD-based join algorithm to solve this problem.

3   Terminology

Definition 1. *(Join-region).* Let $R(A_1 A_2 \dots A_d) \subseteq D_1 \times D_2 \times \dots \times D_d$ be a relation and A_i, which is divided into N intervals, is the join attribute. I_{ij} is the jth interval of A_i. Then the jth *join-region* of A_i is defined as $(D_1 \times \dots \times D_{i-1} \times I_{ij} \times D_{i+1} \times \dots \times D_d)$, denoted as $JR(R, A_i, j)$. $L(R, A_i, j)$ and $U(R, A_i, j)$ represent the lower and upper bound of the join-region $JR(R, A_i, j)$. Those hyper-rectangles that belong to the same join region and are stored on the same node are called a partial join region and is denoted as $R.A_i.j.CMD(x_1, \dots, x_d)$ or $R.A_i.j.NodeNo$. It is obvious to see that every join region has P partial join regions.

For example, given a relation $R(A_1, A_2)$, Figure 2 shows the 2nd join region of R on A_1, $JR(R, A_1, 2)$, in the dark strip which is $(40, 60] \times [0, 40]$. $L(R, A_1, 2)=40$ and $U(R, A_1, 2)=60$. And the two hyper-rectangles that are stored in the Node 0 make up a partial join region denoted as $R.A_1.2.0$.

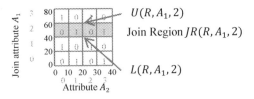

Fig. 2. Relation R and it join region

Obviously, if two join regions from different relations want to get involved in the same join operation, their value ranges have to have some over-lapping fields, that is to say, related.

Definition 2. *(Join related).* Assume there is a theta join between R and S. A_i is the join attribute of R, B_m is the join attribute of S. $JR(R, A_i, j)$ and $JR(S, B_m, n)$ are called *join related*, if one of the conditions below is satisfied:

(1) $\left(\begin{array}{l} L(R, A_i, j) \leq L(S, B_m, n) \leq U(R, A_i, j) \\ \text{or } L(S, B_m, n) \leq L(R, A_i, j \leq U(S, B_m, n) \end{array} \right)$ if join condition is '='

(2) $L(R, A_i, j) < U(S, B_m, n)$ if join condition is '<'

(3) $L(S, B_m, n) < U(R, A_i, j)$ if join condition is '>'

(4) $L(R, A_i, j) \leq U(S, B_m, n)$ if join condition is '≤'

(5) $L(S, B_m, n) \leq U(R, A_i, j)$ I   if join condition is '≥'

Figure 3 illustrates how join regions are join-related. Assume our join condition is '=', then $JR(R, A_i, 3)$ is join related with $JR(S, B_m, 4)$, and $JR(S, B_m, 3)$ respectively.

We could calculate a partial join result from each pair of the join-related regions between R and S, and no unrelated regions should be taken into consideration. The final join result could be obtained by the union of all the partial results.

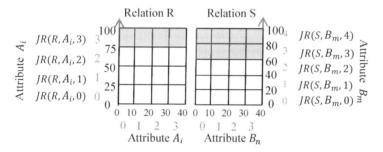

Fig. 3. Example of join related

4   Multi-way Theta-Join Algorithm

Given a multi-way theta-join query on some CMD relations, we want to come up with an optimization plan to calculate it as fast as possible, that is to say, to find out an optimal join operation sequence. The strategy we adopt here is a kind of greedy strategy which selects the join operation that produces the least size of the intermediate result. Our system comprises of one master node which serves as the coordinator and P slaves nodes serve as the parallel storage and computation machines. The overall procedure consists of the following steps:

1. Master estimates the size of the join result for each join operation in the query and selects the join operation which produces the smallest size of the result.
2. Master generates a join instruction vector for the selected join operation and sends it to all the slaves.
3. Slaves execute this join operation in parallel according to the join instruction vector and send back the information of the result relation, e.g. the result relation's size and the value ranges of its attributes;

4. Master updates the size, name, partition information and distribution information of the relations in the remaining join operations, repeat step (1) (2) and (3) until all the join operations are executed.

Step (1) involves a cost model we proposed which will be described in section 4.1, step (2) involves a join instruction vector which will be presented in 4.2 and step (3) involves the details of the execution of a parallel join operation which will be described in section 4.3.

Although the time span may be different due to change of the number of slave nodes we adopt in the experiment, the overall cost would not change accordingly to the number of slaves and finding out the best number of slaves to each dataset is out of this paper's consideration. In the next section, we will talk about the cost model which will determine the order of the join operations.

4.1   Cost Model

To find out the best join operation sequence to minimize the size of intermediate result, we have to define a cost model to evaluate it.

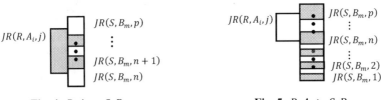

Fig. 4. $R.A_i = S.B_m$ **Fig. 5.** $R.A_i > S.B_m$

Assume that we have two relations R and S to execute the join operation and the join attributes of them are A_i and B_m. First of all, we select all the join regions of S that are related to each of the join region of R. Then we calculate the size of each partial intermediate result that is produced by each join region of R and its join related regions in S. Finally, by adding all the partial intermediate results' sizes together, we can get the total size of the intermediate result of this join operation.

Assume $JR(R, A_i, j)$ is one of the join regions of R, the lower and upper bounds of it are $L(R, A_i, j), U(R, A_i, j)$. There exists the possibility that $L(R, A_i, j) = U(R, A_i, j)$ since one specific value may appear so many times that the tuples in the whole interval have the same value on A_i. If so, we have to take $L(R, A_i, j + 1)$ *and* $U(R, A_i, j + 1)$ into consideration, or even the $L(R, A_i, j + 2)$ and $U(R, A_i, j + 2)$, until $L(R, A_i, k) < U(R, A_i, k)$ is satisfied for some k $(k > j)$. That is to say, $JR(R, A_i, j), JR(R, A_i, j + 1), ..., JR(R, A_i, k)$ may be viewed as a whole new join region, where k is the first integer that satisfies the condition $L(R, A_i, k) < U(R, A_i, k)$. The number of these tuples which have the same value of A_i can be calculated by $\sum_{m=j}^{k-1} |JR(R, A_i, m)| + \rho |JR(R, A_i, k)|$, where $|JR(R, A_i, r)|$ $(j \leq r \leq k)$ denotes the number of the tuples in $JR(R, A_i, r)$, and ρ represents the percentage of the tuples that have same value of A_i equals to $L(R, A_i, k)$. The purpose we introduce this concept is to guarantee the characteristic that every join region's upper bound is bigger than its lower bound so that our cost model can be demonstrated universally without considering some specific situations.

First we consider the situation where the join condition is "=". Figure 4 demonstrates the join regions of S that are related to $JR(R, A_i, j)$ in the join operation $R.A_i = S.B_m$. Here $L(S, B_m, n) < L(R, A_i, j) < U(S, B_m, n)$ and $L(S, B_m, p) < U(R, A_i, j) < U(S, B_m, p)$ and there are some join regions between $JR(S, B_m, n)$ and $JR(S, B_m, p)$. So for those whose join operators are equals, the cost of each join region is:

$$COST(JR(R, A_i, j)) = \sum_{u=n}^{p} (\alpha_u |JR(R, A_i, j)| \times \beta_u |JR(S, B_m, u)|)$$

Here α_u and β_u represent the proportion of each join region that would take part in the join operation and can be estimated by the program based on the data distribution information and the region boundaries. For example, when $u = n$, $\alpha_u = |U(S, B_m, u) - L(R, A_i, j)|_R / |JR(R, A_i, j)|$; when $n < u < p$, $\alpha_u = |U(S, B_m, u) - L(S, B_m, u)|_R / |JR(R, A_i, j)|$; when $n = p$, $\alpha_u = |U(R, A_i, j) - L(S, B_m, u)|_R / |JR(R, A_i, j)|$. β_u can be calculated correspondingly.

Secondly, we take a look at the unequal join operation. Unequal-join's cost model is easier than that of equal-join since it only cares one bound of the join region at a time. Let's take '>' as an example. Assume the join operation we are executing here is $R.A_i > S.B_m$ and $JR(R, A_i, j)$ is one of the join regions of R, then all the join regions of S, whose lower bound is smaller than the upper bound of $JR(R, A_i, j)$ $(U(R, A_i, j))$, should take part in the join operation. The "<" operator is similar. Because much more join regions are involved in the unequal join operation, the size of the intermediate result is obviously larger than that of the equal join, and less possible of an unequal join operation would be selected to execute if there exists an equal join operation.

As shown in Figure 5, the upper bounds of the join regions from $JR(S, B_m, 1)$ to $JR(S, B_m, n - 1)$ is smaller than $L(R, A_i, j)$, and the upper bounds of the join regions from $JR(S, B_m, n)$ to $JR(S, B_m, p)$ is bigger than $L(R, A_i, j)$ while $L(S, B_m, p)$ is the last lower bound that is smaller than $U(R, A_i, j)$. So for those whose join operators are unequal, the cost of each join region is:

$$COST(JR(R, A_i, j)) = |JR(R, A_i, j)| \times \sum_{u=1}^{n-1} |JR(S, B_m, u)|$$
$$+ \sum_{u=n}^{p} (\alpha_u |JR(R, A_i, j)| \times \beta_u |JR(S, B_m, u)|)$$

And the total cost can be calculated with the formula below:

$$TotalCost = \sum_{i=1}^{n} COST(JR(R, A_i, j))$$

In fact, we can estimate the unequal join's total cost by simply use the formula shown below as a more rough method:

$$EstimateTotalCost = \frac{1}{2}(\sum_{j=1}^{k} |JR(S, B_m, n)|)^2$$

Here $JR(S, B_m, n)$ is the join region that can be applied in the join operation. If there are both equal-join and unequal-join, such rough cost could be used since equal joins' costs are more likely to be less than that of unequal-joins'. If there are only unequal-joins left, we should calculate the costs of them carefully with the full formula to decide which one has the least cost among all the join operations.

4.2   Join Instruction Vector Generation

After the master selects a join operation, it will generate a join instruction vector based on the relations' size, partition information and distribution information and send it to all the slaves to instruct them to execute the join operation.

Definition 3. *(Join instruction vector).* Given a join operation, R_b is the relation with a bigger size whose hyper-rectangles would stay at where they are stored during a join operation, while R_t is the smaller one whose hyper-rectangles should be transported among the slaves through the network. A_{b1} is the join attribute of R_b and A_{t1} is the join attribute of R_t. R_r is the result relation. The join instruction vector, which is denoted as JV, is defined as:

$$R_b[A_{b1}A_{b2} ... A_{bn}]R_t[A_{t1}A_{t2} ... A_{tm}] R_r[A_{r1}A_{r2} ... A_{ro}] \quad NMpairs \quad \theta$$

The $NMpairs$ part consists of a series of pairs like $< N_i, M_j >< N_u, M_v > \cdots <$ $N_r, M_s >$ where each pair denotes a pair of join related region's number from R_b and R_t respectively. For example, $< N_i, M_j >$ denotes the join related region pair $JR(R_b, A_{b1}, N_i)$ and $JR(R_t, A_{t1}, M_j)$ from R_b and R_t respectively.

To generate the JV, master has to put the bigger size relation in the position of R_b and put the smaller one in the position of R_t. R_r is named after R_b and R_t, and the attributes of it should have later use in the other join operations. By adopting the join related judging method mentioned in Definition 2, the master can enumerate all the pairs of join related regions between R_b and R_t, and generate the $NMpairs$ of the join JV.

4.3   Parallel Join Operation Algorithm

In this section, we will describe the parallel join operation algorithm in detail. One thing we should keep in mind is that the result we need is only a few attributes, so the attributes, which are not in the result attribute set or have nothing to do with any join operation, should be pruned. From now on, all the attributes that appear in the relations will be the useful attributes. In the following part, we first present the join algorithm in detail, and then we discuss the correctness and time complexity.

Main Idea of the Algorithm

Given a JV, slaves get to know the related relations $R_b(A_{b1}A_{b2}\ldots A_{bn})$ and $R_t(A_{t1}A_{t2}\ldots A_{tm})$ }, join operation is $R_b.A_{b1}\ \theta\ R_t.A_{t1}$, $\theta \in \{<, \leq, =, >, \geq\}$. The overall procedure of executing the join operation consists of the following stages:

1. Data preparation stage: Each slave picks out all the local hyper-rectangles of $R_b's$ join related regions according to the first number in each pair of the $NMpairs$ and organizes them into a B-tree.
2. Data transportation stage: Each slave picks out all the local hyper-rectangles of $R_t's$ join related regions and sends them to all the other nodes. At the same time, each slave receives the hyper-rectangles of $R_t's$ join related regions sent from other nodes. At the end of the transportation stage, each slave has a full version of $R_t's$ join related regions. This stage can run in concurrency with the data preparation stage.
3. Data computation stage: each slave picks out all the hyper-rectangles of the $R_t's$ join related regions one by one and executes the join operation tuple by tuple. When the data preparation stage is completed, this stage can run in concurrency with the data transportation stage.
4. After all the steps above are finished, the slaves send the size and the attributes' value ranges of the new result relation back to the master node.

During the data preparation stage, each slave picks out all the local hyper-rectangles that are corresponding to the join related regions of R_b. By examining the attribute set provided by the JV, it can pick out each tuple's join attribute and all the other useful attributes, and organize the tuples in a B-tree.

In the data transportation stage, each slave picks out all the local hyper-rectangles that are corresponding to the join related regions of R_t and finds the join attribute and the useful attributes. These local hyper-rectangles of R_t can be used immediately after the data preparation stage is finished. At the same time, the slave has to send these local hyper-rectangles to all the other slaves. Every time it receives a foreign hyper-rectangle of R_t, it can apply it to the data computation stage. The hyper-rectangle receiving process can run in concurrency with the data computation stage in a pipeline.

In the data computation stage, for a tuple from a hyper-rectangle of R_t(no matter local or foreign), each slave searches its join attribute in the B-tree to find the upper and lower bounds of the list of tuples of R_b. Thus, a list of tuples in R_b are determined to join with this tuple, and the join results are written into the result relation while this result relation's size and attributes' value ranges have to be updated by the slave.

After the data computation stage is done, the slaves send all the information, e.g. size and attributes' value ranges, of the new result relation back to the master so that master can merge and update the information, e.g. size, attributes' value ranges relation names and partition information, of all the relations and recalculates the costs of the join operations in the remaining join operation set. If all the join operations in the join operation set have been executed, the join query is finished and the result can be presented by the master.

Correctness Analysis

The correctness analysis is pretty easy. Since all the data are stored using CMD storage, the intersection of any two hyper-rectangles is empty. So this characteristic can guarantee the result will not have any redundancy and each operation on any tuple will be performed only once.

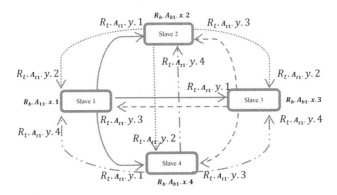

Fig. 6. An example of join operation: between two regions on four nodes

Figure 6 illustrates an example of a join operation between two join regions on four slave nodes. R_b is the relation with a bigger size that should be computed locally and x is the coordinate of one join related region of R_b. Correspondingly, R_t is the other join relation that has to be transported through the network and y is the coordinate of one of its join related regions. We just consider one pair of join regions for simplicity. Let's take a look at what happens on slave1. In the data preparation stage, slave 1 has got the partial join region $R_b.A_{b1}.x.1$.

In the data transportation stage, slave 1 itself has the partial join region $R_t.A_{t1}.y.1$, and can get hyper-rectangles of $R_t.A_{t1}.y.2$ from slave 2, hyper-rectangles of $R_t.A_{t1}.y.3$ from slave 3 and hyper-rectangles of $R_t.A_{t1}.y.4$ from slave 4. By summing up those four partial join regions, slave1 can get a full version of $JR(R_t, A_{t1}, y)$. In the data computation stage, the join operation is executed between the partial join region $R_b.A_{b1}.x.1$ and the full join region $JR(R_t, A_{t1}, y)$. At the same time, slave 2 is executing the join operation between the partial join region $R_b.A_{b1}.x.2$ and the full join region $JR(R_t, A_{t1}, y)$, so are slave 3 and slave 4. By summing up the computation results produced by the four slaves, the full join operation result can be obtained.

The general result can be described in the formula below:

$$Result = \sum_{i=1}^{n}(R_b.A_{b1}.x.i \bowtie R_t.A_{t1}.y.i + R_b.A_{b1}.x.i$$

$$\bowtie (\sum_{j=1}^{i-1} R_t.A_{t1}.y.j + \sum_{j=i+1}^{n} R_t.A_{t1}.y.j))$$

It can be seen from the formula that the algorithm just splits the result into several disjoint parts, so the result produced by the multi-way theta-join algorithm is correct.

Complexity Analysis

The time consumption consists of three parts: data gathering, data transferring and data computing, denoted as T_g, T_t and T_c respectively. Since each relation is almost evenly distributed on the slave nodes, the sizes of R_b and R_t on each node are $\frac{|R_b|}{P}$ and $\frac{|R_t|}{P}$. So $T_g = T_{I/O}(\alpha \frac{|R_b|}{P} + \beta \frac{|R_t|}{P})$, where $T_{I/O}()$ represents the time spent on I/O. $\alpha \frac{|R_b|}{P}$ is the time to gather those partial regions of relation R_b and α is the percentage of the original relation according to the join instruction vector provided by the master which can be calculated by $\alpha = \frac{|join\ related\ region\ unmber|}{|total\ region\ number|}$. $\beta \frac{|R_t|}{P}$ is the time to gather all hyper-rectangles in the partial join related region of relation R_t which will be sent to the other slaves, and β denotes the percentage of its data to be gathered which can be calculated by $\beta = \frac{|join\ related\ region\ unmber|}{|total\ region\ number|}$.

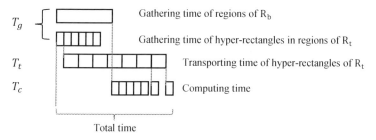

Fig. 7. Illustration of the join operation

One slave node has to transfer its data to all the other slaves, and the total amount of data to be transferred is P-1 times of the one it gathered. And because of the collision of network, the actual time spending is longer than the optimal time, $T_t = T_{comm}(\beta \frac{|R_t|}{P} \times (P-1)) \times \eta$, where $\eta>1$ is the collision factor and $T_{comm}()$ represents the time spent on network.

T_c is time spent on conducting the join operation. Since we use B-tree to store the R_b, so $T_c = T_{comp}(\beta|R_t|)$.

As shown in Figure 7, the whole operation can be executed in a pipeline way. After a join region is gathered it should be sent out immediately to shorten the total time spending. After the gathering of relation R_b is finished, the join computing shall begin even no region has been received since at least some regions are stored locally and these regions should take part in the join operation as well. So the total time spending is

$$T = \frac{T_{I/O}\left(\beta \frac{|R_t|}{P}\right)}{M} + T_{comm}\left(\beta \frac{|R_t|}{P} \times (P-1)\right) \times \eta + \frac{T_{comp}(\beta|R_t|)}{M}$$

Here M denotes the number of R_t's regions that will take part in the join operation.

5   Experiments

We conduct the experiment and compare the results of our CMD-based multi-way theta-join algorithm with the algorithm based on MapReduce proposed by [3].

5.1   Experiment Setup

All our experiments are conducted on a 16-node cluster, with one node serves as the master node and the others act as slave nodes. Each machine has one core of a 2.7GHz processor with 8GB RAM, one 7200 RPM hard disk with 1TB storage volume and is connected to a 1G network router. In total, the cluster therefore has 16 processing cores with 128GB memory and 16TB hard disk. All the nodes are running Ubuntu 12.04-amd64-server operating system. We use Hadoop-1.0.4 version to set up the system for the comparison experiment. The experiment on Hadoop applies B+-tree, which is a better join method than what we used in our algorithm, and the data has been loaded into HDFS in advance, so the time consumption of it is only determined by the computation and storage paradigm of MapReduce. To find out how the nodes number influences our algorithm's efficiency, we conduct each experiment four times on 8 slaves, 10 slaves, 12 slaves and 15 slaves respectively. And to find out the time consumption trend along with the data size, we conduct the experiments on the data sets with tuples of 1M, 5M, 10M, 50M and 100M respectively. The join program based on MapReduce uses the same data set, so the comparison between it and our method is possible.

The experiment applies two kinds of synthetic data sets, which can validate the worst case and the random case of our algorithm. The first kind of datasets has the same value ranges with four attributes which are generated equally from 1 to the size of the dataset. We use a python script to generate these five datasets which have 1 million tuples, 5 million tuples, 10 million tuples, 50 million tuples and 100 million tuples randomly, and the sizes of them are 27 MB, 149 MB, 301 MB, 1.7 GB and 3.4 GB respectively. To simplify our description, we use $R_1(ABCD)$, $R_2(EFGH)$, $R_3(IJKL)$ and $R_4(MNOP)$ to represent the four relations that are involved in our experiment. Since the datasets in the first group have the same value ranges, we use the Table 1 to describe the four datasets together. For example, the second line demonstrates the data distribution of our first experiment whose data sets' size are all 1000000 tuples. And the four attributes in each relations all fall into the range from 1 to 1 million respectively. Since the attributes in datasets 1 all have the same data value ranges, the possibility of a larger result is the largest, which actually indicates the performance of the worst case. The experiment result is repeatable and convincible since the data are generated in uniform distribution.

Table 1. Data set 1

Data size	First attribute	Second attribute	Third attribute	Fourth attribute
1000000	[1,1000000]	[1,1000000]	[1,1000000]	[1,1000000]
5000000	[1,5000000]	[1,5000000]	[1,5000000]	[1,5000000]
10000000	[1,10000000]	[1,10000000]	[1,10000000]	[1,10000000]
50000000	[1,50000000]	[1,50000000]	[1,50000000]	[1,50000000]
1000000000	[1,100000000]	[1,100000000]	[1,100000000]	[1,100000000]

The second group of datasets doesn't have the same value ranges at all, which can affect the experiment result significantly. It can simulate the data skew that happens often in the real world, which will boost up the speed of our algorithm. The data distributions are shown below separately. The sizes of them are 27M, 151M, 304M, 1.7G and 3.4G respectively. Every table's corresponding rows are in the same set of experiment, since they have the same number of data size. For example, for every third row in the following four tables, they all have 5000000 tuples of data, and they are involved in the same experiment.

Table 2. Data set 2

Data size	A	B	C	D
1000000	[1,1000000]	[500000,1000000]	[1,1000000]	[500000,1000000]
5000000	[1,5000000]	[2000000,5000000]	[1,5000000]	[2000000,5000000]
10000000	[1,10000000]	[5000000,10000000]	[1,10000000]	[5000000,10000000]
50000000	[1,50000000]	[20000000,50000000]	[1,50000000]	[20000000,50000000]
100000000	[1,100000000]	[20000000,100000000]	[1,100000000]	[20000000,100000000]
Data size	**E**	**F**	**G**	**H**
1000000	[500000,1500000]	[1,1000000]	[500000,1500000]	[1,1000000]
5000000	[2500000,7500000]	[1,5000000]	[2500000,7500000]	[1,5000000]
10000000	[5000000,10000000]	[1,15000000]	[5000000,15000000]	[1,10000000]
50000000	[25000000,75000000]	[1,50000000]	[25000000,75000000]	[1,50000000]
100000000	[50000000,100000000]	[1,100000000]	[50000000,100000000]	[1,100000000]
Data size	**I**	**J**	**K**	**L**
1000000	[750000,1500000]	[1,1000000]	[750000,1500000]	[1,1000000]
5000000	[4000000,8000000]	[1,5000000]	[4000000,8000000]	[1,5000000]
10000000	[7500000,15000000]	[1,10000000]	[7500000,15000000]	[1,10000000]
50000000	[40000000,80000000]	[1,50000000]	[40000000,80000000]	[1,50000000]
1000000000	[75000000,150000000]	[1,100000000]	[75000000,15000000]	[1,100000000]
Data size	**M**	**N**	**O**	**P**
1000000	[1,1000000]	[750000,2000000]	[1,1000000]	[750000,2000000]
5000000	[1,5000000]	[4000000,10000000]	[1,5000000]	[4000000,10000000]
10000000	[1,10000000]	[7500000,20000000]	[1,10000000]	[7500000,20000000]
50000000	[1,50000000]	[40000000,100000000]	[1,50000000]	[40000000,100000000]
1000000000	[1,100000000]	[75000000,200000000]	[1,100000000]	[75000000,200000000]

5.2   Experiment Result

The query we applied here is *"Select R1.A from R1, R2, R3, R4 where R1.B=R2.F and R1.C<R4.O and R2.E>R4.N and R2.E<R3.I and R3.K=R4.O"*, which has five join operations that could cover all the typical operators. We believe this query could be used to figure out the performance of both our method and other methods fairly due to its randomness of dataset generating and the full coverage of the operators along with the number of join operations.

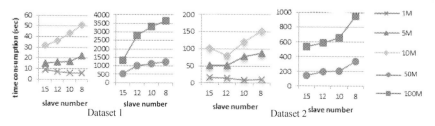

Fig. 8. Execution time of datasets 1 on CMD

Figure 8 demonstrates the impact of node numbers on our algorithm as well as the time consumption trend along with the growth of the data size. It can be seen from the left part of Figure 8 that when the size of input data is small, fewer nodes may have better performance while when the size of input data is larger, more nodes can achieve higher performance. As the number of the node decreases, the time consuming is climbing up. This is because when the data size is small, the time consumption on data transportation is larger than computation, so less transportation among slaves would result in smaller time consumption. When it data size is large, the time of data computation overweighs that of the data transportation, so more slaves could have better performance. The shortest times of our algorithm on dataset 1 are 9s, 52s, 81s, 535s and 1315s while the shortest times of dataset 2 are 6s, 15s, 32s, 148s and 535s. We do not present the time of experiment with single node here because the time consumption was so high that it took much longer time than our algorithm did. Although the trend may not be that dramatic from the figures we present, the time consumptions of less number of nodes were just so frustrating that no one can bear to wait for them to finish their jobs.

Although the data sizes are the same, the time consumptions on processing dataset 2 are much smaller. It is because that fewer regions would be selected since there are fewer overlapping fields between the join attributes, and the data transportation volume and computation volume will drop at the same time.

We did try to conduct the query on the dataset with the least size (1M, which took 9s on our algorithm) on MySQL, but it turns out to run more than a day, so comparison between our method and MySQL will not be presented here due to this huge performance gap. However, methods based on MapReduce may have better performance than MySQL due to its parallelism. We apply the data split method presented in [2], [3] and conduct the join operation in each reducer with the sort-merge algorithm based on a B+ tree, which is faster than B-tree. So, in fact, the algorithm based on MapReduce is faster, experimental environments are the same, and another difference between them is just how to split the data: by HDFS and MapReduce or by CMD.

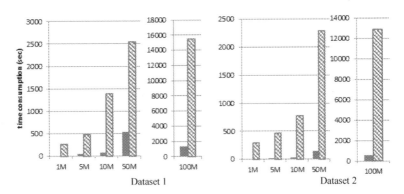

Fig. 9. Comparison between our method and MapReduce on uniform distributed datasets and skewed datasets

Figure 9 compares the best performance of our algorithm on different size of data sets and the best performance we implemented on MapReduce, which was proposed by [2][3]. It can be seen from the Figure 8 that although the time consumptions of MapReduce are acceptable, there is still a huge performance gap between it and our solution especially when the data size grows larger. In fact, we have to separate the 100M line's experiment results from the others' results. Otherwise, the others' bars will become so tiny that nobody can distinguish the time consumptions among them. The cause of this performance gap may lay in the following reasons:

1. MapReduce requires better hardware and larger cluster to fulfill a task this large. In fact, the biggest headache we countered when we conduct the MapReduce experiment is the disk trashing. If the reduce data size is larger than the reducer's memory size, the efficiency will drop dramatically. So we need to find the best parameter to make sure the split data size could fit in the reducers' memory.
2. Although [2] and [3] did propose some data splitting methods to boost the join speed, they were still not powerful enough to avoid the repeat data transportation and computation born with MapReduce itself. In fact, at least two times the size of the original data has to be sent through the network if two relations are involved in a join operation. With more join operations and relations participate the join, even more times of the original data have to be sent through network repeatedly, which results in not only the large latency in network, but also the burden of calculation. By reducing the data transportation and computation, it is little wonder our method is faster than theirs.

In conclusion, as expected and proved by experiment, our algorithm achieves higher performance than both traditional database product and the popular Nosql product.

6   Conclusion

We proposed an efficient algorithm to process multi-way theta-join queries on CMD. This algorithm takes maximum advantage of the CMD storage to split the data evenly. With a new cost model and a related join regions concept, loads of data could be pruned so that lighter burden of networking communication and computation could be achieved. The algorithm we proposed is much faster than that of traditional join methods and the new join methods based on MapReduce by conducting some convincing experiments.

Future directions include spanning the algorithm to higher dimension data sets and a on larger number of clusters. With these works done, the implementation of CMD on real business will come out soon.

Acknowledgement. This work was supported by the NSF of China under Grant No. 61173022, and the Key Program of the NSF of China under Grant No. 61190115.

References

1. Li, J., Srivastava, J., Rotem, D.: CMD: a multidimensional de-clustering method for parallel database systems. In: Proceedings of the 18th International Conference on Very Large Data Bases Conference, Canada (1992)
2. Okcan, A., Riedewald, M.: Processing Theta-Joins using MapReduce. In: SIGMOD 2011, Athens, Greece, June 12-16 (2011)
3. Zhang, X., Chen, L., Wang, M.: Efficient Multiway Theta Join Processing Using MapReduce. Proceedings of the VLDB Endowment 5(11) (August 27-31, 2012)
4. Li, J.-Z., Wei, D.: Parallel CMD- Join Algorithms on Parallel Databases. Journal of Software 9(4) (1998)
5. Li, J.-Z.: A Dynamic and Multidimensional Declustering Method for Parallel Databases. Journal of Software 10(9) (1999)
6. DeWitt, D.: MapReduce: A major step backwards (January 8, 2008), http://databasecolumn.verca.com/2008/01/mapreduce-a-major-step-back.html
7. Dean, J., Ghemawat, S.: Proc. 2004. MapReduce: simplified data processing on large clusters. In: The 6th Symposium on Operating System Design and Implementation, OSDI 2004 (2004)
8. Chang, F., Dean, J., Ghemawat, S.: Proc. 2006. Bigtable: a distributed storage system for structured data. In: The 7th Symp. Operating System Design and Implementation, pp. 205–218. Usenix Assoc. (2006)
9. Apache hadoop, http://hadoop.apache.org
10. Kitsuregawa, M., Tanaka, H., Moto-Oka, T.: Application of Hash to Data Base Machine and Its Architecture. New Generation Comput. 1(1), 63–74 (1983)
11. Boral, H., et al.: Join on a cube: analysis, simulation and implementation. In: Kitsuregawa, M., Tanaka, H. (eds.) Database Machines and Knowledge Base Machines, pp. 61–74. Kluwer, Boston (1988)
12. Schneider, D.A., DeWitt, D.J.: A performance evaluation of parallel in algorithms in a shared-nothing multiprocessor environment. In: Maier, D. (ed.) Proc. of ACM SIGMOD 1989, USA, pp. 110–121. ACM Press, M Baltimore (1989)
13. Kitsuregawa, M., Tanaka, H., Moto-oka, T.: Application of hash to data base machine and its architecture. New Generation Computing 1(1), 25–39 (1983)
14. DeWitt, D.J., Gerber, R.: Multiprocessor hash-based join algorithms. In: Proceedings of VLDB 1985, pp. 151–164. Morgan kaufmann Publishers, Inc., Stockholm (1985)
15. Li, J.: A Dynamic and Multidimensional Declustering Method for Parallel Databases. Journal of Software 10(9) (1999)

MIGSOM: A SOM Algorithm for Large Scale Hyperlinked Documents Inspired by Neuronal Migration

Kotaro Nakayama and Yutaka Matsuo

Center for Knowledge Structuring, The University of Tokyo,
7-3-1 Hongo, Bunkyo-ku, Tokyo Hongo, Japan
http://www.cks.u-tokyo.ac.jp

Abstract. The SOM (Self Organizing Map), one of the most popular unsupervised machine learning algorithms, maps high-dimensional vectors into low-dimensional data (usually a 2-dimensional map). The SOM is widely known as a "scalable" algorithm because of its capability to handle large numbers of records. However, it is effective only when the vectors are small and dense. Although a number of studies on making the SOM scalable have been conducted, technical issues on scalability and performance for sparse high-dimensional data such as hyperlinked documents still remain. In this paper, we introduce *MIGSOM*, an SOM algorithm inspired by new discovery on neuronal migration. The two major advantages of MIGSOM are its scalability for sparse high-dimensional data and its clustering visualization functionality. In this paper, we describe the algorithm and implementation in detail, and show the practicality of the algorithm in several experiments. We applied MIGSOM to not only experimental data sets but also a large scale real data set: Wikipedia's hyperlink data.

Keywords: SOM, Wikipedia, Visualization, Clustering, Link Analysis.

1 Introduction

Elegant mechanisms in biology often impress and fascinate researchers, since they solve complex problems by using simple but rational systems [2]. Although technology has been improved dramatically in terms of computational scale and processing capability, we must admit that the achievements of computer science are far inferior to those of living beings in terms of comprehensiveness and adaptability. That is why the research area of *Biomimicry*, a discipline that studies nature and its models, has emerged in computer science. The basic idea of biomimicry is to observe biological principles contributing to the efficiency and mimic them to advance technology. Biomimicry was originally a trend in the engineering domain, but nowadays many other domains, including computer science and informatics, attempt to apply biomimicry.

In recent years, there has been a renewal of interest in biomimicry in computer science. This trend is related to the dramatic progress of brain science.

S.S. Bhowmick et al. (Eds.): DASFAA 2014, Part I, LNCS 8421, pp. 79–94, 2014.

Researchers have unveiled more about the brain in the last 10 years than in all previous centuries, due to new analytical technologies like fMRI[1]. New technologies allow researchers to analyze the behavior of nervous systems in detail in terms of both time and space. As a result, the mechanism of the central nervous system (CNS) including brain and spinal cord has been uncovered in detail. *Neuronal migration* is one of the research topics that has advanced greatly in this trend. Neuronal migration describes how neurons move from their birth place to their final location in the nervous system. Traditionally, changing connectivity is widely known as a principle of learning in the CNS. Recent neuroscience studies [15–17] have unveiled the important role of neuronal migration in addition to changing connectivity. The detailed mechanism of neuronal migration was a subject of debate among researchers in the past, but latest studies have brought more and more information about this phenomenon, and the important role of neuronal migration in effective CNS network construction has been unveiled.

In this paper, we propose *MIGSOM (MIGrational SOM)*, a new SOM (Self Organizing Map) algorithm inspired by neuronal migration. The SOM, originally proposed by T. Kohonen [10], is widely used in applications such as visualization, clustering, image classification and association memory [14, 1] because of the practicality. One of the most important applications of the SOM is cluster visualization for data where both number and size of the clusters are unclear. In other words, the SOM is used to understand the structure of the input data, and in particular, to identify clusters of input records that have similar characteristics [13]. For instance, Golub et al. has shown that SOMs can be used to analyze classification of tumors based on gene expression patterns, since it is capable of visualizing clusters in detail [6]. Before that, cancer classification was a typical hard task since it is difficult to identify new (cancer) classes. However, Golub et al. proved that SOMs are powerful tools for this task, due to their capability of clustering, representing relations among classes and structure of data (classes as well as sub-classes). It is also interesting to note the SOM's topological localization capability. A number of studies have been made on sensor nodes coordinates based SOM over the past few years [5, 8]. These researches proved that SOMs can be used to plan (optimize) the topological locations of sensors.

An important advantage of the SOM is its iterative learning process with low-memory consumption, since it does not use more than one input record during one iteration. Therefore, the SOM is widely known as a "scalable" algorithm. However, there are a few exceptions to the scalability. Handling large scale sparse data, such as a set of hyperlinked documents, is one of the difficulties on the SOM due to performance loss caused by dimension reduction. MIGSOM, like Kohonen's SOM, is an iterative learning algorithm, but the data structure and process are different and novel. The main characteristics of MIGSOM are 1) the scalability in terms of both number of documents and dimensionality and 2) the clustering performance for web document and 3) the visualization capability. The main contribution of this paper is not limited to the design of a new

---

[1] fMRI (Functional magnetic resonance imaging) is a high-resolution visualizing method for analyzing brain activities by measuring the consumption of oxygen.

algorithm but includes the proof of neuronal migration's potential as a computational model. In Section 2, we first explain how the sparse data problem has been handled in related research.

In Section 3, we give detailed information about MIGSOM. In order to prove the effectiveness and practicality, we show a series of experiments in Sections 4 to 6 including an experiment with a real large scale hyperlinked documents set (Wikipedia's hyperlink data). Finally we draw a conclusion in Section 7.

2   Related Work

The SOM is an unsupervised learning algorithm to map high dimensional data into low dimensional data (usually a 2D regular grid of nodes). Kohonen's SOM [10] is the very first proposal of SOM in history. Although many studies on SOM have been conducted [14, 1, 13, 5, 11, 12, 9] since then, almost all studies are extensions or enhancements based on Kohonen's SOM with the same basic principle. The algorithm of Kohonen's SOM can be briefly summarized as follows:

1) Initialize each node in the map by a random vector. 2) Randomly select one record x from the whole data set X. 3) Compare x to all nodes in the grid and find the most similar node w (winner node). 4) Modify the vectors of nodes around w in distance d. The vectors are averaged by w's vector along with the distance from w. Closer nodes are strongly averaged. 5) Narrow distance d. 6) Go back to step 2.

After carrying out the above steps, the feature vectors in the map will slightly change in each iteration, and a series of iterations creates an organized map. Thus, this process is called self-organizing map. The point to observe here is that only one input record x is needed in each iteration, no matter how large the size of X is. This is the reason why SOM is known to be a scalable algorithm. There are, however, two limitations to this scalability.

The first limitation is caused by the step of selecting a winner node (3rd step). Comparing input vectors with all feature vectors (in nodes) is a time consuming process when the map is large. Map size does not always need to be large even when the number of input records is large. For instance, in an extreme case, SOM can be used with a 2 x 2 (4 nodes) grid to map millions of input records, and it still work if the application does not require detailed clustering. However, in order to visualize a map for a large number of documents, e.g., Web documents, the map size must be big enough according to the number of input records (documents) [11]. The complexity scales dramatically with the number of map units, so that training large maps is time consuming [18].

The second problem is caused by the 4th step. Modifying vectors (actually, vector compositions) is not a big concern for dense data with small dimensionality. However, for sparse data with large dimensionality, such as document term matrices, vector composition causes massive memory consumption. For sparse data, each node in the map initially does not have large dimensionality because vectors can be compressed by storing only non-zero elements. However, each iteration performs a series of vector compositions that make the vectors large and dense, so that memory consumption problems start to occur.

Studies such as Web Mining and document processing (clustering, categorization etc.) often need to handle large scale sparse data such as document-term vectors or document-document vectors (adjacency matrix). This demand has motivated a number of studies on making SOM scalable. These studies can be categorized into two major trends; distributed computing and dimension reduction.

Lawrence et al. proposed an MPI (Message Passing Interface) based distributed computational model of SOM [13]. Although the effectiveness has been demonstrated, it requires a special computing environment such as an IBM SP2 PC cluster.

One of the most famous studies on dimension reduction for SOM is WebSOM [12, 9] proposed by Lagus et al. WebSOM uses dimension reduction methods such as LSI [3] and random mapping in order to convert the sparse large dimensionality into a small dense dimensionality (several hundred dimensions), and then use Kohonen's SOM for clustering. Thanks to these studies, the effectiveness of dimension reduction has been demonstrated, but in contrast, relatively few attempts have been made at renewing SOM itself. We do not disallow the effectiveness of dimension reduction, but dimension reduction reduces information which affects the performance, since dimension reduction likely keeps general information (high frequency) but removes specific information (low frequency) which is useful for clustering. Especially, heterogeneous data like Web corpora, selected dimensions may deviate substantially from the optimal when the data distribution is far from the assumed distribution [4].

3   MIGSOM

MIGSOM is a new SOM algorithm inspired by neuronal migration. The main target of MIGSOM is high-dimensional sparse data such as document-term vectors and document-link vectors (adjacency matrix). As we mentioned, traditional SOMs have had scalability and performance problems for sparse data, due to the trade-off problem on dimension reduction. MIGSOM is designed to satisfy these two demands at the same time. In this section, we first explain the basis of neuronal migration. After that, we describe our proposed method.

3.1   Neuronal Migration

It is commonly accepted in brain science that neurons are born in limited places and repeatedly move to find a suitable place. This behavior is originally discovered in 1970s and it is named "neuronal migration." Recently, the detailed mechanism and the important role of neuronal migration have been unveiled.

There are two types of neuronal migration; translocation and locomotion [15]. Translocation is likely found in early stages of the central nervous system. In translocation, Neurons find a suitable place and move using their own axons. Locomotion, in contrast, is likely found in later stages of the central nervous system. In locomotion, neurons are guided by glial cells and move by following the glial cells. In fact, translocation uses glial cells too, but the way of usage is slightly different. Glial cells (aka neuroglia) are a type of cells in the central

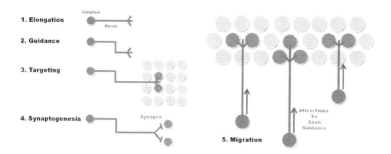

Fig. 1. Translocation

nervous system. Glial cells have long been thought to be trivial cells that just provide energy to the neurons, but recent studies have brought more and more evidences proving the important roles of Glial cells in the learning process.

Both migration types have distinct characteristics and are interesting because of their great potential for computer science. However, we pay special attention to translocation, in this research work because translocation is simpler and more principle.

We show the detailed process of translocation in Figure 1. Neurons start to expand their axons after they are born. Then, they explore their peripheral area by using their axons to decide the direction to expand the axons. After finding a suitable place to establish, they create synaptic connections to the neurons around the area. Finally, they move to the location by shortening the axons.

3.2 MIGSOM Basis

MIGSOM, like Kohonen's SOM, has a map (2D grid) to represent geometric relationships among records. This means that similar records are deployed (mapped) closer to each other in the grid. There are two major differences between Kohonen's SOM and MIGSOM; its representation and its learning process. In traditional SOMs, each node in the grid has its own vector, but MIGSOM's nodes, in contrast, correspond to input records and do not have vectors independent from the input records. For instance, for document categorization based on a document-term matrix, each node corresponds to a document vectorized by terms. Precisely, there are two types of nodes; neuron cells and glial cells. Each neuron cell corresponds to a document (input record) and each glial cell has a randomly generated vector. Glial cells work as intermediators to guide neurons having similarity just like glial cells in real nervous system.

MIGSOM's learning process is also different from that of traditional SOMs. MIGSOM does not modify vectors of nodes but *moves* (migrates) the neurons and glial cells. We illustrate the detailed process of iteration (training) of MIGSOM in Figure 2.

MIGSOM begins the training procedure by randomly selecting a node g from the set of nodes G and adopts a cell unit (a neuron or a glial cell) on the node g as training data (Figure 3 upper). Secondly, it initializes the motion vector

---

Algorithm $train()$:

1   Randomly select g from G

2    $\vec{m_g} = \vec{O}$ #Initialize by null vector

3    $N = GaussianSelection(g)$

4   **for each** $n \in N$

5        $d = distance(n, g)$ #Distance in map

6        $p = tanh(d)$

7        $\vec{m_g} = \vec{m_g} + U(n, g) \cdot p \cdot Sim(n, g)$

8   if $|\vec{m_g}| > t$

9        $Translocate(g, \vec{m_g})$

- G: a set of nodes in the grid.
- $\vec{m_g}$: motion vector of a cell unit on g.
- $GaussianSelection(g)$: function that selects a set of random nodes around g following Gaussian distribution.
- $distance(n, g)$: function that returns the Euclidean distance between n and g on the map.
- $tanh(dist)$: hyperbolic tangent function of distance to calculate the force of reeling in (attracting force). Farther nodes have stronger attracting force.
- $U(n, g)$: a unit vector from g to n.
- $Sim(n, g)$: a function that returns similarity between n's vector and g's vector.
- $Translocate(g, \vec{m_g})$: Translocation function.

Fig. 2. MIGSOM's training algorithm

of the training cell unit $\vec{m_g}$ with a null vector. In the third step, it randomly selects nodes around g. For this purpose, we adopted the distance dependent connectivity model based on Gaussian distribution proposed by Hellwig [7]. This means that closer nodes have a higher probability to be selected than more distant nodes.

For each node n in the set of randomly selected nodes N, MIGSOM calculates the distance between g and n (Figure 3 lower). At the same time, it calculates the similarity between vectors of cell units on n and g. In step 7, MIGSOM modifies the motion vector according to the similarity and distance. Finally, it translocates the cell unit on g to the direction of $\vec{m_g}$ just like an axon based guidance for finding a fitting place where similar neurons exist. To avoid oscillation problem, we employed a threshold t; a cell unit migrates only if $|\vec{m_g}|$ (norm of $\vec{m_g}$) is larger than the threshold t. We restricted the movement to 8 neighbor nodes next to g (Figure 3 upper). At same time, the cell unit on the targeted node moves to g. Strictly speaking, translocation is a swapping of cell units between two nodes in a map. After conducting a number of iterations, small clusters of neurons will be formed and the clusters attract more similar neurons just like a set of neurons which guides neuronal migration in Figure 1 (lower). Eventually, the number of migration will be greatly decreased when the map becomes a stable state and the procedure can be finalized.

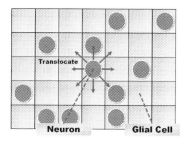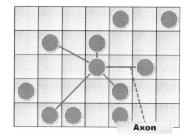

Fig. 3. MIGSOM basis (Movements of neurons are the principle of training. Neurons use their axons to find similar cells. The connectivity is based on Gaussian selection)

3.3   Axonal Cache

The scalability of MIGSOM relies on the *locality of vector comparison*. As mentioned before, traditional SOMs' scalability problem is cased by a massive amount of vector comparisons. MIGSOM, like traditional SOMs, also needs a large number of vector comparisons in step 7, since it calculates the similarity between n's vector and g's vector. However, since N is a set of nodes that tend to be close to g, the possibility of being compared with nodes that have already been compared in the past is high. We call this characteristic *locality of vector comparison*. The locality leads to an important feature on MIGSOM; *cacheability*. Caching the results of vector comparisons greatly contributes to the scalability of MIGSOM, because the same vector pairs are likely to be compared repeatedly. Once the similarity for a vector pair is calculated, the similarity is stored for future iterations. This mechanism is easily understandable if we note that axons store the similarity between two cell units. Therefore, we use the term *axonal cache* to refer to this mechanism.

The implementation of the axonal cache is quite simple. The axonal cache treats caches by following FIFO (First In First Out) principle. It stores all vector comparison results in a queue style cache until the cache size reaches the memory limitation. Once it has reached the size limit, it drops the oldest value from the cache. The advantage of this implementation is that both acquisition and maintenance of values on the cache are $O(1)$. A drawback of this implementation is that it drops values even if the value is repeatedly used so that recalculation of similarity is needed when the dropped value is requested again. In an environment with enough memory, however, frequency of recalculation is low, thus total cost will be lower than using a $O(log\ n)$ algorithm every time. There are alternative caching algorithms that work better than simple FIFO, however it is reasonable enough to confirm our idea as the first step.

Cacheability is one of the advantages of MIGSOM. Traditional SOMs are not able to cache the results of vector comparisons, since vector composition is the principle mechanism of learning, and it changes the value of vector comparison. MIGSOM, in contrast, never combines vectors during the learning process, thus it is able to cache the result of vector comparisons.

4   Performance Experiments

In this section, we show the result of an experiment that we have conducted to measure the effectiveness of our proposed method in terms of performance by comparing it with Kohonen's SOM and with LSI, one of the standard methods in dimension reduction on SOM [12, 9]. Table 1 shows the experimental environment. The detailed procedure of the experiment is as follows:

1) For all combinations of input records, extract an evaluation pair $p = p_1, p_2$ then make a set of all evaluation pairs P. 2) For each evaluation pair p, find corresponding nodes for both p_1 and p_2 in the map and keep them as n_1 and n_2. 3) Calculate cosine similarity for p_1 and p_2. 4) Calculate squared Euclidean distance in the map for n_1 and n_2. 5) Calculate Pearson correlation between cosine similarity set and squared Euclidean distance set.

In short, we measured how similar input records are mapped to close points in the map. In order to investigate the relation between map size and performance, we slightly changed the number of input records (100, 150, 200, 250) and map size in the performance experiment. The dimension of the vectors is equivalent to the data size N. For each data size, we generated 10 datasets (Random sparse matrices generated by GNU Octave) and used the average of the results to make the result accurate. We show the result in Figure 4. In each diagram, the horizontal axis represents the CPU cycles (Processing time) and the vertical axis represents the Pearson correlation score.

First of all, we noticed that Kohonen's algorithm converges faster than MIGSOM. While the baseline method is able to converge after about 500 (N: 100-150) to 1,000 CPU cycles (N: 200-250), MIGSOM needs about 1,600 to 1,800 CPU cycles to converge. However, in terms of correlation, MIGSOM achieved results that were as good as Kohonen's or even slightly better, while the LSI + SOM method clearly decreased performance.

The Kohonen's method scored about 0.6 - 0.7 when the data size is 100, but the correlation score decreased slightly with increasing data size (and map size). For a data size of 250, the Kohonen's method achieved only less than 0.4. MIGSOM, on the other hand, has achieved better results for bigger data size with limited influence of data size related correlation decrease. Although it requires longer time to converge, it scored about 0.5 for a data size of 250 while Kohonen's scored less than 0.4.

Let us now explain an observation on the effectiveness of migrational acceleration. Figure 5 shows a series of diagrams illustrating α's effect on convergence. We changed α from 0.0 (no acceleration) to 0.8 to find out how *alpha* affects the correlation. As we can see, MIGSOM with higher α (0.4, 0.6 and 0.8) seems to converge earlier than in case α is 0.2. Furthermore, 0.6 and 0.8 seem to have a slightly better result than 0.4 although 0.6 and 0.8 do not differ significantly. We have conducted a small experiment thus, it is early to say which α is better since it is possible that various factors such as map size and data size affect the results. However, it seems reasonable to suppose that migrational acceleration contributes to earlier convergence.

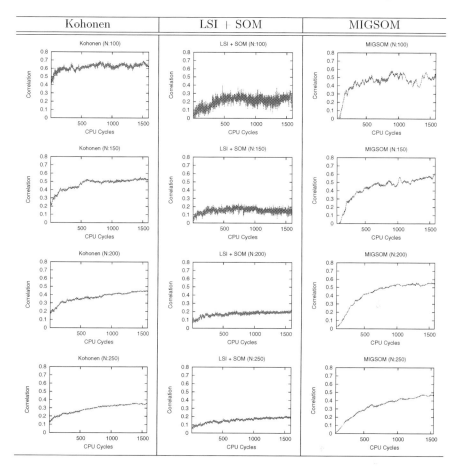

Fig. 4. Performance comparison (MIGSOM achieved comparable performance to Kohonen's while LSI + SOM method clearly decreased the performance)

Since our purpose is to develop a scalable SOM algorithm that is able to avoid performance loss caused by dimension reduction, we can conclude that MIGSOM's mapping functionality, the ability to map similar input records to closer places, fulfills the requirement.

4.1 Migrational Acceleration

After designing MIGSOM, we developed a prototype implementation with learning monitoring functionality and conducted a small-scale preliminarily experiment to find out whether the algorithm works correctly. When we tracked the behaviors of neurons carefully in a preliminarily experiment, we realized that sometimes neurons move back and forth between the same places (oscillating motion) without being able to move to a better place. After careful investigation, we found out that this problem was caused by random connections. In each

Table 1. Experimental environment

CPU	Xeon 2.27 GHz
Memory	12GB
OS	Ubuntu Linux 10.04
MIGSOM implementation	Java (JDK 6)
Numerical library	Apache Commons-Math

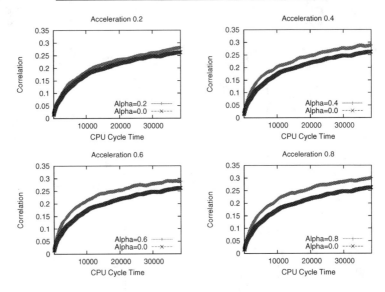

Fig. 5. Effectiveness of acceleration model (Migrational acceleration contributes to earlier convergence)

iteration, MIGSOM clears the motions of neuron and recalculate it from scratch. This means that the motion information in previous iterations is dropped even if it contains useful information. In order to prevent the dropping useful information extracted in previous iterations, we modified the second step of MIGSOM's training process as follows;

$$2 \quad \vec{m_g} = \alpha \cdot \vec{m_g}$$

α ($0 <= \alpha <= 1.0$) is a coefficient that determines how strong previous motions affect the current motion. If we set α to 0.0, the training process becomes exactly the same procedure as before. This modification acts on to use the previous motion instead of initializing the motion by a null vector in each iteration.

5   Scalability Experiment

The second experiment measured the scalability of MIGSOM in terms of both map size and dimensionality. Figure 6 (upper) shows the memory usage for dense data (dimensionality of 100) in each method when the performance reached the

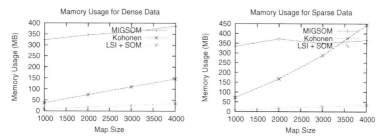

Fig. 6. Memory usage (Kohonen's SOM experiences a memory consumption burst for sparse data)

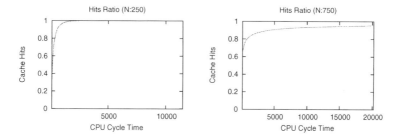

Fig. 7. Hit ratio of axonal cache (Hits ratio do not reach to 100% but increase as time because of the locality)

top score. We slightly changed the data size from 1,000 to 4,000 as well as the map size in proportion. This time, we generated 10 random sparse matrices and again used the average of the results. We can clearly see that all methods have linear memory consumption and the LSI + SOM method has an advantage in memory consumption as we had expected. Outstanding is Figure 6 (lower). It shows the memory size for sparse data used in each method. As the graph indicates, MIGSOM has efficient memory consumption for sparse data (upper) while the memory usage of Kohonen's SOM dramatically increases with increased map size caused by the vector composition. The LSI + SOM method also has significant advantage in sparse data, so that dimension reduction is necessary to handle large scale data in a SOM. However, the advantage comes with a non-trivial performance sacrifice as we showed in the previous section. MIGSOM has larger memory consumption than the LSI + SOM method, but the gradient is closer to linear. This means that MIGSOM is capable of handling massive amounts of data while Kohonen's SOM is not.

Additionally, we conducted an experiment on the hit ratio of the axonal cache (Figure 7). As shown in the upper side of Figure 7 (for N = 250), the cache hit ratio increases as time progress, but this is just due to the nature of caching. What we need to notice here is the behavior for larger data such as shown in the lower side of Figure 7 (for N = 750). At the beginning, the ratio grows rapidly, but after a while the acceleration ratio shifts to slight increase. This phenomenon is caused by the locality of vector comparison. When the cache size reaches the limit, it starts to drop values causing the cache hit ratio to stop growing. However, at the same time, MIGSOM's training process clusters similar records in the map, so that the locality of vector comparison becomes stronger and cache ratio slightly improves. To summarize above, we confirmed that locality of vector comparison exists and that the axonal cache is effectively working as expected.

In next section, we demonstrate the scalability of MIGSOM in detail by applying it to large scale real data.

6   A Case Study: Wikipedia SOM

In order to prove the practicality of MIGSOM, we applied MIGSOM to Wikipedia's hyperlink data and visualized the data for clustering analysis[2]. Because of the scale, coverage and practicality, Wikipedia has become an invaluable corpus for various research areas such as Web mining, natural language processing, artificial intelligence and database. A Wikipedia article corresponds to a conceptual entity and articles are connected by a number of hyperlinks. Wikipedia's dense link structure has been widely proved in previous studies and allows researchers to extract relations among articles (entities). An adjacency matrix generated from Wikipedia's hyperlink network is a typical large scale sparse matrix, thus it is suitable for testing our proposed method.

In this case study, we used an English Wikipedia dump from April 2009 (3+ million articles, 128+ million internal hyperlinks) as the dataset[3]. First, we vectorized all articles by hyperlinks appearing in the article, then we created an adjacency matrix (about 3 million by 3 million) from the hyperlink network. After that, we filtered out two types of articles; those having less than 10 backward links (typically noisy articles), and those that cannot be used for visualization (do not contain any images). About 118,000 articles remained (thus we created a 118,000 by 3 million adjacency matrix).

Next, we prepared a 2D map with 150,000 nodes (300 rows by 500 columns) and randomly deployed input records (articles) on it. For about 3,200 nodes not used for article deployment, we deployed glial cells with random vectors and executed MIGSOM learning. Although it is debatable how many glial cells must be deployed for better learning, we empirically observed that about 2% - 5% of input records gives us reasonable performance in terms of both clustering and computational time. Figure 8 illustrates the progress of MIGSOM's learning process through time. To make the analysis easy, we colored all nodes having an article by k-means clustering algorithm (into 10 clusters) as preparation. We used PIL (Python Imaging Library) to render the processed map to generate an unified map image. As the set of figures indicates, similar articles are placed closer over time. In the beginning (top left), the articles were randomly deployed, but we can notice that several small blocks of articles having the same color

---

[2] Wikipedia SOM Visualizer: http://sigwp.org/wikisom
[3] http://download.wikimedia.org/

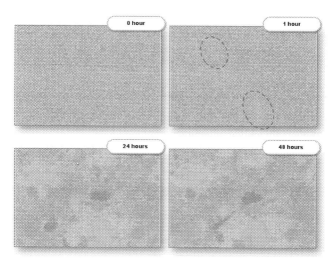

Fig. 8. Progress of the learning process Several small clusters appeared in 1 hour

(belong to the same cluster in k-means) appeared in 1 hour (top right). After 24 hours (bottom left), nodes having the same color are clearly placed to close positions on the map. It may be worth pointing out that the map is deployed on annulus space, a space where one side of the map connects to the other side of the map. Because of the annulus space, all nodes in the map have a uniformed number of neighbors and selection probability.

We have developed an Ajax[4] based interactive visualization interface to make it easy to analyze the clustering result. Thanks to the asynchronous communication technique, the server interactively gives detailed (high resolution) map image only for the requested segments to the client side. Figure 9 shows screen shots of the visualization system. The interactive Ajax analysis functionality allowed us to observe the relationships among clusters in both detail and whole seamlessly. From an observation, we realized that various clusters like Sports / Culture, Science, Nature and Cars are constructed and similar clusters are located closer on the map. A number of observations on boundaries among clusters gave us more findings. For instance, Geography cluster has sub-clusters such as roads, places and landmarks and they are strongly connected by densely binded boundaries. On the other hands, the boundary between Geography cluster and Sports are quite sparse and widely separated.

As we mentioned, one of the most important applications of SOMs is cluster visualization for data where both number and size of the clusters are unclear. For applications such as cancer classification, for instance, the capability of clustering and representing relations among classes and structure of data (classes as well as sub-classes) is an important criterion. We believe that it is no exaggeration to say that the clustering capability of MIGSOM has been proved in our experiment

---

[4] Ajax: Asynchronous JavaScript and XML.

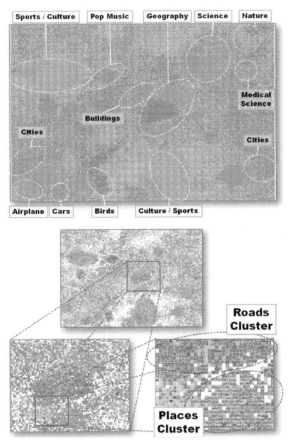

Fig. 9. Discovered clusters (http://sigwp.org/wikisom for demo. Currently, the demo is provided only for review. Please do not access other pages to protect the anonymity of this paper)

described above and MIGSOM is promising as a tool for cluster visualization. Furthermore, MIGSOM's scalability for sparse data has been proved while the Kohonen's SOM could not even analyze the data due to the memory consumption problem mentioned in the previous section.

Although our experiments unveiled the practicality and potential of MIGSOM, it still has potential for improvement, especially in terms of processing time. Fortunately, we have already found several probable solutions for this issue. The first solution is an enhancement of the axonal cache mechanism. The current version of MIGSOM's axonal cache algorithm tries to reconnect all the connections of a neuron. However, it is a reasonable strategy to store important connections (axons to strongly similar neurons) and reconnect only unimportant connections (axons to not similar neurons) to reduce processing time. The second solution is parallel computing. Since MIGSOM's learning process in an iteration is independent from other iterations, it can easily be paralleled.

7   Conclusion

In this paper, we proposed a new SOM algorithm named MIGSOM inspired by neuronal migration. Although much still remains to be done, our conviction that neuronal migration has a potential as an efficient computational model has been confirmed. The detailed experiments showed us the performance of MIGSOM that is comparable (even better) to Kohonen's SOM (MIGSOM achieved 10% better in the performance comparison (N: 250)) while the LSI + SOM method clearly decreased the performance. Furthermore, MIGSOM's advantage on scalability due to the linear memory consumption for sparse data has been shown in an experiment. The scalability is also supported by an experiment on Wikipedia's hyperlink data, which proves the practicality of MIGSOM as a clustering analysis tool.

Although the advantages of MIGSOM have been proved, improving the algorithm for reducing processing time is our next primary challenge. Debates for parameters such as ratio of neurons vs glial cells also still remain to be done. As a future work, designing an algorithm based on locomotion, another neuronal migration model, is also an interesting research direction since it is also an essential principle for constructing efficient information network on central nervous systems.

Acknowledgments. This work was supported by MIC SCOPE (Strategic Information and Communications R&D Promotion Programme).

References

1. Beal, J.: Self-managing associative memory for dynamic acquisition of expertise in high-level domains. In: Proc. of International Joint Conference on Artificial Intelligence (IJCAI), pp. 998–1003 (2009)
2. Benyus, J.M.: Biomimicry: Innovation Inspired by Nature. Harper Perennial (September 2002)
3. Deerwester, S., Dumais, S.T., Furnas, G.W., Landauer, T.K., Harshman, R.: Indexing by latent semantic analysis. Journal of the American Society for Information Science 41(6), 391–407 (1990)
4. Ding, C.H.Q., He, X., Zha, H., Simon, H.D.: Adaptive dimension reduction for clustering high dimensional data. In: Proc. of IEEE International Conference on Data Mining (ICDM), pp. 147–154 (2002)
5. Giorgetti, G., Gupta, S.K.S., Manes, G.: Wireless localization using self-organizing maps. In: Proc. of International Conference on Information Processing in Sensor Networks (IPSN), pp. 293–302. ACM, New York (2007)
6. Golub, T.R., Slonim, D.K., Tamayo, P., Huard, C., Gaasenbeek, M., Mesirov, J.P., Coller, H., Loh, M.L., Downing, J.R., Caligiuri, M.A., Bloomfield, C.D.: Molecular classification of cancer: class discovery and class prediction by gene expression monitoring. Science 286, 531–537 (1999)
7. Hellwig, B.: A quantitative analysis of the local connectivity between pyramidal neurons in layers 2/3 of the rat visual cortex. Biological Cybernetics 82(2), 111–121 (2000)

 8. Jin, M., Xia, S., Wu, H., Gu, X.: Scalable and fully distributed localization with mere connectivity. In: Proc. of IEEE International Conference on Computer Communications (INFOCOM), pp. 3164–3172 (2011)
 9. Kaski, S., Honkela, T., Lagus, K., Kohonen, T.: Websom - self-organizing maps of document collections. Neurocomputing 21(1-3), 101–117 (1998)
10. Kohonen, T.: The self-organizing map. Proceedings of the IEEE 78(9), 1464–1480 (1990)
11. Kohonen, T.: Self organization of a massive document collection. IEEE Transactions on Neural Networks 11(3), 574–585 (2000)
12. Lagus, K., Kaski, S., Kohonen, T.: Mining massive document collections by the websom method. Information Science 163(1-3), 135–156 (2004)
13. Lawrence, R.D., Almasi, G.S., Rushmeier, H.E.: A scalable parallel algorithm for self-organizing maps with applications to sparse data mining problems. Data Mining and Knowledge Discovery 3(2), 171–195 (1999)
14. Lefebvre, G., Laurent, C., Ros, J., Garcia, C.: Supervised image classification by som activity map comparison. In: Proc. of International Conference on Pattern Recognition (ICPR), pp. 728–731. IEEE Computer Society, Washington, DC (2006)
15. Nadarajah, B., Brunstrom, J.E., Grutzendler, J., Wong, R.O., Pearlman, A.L.: Two modes of radial migration in early development of the cerebral cortex. Nature Neuroscience 4(2), 143–150 (2001)
16. Neal, J., Takahashi, M., Silva, M., Tiao, G., Walsh, C.A., Sheen, V.L.: Insights into the gyrification of developing ferret brain by magnetic resonance imaging. Journal of Anatomy 210(1), 66–77 (2007)
17. Ohshima, T., Hirasawa, M., Tabata, H., Mutoh, T., Adachi, T., Suzuki, H., Saruta, K., Iwasato, T., Itohara, S., Hashimoto, M., Nakajima, K., Ogawa, M., Kulkarni, A., Mikoshiba, K.: Cdk5 is required for multipolar-to-bipolar transition during radial neuronal migration and proper dendrite development of pyramidal neurons in the cerebral cortex. Development 134, 2273–2282 (2007)
18. Vesanto, J., Alhoniemi, E.: Clustering of the self-organizing map. IEEE Transactions on Neural Networks 11(3), 586–600 (2000)

Scalable Numerical Queries by Algebraic Inequality Transformations

Thanh Truong and Tore Risch

Department of Information Technology, Uppsala University
Box 337, SE-751 05, Sweden
thanh.truong@it.uu.se, tore.risch@it.uu.se

Abstract. To enable historical analyses of logged data streams by SQL queries, the Stream Log Analysis System (SLAS) bulk loads data streams derived from sensor readings into a relational database system. SQL queries over such log data often involve numerical conditions containing inequalities, e.g. to find suspected deviations from normal behavior based on some function over measured sensor values. However, such queries are often slow to execute, because the query optimizer is unable to utilize ordered indexed attributes inside numerical conditions. In order to speed up the queries they need to be reformulated to utilize available indexes. In SLAS the query transformation algorithm AQIT (Algebraic Query Inequality Transformation) automatically transforms SQL queries involving a class of algebraic inequalities into more scalable SQL queries utilizing ordered indexes. The experimental results show that the queries execute substantially faster by a commercial DBMS when AQIT has been applied to preprocess them.

1 Introduction

We first introduce a real-world scenario application under investigation in the Smart Vortex project [15], which requires queries involving numerical expressions. A factory operates some machines. On each machine, there are a number of sensors to measure different physical properties, e.g. power consumption, pressure, temperature, etc. The sensors generate logs of measurements per machine that carry a time stamp ts, a machine identifier m, a sensor identifier s, a measured value mv, and a measurement class mc for the kind of measurements made by the sensor. Examples of measurement classes are oil pressures of hydraulic filters and pressures of gear pumps. The logs are analyzed by bulk loading them into a relational DBMS. To speed up performance when analyzing sensors of the same kind on many different machines, there is one table for each *measurement class* of each kind of physical property. To avoid repetition of unchanged sensor readings, each measured value mv on machine m is associated with a valid time interval bt and et indicating the begin time and end time for mv, computed from the log time stamp ts when the data is bulk loaded. Hence, the measurement of class $mc=MC$ on machines m will be stored in the table *measuresMC(m, s, bt, et, mv)*. These tables will contain large volumes of log data from many sensors of the same kind on different machines.

S.S. Bhowmick et al. (Eds.): DASFAA 2014, Part I, LNCS 8421, pp. 95–109, 2014.
© Springer International Publishing Switzerland 2014

After the data streams have been loaded into *measuresMC()*, the user can issue off-line historical queries to find errors on machines in the past by looking for abnormal values of *mv*. This often requires search conditions containing inequalities inside numerical expression. In our scenario, in order to improve the performance of inequality queries over *mv*, a B-tree index is added on each *measuresMC.mv*, denoted *idx(measuresMC.mv)*. The following are typical numerical query conditions on tables *measuresA*, and *measuresB* to identify faulty behaviors of machines:

- *C1:* Were the measurements of class A higher than a threshold $v_0 = 15.6$? We express the condition as $C1(mv): mv > v_0$.
- *C2:* Were the measurements of class A higher than $r_1 = 300$ above the expected value $v_1 = 15.6$? We express the condition as $C2(mv): mv - v_1 > r1$.
- *C3:* Were the measurements of class B outside the range $r_2 = 11$ from the ideal value $v_1 = 20$? We express the condition as $C3(mv): |mv - v_1| > r_2$.
- *C4:* Were the measurements of class B outside the range $r_3 = 20\%$ from $v_1 = 20$? We express the condition as $C4(mv): \left| \dfrac{mv - v_1}{v_1} \right| > r_3$.

The above conditions can be expressed in SQL. Relational databases can handle SQL query conditions of type *C1* efficiently, since there is an ordered index *idx(measuresA.mv)*. However, in *C2-C4* the inequalities are not defined directly over the attribute *mv* but through some numerical expressions, which makes the query optimizer not utilizing the indexes and hence the queries will execute slowly. We say that the indexes *idx(measuresA.mv)* and *idx(measuresB.mv)* are *not exposed* in *C2-C4*. To speed up such queries, the DBMS vendors recommend that the user reformulates them [11] which often requires rather deep knowledge of low-level index operations.

To automatically transform a class of queries involving inequality expressions into more efficient queries where indexes are exposed, we have developed the query transformation algorithm *AQIT* (*Algebraic Query Inequality Transformation*). We show that AQIT substantially improves performance for queries with conditions of type *C2-C4*, exemplified by analyzing logged abnormal behavior in our scenario. Without the proposed query transformations the DBMS will do a full scan, not utilizing any index.

AQIT transforms queries with inequality conditions on single indexed attributes to utilize range search operations over B-tree indexes. In general, AQIT can transform inequality conditions of form $F(mv) \, \psi \, \varepsilon$, where *mv* is a variable bound to an indexed attribute *A*, *F(mv)* is an expression consisting of a combination of *transformable* functions *T*, currently $T \in \{+, -, /, *, power, sqrt, abs\}$, and ψ is an inequality comparison $\psi \in \{\leq, \geq, <, >\}$. AQIT tries to reformulate inequality conditions into equivalent conditions, $mv \, \psi' \, F'(\varepsilon)$ that makes the index on attribute *A*, *idx(A)* exposed to the query optimizer. AQIT has a strategy to automatically determine ψ' and $F'(\varepsilon)$. If AQIT fails to transform the condition, the original query is retained. For example, AQIT is currently not applicable on multivariable inequalities, which are subjects for future work.

In summary, our contributions are:

1. We introduce the algebraic query transformation strategy AQIT on a class of numerical SQL queries. AQIT is transparent to the user and does not require manual reformulation of queries. We show that it substantially improves query performance.

2. The prototype system SLAS (Stream Log Analysis System) implements AQIT as a SQL pre-processor to a relational DBMS. Thus, it can be used on top of any relational DBMS. Using SLAS we have evaluated the performance improvements of AQIT on log data from industrial equipment in use.

This paper is organized as follows. Section 2 discusses related work. Section 3 presents some typical SQL queries where AQIT improves performance. Section 4 gives an overview of SLAS and its functionality. Section 5 presents the AQIT algebraic transformation algorithm on inequality expressions. Section 6 evaluates the scalability of applying AQIT for a set of benchmark queries based on the scenario database, along with a discussion of the results. Section 7 gives conclusions and follow-up future work.

2   Related Work

The recommended solution to utilize an index in SQL queries involving arithmetic expressions is to manually reformulate the queries so that index access paths are exposed to the optimizer [5] [11] [13]. However, it may be difficult for the database user to do such reformulations since it requires knowledge about indexing, the internal structure of execution plans, and how query optimization works. There are a number of tools [16] [11], which point out inefficient SQL statements but do not automatically rewrite them. In contrast, AQIT provides a transparent transformation strategy, which automatically transforms queries to expose indexes, when possible. If this is not possible, the query is kept intact.

Modern DBMSs such as Oracle, PostgreSQL, DB2, and SQL Server support *function indexes* [10] [8], which are indexes on the result of a complex function applied on row-attribute values. When an insertion or update happens, the DBMS computes the result of the function and stores the result in an index. The disadvantage of function indexes compared to the AQIT approach is that they are infeasible for ad hoc queries, since the function indexes have to be defined beforehand. In particular, function indexes are very expensive to build in a populated database, since the result of the expression must be computed for every row in the database. By contrast, AQIT does not require any pre-computations when data is loaded or inserted into the database. Therefore AQIT makes the database updates more efficient, and simplifies database maintenance.

Computer algebra systems like Mathematica [1] and Maple [4] and constraints database systems [7] [9] also transform inequalities. However, those systems do not have knowledge about database indexes as AQIT. The current implementation is a DBMS independent SQL pre-processor that provides the index specific query rewritings.

FunctionDB [2] also uses an algebraic query processor. However, the purpose of FunctionDB is to enable queries to continuous functions represented in databases, and it provides no facilities to expose database indexes.

Extensible indexing [6] aims at providing scalable query execution for new kinds of data by introducing new kinds of indexes. However, it is up to the user to reformulate the queries to utilize a new index. By contrast, our approach provides a general mechanism for utilizing indexes in algebraic expressions, which complements extensible indexing. In the paper we have shown how to expose B-tree indexes by algebraic rewrites. Other kinds of indexes would require other algebraic rules, which is a subject of future work.

3   Example Queries

A relational database that stores both meta-data and logged data from machines has the following three tables:

machine(m, mm) represents meta-data about each machine installation identified by *m* where *mm* identifies the machine model. There is a secondary B-tree index on *mm*.

sensor(m, s, mc, ev, ad, rd) stores meta-data about each sensor installation *s* on each machine *m*. To identify different kinds of measurements, e.g oil pressure, filter temperature etc., the sensors are classified by their *measurement class, mc*. Each sensor has some tolerance thresholds, which can be an absolute or relative error deviation, *ad* or *rd*, from the expected value *ev*. There are secondary B-tree indexes on *ev, ad,* and *rd*.

measuresMC(m, s, bt, et, mv) enables efficient analysis of the behavior of different kinds of measurements over many machine installations over time. The table stores measurements *mv* of class *MC* for sensor installations identified by machine *m* and sensor *s* in valid time interval *[bt,et)*. By storing *bt* and *et* temporal interval overlaps can be easily expressed in SQL [3][14]. There are B-tree indexes on *bt, et,* and *mv*.

We use the abnormality thresholds *@thA* for queries determining deviations in table *measuresA*, *@thB* for queries determining absolute deviation in table *measuresB*, and *@thRB* for queries determining relative deviation in table *measuresB*. We shall discuss these thresholds in Section 6 in greater details.

The following queries *Q1, Q2,* and *Q3* identify abnormalities:

- **Query Q1** finds when and on what machines, the pressure reading of class A was higher than *@thA* from its expected value:

```
1 SELECT va.m, va.bt, va.et
2 FROM    measures A va, sensor s
3 WHERE    va.m = s.m AND va.s = s.s AND va.mv > s.ev + @thA.
```

AQIT has no impact for query *Q1* since the index *idx(measuresA.mv)* is already exposed.

- **Query Q2** identifies abnormal behaviors based on absolute deviations: When and for what machines did the pressure reading of class B deviate more than *@thB* from its expected value? AQIT translates the query into the following SQL query *T2:*

```
Q2:                                           T2:
1 SELECT vb.m, vb.bt, vb.et                   SELECT vb.m, vb.bt, vb.et
2 FROM  measuresB vb, sensor s                FROM   measuresB vb, sensor s
3 WHERE vb.m =s.m AND vb.s=s.s                WHERE  vb.m=vb.m AND vb.s=s.s  AND
AND                                             ((vb.mv > @thB + s.ev) OR
4 abs(vb.mv - s.ev) > @thB                      (vb.mv < - @thB + s.ev))
5
```

In *T2* lines 4-5 expose the ordered index *idx(measuresB.mv)*.

- **Query Q3** identifies two different abnormal behaviors of the same machine at the same time based on two different measurement classes and relative deviations: When and for which machines were the pressure readings of class A higher than *@thA* from its expected value at the same time as the pressure reading of class B were deviating *@thRB* % from its expected value? After the AQIT transformation *Q3* becomes *T3*:

```
Q3:                                           T3:
1 SELECTva.m,greaest(va.bt,vb.bt)             SELECT va.m,greatest(va.bt, vb.bt),
2      least(va.et, vb.et)                       least(va.et, vb.et)
3 FROM  measuresA va,measuresB vb,            FROM    measuresA va, measuresB vb,
4      sensor sa, sensor sb                          sensor sa, sensor sb
5 WHERE va.m=sa.m AND va.s=sa.s AND           WHERE va.m=sa.m AND va.s=sa.s AND
6 vb.m=sb.m AND vb.s=sb.s         AND 7             vb.m=sb.m AND vb.s=sb.s AND
va.m=vb.m                         AND               va.m =vb.m                AND
8 va.bt<=vb.et AND va.et>=vb.bt AND            va.bt<=vb.et AND va.et>=vb.bt AND
9 va.mv - sa.ev > @thA            AND          va.mv >@thA + sa.ev           AND
10 abs((vb.mv-sb.ev)/sb.ev)>@thRB             ((vb.mv>(1+@thRB)*sb.ev AND sb.ev >0)
11                                            OR (vb.mv<(1+@thRB)*sb.ev AND sb.ev<0)
12                                            OR (vb.mv<(-@thRB+1)*sb.ev ANDsb.ev>0)
13                                            OR (vb.mv>(-@thRB+1)*sb.ev AND sb.ev<0))
```

Lines 8 in *Q3* selects temporal overlap of the time interval *[va.bt, va.et]* with *[vb.bt, vb.et]*. The functions *greatest(va.bt, vb.bt)* and *least(va.et, vb.et)* return the maximum and minimum values of their two arguments, respectively. These functions are supported by Oracle, MySQL, DB2 and PostgreSQL but not by SQL Server [14]. Therefore, we defined *greatest(x, y)* and *least(x, y)* as user defined functions for SQL Server.

In *T3* line 9 exposes *idx(measuresA.mv)* and lines 10-13 expose *idx(measuresB.mv)*.

4   Stream Log Analysis System (SLAS)

Fig. 1 illustrates the architecture of SLAS. It uses a data stream management system, *DSMS*, to process *raw streams* of measurements from different *machines*. The *log writer* receives from the DSMS a stream of tuples with format *(mc, m, s, ts, mv)* specified as a continuous query. The log writer produces once per system-determined time interval a CSV file of tuples *(m, s, bt, et, mv)* for each measurement class *mc* to be loaded into the corresponding table *measuresMC*. Here, *[bt,et)* is the valid time interval for *mv*, computed from *ts*. When the log writer has written a CSV file it notifies the *log loader* for measurement class *mc*, which bulk loads the new log file rows into the corresponding measurement log table *measuresMC*.

In order to limit and customize the amount of log data stored in the DBMS the *log deleter* continuously deletes log data from the DBMS according to user specified configuration parameters.

The user can analyze the stored data streams by issuing historical SQL queries over loaded log data through the *AQIT processor*. The strategy used by AQIT to improve numerical SQL queries is the focus of this paper.

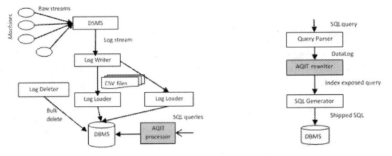

Fig. 1. Stream Log Analyse System   **Fig. 2.** AQIT Preprocessor

Fig. 2 illustrates the query processing of AQIT. An SQL query is first parsed into an internal query in a Datalog dialect [12]. The *AQIT rewriter* transforms the Datalog query into an equivalent *index exposed query*. The *SQL Generator* transforms the index exposed Datalog query into an equivalent *shipped SQL* query sent to the backend DBSM through JDBC for optimization and evaluation.

5   Algebraic Query Inequality Transformation

To explain the AQIT transformations we need the following definitions:

Definition 1. A *source predicate r(…)* of a query is a predicate that represents access to a relation named *r*.

Definition 2. If there is a B-tree index *idx(r.a)* on some attribute *a* of a source predicate   *r(…a…)*, we say that *r* is an *indexed predicate*.

Definition 3. If there is an occurrence of a variable *v* representing *idx(r.a)* in an indexed predicate *r(…v…)* of a query, we say that *v* is an *indexed variable* in the query.

Definition 4. If there is an inequality $\psi(v,x)$ where *v* is an indexed variable, we say that the indexed variable *v* is *exposed* by the inequality predicate ψ.

In this section, we use *Q1* and *Q2* to show how AQIT works. First the parser translates *Q1*, and *Q2* into the following Datalog queries *DQ1* and *DQ2*:

```
DQ1(m,bt,et) ←
measuresA(m,s,bt,et,mv)        AND
sensor(m,s,_,_,ev,_,_)         AND
v1 = ev + @thA                 AND
mv > v1
```
```
DQ2(m,bt,et) ←
measuresA(m,s,bt,et,mv)        AND
sensor(m,s,_,_,ev,_,_)         AND
v1 = mv - ev                   AND
v2 = abs(v1)                   AND
v2 > @thA
```

Here, the source predicates *measuresA(m,s,bt,et,mv)* and *measuresB(m,s,bt.et,mv)* represent relational tables for two different measurement classes. For both tables there is a B-tree index on *mv* to speed up comparison and proximity queries, and therefore *measuresA()* and *measuresB()* are indexed predicates and the variable *mv* is an indexed variable. In *Q1*, the index *idx(measuresA.mv)* is already exposed because there is a comparison between *measuresA.mv* and variable *v1*, so AQIT will have no effect.

In *Q2*, the index *idx(measuresB.mv)* is not exposed by the inequality predicate *v2 > @thB* since the inequality is defined over a variable *v2*, which is not bound to the indexed attribute *measuresB.mv*. Here AQIT transforms the predicates to expose the index *idx(measuresB.mv)* so in *T2 idx(measuresB.mv)* is exposed in both *OR* branches.

5.1   AQIT Overview

The AQIT algorithm takes a Datalog predicate as input and returns another semantically equivalent predicate that exposes one or several indexes, if possible. AQIT is a *fixpoint* algorithm that iteratively transforms the predicate to expose hidden indexes until no further indexes can be exposed. The full pseudo code can be found in [17].

The transformations are made iteratively by the function *transform_pred()* in Listing 1. At each iteration, it invokes three functions, called *chain()*, *expose()*, and *substitute()*. *chain()* finds some path between an indexed variable and an inequality predicate that can be exposed, *expose()* transforms the found path so that the index becomes exposed, and *substitute()* replaces the terms in the original predicate with the new path.

```
function transform_pred(pred):
input:    A predicate pred
output:   A transformed predicate or the original pred
begin
  if pred is disjunctive then
      set failure = false
      /*result list of transformed branches*/
      set resl = null
      do /*transform each branch*/
          set b = the first not transformed branch in pred
          set nb = transform_pred(b)/*new branch*/
          if  nb not null then add nb to resl
          else  set failure = true
      until failure or no more branch of pred to try
        if not failure then
            /*return a disjunction from resl*/
            return  orify(resl)
        end if
  else if pred is conjunctive then
      set path = chain(pred)
      if path not null then
          set exposedpath = expose(path)
          if  exposedpath not null then
              return substitute(pred, path, exposedpath)
          end if
      end if
  end if
  return pred
end
```

Listing 1. Transform Predicate

Chain. The *chain()* algorithm tries to produce a path of predicates that links one indexed variable with one inequality predicate. If there are multiple indexed variables a simple heuristic is applied. It sorts the indexed variables decreasingly based on selectivities of the indexed attributes, which can be obtained first from the backend DBMS. The path must be a conjunction of *transformable terms* that represent expressions transformable by AQIT. Each transformable term in a path has a single common variable with adjacent terms. Such a chain of connected predicates is called an *index inequality path (IIP)*. Query *DQ2* has the following IIP called *Q2-IIP* from the indexed variable *mv* to the inequality *v2 > @thB*, where the functions '−' and '*abs*' are transformable:

`Q2-IIP: measuresB(m, s, bt, et, mv)→ v1=mv - ev → v2=abs(v1)→ v2>@thB`

In this case *Q2-IIP* is the only possible IIP, since there are no other unexposed index variables in the query after *Q2-IIP* has been formed. The following graph illustrates *Q2-IIP*, where nodes represent predicates and arcs represent the common variable of adjacent nodes:

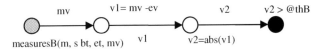

Fig. 3. Q2-IIP

An IIP starts with an indexed *origin* predicate and ends with an inequality *destination* predicate. The origin node in an IIP is always an indexed predicate where the outgoing arc represents one of the indexed variables.

chain() is a backtracking algorithm trying to extend partial IIPs consisting of transformable predicates from an indexed variable until some inequality predicate is reached, in which case the IIP is *complete*. The algorithm will try to find one IIP per indexed variable. If there are several common variables between transformable terms, *chain()* will try each of them until a complete IIP is found. If there are other not yet exposed ordered indexes for some source predicates, the other IIPs may be discovered later in the top level fixpoint iteration.

The *chain()* procedure successively extends the IIP by choosing new transformable predicates *q* not on the partial IIP such that one of *q*'s arguments is the variable of the right-most outgoing arc (*mv* in our case) of the partial IIP. For *DQ2* only the predicate *v1=mv-ev* can be chosen, since *mv* is the outgoing arc variable and '−' is the only transformable predicate in *DQ2* where *mv* is an argument. When there are several transformable predicates, *chain()* will try each of them in turn until the IIP is complete or the transformation fails.

An IIP through a disjunction is treated as a disjunction of IIPs with one partial IIP per disjunct in Listing 1. In this case the index is considered utilized if all partial IIPs are complete.

Expose. The *expose()* procedure is applied on each complete IIP in order to expose the indexed variable. The indexed variable is already exposed if there are no intermediate nodes between the origin node and the destination node in the IIP. For example, the IIP for *Q1* is `Q1-IIP: measuresA(m, s, bt, et, mv)→ mv>v1`. Here the indexed variable *mv* is already exposed to the inequality. Therefore, in this case *expose()* returns the input predicate unchanged.

The idea of *expose()* is to shorten the IIP until the index variable is exposed by iteratively combining the two last nodes through the algebraic rules in

Table 4 into larger destination nodes while keeping the IIP complete. To keep the IIP complete the incoming variable of the last node must participate in some inequality predicate. As an example, the two last nodes in *Q2-IIP* in Fig. 3 are combined into a disjunction in Fig. 4. Here the following algebraic rule is applied: *R10:* $|x| > y \Rightarrow (x > y \lor x < -y)$.

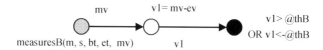

Fig. 4. Q2-IIP after the first reduction

The algebraic rule *R10* exposes a variable x hidden inside *abs()* of an inequality. The following table shows how R10 is applied on the two last nodes in Fig. 3 to form the new predicate in Fig. 4.

Table 1. Applying R10

Before	After
v2 = abs(v1) AND v2 > @thB	(v1 > @thB OR v1 < -@thB)

By iteratively exposing each variable on the IIP, the indexed variable (and the index) will possibly be exposed. For example, Q2-IIP in Fig. 4 is reduced into Fig. 5 by applying the algebraic rules *R3:* $x - y > z \Rightarrow x > y + z$ and   *R4:* $x - y < z \Rightarrow x < y + z$.

Fig. 5. Q2-IIP after the second reduction

The following two tables show how rules R3 and R4 have been applied:

Table 2. Applying R3

Before	After
v1 = mv –ev AND v1 > @thB	v3 = ev + @thB AND mv > v3

Table 3. Applying R4

Before	After
v1 = mv –ev AND v1 < -@thB	v4 = ev -@thB AND mv < v4

The new variables *v3* and *v4* are created when applying the rewrite rules to hold intermediate values.

In Fig. 5, there are no more intermediate nodes and the index *idx(measuresB.mv)* is exposed, so *expose()* succeeds.

expose() may fail if there is no applicable algebraic rule when trying to combine some two last nodes, in which case the *chain()* procedure will be run again to find a next possible IIP until as many indexed variables as possible are exposed.

Substitute. When *expose()* has succeeded, *substitute()* updates the original predicate by replacing all predicates in the original IIP, except its origin, with the new destination predicate in the transformed IIP [17]. For *Q2* this will produce the final transformed Datalog query:

```
DQ2 (m,bt,et) ← measuresB (m,s,bt,et,mv) AND
          sensor (m, s, _,_,ev,_,_)    AND
          v3  =  ev + @thB             AND
          v4 = ev -@thB                AND
          (mv < v4 OR mv >  v3)
```

The Datalog query is the translated by the SQL Generator into SQL query *T2*.

5.2   Inequality Transformation Rules

Table 4 the algebraic rewrite rules currently used by AQIT are listed. The list can be extended for new kinds of algebraic index exposures. In the rules, *x*, *y*, and *z* are variables and ψ denotes any of the inequality comparisons $\geq$, $\leq$,<, or >, while ψ^{-1} denotes the inverse of ψ. *CP* denotes a positive constant (*CP* > 0), while *CN* denotes a negative constant (*CN* < 0). Each rule shows how to expose the variable *x* hidden inside an algebraic expression to some inequality expression.

Table 4. Algebraic inequality transformations

R1	$(x + y)\ \psi z$	⇔	$x\ \psi (z - y)$
R2	$(y + x)\ \psi z$	⇔	$x\ \psi (z - y)$
R3	$(x - y)\ \psi z$	⇔	$x\ \psi (z + y)$
R4	$(y - x)\ \psi z$	⇔	$x\ \psi^{-1} (y - z)$
R5	$(x * CP)\ \psi z$	⇔	$(x\ \psi z/CP)$
R6	$(x * CN)\ \psi z$	⇔	$(x\ \psi^{-1}\ z/CN)$
R7	$x/y\ \psi z \wedge\ y! = 0$	⇔	$(x\ \psi y*z \wedge y > 0) \vee (x\ \psi^{-1} z*y \wedge y < 0)$
R8	$y/x\ \psi z$	⇔	$(y/z\ \psi x \wedge x*z > 0) \vee (y/z\ \psi^{-1} x \wedge x*z < 0)$ $\vee\ (y = 0 \wedge 0\ \psi\ z)$
R9	$|x| \leq y$	⇔	$(x \leq y \wedge x\ \geq - y)$
R10	$|x| \geq y$	⇔	$(x\ \geq y \vee x \leq - y)$
R11	$\sqrt{x}\ \psi y$	⇔	$x\ \psi y^2$
R12	$x^y\ \psi z$	⇔	$(x\ \psi \sqrt[y]{z} \wedge y\ > 0) \vee (x\ \psi^{-1} \sqrt[y]{z}\ \wedge y < 0)$ $\vee\ (x\ \psi z \wedge y = 0)$
R13	$(x + y)/x\ \psi\ z$	⇔	$(1 + y/x)\ \psi\ z$
R14	$|(x - y) / y\ | > z$	⇔	$(x > (z + 1)* y \wedge y > 0) \vee (x < (z + 1)* y\ \ y < 0)$ $\vee (x < (- z+ 1)* y \wedge y > 0) \vee (x > (- z + 1)* y\ \ y < 0)$

6   Experimental Evaluation

We experimentally compared the performance of a number of typical queries finding different kinds of abnormalities based on 16000 real log files from two industrial machines. To simulate data streams from a large number of machines, 8000 log files were constructed by pairing the real log files two-by-two and then time-stamping their events based on off-sets from their first time-stamps. This produces realistic data logs and enables scaling the data volume by using an increasing number of log files.

6.1   Setup

To investigate the impact of AQIT on the query execution time, we run the SLAS system with SQL Server™ 2008 R2 as DBMS on a separate server node. The DBMS was running under Windows Server 2008 R2 Enterprise on 8 processors of AMD Opteron ™ Processor 6128, 2.00 GHz CPU and 16GB RAM. The experiments were conducted with and without AQIT preprocessing.

6.2   Data

Fig. 6 (a) is a scatter plot from a small sampled time interval of pressure readings of class A. This is an example of an asymmetric measurement series with an initial warm-up period of 581.1 seconds.

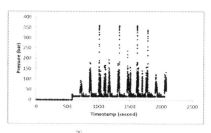
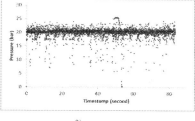

Fig. 6. Pressure measured of class A (a) and class B (b)

The abnormal behavior in this case is that the measured values are larger than the expected value (17.02) within a threshold. When the deviation threshold is 0 all measurements are abnormal, while when the threshold is 359.44 no measurements are abnormal. For example, *Q1* finds when a sensor reading of class A is abnormal based on threshold @*thA* that can be varied.

Fig. 6 (b) plots pressure readings of measurements of class B over a small sampled time interval. Here the abnormality is determined by threshold @*thB*, indicating absolute differences between a reading and the expected value (20.0), as specified in *Q2*. When the threshold is 0 all measurements are abnormal, while when the threshold is 20.0 no measurements are abnormal.

In addition, the abnormality of measurements of class B is determined by threshold @*thRB* as in *Q3*, indicating relative difference between a reading and the expected value. When the relative deviation threshold is 0%, no measurements are abnormal, while when the threshold is 100% all measurements are abnormal.

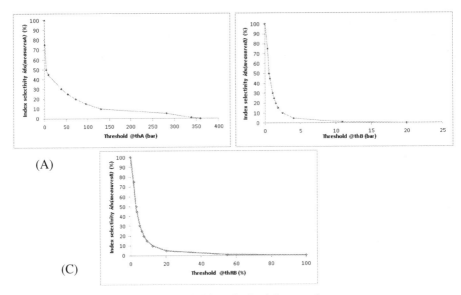

Fig. 7. Thresholds and selectivity mappings

6.3   Benchmark Queries

We measured the impact of index utilization exposed by AQIT by varying the abnormality thresholds *@thA* for queries determining deviations in *measuresA*, and the thresholds *@thB* and *@thRB* for queries determining deviations in *measuresB*. The larger the threshold values the fewer abnormalities will be detected. We also defined three other benchmark queries *Q4*, *Q5*, and *Q6*. All the detailed SQL and Datalog formulations before and after AQIT for the benchmark queries are listed in [18].

- *Q4* identifies when the pressure readings of class B deviates more than *@thB* for the machines in a list *machine-models* of varying length. Here, if a query spans many machine models the impact of AQIT should decrease since many different index keys are accessed.

```
Q4: SELECT vb.m, vb.bt, vb.et
FROM measuresB vb, sensor s,
        machine ma
WHERE vb.m = s.m AND va.s=s.s AND
        vb.m = ma.m             AND
        ma.mm in@machine-models AND
        abs(vb.mv - s.ev) > @thB
```

```
T4: SELECT vb.m, vb.bt, vb.et
FROM measuresB vb, sensor s,
        machine ma
WHERE vb.m = s.m AND va.s=s.s AND
        vb.m = ma.m             AND
        ma.mm in @machine-models AND
        (vb.mv > @thB + s.ev OR vb.mv < -
        @thB + s.ev
```

- *Q5* identifies when the pressure reading of class B deviates more than *@thB* for two specific machine models using a temporal join. The query involves numerical expressions over two indexed variables, which are both exposed by AQIT. See [18] for details.
- Query *Q6* is a complex query that identifies a sequence of two different abnormal behaviors of the same machine happening within a given time interval, based on two different measurement classes: On what machines the pressure readings class B were out-of-bounds more than *@thB* within 5 seconds after the pressure readings of class A were higher than *@thA* from the expected value. Here, both *idx(measuresA.mv)* and *idx(measuresB.mv)* are exposed by AQIT. See [18] for details.

6.4 Performance Measurements

To measure performance based on different selectivities of indexed attributes, in Fig. 7 we map the threshold values to the corresponding measured index selectivities of *idx(measuresA.mv)* and *idx(measuresB.mv)*. 100% of the abnormalities are detected when any of the thresholds is 0 and thresholds above the maximum threshold values (*@thA*=359.44, *@thB*=20.0, and *@thRB*=100%) detect 0% abnormalities.

Experiment A varies the database size from 5GB to 25GB while keeping the selectivities (abnormality percentages) at 5% and a list of three different machine models in *Q4*.

Fig. 8 (a) shows the performance of example queries *Q2, Q3, Q4, Q5*, and *Q6* (without AQIT) and their corresponding transformed queries *T2, T3, T4, T5*, and *T6* (with AQIT) when varying the database size from 5 to 25 GB. The original queries without AQIT are substantially slower since no indexes are exposed and the DBMS will do full scans, while for transformed queries the DBMS backend can utilize the exposed indexes.

Experiment B varies index selectivities of *idx(measuresA.mv)* and *idx(measuresB.mv)* while keeping the database size at 25 GB and selecting three different machine models in *Q4*. We varied the index selectivities from 0% to 100%. Fig. 8 (b) presents execution times of the all benchmark queries with and without AQIT.

Without AQIT, the execution times for *Q2 - Q6* stay constant when varying the selectivity since no index is utilized and the database tables are fully scanned.

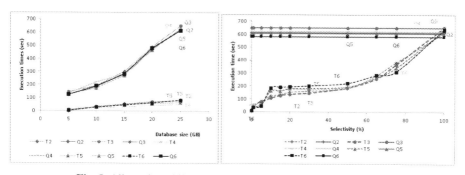

Fig. 8. All queries while changing DB size (a) and selectivities (b)

Fig. 8 (b) shows that AQIT has more effect the lower the selectivity, since index scans are more effective for selective queries. For non-selective queries the indexes are not useful. When all rows are selected the AQIT transformed queries are slightly slower than original ones; the reason being that they are more complex. In general AQIT does not make the queries significantly slower.

Experiment C varies the number machine models in *Q4* from 0 to 25 while keeping the database size at 25 GB and the selectivity at 5%, as illustrated by Fig. 9. It shows that when the list is small the transformed query *T4* scales much better than the original query *Q4*. However, when the list of machine increases, *T4* is getting slower. The reason is that the index *idx(measuresB.mv)* is accessed once per machine model, which is faster for fewer models.

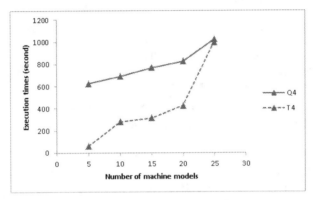

Fig. 9. Execution times of Query 4 when varying the list of machine models

The experiments A, B, and C show that AQIT improves the performance of the benchmark queries substantially and will never make the queries significantly slower. In general AQIT exposes hidden indexes while the backend DBMS decides whether to utilize them or not.

7   Conclusion and Future Work

In order to improve the performance of queries involving complex inequality expression, we investigated and introduced the general algebraic query transformation algorithm AQIT. It transforms a class of SQL queries so that indexes hidden inside numerical expressions are exposed to the back-end query optimizer.

From experiments, which were made on a benchmark consisting of real log data streams from industrial machines, we showed that the AQIT query transformation substantially improves query execution performance.

We presented our general system architecture for analyzing logged data streams, based on bulk loading data streams into a relational database. Importantly, looking for abnormal behavior of logged data streams often requires inequality search conditions and AQIT was shown to improve the performance of such queries.

We conclude that AQIT improves substantially the query performance by exposing indexes without making the queries significantly slower.

Since inequality conditions also appear in spatial queries we plan to extend AQIT to support transforming spatial query conditions as well user defined indexing. We also acknowledge that the inequality conditions could be more complex with multiple variables and complex mathematical expression, which will require other algebraic rules.

Acknowledgements. The work was supported by the Smart Vortex EU project [15].

References

1. Andrew, A.D., Cain, G.L., Crum, S., Morley, T.D.: Calculus Projects Using Mathematica. McGraw-Hill (1996)
2. Arvind, T., Samuel, M.: Querying continuous functions in a database system. In: Proc. SIGMOD 2008, Vancouver, Canada, pp. 791–804 (2008)
3. Celko, J.: SQL for Smarties, 4th edn. Advanced SQL Programming (2011) ISBN: 978-0-12-382022-8
4. Chang, C.M.: Mathematical Analysis in Engineering. Cambridge University Press (1994)
5. Dageville, B., Das, D., Dias, K., Yagoub, K., Zaït, M., Ziauddin, M.: Automatic SQL Tuning in Oracle 10g. In: Proc. VLDB 2004, Toronto, Canada, pp. 1098–1109 (2004)
6. Eltabakh, M.Y., Eltarras, R., Aref, W.G.: Space-Partitioning Trees in PostgreSQL: Realization and Performance. In: Proc ICDE, Atlanta, Georgia, USA, pp. 100–112 (April 2006)
7. Gabriel, K., Leonid, L., Jan, P.: Constraint Databases, pp. 21–54. Springer, Heidelberg, ISBN 978-3-642-08542-0
8. Gray, J., Szalay, A., Fekete, G.: Using Table Valued Functions in SQL Server 2005 to Implement a Spatial Data Library,Technical Report, Microsoft Research Advanced Technology Division (2005)
9. Grumbach, S., Rigaux, P., Segoufin, L.: The DEDALE system for complex spatial queries. In: Proc SIGMOD 1998, Seattle, Washington, pp. 213–224 (1998)
10. Hwang, D.J.-H.: Function-Based Indexing for Object-Oriented Databases, PhD Thesis, Massachusetts Institute of Technology, 26–32 (1994)
11. Leccotech. LECCOTECH Performance Optimization Solution for Oracle, White Paper (2003), http://www.leccotech.com/
12. Litwin, W., Risch, T.: Main Memory Oriented Optimization of OO Queries using Typed Datalog with Foreign Predicates. IEEE Transactions on Knowledge and Data Engineering 4(6) (December 1992)
13. Oracle Inc. Query Optimization in Oracle Database 10g Release 2. An Oracle White Paper (June 2005)
14. Snodgrass, R.T.: Developing Time-Oriented Database Applications in SQL. Morgan Kaufmann Publishers, Inc., San Francisco (1999) ISBN 1-55860-436-7
15. Smart Vortex Project , http://www.smartvortex.eu/
16. Quest Software.Quest Central for Oracle: SQLab Vision (2003), http://www.quest.com
17. http://www.it.uu.se/research/group/udbl/aqit/PseudoCode.pdf
18. http://www.it.uu.se/research/group/udbl/aqit/Benchmark_queries.pdf

SAQR: An Efficient Scheme for Similarity-Aware Query Refinement

Abdullah Albarrak, Mohamed A. Sharaf, and Xiaofang Zhou

The University of Queensland
Queensland, Australia
{a.albarrak,m.sharaf,zxf}@uq.edu.au

Abstract. Query refinement techniques enable database systems to automatically adjust a submitted query so that its result satisfies some specified constraints. While current techniques are fairly successful in generating refined queries based on cardinality constraints, they are rather oblivious to the (dis)similarity between the input query and its corresponding refined version. Meanwhile, enforcing a *similarity-aware* query refinement is a rather challenging task as it would require an exhaustive examination of the large space of possible query refinements. To address this challenge, we propose a novel scheme for efficient *Similarity-aware Query Refinement (SAQR)*. SAQR aims to balance the tradeoff between satisfying the cardinality and similarity constraints imposed on the refined query so that to maximize its overall benefit to the user. To achieve that goal, SAQR implements efficient strategies to minimize the costs incurred in exploring the available search space. In particular, SAQR utilizes both similarity-based and cardinality-based pruning techniques to bound the search space and quickly find a refined query that meets the user expectations. Our experimental evaluation shows the scalability exhibited by SAQR under various workload settings, and the significant benefits it provides.

1 Introduction

The explosion of available data in many web, scientific and business domains dictates the need for effective methods to quickly guide the users in locating information of high interest. *Query refinement* is one such method that has been widely employed to assist users in that direction [11,6]. In particular, query refinement enable database systems to automatically adjust a submitted query so that the query result satisfies some pre-specified requirements. Setting some constraint on the *cardinality* of the query result is one example of such requirements, which provides practical solution to the problem of queries returning too many or too few answers. Towards this, *cardinality-based* query refinement has recently attracted several research efforts (e.g., [10,2,16,11]).

Specifically, cardinality-based query refinement techniques aim to quickly navigate a large search space of possible refined queries and return one refined query such that the cardinality of its result is very close to a pre-specified cardinality constraint. That is, minimize the deviation between the pre-specified cardinality and the achieved one. Towards this, it has been shown that simple *local search* techniques based on greedy heuristics (e.g., *Hill Climbing*) often provide efficient and effective solutions to the

S.S. Bhowmick et al. (Eds.): DASFAA 2014, Part I, LNCS 8421, pp. 110–125, 2014.

cardinality-based query refinement problem [2]. In particular, the relationship between the possible refined queries and their corresponding cardinalities exhibit a *monotonic* pattern [2], which makes it practical to apply local search methods.

These techniques, however, are oblivious to the (dis)similarity between the input query and its corresponding refined version. That is, to meet the cardinality constraint, the generated refined query might often be very far (i.e., dissimilar) from the input query. While the (dis)similarity between two queries can be quantified using several alternative measures (e.g., [15,13,6,12,16]), it should be clear that, irrespective of the adopted similarity measure, a refined query that is very different from the input one will have a very limited benefit to the end user and is often rendered useless.

To address the limitation of current cardinality-based query refinement techniques, in this work, we propose the similarity-aware query refinement problem, in which the user satisfaction is measured in terms of both: 1) meeting some specified cardinality constraint on the refined query, and 2) maximizing the similarity between the submitted input query and its corresponding refined one. Achieving such goal is a rather challenging task as it would require an exhaustive examination of the large space of possible query refinements. In particular, the monotonic property exhibited under the cardinality-based refinement no longer holds for our dual-criteria objective, in which cardinality is combined with similarity. Hence, current query refinement techniques that are based on local search heuristics have higher chances of meeting a local minima and falling short in finding a refined query that strikes a fine balance between minimizing the deviation in cardinality and maximizing the similarity.

Motivated by that challenge, we propose a novel scheme called *SAQR*, which aims to balance the tradeoff between satisfying the cardinality and similarity constraints imposed on the refined query so that to maximize its overall benefit to the user. In particular, the contributions of this paper are summarized as follows:

- Defining the similarity-aware query refinement problem, which captures the user's constraints on both cardinality and similarity.
- Proposing a new scheme, SAQR, which utilizes pruning techniques based on both similarity and cardinality to efficiently formulate a refined query that meets the user's expectations.
- Employing a hierarchical representation of the refined queries search space, which allows for significant reduction in the cost incurred by SAQR while maintaining the quality of its solution.
- Demonstrating extensive evaluation results, which consistently show the significant gains provided by SAQR compared to existing query refinement techniques.

Roadmap. This paper is organized as follows: Section 2 provides preliminary background, whereas the problem definition is stated in Section 3. Our SAQR scheme for similarity-aware refinement appears in Sections 4 and 5. Experimental results are presented in Section 6 and related work is discussed in Section 7.

2   Preliminaries

The input to the refinement process is an initial Select-Project-Join (SPJ) query I, which is to be transformed into a refined query R. Query I is a conjunctive query defined in

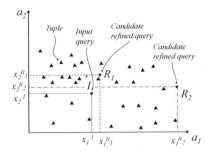

Fig. 1. Example - refining an input query I in a two-dimensional space

terms of d range predicates $P_1^I, P_2^I, ..., P_d^I$. Each range predicate P_i^I is in the form $l_i^I \le a_i \le u_i^I$, where a_i is the i-th attribute (i.e., dimension), and l_i^I and u_i^I are the lower and upper limits of query I along dimension a_i, respectively. This results in a range query represented as a d-dimensional box (also known as hyper-rectangle or orthotope).

The domain of each dimension a_i is limited by a lower bound L_i and an upper bound U_i. Hence, a predicate P_i^I can be further expressed as: $L_i \le l_i^I \le a_i \le u_i^I \le U_i$. We note that a dimension a_i that is not included in I, is equivalent to $L_i \le a_i \le U_i$. Moreover, join queries over multiple relations are expressed as equality predicates, which are not to be refined. A refined query R for an initial query I is achieved by modifying the lower and upper limits for some of the predicates in R. That is, for a predicate P_i^I in query I, a refined predicate P_i^R in R takes the form $l_i^R \le a_i \le u_i^R$.

Similar to [10], we assume that a double-sided predicate is equivalent to two separate single-sided predicates. Hence, a predicate $l_i^I \le a_i \le u_i^I$ is equivalent to: $a_i \le u_i^I$, and $-a_i \le -l_i^I$. Accordingly, refining a numerical, single-sided predicate $a_i \le x_i^I$ takes place by means of one of two operations as follows:

1. *predicate expansion*: in which $a_i \le x_i^R$, where $x_i^R > x_i^I$, or
2. *predicate contraction*: in which $a_i \le x_i^R$, where $x_i^R < x_i^I$.

Thus, under those operations, the value $|x_i^I - x_i^R|$ is the amount of refinement applied to predicate P_i^I. For example, Figure 1 shows two alternative refined queries R_1 and R_2, which are generated by expanding the input query I across its two dimensions a_1 and a_2. Note that in the presence of categorical predicates, a multi-level hierarchy is typically used to rank the different categorical values. Hence, refining a categorical predicate is simply mapped to moving up or down within the hierarchy.

Clearly, the number of possible refined queries is exponential in the number of dimensions and forms a combinatorial search space. For instance, consider a query I over a d-dimensional database, in which each dimension a_i is discrete and the number of distinct attributes in each dimension a_i is n. For query I, the set of possible refined queries form a query space $\mathbb{R}$, where the size of that space $|\mathbb{R}| = n^d$.

Given an objective for query refinement (e.g., satisfying a certain cardinality constraint), exploring the large search $\mathbb{R}$ in search for the optimal parameter settings (i.e., optimal R) becomes a non-trivial and challenging task. For instance, it has been shown that query refinement to meet cardinality constraints is an NP-Hard problem [2].

To circumvent the high-complexity of query refinement, several search heuristics have been proposed (e.g., [11,10,2,6,16]), in which the cardinality of a query result has been the main goal for query refinement as discussed next.

Cardinality-Based Query Refinement. Given a database B, an input conjunctive query I, and a cardinality constraint K over the result of I, the goal of cardinality-based query refinement is to find R that satisfies the cardinality constraint over B [2]. Ideally, the cardinality K_R of R should be equal to the cardinality constraint K. In reality, however, achieving the exact cardinality K_R is unrealistic either because of the data distribution or to reduce the search complexity. In either cases, the database management system returns the refined query R and the goal is to minimize the amount of *deviation* from the target cardinality, which is defined as follows: $\Delta_R^K = \frac{|K - K_R|}{max(K, K_R)}$ where K_R is the cardinality of R, K is the target cardinality constraint, and Δ_R^K is the deviation in cardinality. Note that Δ_R^K is normalized and in the range $[0 - 1]$.

As mentioned above, minimizing the deviation in cardinality (i.e., Δ_R^K) is an NP-Hard problem. This has motivated the proposal of several practical heuristics for solving the cardinality-based query refinement problem. As expected, the main idea underlying those heuristics is to limit the search space to a small set of possible candidate queries $\mathbb{R}_c$, such that $\mathbb{R}_c \subseteq \mathbb{R}$, and $|\mathbb{R}_c| \ll |\mathbb{R}|$. It has been shown that simple *local search* techniques based on greedy heuristics (e.g., Hill Climbing) often provide efficient solutions to the cardinality-based query refinement problem [2,6]. Similarly, those techniques have also been shown to provide effective solutions that minimize the deviation in cardinality, despite of of being susceptible to hitting local minima. The reason is that, there are typically many possible combinations of refinements that can result in meeting the cardinality constraint. This increases the possibility that a local search method will find one of those combinations without getting stuck at some local minimum.

For each candidate query $R_c \in \mathbb{R}$, a *probe* of the database is required to estimate the cardinality of R_c. Current techniques use alternative methods to perform such probe such as sampling [11,10], and histograms [2,6]. Irrespective of the employed estimation method, a call has to be issued to the database *evaluation layer*, where the cardinality of R_c is estimated by running it on a small set of data (i.e., a sample or histogram table). This makes the probing operation inherently expensive and is a strong motivation for reducing $|\mathbb{R}_c|$. Accordingly, the incurred *cost* of the query refinement process is measured in terms of the number of probes to the evaluation layer [2], and is defined as: $C_R = |\mathbb{R}_c|$. In this work, we follow the same approach as in [2,10], in which we consider the probing operation for cardinality estimation as a *blackbox* and our goal is to minimize the number of such probes. Differently from those approaches, however, our goal is to achieve a similarity-aware query refinement, as described in the next section.

3   Similarity-Aware Query Refinement

While the techniques mentioned in the previous section are very effective in solving the cardinality-based query refinement problem, they are generally oblivious to the (dis)similarity between the input query I and its corresponding refined version R. That is, the generated refined query R is often very different (i.e., dissimilar) from the input query I. Clearly, while the user might be satisfied that the refined query returns

Table 1. Summary of Symbols

Symbol	Description	Symbol	Description
I	input query	$1 - \alpha$	weight of deviation in cardinality
$a_i \leq x_i^I$	predicate on attribute a_i in input query I	Δ_R	total deviation of R
R	a refined query	C_R	total cost of finding R
$a_i \leq x_i^R$	predicate on attribute a_i in refined query R	$\mathbb{R}$	space of possible refined queries
K	cardinality constraint on R	R_i	a possible refined query in $\mathbb{R}$
K_R	cardinality of R	R_{opt}	the optimal refined query in $\mathbb{R}$
Δ_R^K	deviation in cardinality of R	Δ^D-list	a partial list of Δ_i^D of $R_i \in \mathbb{R}$
Δ_R^D	deviation in similarity of R	Δ^K-list	a partial list of Δ_i^K of $R_i \in \mathbb{R}$
α	weight of deviation in similarity	δ	resolution of the search space

less/more tuples, they would also expect the refined query to be very close (i.e., similar) to their input query. A refined query that is very different from the input one will have a very limited benefits to the end user and is often rendered useless. To further clarify that point, consider the following example:

Example 1. A user is looking for an airline ticket from New York City to Denpasar Bali. Typically, users have preferences for the trip cost (i.e., `Price`) and duration (i.e., `Duration`). More importantly, they would also have certain expectations for the result cardinality (e.g., 30 tickets to choose from). Alternatively, a constraint on the result's cardinality might be set as an application-based objective (e.g., application screen must project at most 50 results). In either case, the user's input query I might look like this:

```
SELECT ticket FROM Table
WHERE Price ≤ $1800    AND Duration ≤ 12H
AND Departure ='NYC'    AND Destination = 'DPS'
AND Date = '20-04-2014';
```

Figure 1 shows query I in a 2-dimensional space in which a_1 represents `Duration`, whereas a_2 represents `Price`. Clearly, since most flights from **NYC** to **DPS** take more than 12 hours, query I might return very few results. To find more tickets, query I can be expanded into R_2 : `WHERE Price ≤ $1950 AND Duration ≤ 45H`, as shown in Figure 1. However, as mentioned earlier, R_2 is essentially different from the user's preferences as it returns flights with excessively long durations. Alternatively, expanding query I into R_1: `WHERE Price ≤ $2200 AND Duration ≤ 20H` might be largely accepted by the user, while returning the same cardinality as R_2.

Problem Statement. To address the limitation of current cardinality-based query refinement, in this work, we propose the similarity-aware query refinement problem, in which the user satisfaction is measured in terms of both: 1) meeting some pre-specified cardinality constraint K on R, and 2) minimizing the distance between R and I under some given distance function $D()$.

Ideally, the distance between R and I (i.e., $D(R, I)$) should be equal to zero (i.e., maximum similarity such that $R \equiv I$). In reality, however, achieving that exact similarity is unrealistic, unless $K = K_I$. That is, I already meets its cardinality constraint K and no further refinement is required. Hence, in this work, we adopt a dual-criteria objective model that particularly captures and quantifies the overall *deviation* in meeting the user expectations for both similarity and cardinality, which is formally defined as:

$$\Delta_R = \alpha \Delta_R^D + (1 - \alpha) \Delta_R^K \tag{1}$$

In Eq. 1, Δ_R^K is the deviation in cardinality, which is computed as described in the previous section. Δ_R^D is the deviation in similarity, which is captured by means of a distance function $D()$, as it will be described below. The parameter α simply specifies the weight assigned to the deviation in similarity, and in turns, $1 - \alpha$ is the weight assigned to the deviation in cardinality. Putting it together, the similarity-aware query refinement problem is specified as follows:

Definition 1. *Given a database B, an input conjunctive query I, a distance function $D()$, and a cardinality constraint K over the result of I, the goal of similarity-aware query refinement is to find R that has the minimum overall deviation Δ_R.*

Note that the weight α can be user-defined so that to reflect the user's preference between satisfying the cardinality and similarity constraints. Alternatively, it can be system-defined and is set automatically to meet certain business goals or objectives that are defined by the application. On the one hand, setting $\alpha = 0$ is equivalent to the cardinality-based query refinement problem as described in the previous section. On the other hand, setting $\alpha = 1$ is equivalent to the extreme case described above, in which $R \equiv I$. In the general case, in which $0 < \alpha < 1$, both the cardinality and similarity constraints are considered according to their respective weights and the overall deviation is captured by Δ_R. Hence, a small value of Δ_R indicates a small deviation in meeting the constraints, and more satisfaction by the refined query R. Interestingly, those two constraints are typically at odds. That is, maximizing similarity while minimizing deviation in cardinality are two objectives that are typically in conflict with each other. Moreover, the parameter α specifies by how much those two objectives (i.e., Δ_R^K and Δ_R^D) contribute to the overall deviation Δ_R. For instance, in Figure 1, if similarity is neglected (i.e., α equals zero), then both R_1 and R_2 are two accepted refined queries since similarity has no weight. However, in the presence of α (e.g., $\alpha = 0.5$), then R_1 is probably preferred over R_2 since R_1's similarity deviation from the query I is less than R_2's deviation. Measuring that deviation is discussed next.

Similarity Measure. Measuring the (dis)similarity between two *point* queries is very well-studied in the literature, where typically a variant of the L_p norm metric is used. Meanwhile, there is a lack of a standard for measuring the distance between two box queries (i.e., I and R), which are the building blocks for the query refinement process. In this paper, we broadly classify metrics for measuring the distance between two box queries as: 1) data-oriented, 2) predicate-oriented, and 3) value-oriented.

In the data-oriented measures (e.g., [13,6]), the distance between I and R is based on the data points (i.e., tuples) that are included in the result of each query. For instance, to measure the distance between I and an expanded R, [6] computes the distance between I and all the points in $R - I$ (i.e., the extra points added due to expansion). Clearly, data-oriented methods incur a large overhead, which potentially renders a query refinement process infeasible. In the predicate-oriented measures (e.g., [15]), the distance between I and R is mapped to that of measuring the *edit distance* needed to transform I into R. Though of its simplicity, a predicate-oriented measure is very coarse for the purpose of query refinement as it falls short in distinguishing between the different possible modifications that can be applied to each predicate.

Finally, in the value-oriented measures (e.g., [12,16]), the distance between I and R is based on the amount of refinement experienced by each predicate. Formally,

$$D(I, R) = \frac{1}{d} \sum_{i=1}^{d} \frac{|x_i^R - x_i^I|}{U_i - L_i} \tag{2}$$

We note that, in comparison to the predicate-based methods, Eq. 2 considers the amount of applied refinement (i.e., $|x_i^R - x_i^I|$). We also note that Eq. 2 provides a reasonable approximation of the data-oriented measures at a negligible cost. Accordingly, we adopt the value-oriented method as our measure for distance in the similarity-aware query refinement problem, for which our proposed solutions are described next.

4   The *SAQR-S* Scheme

In this section, we present our SAQR-Similarity scheme (SAQR-S for short), which leverages the distance constraint to effectively prune the search space. In the next section, we present SAQR-CS, which extends SAQR-S by exploiting the cardinality constraint for further pruning of the search space and higher efficiency.

Our similarity-aware query refining problem, as defined in Eq. 1, is clearly a preference query over the query space $\mathbb{R}$ and naturally lends itself as a special instance of *Top-K* or *Skyline* queries. In particular, our goal is to search the query space $\mathbb{R}$ for the one refined query R_{opt} that minimizes the aggregate function defined in Eq. 1.

Query R_{opt} is equivalent to Top-1 query over the aggregation of two attributes: 1) similarity deviation (i.e., Δ^D), and 2) cardinality deviation (i.e., Δ^K). R_{opt} should also fall on the skyline of a 2-dimensional space over those two attributes [5,14]. However, efficient algorithms for preference query processing (e.g., [9]), are not directly applicable to the similarity-aware query refinement problem due to the following reasons:

1. For any query $R_i \in \mathbb{R}$, the values of Δ_i^D and Δ_i^K are not physically stored and they are computed on demand based on the input query I and the constraint K.
2. In addition to the cardinality constraint K, computing Δ_i^K for any query $R_i \in \mathbb{R}$, requires an expensive probe to estimate the cardinality K_i of query R_i.
3. The size of the query search space $|\mathbb{R}|$ is prohibitively large and potentially infinite.

To address the limitations listed above, in this paper, we propose the SAQR scheme for similarity-aware query refinement. In particular, SAQR adapts and extends algorithms for Top-K query processing towards efficiently and effectively solving the similarity-aware query refinement problem. Before describing SAQR in details, we first outline a baseline solution based on simple extensions to the *Threshold Algorithm (TA)* [4].

Conceptually, to adapt the well-know TA to the query refinement model, each possible refined query $R_i \in \mathbb{R}$ is considered an object with two *partial scores*: 1) deviation in similarity (i.e., $\alpha \Delta_i^D$), and 2) deviation in cardinality (i.e., $(1 - \alpha)\Delta_i^K$). Those two partial scores are maintained in two separate lists: 1) Δ^D-list, and 2) Δ^K-list, which are sorted in descending order of each score.

Under the classical TA algorithm, the two lists are traversed sequentially in a round-robin fashion. While traversing, the query with the minimum deviation seen so far is maintained along with its deviation. A lower bound on the total deviation (i.e., threshold) is computed by applying the deviation function (Eq. 1) to the partial deviations of the last seen queries in the two different lists. TA terminates when the minimum deviation seen so far is below that threshold or when the lists are traversed to completeness.

Clearly, such straightforward conceptual implementation of TA is infeasible to the similarity-aware query refinement problem due to the three reasons listed above. To address the first problem (i.e., absence of partial deviation values), SAQR-S generates the Δ^D-list on the fly and on-demand based on the input query I. In particular, given query I, it progressively populates the Δ^D-list with the distance between I and the nearest possible refined query $R_i \in \mathbb{R}$.

To control and minimize the size of the possible search space, a value δ is defined and the nearest query is defined in terms of that δ. In particular, given an input query I, a first set of nearest queries is generated by replacing each predicate $a_i \leq x_i^I$ with two predicates $a_i \leq x_i^I \pm \delta$. The same process is then repeated recursively for each set of generated queries. Notice that the search space is typically normalized, hence, $0 < \delta < 1$. Clearly, using δ allows for simply *discretizing* the rather continuous search space $\mathbb{R}$. Hence, $\mathbb{R}$ can be perceived as a uniform grid of granularity δ (i.e., each cell is of width δ). We note that at any point of time, the Δ^D-list is always sorted since the values in that list are generated based on proximity.

One approach for populating the Δ^K-list is to first generate the distance Δ^D-list and then compute the corresponding Δ_i^K value for each query R_i in the Δ^D-list. Those values are then sorted in descending order and the TA algorithm is directly applied on both lists. Clearly, that approach has the major drawback of probing the database for estimating the cardinalities of all the possible queries in the new discretized search space. Instead, we leverage the particular *Sorted-Random (SR)* model of the Top-K problem to minimize the number of those expensive cardinality estimation probes.

The SR model is particularly useful in the context of web-accessible external databases, in which one or more of the lists involved in an objective function can only be accessed in random and at a high-cost [9,5]. Hence, in that model, the sorted list (i.e., S) basically provides an initial set of candidates, whereas random lists (i.e., R) are probed on demand to get the remaining partial values of the objective function. In our model, the Δ^D-list already provides that sorted sequential access, whereas Δ^K-list is clearly an external list that is accessed at the expensive cost of probing the database. Under that setting, while the Δ^D-list is generated incrementally, two values are maintained (as in [9]): 1) Δ_m: the minimum deviation Δ among all queries generated so far, and 2) Δ_M: a threshold on the minimum possible deviation Δ for the queries to be generated. These two values enable efficient navigation of the search space $\mathbb{R}$ by pruning a significant number of the queries. This is achieved by means of two simple techniques:

1. Short circuiting: when a query R_i is generated, its $\alpha\Delta_i^D$ value is compared to the minimum deviation seen so far (i.e., Δ_m) and the cardinality of R_i is only estimated if $\alpha\Delta_i^D < \Delta_m$. Otherwise, the cardinality probing is short circuited and the next query is generated.

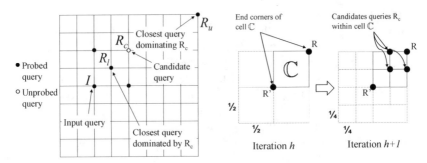

(a) Estimating upper and lower bounds of R_c (b) $\mathbb{C}$ is partitioned for further refinement

Fig. 2. Cardinality based pruning and the pyramid structure in SAQR-CS

2. Early termination: when a query R_i is generated, the threshold Δ_M is updated to $\alpha\Delta_i^D$. That is, assuming that some ungenerated query will have a $\Delta^K = 0$, the deviation of that query from I still cannot exceed Δ_M. An early termination is reached when $\Delta_m \leq \Delta_M$.

5   The *SAQR-CS* Scheme

The SAQR-S scheme, presented in the previous section, basically leverages the deviation in distance in order to bound the search space and reduce the number of generated candidate refined queries. However, it still has two major drawbacks: 1) It probes the database for estimating the cardinality for every candidate query R_i that survives the short circuit test, and 2) The overall search space is still large despite of the discretization process. In this section, we propose the extended SAQR-Cardinality/Similarity scheme (SAQR-CS for short), which provides the following features:

1. SAQR-CS exploits the *monotonicity* property of the cardinality constraint so that to provide significant reductions in the search space, and
2. SAQR-CS employs a *hierarchical* representation of the search space that allows for adaptive navigation and further reductions in the total cost.

The Monotonicity of Cardinality. Consider a candidate query R with d conjunctive predicates $P_1 \wedge P_2 \wedge ... \wedge P_d$, such that each predicate P_i is defined as: $a_i \leq x_i^R$. Further, assume that n_i is the number of distinct values for dimension a_i. The space of possible cardinalities of query R can be modeled as d-dimensional $n_1 \times ... \times n_d$ grid $\mathcal{G}$ [2]. The value of $\mathcal{G}[p_1, ..., p_d]$ for $1 \leq p_i \leq n_i$ is precisely the cardinality of the query R when each predicate P_i is instantiated with the p_i-th smallest distinct value of attribute a_i. Therefore, $\mathcal{G}$ satisfies the following monotonicity property: $\mathcal{G}[p_1, ..., p_d] \leq \mathcal{G}[q_1, ..., q_d]$ when $p_i \leq q_i$ for every attribute a_i [2].

Cardinality-Based Pruning. SAQR-CS exploits the monotonicity property of the cardinality constraint so that to provide significant reductions in the search space. In particular, if a candidate refined query R_c passed the short circuit test, SAQR-CS estimates a lower bound K_c^l and an upper bound K_c^u on the cardinality of query R_c (i.e.,

$K_c^l \leq K_c \leq K_c^u$). Estimating those bounds is very efficient since it is completely based on the candidate queries that have been already probed and added to the Δ^D-list, thus requires no further probes.

When a new query R_c is generated, it sets the bounds K_c^l and K_c^u as follows (See Figure 2(a)):

- $K_c^l = K_l$, where K_l is the cardinality of query R_l, which is the closest probed query *dominated* by R_c. That is, when $x_i^l \leq x_i^c$ for every attribute a_i.
- $K_c^u = K_u$, where K_u is the cardinality of query R_u, which is the closest probed query *dominating* R_c. That is, when $x_i^c \leq x_i^u$ for every attribute a_i.

Given K_c^l and K_c^u, SAQR-CS estimates the benefit of probing the database to get the exact cardinality K_c. In particular, since K_c can take any value in $[K_c^l$ - $K_c^u]$, it is required to check if any value in that range can provide an overall deviation smaller than the one achieved so far (i.e., $\Delta_c < \Delta_m$). To perform that test, we simply substitute in Eq. 1 the deviation in similarity provided by R_c (i.e., Δ_c^D) so that to find the range of values $[\bar{K}_c^l - \bar{K}_c^u]$ such that if $\bar{K}_c^l \leq K_c \leq \bar{K}_c^u$, then $\Delta_c < \Delta_m$. By substitution and manipulating Eq. 1, it is easy to calculate that range as follows:

$$\bar{K}_c^l = K \times \frac{(1 - \alpha(1 + \Delta_c^D) - \Delta_m)}{1 - \alpha)} \text{ , and } \bar{K}_c^u = K \times \frac{(1 - \alpha)}{1 - \alpha(1 - \Delta_c^D) - \Delta_m}$$

Thus, only if there is an overlap between $[K_c^l$ - $K_c^u]$ and $[\bar{K}_c^l - \bar{K}_c^u]$, R_c might provide a smaller deviation and the database is probed to retrieve its actual cardinality.

Hierarchical Representation of the Search Space. Clearly, the effectiveness of the bounds described above in pruning the search space depends on the tightness of the bounds K_c^l and K_c^u. However, achieving such tight bounds is not always possible when the candidate refined queries are generated in order of their proximity to the input query I on a uniform grid with a constant width δ as in the previous section.

For instance, under that approach, a generated candidate query R_c that is positioned between the input query I and the origin for the search space, will often have a loose lower bound $K_c^l = 0$. Similarly, if R_c is positioned between I and the limits of the search space, then it will have a loose upper bound of $K_c^u = |B|$ (similar to the bound provided by R_u as in Figure 2(a)).

To achieve tighter bounds, SAQR-CS employs a hierarchical representation of the search space based on the *pyramid* structure [1] (equivalent to a partial quad-tree [8]). The pyramid decomposes the space into H levels (i.e., pyramid height). For a given level h, the space is partitioned into 2^{dh} equal area d-dimensional grid cells. For example, at the pyramid root (level 0), one cell represents the entire search space, level 1 partitions space into four equal-area cells, and so forth.

To create the pyramid representation, SAQR-CS generates candidate queries recursively using a dynamic δ, where the value of δ_h in any iteration h is equal to $\frac{1}{2^h}$. Queries generated in iteration h are processed similar to SAQR-S, in addition to: 1) applying the cardinality-based pruning outlined above, and 2) maintaining the minimum deviation Δ_m across iterations. The pyramid representation provides the following advantages:

- Effective pruning: it allows to compute tight cardinality bounds for candidate query R_c based on the known cardinalities of other queries that are either at the same level or higher levels in the pyramid representation.

– Efficient search: it allows to to quickly explore the search space at a coarser granularity and only further refine those cells that might contain R_{opt}.

To further explain the second point listed above, in the end of each iteration h, SAQR-CS decides which cells are to be further explored at a finer granularity in iteration $h + 1$ (i.e., expanded with a smaller δ). To do this, notice that in every iteration h, each generated candidate query represents some intersection point between the lines of the grid located at level h of the pyramid. As shown in Figure 2(b), let R be one of those queries generated in iteration h such that each predicate P_i^R is defined as $a_i \leq x_i^R$. Similarly, let R' be another query generated in that same iteration such that each predicate $P_i^{R'}$ is defined as $a_i \leq x_i^{R'}$, where $x_i^{R'} = x_i^R - \delta_h$ for every attribute a_i. Hence, R and R' define the corners of a d-dimensional cell $\mathbb{C}$ of width δ_h, such that R is the nearest dominating query for all the queries enclosed in that subspace defined by $\mathbb{C}$. Similarly, R' is the nearest query dominated by all the queries enclosed in $\mathbb{C}$. At the end of each iteration, SAQR-CS assesses the possibility that any query in $\mathbb{C}$ can provide a deviation smaller than the one seen so far. To perform that assessment, SAQR-CS further utilizes the previously outlined monotonicity property but at a cell-level (i.e., multiple queries) rather than the query-level. Similar to the query-level pruning technique, SAQR-CS needs to compute two ranges of bounds:

– Range $[K_{\mathbb{C}}^l - K_{\mathbb{C}}^u]$: defines the lower and upper bounds on the cardinality of any query $R_c \in \mathbb{C}$.
– Range $[\bar{K}_{\mathbb{C}}^l - \bar{K}_{\mathbb{C}}^u]$: defines the lower and upper bounds on the cardinality of any query $R_c \in \mathbb{C}$ such that if $\bar{K}_{\mathbb{C}}^l \leq K_{\mathbb{C}} \leq \bar{K}_{\mathbb{C}}^u$, then $\Delta_c < \Delta_m$

Hence, only if those two ranges overlap, then there is some query $R_c \in \mathbb{C}$ that might provide an overall deviation that is smaller than the deviation achieved so far (i.e., $\Delta_c < \Delta_m$) and $\mathbb{C}$ is explored at a finer granularity. Clearly, the bounds in the range $[K_{\mathbb{C}}^l - K_{\mathbb{C}}^u]$ are easily calculated given the end corners of $\mathbb{C}$. That is, $K_{\mathbb{C}}^l = K_{R'}$, whereas, $K_{\mathbb{C}}^u = K_R$. To calculate the range $[\bar{K}_{\mathbb{C}}^l - \bar{K}_{\mathbb{C}}^u]$, however, SAQR-CS requires some estimate of distance so that to substitute in Eq. 1 and obtain the value for that range. Recall that at the query-level, that distance is basically the deviation in similarity provided by the query under examination. At the cell-level, however, each cell $\mathbb{C}$ contains a set of candidate queries with different distances from the input query I. Hence, SAQR-CS assumes a *conservative* approach and estimates that distance as the smallest perpendicular distance between any surfaces of $\mathbb{C}$ and the input query I, denoted as $\bar{\Delta}_{\mathbb{C}}^D$. Given that distance estimation, the range $[\bar{K}_{\mathbb{C}}^l - \bar{K}_{\mathbb{C}}^u]$ is calculated similar to the one at the query-level. That is:

$$\bar{K}_{\mathbb{C}}^l = K \times \frac{(1 - \alpha(1 - \bar{\Delta}_{\mathbb{C}}^D) - \Delta_m)}{1 - \alpha} \text{ , and } \bar{K}_{\mathbb{C}}^u = K \times \frac{(1 - \alpha)}{1 - \alpha(1 - \bar{\Delta}_{\mathbb{C}}^D) - \Delta_m}$$

We note that while there are alternatives for estimating the distance $\bar{\Delta}_{\mathbb{C}}^D$ (e.g., average, and maximum distance), our conservative approach is expected to minimize deviation as it assumes best case scenario for each cell. Other alternatives, such as an average distance, are expected to reduce the number of explored cells, and in turn the number of database probes, at the expense of an increase in the provided deviation.

Table 2. Evaluation Parameters

Parameter	Range	Default
Deviation weight (α)	0.0–1.0	0.5
Dimensions (d)	1–4	2
Grid resolution (δ)	$1/2^5$–$1/2$	$1/2^5$
Database Size ($|B|$)	–	60K tuples

In the end of each iteration, a cell $\mathbb{C}$ is eliminated from the search space if it exhibits no overlap between the ranges $[K_{\mathbb{C}}^l - K_{\mathbb{C}}^u]$ and $[\bar{K}_{\mathbb{C}}^l - \bar{K}_{\mathbb{C}}^u]$. SAQR-CS terminates when there no further cells to be explored, or when it reaches some maximum preset height H for the pyramid structure (or equivalently, when reaching a preset value of δ). Putting it together, SAQR-CS provides the following advantages:

- Minimizing Cost: by exploiting a pyramid representation in conjunction with cardinality-based pruning, SAQR-CS is able to recursively prune large portions of the search space and avoid unnecessary database probes.
- Maximizing Accuracy: by exploring the most promising portions of the search space at finer granularity, SAQR-CS is able to achieve high accuracy in finding a refined query that minimizes the overall deviation from the input query.

6   Experimental Evaluation

We have implemented SAQR as a Java front-end on top of the MySQL DBMS. Due to the limited space, in the following we present a representative sample of our experimental results under the settings summarized in Table 2 and described below.

Schemes. In addition to *SAQR-S* and *SAQR-CS*, we evaluate the performance of the *Hill Climbing (HC)* local search scheme. HC is proposed in [2] to automatically refine queries for DBMS testing. HC discretizes the search space and navigates it in a greedy manner until no further reduction in deviation is attainable. However, in this work we have extended HC [2] to use our similarity-aware objective function (i.e., Eq 1). To achieve a fair comparison between the different schemes, HC and SAQR-S use the same cell width δ value. Meanwhile, SAQR-CS is tuned so that the cell width at the lowest layer of the pyramid structure is also equal to δ.

Databases. In our experiments, we use real scaled TPC-D database of 60K tuples. The database is created similar to [10], using the publicly available tool [3] which provides the capability of generating databases with different scales. In our experiments, all the numerical columns in the generated database are normalized in the range [0-1].

Queries. To cover a large spectrum of query contraction and expansion scenarios, we generated a set of 100 <query, cardinality> pairs. Each pair is an input query together with its cardinality constraint generated according to a uniform distribution.

Performance Measures. Performance is evaluated in terms of:

- **Average cost** (C_R): That is the average number of probes (calls) made to the database *evaluation layer* for refining all the queries in the workload.

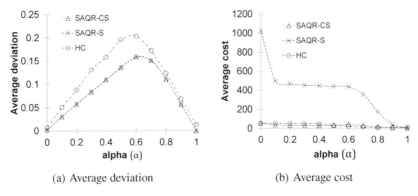

(a) Average deviation

(b) Average cost

Fig. 3. Deviation weight (α) for similarity varied from 0–1

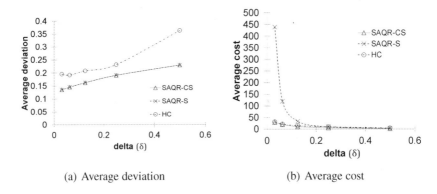

(a) Average deviation

(b) Average cost

Fig. 4. Average deviation and cost when grid resolution (δ) is varied from $\left(\frac{1}{2}\right)^5$ to $\left(\frac{1}{2}\right)$

- **Average deviation (Δ_R):** That is the average deviation experienced by all the queries in the workload, where the deviation perceived by each query is computed according to Eq. 1.

Impact of Deviation Weight. In Figure 3, we use the default experiment settings while varying the deviation weight parameter α from 0.0 to 1.0. Figure 3(a) shows that all schemes follow the same pattern: deviation increases with α until it reaches a critical point after which it rebounds. This behavior happens because minimizing deviation in cardinality while maximizing similarity are two objectives that are typically in conflict. The degree of conflict is determined by α, for example at $\alpha = 1$ there is no conflict and the total deviation is based on similarity. Hence, SAQR easily finds a refined query that satisfies K, though it might be very dissimilar from the input. For α around 0.5, both objectives are of equal importance, making it hard to find a refined query that satisfies them both simultaneously leading to higher deviation (compared to $\alpha = 1$). However, SAQR effectively balances that tradeoff and finds a near-optimal solution. For instance, when α equals 0.5, SAQR-CS achieves 30% better deviation than HC.

Figure 3(b) shows the high efficiency of both HC and SAQR-CS compared to SAQR-S. However, at higher values of α, the cost of SAQR-S is relatively low. This is because

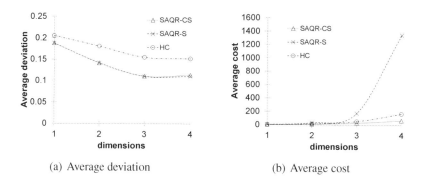

(a) Average deviation   (b) Average cost

Fig. 5. Dimensions (d) from 1–4

SAQR-S traverses the grid cells starting from the closest point to the input query (smallest distance) to the furthest one. Thus, assigning higher weight to deviation in similarity increases the chance for SAQR-S to satisfy the early termination condition.

Impact of Grid Resolution. In this experiment, we examine the impact of the grid resolution δ on the performance metrics. As shown in Figure 4(a), all three schemes exhibit a direct correlation between δ and the average deviation. This relation is natural because when δ is increased, the search space is highly approximated, which essentially increases the probability of missing the exact target constraints, and vice versa. Figure 4(b) shows that the average cost drops as δ is increased. For instance, the average cost provided by SAQR-S is reduced by 72% when δ increased from $(\frac{1}{2})^5$ to $(\frac{1}{2})^4$, while the reduction in the costs of HC and SAQR-CS is 31% and 33%, respectively. This is because the total number of cells in the grid is decreased, leading to less number of cells to be scanned. Looking at both Figures 4(a) and 4(b) uncovers the trade-off between the deviation and cost metrics which is controlled by δ.

Impact of Dimensionality. In this experiment we measure the effect of dimensionality d on performance, while $\delta = (\frac{1}{2})^3$. Figure 5(a) shows that SAQR achieves better deviation than HC for all values of d. For instance, when all queries include four dimensions, SAQR-CS and SAQR-S reduced the deviation by 28% when compared to HC. The reason for the poor performance of HC is that when the number of dimensions is increased, there are more possible refined queries, which increases the chances for HC to get stuck at local minima and miss the optimal solution. Figure 5(b) shows that, in general, SAQR-CS outperforms all other schemes. For instance, for four-dimensional queries, SAQR-CS reduced the cost compared to HC and SAQR-S by 60% and 95%, respectively. Clearly, these reductions are due to the cardinality-based pruning and pyramid structure techniques employed by SAQR-CS.

7   Related Work

The work on query refinement can be broadly classified into two categories: databases testing [10,2] and databases exploration [12,11,6,16]. In database testing, [2] addressed the complexity of the query refinement problem and developed heuristics to find approximate solutions for generating targeted queries that meet certain cardinality constraints. Similarly, [10] proposed a scheme based on *intermediate subexperssions* cardinality constraints. For database exploration, [12] presented a scheme based on data mining techniques to overcome the empty-result query problem. The basic idea is to discover and exploits the correlation between the attributes to produce a set of decision rules which is used in finding the suitable refined query accordingly. [7] addressed the same problem and proposed a framework to systematically refine queries to produce non-empty answer-set, without explicitly specifying any cardinality constraints.

More related to our work is the SAUNA scheme [6], in which starting with a user's initial query and a cardinality constraint, SAUNA automatically produces a refined query which satisfies the constraint on the number of answers and preserves the aspect ration of the initial query. Moreover, [11] proposed an interactive, user-based model (SnS) for query refinement. However, SnS requires incorporating the user's involvement to manually guide the refinement process. Finally, the QReIX tool automatically suggests alternative queries based on the given cardinality constraints while providing indicators of the closeness between the input query and the new generated one [16].

8   Conclusions

Motivated by the need for efficient similarity-aware query refinement, in this paper, we propose a novel scheme called SAQR. SAQR balances the tradeoff between satisfying the cardinality and similarity constraints imposed on the refined query so that to maximize its overall benefit to the user. To achieve that goal, SAQR implements efficient strategies to minimize the costs incurred in exploring the available search space. In particular, SAQR utilizes both similarity- and cardinality-based pruning techniques to bound the search space and quickly find a refined query that meets the user expectations. Our experiments show that SAQR achieves scalability comparable to light-weight local search strategies, while significantly improving the quality of query refinement.

Acknowledgment. We would like to thank the anonymous reviewers for their valuable comments. We also thank Hina Khan for her feedback and suggestions. This work is partially supported by Australian Research Council grant DP110102777.

References

1. Aref, W.G., Samet, H.: Efficient processing of window queries in the pyramid data structure. In: PODS (1990)
2. Bruno, N., Chaudhuri, S., Thomas, D.: Generating queries with cardinality constraints for dbms testing. IEEE Trans. Knowl. Data Eng. 18(12), 1721–1725 (2006)
3. Chaudhuri, S., Narasayya, V.: Program for generating skewed data distributions for tpc-d, ftp://ftp.research.microsoft.com/users/surajitc/TPCDSkew/
4. Fagin, R., et al.: Optimal aggregation algorithms for middleware. In: PODS (2001)
5. Ilyas, I.F., Beskales, G., Soliman, M.A.: A survey of top-k query processing techniques in relational database systems. ACM Comput. Surv. 40(4) (2008)

6. Kadlag, A., Wanjari, A.V., Freire, J.-L., Haritsa, J.R.: Supporting exploratory queries in databases. In: Lee, Y., Li, J., Whang, K.-Y., Lee, D. (eds.) DASFAA 2004. LNCS, vol. 2973, pp. 594–605. Springer, Heidelberg (2004)
7. Koudas, N., et al.: Relaxing join and selection queries. In: VLDB (2006)
8. Levandoski, J.J., et al.: Lars: A location-aware recommender system. In: ICDE (2012)
9. Marian, A., Bruno, N., Gravano, L.: Evaluating top-k queries over web-accessible databases. ACM Trans. Database Syst. 29(2), 319–362 (2004)
10. Mishra, C., et al.: Generating targeted queries for database testing. In: SIGMOD (2008)
11. Mishra, C., Koudas, N.: Interactive query refinement. In: EDBT (2009)
12. Muslea, I.: Machine learning for online query relaxation. In: KDD (2004)
13. Pan, L., et al.: Probing queries in wireless sensor networks. In: ICDCS (2008)
14. Tao, Y., Xiao, X., Pei, J.: Efficient skyline and top-k retrieval in subspaces. IEEE Trans. Knowl. Data Eng. 19(8), 1072–1088 (2007)
15. Tran, Q.T., et al.: How to conquer why-not questions. In: SIGMOD (2010)
16. Vartak, M., Raghavan, V., Rundensteiner, E.A.: Qrelx: generating meaningful queries that provide cardinality assurance. In: SIGMOD Conference (2010)

Concurrent Execution of Mixed Enterprise Workloads on In-Memory Databases

Johannes Wust[1], Martin Grund[2], Kai Hoewelmeyer[1],
David Schwalb[1], and Hasso Plattner[1]

[1] Hasso Plattner Institute, 14440 Potsdam, Germany
johannes.wust@hpi.uni-potsdam.de
[2] University of Fribourg, 1700 Fribourg, Switzerland

Abstract. In the world of enterprise computing, single applications are often classified either as transactional or analytical. From a data management perspective, both application classes issue a database workload with commonly agreed characteristics. However, traditional database management systems (DBMS) are typically optimized for one or the other. Today, we see two trends in enterprise applications that require bridging these two workload categories: (1) enterprise applications of both classes access a single database instance and (2) longer-running, analytical-style queries issued by transactional applications. As a reaction to this change, in-memory DBMS on multi-core CPUs have been proposed to handle the mix of transactional and analytical queries in a single database instance. However, running heterogeneous queries potentially causes situations where longer running queries block shorter running queries from execution. A task-based query execution model with priority-based scheduling allows for an effective prioritization of query classes. This paper discusses the impact of task granularity on responsiveness and throughput of an in-memory DBMS. We show that a larger task size for long running operators negatively affects the response time of short running queries. Based on this observation, we propose a solution to limit the maximum task size with the objective of controlling the mutual performance impact of query classes.

1 Introduction

In-memory databases that leverage column-oriented storage have been proposed to run analytical queries directly on the transactional database [1]. This enables building analytical capabilities on top of the transactional system, leading to reduced system complexity and overall cost. Although queries are typically executed very fast on in-memory databases, execution time is still bound by bottleneck resources, such as CPU cycles or main memory access. Therefore, running a mixed workload of short and long-running queries with varying service-level-objectives (SLOs) can lead to resource contention in the DBMS.

The main challenge we face in this context is that we have to execute query classes that differ in their execution model, their execution time, as well as their service level objectives. In [2], we have provided an overview of three different

S.S. Bhowmick et al. (Eds.): DASFAA 2014, Part I, LNCS 8421, pp. 126–140, 2014.

query classes in enterprise applications that differ in these categories by being either executed sequentially or in parallel, by having a short or relatively longer response time, and by having either response time or throughput as optimization goal. With TAMEX [2], we have demonstrated that we can achieve an almost constant response time for high priority queries applying a task-based query execution and a non-preemptive priority scheduling policy. The underlying assumption is that we can split queries into equally sized small tasks.

In TAMEX's execution model, we generate smaller tasks typically by horizontally splitting input relations to exploit data level parallelism. However, this comes at a price, as splitting input data sets and merging of results creates an overhead. In this paper, we demonstrate that the known principle of reducing task granularity in a highly loaded system to reduce overhead and maintain throughput [3] sacrifices responsiveness of the system, as larger tasks lead to a longer CPU occupation and high priority tasks need to wait longer until they can start execution. Therefore, the objective of this paper is to provide a better understanding of the trade-off of achieving a highly responsive database system by splitting selected parts of long running complex queries into sequences of smaller tasks, while maintaining throughput of longer running queries. Our approach is to map this balancing problem to a queueing network model to better interpret our experimental results and learn from theory to find an appropriate solution. Based on our findings, we introduce a maximum task size for query execution to control the mutual performance impact of query classes.

The contributions of this paper are: (1) An experimental evaluation of query throughput for varying degrees of intra-operator parallelism, (2) a demonstration of the performance impact of the task sizes of a lower priority tasks on the response time of high priority tasks based on a queueing network model, and (3) a query execution framework, that dynamically determines the task granularity for incoming queries to maintain a system wide maximum task size.

The paper is structured as follows: Section 2 briefly describes the assumed system model, task-based query execution with TAMEX, as well the assumed workloads of enterprise applications. In Section 3, we evaluate the impact of task granularity on the throughput of analytical queries. Section 4 discusses the impact of task granularity of low priority tasks on the response time of high priority tasks using a queueing network model. In Section 5, we propose the system parameter for the maximum task size to maintain a robust response time for time critical systems and show the effectiveness in an evaluation. Section 6 discusses related work and we close with some concluding remarks.

2   System Model and Task-Based Query Execution

This section gives a brief overview of the underlying system model of an In-Memory Database Management System (IMDBMS), as well as the task-based query execution framework TAMEX [2]. Furthermore, we introduce heterogeneous query classes in enterprise workloads.

2.1   System Model

We assume an IMDBMS following the system model described in [4], where data is physically stored decomposed in a column-oriented structure. To achieve high read and write performance, an insert-only approach is applied and the data store is split in two parts, a read optimized main partition and a write optimized differential store [5]. We apply a multi version concurrency control (MVCC) based on transaction IDs (TID) to determine which records are visible to each transaction when multiple transactions run in parallel. See [4] for more details. As our proposed approach for query execution is largely agnostic to specific architectural details of the database, it can be easily generalized and applied to other architectures. However, our approach assumes that the execution of a single query can be separated into a set of atomic tasks, which can be executed in parallel if necessary, as we will explain in the next section.

2.2   Heterogeneous Workloads

In data management for enterprise applications, it is commonly distinguished between analytical and transactional applications. It is widely agreed that transactional queries are considered as short-running with high response time requirements, as they are often coupled to customer interactions or other time-critical business processes. In contrast, analytical queries are typically classified as long-running with comparably less critical response-time requirements.

As motivated in the Introduction, we see a trend towards a mix of analytical and transactional workloads. However, a mixed workload is not a mere combination of analytical and transactional queries. Additionally, we see queries with a complexity typically found in analytical queries, but with stricter response-time requirements, so called *transactional analytics*. The two main reasons for these queries are the introduction of new, often interactive applications, that mix transactional and analytical queries as discussed in [6], as well as the calculation of dynamic views instead of reading materialized aggregates as proposed in [1] to simplify the data schema. A more detailed analysis of the query and data characteristics of enterprise applications can be found in [6,7].

For the experiments in this paper, we have selected a representative query of each of these three classes. The queries access a sales header and corresponding item table with up to 100 million for the items table and half the rows for the header table. The transactional query, referred to as Q1, is a single index lookup on the sales header table. The query of class transactional analytics, referred to as Q2, is a join on the results of two scans that filter the sales item table for two products. The query is used to calculate cross-selling opportunities as described in [8]. For Q3, the analytical query, we chose a query that returns all products that have been sold in a certain time period and at a certain quantity, mainly consisting of a large join of the sales header and item table, that have been filtered for the time period and quantity.

2.3   Task-Based Query Execution with TAMEX

To efficiently process heterogeneous query workloads, we have implemented the task-based query execution framework TAMEX based on HYRISE [9]. A detailed description of TAMEX is provided in [2].

We understand task-based query execution as the transformation of the logical query plan into a set of atomic tasks that may have data dependencies, but otherwise can be executed independently. We consider such atomic tasks as the smallest scheduling unit. Compared to scheduling whole queries, a task-based approach provides two main advantages: better load balancing on a multiprocessor system, as well as more control over progress of query execution based on priorities. Assuming a sufficiently small task size, processing units can be freed quickly to execute incoming high priority queries.

TAMEX adopts this concept by transforming incoming queries into a directed acyclic graph of tasks – representing the physical operators – and schedules these tasks based on priorities. To keep the scheduling overhead for short running queries low, they are executed as a single task. To support intra-operator parallelism for more complex queries, database operators can be split into a sequence of smaller tasks by partitioning the input and merging the results appropriately. The task scheduler assigns all ready tasks to a priority queue; all tasks with unmet dependencies are placed into a wait set until they become ready. Worker threads of a thread pool take the tasks from the queue and execute them. To avoid frequent context switches, tasks cannot be preempted. Once a task is finished, the next task to be executed is chosen based on priorities and release time as second criterion. That way, incoming high priority tasks can start executing on all processing units, once the currently running tasks have finished. In our setup each thread is assigned to a physical processing unit and executes one and only one task at a time. In case of highly contending workloads, the number of workers could be dynamically adapted, to keep all processing units busy, as proposed in [10].

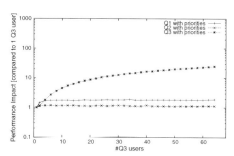

Fig. 1. Performance impact of varying the number Q_3 users with TAMEX [2]

We have shown the effectiveness of query priorities in a task-based execution framework to maintain a low response time for high priority queries in [2]. Figure 1 shows the response times for a transactional (Q1) and a transactional analytics (Q2) query while constantly increasing the number of parallel analytical (Q3) users.

3   Task Granularity for Analytical Queries

Parallelism of query execution can happen on three different levels: *inter-query*, *inter-operator*, as well as *intra-operator*. In TAMEX, the concurrent execution of separate queries is applied to any of the query classes introduced in Section 2.2. The other two forms of parallelism, commonly referred to as *intra-query* parallelism, are mainly relevant for longer-running analytical style queries, as the overhead of splitting data and merging results does not pay off for short-running transactional queries that operate on small data sets.

A major challenge is choosing the best task granularity given a fixed number of cores. Therefore, the goal of this section is to demonstrate the impact of task sizes on the response-time and throughput of analytical queries. In the following, task size and task runtime are used interchangeably. We discuss the scaling of individual operators and the throughput of concurrently executing analytical queries depending on the number of task instances we generate for a physical operator.

3.1   Scaling of Individual Database Operators

This subsection discusses the scaling behavior of two fundamental database operators, a Scan and Join operator, in our in-memory storage engine as described in Section 2.3. The join is implemented as a parallel radix-clustered hash-join, similar to the algorithm described in [11]. All experiments were executed on a server with 4 Intel(R) Xeon(R) X7560 CPUs with 8 cores each and 512GB RAM and TAMEX was started with a fixed number of 32 threads, hence, at most 32 operator instances can run in parallel even if the operator was split in more instances. For the first experiment, we analyzed each operator by splitting the overall physical operation into an increasing number of task instances that are then executed in parallel. To evaluate the results, we measured total response time and mean execution time of the tasks.

Figure 2 shows the result for the Scan operator. In addition to scaling the number of instances, the series in the graph shows an increasing number of records in the input table. As we see in Figure 2(a), for a scan on a large table, the optimal degree of parallelism is the number of cores, as we can fully leverage all available cores. Further increasing the number of instances also increases the runtime. The spike after 32 instances happens due to bad load balancing, as little more than 32 instances leave several processors idle after 32 tasks have been executed. For very large number of tasks, we can see that the overhead for splitting and merging the results becomes noticeable but still moderate; the experiment demonstrates, that splitting the scan operator in more tasks than available hardware contexts is a viable option without causing too much overhead. Figure 2(b) shows the mean task size for the corresponding task instances of the Scan operator. Splitting the scan operator in a higher number of instances constantly reduces the mean task size.

Figure 3 shows the results for a similar experiment for the radix join. The number of rows indicates the combined size of processed rows of both tables, whereas

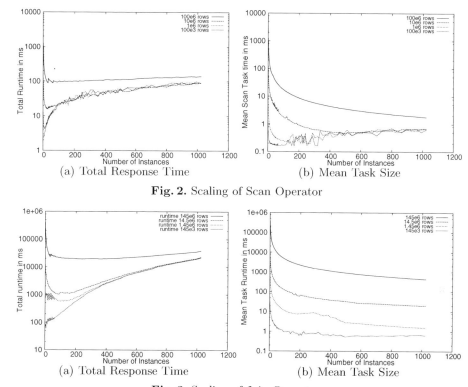

Fig. 2. Scaling of Scan Operator

Fig. 3. Scaling of Join Operator

the larger table was twice the size of the smaller table. Figure 3(a) indicates that splitting the join operator to a large number of tasks significantly increases the runtime. This can be explained with an increased overhead for splitting the join operator, as each join operator requires access to the corresponding partitions of the hashed relations. Figure 3(b) shows the mean task size for instances of the tasks corresponding to the join operator. It is worth mentioning, that we could reach even smaller task sizes for the join. However, we would require more bits for the radix partitioning which would lead to a higher runtime for the data set at hand.

3.2   Query Throughput for Varying Task Granularity

Based on the results of the individual operators, we can conclude that the best parallelization strategy for a lightly loaded system is to divide each database operator into a number of tasks that equals the number of available hardware threads. In this subsection, we discuss the impact of choosing the number of instances on query response time and query throughput, when several queries are executed in parallel. Again, we limit our experiment to a homogeneous workload and discuss the behavior in the presence of heterogeneous workloads in subsequent Sections. We have tested the parallel execution of Q3 queries on 10 millions

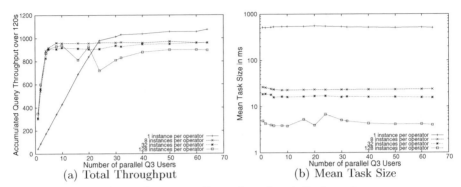

(a) Total Throughput   (b) Mean Task Size

Fig. 4. Concurrent Execution of analytical queries

records for the items table (see Section 2.2). In addition to varying the number of users that constantly issue Q3 queries without think time, the series in the graph shows different strategies for choosing the task granularity. We have varied the number of instances we split each scan and join operator from 1, basically no intra-operator parallelism, to 128, which means that we generate more tasks as available processing units.

Figure 4(a) shows the resulting throughput accumulated over all users. A higher number of instances leads to a higher throughput for a low number of users, as the work can be better balanced across the cores. However, for a number of users greater than 32, the system is constantly under full load, and choosing larger task sizes decreases the overhead of splitting inputs and merging results for individual database operators and thus, increases overall throughput. Figure 4(b) shows that the mean task size differs for each strategy and is largely independent of the load on the system. We can conclude that larger task sizes lead to less overhead and higher throughput in a loaded system.

4   Response Time of High Priority Tasks

We have considered a homogeneous, analytical workload in the previous section. In our experiments with TAMEX [2], we have seen that the task granularity of large analytical queries can impact the response time of short running queries. To better understand this effect, we have modeled the problem as a queueing network. In terms of a queueing network, the problem we are facing is to determine the expected response times for high priority tasks in the presence of other task classes with different service times, when all tasks are executed by a service station consisting of a single queue with non-preemptive priority scheduling discipline and multiple identical server stations. As discussed extensively in [12], "multiserver systems [are] a notoriously difficult problem in queueing theory". For the non-preemptive case, the only theoretical work we found on response times of different priority classes [13] restricts the problem to two priority classes with the same mean service time. Lacking directly applicable theoretical results, we follow two paths to learn from theory: We briefly review the expected response time for the well-understood case of a single server system to demonstrate the interdependencies of different service classes and obtain approximate results for the multiserver case by simulation.

For the theoretical consideration, we model our problem as an open queueing model network. For illustration, we restrict the number of priority classes to two, a class of high priority tasks h and a class of low priority tasks l. The tasks of the two classes arrive according to a Poisson stream with rate λ_h and λ_l, and an exponentially distributed service rate of μ_h and μ_l. According to queueing theory, the response or sojourn time for tasks of class h consists of the mean service time $1/\mu_h$, as well as the waiting time in the queue. $\rho_h = \lambda_h/\mu_h$ and ρ_l denote the average number of tasks in service. As explained in detail in [14], the mean sojourn time $E(S_h)$ of the highest priority class is given by the sum of the time a task has to wait in the queue until other high priority tasks $E(L_h)$, currently in the system, have finished, its own service time, as well as the residual service time of the task that is currently served. (Note that in case of an exponential distribution the residual service time is equally distributed as the service time and has a mean of $1/\mu$):

$$E(S_h) = E(L_h)/\mu_h + 1/\mu_h + \rho_l/\mu_l \tag{1}$$

Applying Little's law leads to

$$E(S_h) = \frac{1/\mu_h + \rho_l/\mu_l}{1 - \rho_h} \tag{2}$$

Equation 2 shows that an increased mean service time of $1/\mu_l$ for low priority tasks leads to an increased sojourn time for high priority tasks. This holds even if the average number of low priority tasks in the system is kept constant by decreasing the number of tasks proportionally with increasing task sizes. In a system that is heavily loaded with low priority tasks ($\rho_l >> \rho_h$), Equation 2 shows that the sojourn time of high priority tasks is dominated by ρ_l/μ_l, and given a fixed amount of work, depends only on the granularity of the low priority tasks, which is reflected in the mean service time $1/\mu_l$.

To confirm the hypothesis that the service time of the lower priority tasks also significantly influences the response time of the high priority class in the case of multiple servers, we have simulated a queueing network with Java Modelling Tools [15]. As discussed above, the response time of a high priority task depends on the number of high priority tasks that are in the system as well as the residual time of a task that is currently served. We focus our analysis on the impact of the residual time of lower priority tasks, and therefore limit the number of high priority tasks to a single task to eliminate this waiting time. We have modeled two task classes: a high priority class as a closed class with population 1 and the low priority class as an open class with indefinite population and an exponentially distributed inter-arrival time, leading to a Poisson distribution of the number of arriving tasks per interval. The high priority task is modeled as a user that constantly issues a single request, waits for its completion, and sends the request again after a short waiting time. The open class of low priority tasks flow from a source that generates tasks according to the given distribution to the service station and are taken out of the network after being served by a sink node. The service station consists of a queue with a non-preemptive scheduling policy which serves tasks of both classes and 8 equal servers with exponentially distributed service times $1/\mu_h$, and $1/\mu_l$ for both classes. We measure the time that the high priority task needs to wait in the queue until it can be served depending on the average service time $1/\mu_l$. We vary the load of the system by running our experiment with different ρ_l.

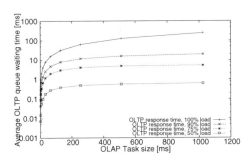

Fig. 5. Simulation of response time of high priority classes depending on task size of low priority classes

Figure 5 shows the resulting average waiting time in the queue for the high priority task depending on the average task size of the low priority tasks. The average service time $1/\mu_h$ for a high priority task is 1 ms and $1/\mu_l$ is increased from 1 ms up to 1000 ms. By increasing the task size of the low priority tasks, we adjusted the arrival rate to keep ρ_l constant at 50%, 75%, 90%, and 100%. We can see that the task size of the low priority task has a significant influence on the waiting time of the transactional query if the system is highly loaded. A task size of 1000 ms for the low priority task leads to an average waiting time of 20 ms for a high priority task and almost 250 ms for a fully loaded system. In a lightly loaded system, the likelihood that an incoming high priority task can start immediately as a processing unit is idle is high, so the waiting time is only marginally affected by the task size of low priority tasks.

From the simulation of the queueing network, we conclude for our task-based query execution system that the task sizes of low priority queries can have a significant impact on the response times of high priority tasks, if the system is loaded. We conclude that we can effectively maintain a high response time for high priority queries and limit the impact of low priority queries by limiting the task size of queries in the system.

5   Enforcing a Maximum Task Size

In this Section, we propose the introduction of a system parameter for the Maximum Tasks Size (MTS) to maintain a consistent response time of the system. Based on the MTS, the number of instances for each operator of a query is

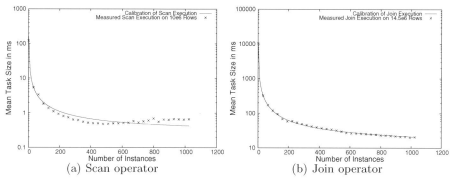

Fig. 6. Measured task size compared with predicted size

chosen. To limit the dependency on the quality of predictions of the size of intermediate results, we determine the number of instances dynamically during runtime, when the intermediate results are known. We first describe our solution and subsequently demonstrate the effect of varying the MTS.

5.1 Query Execution with a Maximum Task Size

The problem to determine the expected mean run time of tasks depending on the number of instances is related to predicting the runtime of a query, but simpler to solve. The two main challenges of predicting the total execution time of a query are calibrating the cost for each supported database operator and estimating the cardinalities of intermediate results [16]. As demonstrated in [16], the cost for each operator can be calibrated off line. In our case, estimating the cardinalities of intermediate results can be avoided by deferring the determination of the task granularity until the intermediate results are known. In other words, we split a database operator in executable tasks at the point in time when the sizes of all inputs are known, based on input data characteristics and the calibrated cost of the corresponding operator.

For our implementation, we have calibrated the operator cost for the scan and join operators as introduced in Section 3.1. For simplicity, we assume a fixed data set. To predict the runtime $rt(n)$ of a task when the operator is split in n tasks, we assume the functional dependency of Equation 3, where rt_{seq} is the sequential runtime of the operator and ov the overhead for creating an additional task.

$$rt_{par}(n) = \frac{rt_{seq}}{n} + ov \qquad (3)$$

After calibrating rt_{seq} and ov to our system, Equation 3 provides a good approximation of the runtime of a single task, as shown in Figure 6(a) for the scan operator, and Figure 6(b) for the join operator, when comparing the size to the mean task size for the corresponding operator on a table with 10 million rows. We use the inverse function to determine the number of instances for a provided MTS.

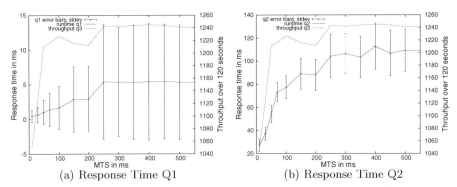

(a) Response Time Q1   (b) Response Time Q2

Fig. 7. Database performance with a varying Maximum Task Size

In the query processing of TAMEX, as briefly described in Section 2.3 and in more detail in [2], we apply the dynamic splitting of a task according to the provided MTS, once a task is ready to be pushed in the ready queue. At that point of time, all inputs to the operator are known and we determine the number of instances to meet the provided MTS and push the corresponding number of tasks in the ready queue.

5.2 Evaluation

To evaluate our concept of MTS, we have implemented the concept of MTS in TAMEX and set up an experiment with three different classes of users, each issuing queries of type Q1, Q2, and Q3 on 10 million records for the items table, as introduced in Section 2.2. Q1 has highest priority followed by Q2 and Q3. Each user constantly issues a single query and waits for the response. For the Q2 user, we have set a think time of 1 second, before the user issues a new query to simulate interactive behavior. Notice, that according to the query execution model of TAMEX, as described in Section 2.3, the Q1 query is executed as a single task. Q2 and Q3 consist of a graph of scan and join operators that are dynamically split into task according to a varying MTS. Figure 7 shows the results separately for Q1 and Q2. In line with the results predicted by our Queuing network model described in Section 4, the response time for queries of classes transactional and transactional analytics increase heavily with an increased task size of lower priority analytical queries. Additionally, also the deviation of the runtime increases with an increased MTS. As we can see in Figure 7(b), the impact on Q2 is higher in absolute times, as Q2 consists of several tasks and each task potentially needs to wait longer. On the other hand, and in line with the experiments of Section 3.2, an increasing the task size, also increases the analytical throughput.

5.3 Choosing the Maximum Task Size

As demonstrated in this paper, choosing the granularity of tasks has a great impact on the competing objectives of high response time for priority queries and high throughput for reporting queries. The MTS is an effective means to control the performance behavior in favor of one of the two objectives. Hence, how to choose the MTS mainly depends on the priorities of the applications that use the database. A static strategy could set the MTS to a small value during operating hours and higher during the night, to favor reports or maintenance tasks. It is also conceivable to dynamically change the MTS, for example if a threshold for the maximum average response time for a certain query class is exceeded.

6   Related Work

We have identified related work in three fields: heterogeneous workload management for database management systems, query execution on multiprocessors and task-based scheduling in other disciplines in computer science, such as real-time systems and cluster scheduling.

6.1   Heterogeneous Workload Management for Database Management Systems

Workload management for heterogeneous queries has been reviewed frequently in the context of web requests [17,18] and business intelligence applications [19,20]. In contrast to our research, most existing work on workload management is specific to disk-based DBMS and considers a query as the level for scheduling. In general, we can divide the proposed approaches for managing workloads of different query classes into two classes: *external* and *internal*. External workload management controls the number of queries that access the database (admission control); internal workload management systems typically control available resources, such as CPU or main memory, and assign them to queries. Niu et al. [21] give a more detailed overview of workload management systems for DBMS.

Early work on internal workload management for disk-based systems has been published by Carey et al. [19,22]. A more recent work by McWherter et al. [18] shows the effectiveness of scheduling bottleneck resources using priority-based algorithms in a disk-based DBMS.

More recent work has proposed solutions for adaptive admission control based on query response time. Schroeder et al. [23] propose an external queue management system that schedules queries based on defined service-levels per query-class and a number of allowed queries in the database, the so-called multiprogramming level. Niu et al. [21] propose a solution that manages a mixed workload of OLTP and OLAP queries by controlling the resources assigned to OLAP queries depending on the response times of OLTP queries. Krompass et al. [24] extended this approach for multiple objectives. The work of Kuno et al. [20] and Gupta et al. [25] propose mixed workload schedulers with admission control based on query run-time prediction. Although external workload management systems are applicable to in-memory databases, they fall short in our scenario, as queries need to get access to a large number of processing units quickly, e.g. to answer complex interactive queries. Psaroudakis et al. [10] discuss scheduling for a mix of analytical and transactional queries in the context of in-memory databases. They demonstrate the principle of reducing the task size of analytical queries to improve throughput which is in line with our results presented in Section 3. However, they do not discuss the effects of varying the task size on other query classes in detail.

6.2   DBMS and Query Execution on Multiprocessors

Our framework is designed for parallel query execution on multiprocessor CPUs. Hardavellas et al. discuss opportunities and limitations of executing of database servers on multiprocessors [26], concluding that data cache misses are the performance bottleneck resource in memory-resident databases. Zhou et al. [27] investigated the impact of simultaneous multithreading. Furthermore, Krikellas et al. [28] and Wu et al. [29] discuss the question of how a single query plan can be scheduled on a number of given threads. Wu et al. [29] also propose an extension to dynamic arrival of queries. Both approaches apply heuristics to generate a schedule with minimal makespan for a given query, but the algorithms seem to impose significant overhead in case of a larger number of tasks and threads.

Bouganim et al. [3] and as Rahm et al. [30] have studied load balancing strategies for parallel database systems. In the area of task-based scheduling for queries, Lu et al. [31] propose a task-based approach for query execution with dynamic load balancing by task stealing. In contrast to our approach, they schedule tasks of larger size that can be further split during runtime in case of load imbalances.

6.3   Task-Based Scheduling

Splitting larger jobs into tasks and scheduling these tasks on a number of processing units is a well researched problem in computer science. We limit the discussion to surveys, as well as related work that demonstrates the effects of scheduling smaller, non-preemptive tasks.

Casavant et al. provide a taxonomy for scheduling algorithms [32]. El-Rewini et al. [33] and Feitelson et al. [34] give an overview of algorithms to assign tasks to processing elements. In TAMEX, we adopt a FCFS policy for tasks of the same priority. More sophisticated scheduling algorithms require the knowledge of execution times for all tasks, which we only estimate for tasks that are subject to being split for intra-operator parallelism, to keep overall overhead low.

For cluster scheduling, Ousterhout et al. [35] present a task-based execution framework, called Tiny Tasks. Similar to our approach, tasks are non-preemptive and small task sizes alleviate long wait times for interactive, high priority tasks. In scheduling real-time tasks, Buttazzo et al. [36] provide a survey on limited preemptive scheduling algorithms. They highlight the main advantage of a non-preemptive discipline as reducing overhead (e.g. scheduling cost or cache-related cost) and the disadvantage of potentially blocking higher priority tasks. Our approach of splitting larger jobs in small non-preemptive tasks shares the motivation of limited preemptive scheduling to keep the database system responsive while keeping the cost of scheduling and context switches low.

7   Conclusion and Future Work

In this paper, we have demonstrated the impact of task granularity on responsiveness and throughput of an in-memory DBMS. We have shown that a larger task size for long running operators negatively affects the response time of short running queries and have proposed a system-wide maximum task size which has been demonstrated to effectively control the mutual performance impact of query classes.

References

1. Plattner, H.: A common database approach for OLTP and OLAP using an in-memory column database. In: SIGMOD Conference, pp. 1–2. ACM (2009)
2. Wust, J., Grund, M., Plattner, H.: Tamex: A task-based query execution framework for mixed enterprise workloads on in-memory databases. In: IMDM, INFORMATIK (2013)
3. Bouganim, L., Florescu, D., Valduriez, P.: Dynamic Load Balancing in Hierarchical Parallel Database Systems. In: VLDB, pp. 436–447. Morgan Kaufmann (1996)
4. Plattner, H.: SanssouciDB: An In-Memory Database for Processing Enterprise Workloads. In: BTW, pp. 2–21. GI (2011)
5. Krüger, J., Kim, C., Grund, M., Satish, N., Schwalb, D., Chhugani, J., Plattner, H., Dubey, P., Zeier, A.: Fast Updates on Read-Optimized Databases Using Multi-Core CPUs. PVLDB 5(1), 61–72 (2011)
6. Krueger, J., Tinnefeld, C., Grund, M., Zeier, A., Plattner, H.: A case for online mixed workload processing. In: DBTest (2010)
7. Wust, J., Meyer, C., Plattner, H.: Dac: Database application context analysis applied to enterprise applications. In: ACSC (2014)
8. Wust, J., Krüger, J., Blessing, S., Tosun, C., Zeier, A., Plattner, H.: Xsellerate: supporting sales representatives with real-time information in customer dialogs. In: IMDM, pp. 35–44. GI (2011)
9. Grund, M., Krüger, J., Plattner, H., Zeier, A., Cudré-Mauroux, P., Madden, S.: HYRISE - A Main Memory Hybrid Storage Engine. PVLDB 4(2), 105–116 (2010)
10. Psaroudakis, I., Scheuer, T., May, N., Ailamaki, A.: Task Scheduling for Highly Concurrent Analytical and Transactional Main-Memory Workloads. In: ADMS Workshop (2013)
11. Kim, C., Sedlar, E., Chhugani, J., Kaldewey, T., Nguyen, A.D., Di Blas, A., Lee, V.W., Satish, N., Dubey, P.: Sort vs. Hash Revisited: Fast Join Implementation on Modern Multi-Core CPUs. PVLDB 2(2), 1378–1389 (2009)
12. Wierman, A., Lafferty, J., Scheller-wolf, A., Whitt, W.: Scheduling for today's computer systems: Bridging theory and practice. Technical report, Carnegie Mellon University (2007)
13. Kella, O., Yechiali, U.: Waiting times in the non-preemptive priority m/m/c queue. In: Stochastic Models (1985)
14. Adan, I., Resing, J.: Queueing Theory: Ivo Adan and Jacques Resing. Eindhoven University of Technology (2001)
15. Bertoli, M., Casale, G., Serazzi, G.: Java Modelling Tools: an Open Source Suite for Queueing Network Modelling andWorkload Analysis. In: QEST (2006)
16. Wu, W., Chi, Y., Zhu, S., Tatemura, J., Hacigümüs, H., Naughton, J.F.: Predicting query execution time: Are optimizer cost models really unusable?. In: ICDE, pp. 1081–1092. IEEE Computer Society (2013)

17. Biersack, E.W., Schroeder, B., Urvoy-Keller, G.: Scheduling in practice. SIGMET-RICS Performance Evaluation Review 34(4), 21–28 (2007)
18. McWherter, D.T., Schroeder, B., Ailamaki, A., Harchol-Balter, M.: Priority Mechanisms for OLTP and Transactional Web Applications. In: ICDE, pp. 535–546 (2004)
19. Brown, K., Carey, M., DeWitt, D., Mehta, M., Naughton, F.: Resource allocation and scheduling for mixed database workloads (January 1992), cs.wisc.edu
20. Kuno, H., Dayal, U., Wiener, J.L., Wilkinson, K., Ganapathi, A., Krompass, S.: Managing Dynamic Mixed Workloads for Operational Business Intelligence. In: Kikuchi, S., Sachdeva, S., Bhalla, S. (eds.) DNIS 2010. LNCS, vol. 5999, pp. 11–26. Springer, Heidelberg (2010)
21. Niu, B., Martin, P., Powley, W.: Towards Autonomic Workload Management in DBMSs. J. Database Manag. 20(3), 1–17 (2009)
22. Carey, M.J., Jauhari, R., Livny, M.: Priority in DBMS Resource Scheduling. In: VLDB, pp. 397–410. Morgan Kaufmann (1989)
23. Schroeder, B., Harchol-Balter, M., Iyengar, A., Nahum, E.M.: Achieving Class-Based QoS for Transactional Workloads. In: ICDE, p. 153 (2006)
24. Krompass, S., Kuno, H.A., Wilkinson, K., Dayal, U., Kemper, A.: Adaptive query scheduling for mixed database workloads with multiple objectives. In: DBTest (2010)
25. Gupta, C., Mehta, A., Wang, S., Dayal, U.: Fair, effective, efficient and differentiated scheduling in an enterprise data warehouse. In: EDBT, pp. 696–707. ACM (2009)
26. Hardavellas, N., Pandis, I., Johnson, R., Mancheril, N., Ailamaki, A., Falsafi, B.: Database Servers on Chip Multiprocessors: Limitations and Opportunities. In: CIDR, pp. 79–87 (2007), www.cidrdb.org
27. Zhou, J., Cieslewicz, J., Ross, K.A., Shah, M.: Improving Database Performance on Simultaneous Multithreading Processors. In: VLDB, pp. 49–60. ACM (2005)
28. Krikellas, K., Cintra, M., Viglas, S.: Scheduling threads for intra-query parallelism on multicore processors. In: EDBT (2010)
29. Wu, J., Chen, J.J., Wen Hsueh, C., Kuo, T.W.: Scheduling of Query Execution Plans in Symmetric Multiprocessor Database Systems. In: IPDPS (2004)
30. Rahm, E., Marek, R.: Dynamic Multi-Resource Load Balancing in Parallel Database Systems. In: VLDB, pp. 395–406. Morgan Kaufmann (1995)
31. Lu, H., Tan, K.-L.: Dynamic and Load-balanced Task-Oriented Datbase Query Processing in Parallel Systems. In: Pirotte, A., Delobel, C., Gottlob, G. (eds.) EDBT 1992. LNCS, vol. 580, pp. 357–372. Springer, Heidelberg (1992)
32. Casavant, T.L., Kuhl, J.G.: A Taxonomy of Scheduling in General-Purpose Distributed Computing Systems. IEEE Trans. Software Eng. 14(2), 141–154 (1988)
33. El-Rewini, H., Ali, H.H., Lewis, T.G.: Task Scheduling in Multiprocessing Systems. IEEE Computer 28(12), 27–37 (1995)
34. Feitelson, D.G., Rudolph, L., Schwiegelshohn, U., Sevcik, K.C., Wong, P.: Theory and Practice in Parallel Job Scheduling. In: Feitelson, D.G., Rudolph, L. (eds.) IPPS-WS 1997 and JSSPP 1997. LNCS, vol. 1291, pp. 1–34. Springer, Heidelberg (1997)
35. Ousterhout, K., Panda, A., Rosen, J., Venkataraman, S., Xin, R., Ratnasamy, S., Shenker, S., Stoica, I.: The case for tiny tasks in compute clusters. In: HotOS 2013, p. 14. USENIX Association, Berkeley (2013)
36. Buttazzo, G.C., Bertogna, M., Yao, G.: Limited Preemptive Scheduling for Real-Time Systems. A Survey. IEEE Trans. Industrial Informatics 9(1), 3–15 (2013)

On Data Partitioning in Tree Structure Metric-Space Indexes

Rui Mao[1], Sheng Liu[1], Honglong Xu[1], Dian Zhang[1,*], and Daniel P. Miranker[2]

[1] Guangdong Province Key Laboratory of Popular High Performance Computers
College of Computer Science and Software Engineering, Shenzhen University
Shenzhen, Guangdong 518060, China
{mao,liusheng,2100230221,zhangd}@szu.edu.cn
[2] Department of Computer Sciences, University of Texas at Austin
Austin, TX 78712, USA
miranker@cs.utexas.edu

Abstract. Tree structure metric-space indexing methods recursively partition data according to their distances to a set of selected reference points (also called pivots). There are two basic forms of data partitioning: ball partition and General Hyper-plane (GH) partition. Most existing work only shows their superiority experimentally, and little theoretical proof is found. We propose an approach to unify existing data partitioning methods and analyze their performance theoretically. First, in theory, we unify the two basic forms of partitioning by proving that there are rotations of each other. Second, we show several theoretical or experimental results, which are able to indicate that ball partition outperforms GH partition. Our work takes a step forward in the theoretical study of metric-space indexing and is able to give a guideline of future index design.

Keywords: data partitioning, metric-space access method, pivot space model.

1 Introduction

Metric-space indexing, also known as distance-based indexing, is a general purpose approach to support similarity queries [9, 17, 29, 37]. It only requires that similarity is defined by a metric-distance function. Tree structure is one of the most popular metric-space indexing structures. In their top-down construction, tree structure metric-space indexing methods build index trees by recursively applying two basic steps: pivot selection and data partition. In pivot selection, only a small number of reference points (pivots), are selected from the database. The distances from data points to the pivots form a projection from the metric space to a low dimensional space, the pivot space [9, 24]. In data partitioning, data points are partitioned by their distances to the pivots, similar to the partitioning methods of multi-dimensional indexing [29].

Basically, there are two kinds of data partition strategies, ball partition and General Hyper-plane (GH) partition. However, most traditional methods are based on

---

* Corresponding author.

S.S. Bhowmick et al. (Eds.): DASFAA 2014, Part I, LNCS 8421, pp. 141–155, 2014.

heuristics. Theoretical analysis is usually overlooked. As a result, a large amount of these work only show their superiority from some carefully chosen datasets.

To solve this problem, we propose an approach that is able to unify ball and GH partitions, and compare their index performance theoretically and experimentally.

First, we show that ball partition and GH partition can be unified under the pivot space model [24] (Section 3.1), which was also proposed by our group. In the pivot space, the ball partition boundary is a straight line parallel to the axe, while GH partition boundary is a straight line with slope 1. That is, they are rotations of each other. Moreover, to make the analysis comprehensive, we leverage the extension versions of typical partition approach for comparison. In detail, we propose the Complete General Hyper-plane Tree (CGHT), which is an extension of the primitive form of GH partition, the GHT [33]. Results show that CGHT outperforms GHT.

Second, we propose an approach to analyze and predict an index's performance by means of the number of data points in a neighborhood (defined as r-neighborhood in Section 4) of the partition boundary. That is, less number of points in the r-neighborhood means better query performance. Further, we show two theoretical and two experimental results, which indicate that ball partition outperforms GH partition in query performance. They are: (1) among all the rotations, we theoretically prove that ball partition has the minimal width of r-neighborhood. Although the number of points in the r-neighborhood, or the size of the r-neighborhood, is jointly dominated by width and data density along the width, a minimal width is an indication of smaller size; (2) if data follow normal distribution in the 2-dimensional pivot space, we theoretically prove that ball partition has smaller r-neighborhood size than GH partition; (3) we experimentally show that MVPT has smaller r-neighborhood size than CGHT on a comprehensive test suite; (4) we experimentally show that MVPT has better query performance than CGHT on the comprehensive test suite.

The remaining of the paper is organized as follows. Related work is introduced in Section 2. In Section 3, we show how ball partition and GH partition are unified, followed by 2 theoretical indications of MVPT's better query performance than CGHT in Section 4. Section 5 elaborates the two experimental indications of MVPT's better query performance. Finally, conclusions and future work are discussed in Section 6.

2   Related Work

Ball partition and GH partition are the two basic kinds of data partition strategies in tree structure metric-space indexes.

Ball partition considers pivots one at a time, e.g. Vantage Point Tree (VPT) [33, 36] and Multi-Vantage Point Tree (MVPT) [5]. That is, data is first partitioned according to their distances to the first pivot. Each partition boundary forms a circle. Then, each partition area is further partitioned according to the data distances to the second pivot. Such process repeats until every pivot is used. For example, given k pivots and m partitions for each pivot, the total number of partitions is m^k.

GH partition is similar to clustering. Each data point is assigned to its closest pivot (or cluster center). In other words, the distance difference from a data point to two pivots is considered, and the metric space is partitioned by a surface equidistant from the two pivots. General Hyper-plane Tree (GHT) [33], GNAT [6] and SA-tree [28] belong to GH partition.

The BSP [16], M-Tree [9] and slim-tree [10] family also adopt the GH partition with multiple pivots when data is partitioned. However, each partition is represented as a ball, i.e. a center and a covering radius, similar to ball partition.

3   Unification of Ball Partition and GH Partition

In this section, we first give a brief introduction to our theoretical background, the pivot space model. Then, we consider VPT and GHT, the primitive forms of ball partition and GH partition. However, VPT has only one pivot while GHT has two pivots. To make the numbers of pivots and partitions equal as a fair comparison, we propose the Complete General Hyper-plane Tree (CGHT). Finally, we compare CGHT with MVPT, with same numbers of pivots and partitions, and unify them in the pivot space.

3.1   Theoretical Background: The Pivot Space Model

This subsection presents a brief introduction to the pivot space model. The readers can refer to [24] for more details.

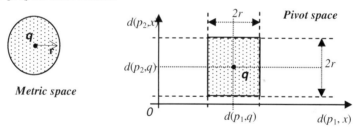

Fig. 1. In a metric space, the ball of a range query is covered by a square in the pivot space

Some notations that will be used in the paper are listed as follows. Let R^m denotes a general real coordinate space of dimension m. Let (M, d) be a metric space, where M is the domain of the space, and d is a metric distance function. Let $S = \{x_i \mid x_i \in M, i = 1, 2, ..., n\}$, be the database, and x_i is the data point in the database, $n \geq 1$. S is a finite indexed subset of M. Let $P = \{p_j \mid j = 1, 2, ..., k\}$ be a set of k pivots. $P \subseteq S$. Duplicates are not allowed.

There are three steps to answer a range query $R(q, r)$ in a metric space [24]:

Step 1: (1) Map the data into R^k. (2) Map the query object into R^k. (3) Determine a region in R^k that completely covers the range query ball.

Given the set of pivots, each point x in S can be mapped to a point x_p in the non-negative orthant of R^k. The j-th coordinate of x_p represents the distance from x to p_j:

$$x_p = (d(x, p_1), ..., d(x, p_k)),$$

The pivot space of S, $F_{P,d}(S)$, is defined as the image set of S:

$$F_{P,d}(S) = \{x_p \mid x_p = (d(x,p_1), \ldots, d(x,p_k)), x \in S\}.$$

The image shape of a range query ball $R(q, r)$ in a general metric space is not clear in a corresponding pivot space. However, it can be proved, from the triangle inequality, the image of the query ball is completely covered by a hypercube of edge length $2r$ in the pivot space [9]. Actually, the hypercube is a ball of radius r in the new metric space specified by the pivot space and the L^x distance. We call the hypercube the query cube. Fig. 1 shows an example where 2 pivots are selected [24]. All the points in the query ball are mapped into the query square in the 2-d pivot space plane. Points outside the square can be safely discarded according to the triangle inequality.

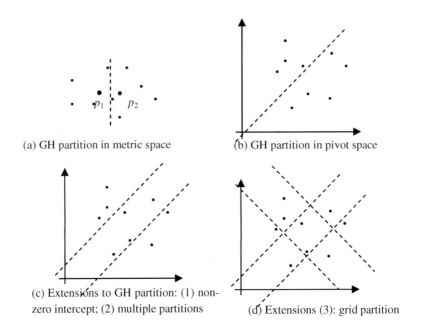

(a) GH partition in metric space

(b) GH partition in pivot space

(c) Extensions to GH partition: (1) non-zero intercept; (2) multiple partitions

(d) Extensions (3): grid partition

Fig. 2. GH partition and extensions

Step 2: Exploit multi-dimensional techniques to obtain all the points in the region determined in Step 1.

Fundamentally, Step 2 uses divide-and-conquer, based on the data coordinates. Many partition methods in multi-dimensional indexing can be applied here. For example, MVPT's partition is similar to a kd-tree [2].

Step 3: For each point obtained in Step 2, we compute its distance to the query object to eliminate the false positives.

3.2   The Complete General Hyper-plane Tree

We start with the primitive form of ball partition and GH partition for comparison. VPT [33, 36] is the primitive form of ball partition. It first selects a pivot, and then draws a circle centered at the pivot to partition the data. The radius of the circle is chosen so that the partitioning is balanced. GHT [33] is the primitive form of GH partition. It first selects two pivots. Then it uses a hyper-plane, which is equidistant to the two pivots, to partition the data. Obviously, VPT and GHT use different number of pivots. Since GH partition uses at least two pivots, we consider Multiple Vantage Point Tree (MVPT) [5] instead of VPT as a representative of ball partition methods. MVPT first selects a few pivots. Then data is partitioned based on their distances to the first pivot. Such partitioning repeats until all the pivots are used.

A MVPT with 2 pivots has the same number of pivots with a GHT. However, at each index node, a MVPT with 2 pivots generates at least 4 partitions (2 by 2), while a GHT generates only 2. To make the number of partitions equal, we extends GHT and propose the CGHT, which has the same numbers of pivots and partitions with MVPT with 2 pivots.

In the following, for simplicity, we assume only two pivots are selected, $k = 2$. Let d_1 and d_2 be the distances from a data point to the two pivots, respectively.

Given the two pivots, a GH partition separates the data with a hyper-plane formed by points equidistant to the two pivots (Fig. 2 (a)) [33]. In the pivot space, the partition line is specified by $d_1 - d_2 = 0$. Its slope is 1 and its intercept is 0 (Fig. 2(b)). Examining Fig. 2(b), we can extend GHT as follows.

Extension (1): use non-zero intercept (Fig. 2(c)).
Extension (2): use multiple partitions (Fig. 2(c)).
Extension (1) has been studied by Zhang and Li, and Lokoc and Skopal [23, 38].

In GHT, distances of each data point to two pivots are computed. So the two variables, d_1 and d_2, are known. However, only one combined variable, $d_1 - d_2$, is exploited in partitioning. In statistics, if d_1 and d_2 are two independent random variables, $d_1 - d_2$ and $d_1 + d_2$ are deemed independent [8]. Although d_1 and d_2 might not be actually independent, we believe that variable $d_1 + d_2$ is able to provide additional pruning ability without extra distance calculations. This idea leads to Extension (3):

Extension (3): use $d_1 + d_2$ to partition.

---

p_1: pivot 1
p_2: pivot 2
plus_max[]: array of all children's upper bounds of $d_1 + d_2$
plus_min[]: array of all children's lower bounds of $d_1 + d_2$
minus_max[]: array of all children's upper bounds of $d_1 - d_2$
minus_min[]: array of all children's lower bounds of $d_1 - d_2$
children[]: array of pointers to all children

Fig. 3. Internal node structure of CGHT

Fig. 2 (d) shows the case that all the 3 extensions are combined. As Extension (3) exploited all the information provided by the two pivots, we name this new structure the Complete General Hyper-plane Tree (CGHT). Here "Complete" means the two distance computation to the two pivots are taken into full use. That is, during the search procedure, in addition to pruning based on one variable $d_1 - d_2$, one can also do pruning base on another variable, $d_1 + d_2$. The structure of one CGHT internal node is shown in Fig. 3. The structure of one CGHT leaf node is the same as that of GHT [33].

---

(1) if data set is small, construct a leaf node; otherwise:
(2) pivot selection
(3) compute $d_1 + d_2$ and $d_1 - d_2$ for all data points
(4) run clustering partition on $d_1 + d_2$ and $d_1 - d_2$ values.
(5) recursively bulkload every partitions generated in (4).

Fig. 4. Steps to bulkload a CGHT

Bulkloading of CGHT follows the general steps in constructing a pivot-based index structure (Fig. 4). First, two pivots are selected, using any pivot selection algorithm. Second, values of $d_1 + d_2$ and $d_1 - d_2$ are computed for each data point. Third, most data partition algorithms (from the three categories of metric space indexing methods) can be applied to partition the data, based on $d_1 + d_2$ and $d_1 - d_2$ values. These steps are run recursively on each partition generated.

---

(1) if leaf node, do the GHT leaf node search; otherwise
(2) compute $d_1 + d_2$ and $d_1 - d_2$ values for the query object
(3) for each child i,
    if not [(plus_min[i]–$2r \leq d_1+d_2 \leq$ plus_max[i]+$2r$) and
            (minus_min[i]–$2r \leq d_1–d_2 \leq$ minus_max[i]+$2r$)]
        then this child can be pruned.
(4) recursively search child that cannot be pruned in (3).

Fig. 5. Range query steps of CGHT

Range query algorithm of CGHT is summarized in Fig. 5. The correctness of the pruning step (3), can be proven by the triangle inequality.

3.3   Unification of CGHT and MVPT

The CGHT partition lines with slope 1 in the pivot space (Fig. 2(d)) are actually hyperbolas in metric space (Fig. 6(a)), while those with slope -1 are actually ellipses. For the MVPT partition circles in the metric space (Fig. 6(b)), in the pivot space (Fig. 6(c)), they are actually straight lines parallel to the axes.

Comparing Fig. 2(d) with Fig. 6(c), it is obvious that one can turn MVPT into CGHT by a 45° rotation, or vice versa.

Now ball partition and GH partition are unified. A following reasonable question is what rotation angle is optimal with respect to query performance. We show 2 theoretical indications of ball partition's better query performance in Section 4, and 2 experimental indications in Section 5.

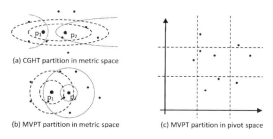

(a) CGHT partition in metric space

(b) MVPT partition in metric space

(c) MVPT partition in pivot space

Fig. 6. CGHT and MVPT partitions

4   Theoretical Comparison of Ball Partition and GH Partition

In this section, we analyze ball partition and GH partition's query performance theoretically. We first define the r-neighborhood of the partition boundary, the number of points in which is a dominant factor of the query performance. Then, the total number of points in the neighborhood (size of r-neighborhood) is investigated for both ball partition and GH Partition from two aspects: (1) width of r-neighborhood, a dominative factor of r-neighborhood size; (2) r-neighborhood size for a particular data distribution, 2-dimensional normal distribution, in the pivot space. Our results indicate that MVPT outperforms CGHT.

4.1 r-Neighborhood of Partition Boundaries

Since the key of tree structure indexing is to prune as much data as possible, the pruning power is one of the dominant factors of query performance. Given a range query $R(q, r)$, if q is so close to a partition boundary that the query ball centered at q with radius r intersect the boundary, either sides of the boundary can be pruned, leading to low pruning power, or low query performance. To increase the pruning power, one has to lower the probability that queries fall too close to partition boundaries. As a result, we define the "r-neighborhood". In plain terms, an r-neighborhood is the neighborhood of a partition boundary, such that a range query R(q, r) may intersect the boundary if q falls into the r-neighborhood.

Definition 1. The **r-neighborhood** of a partition boundary L, denoted as $N_r(L)$, is a neighborhood of L. If a query object q falls into it, the range query ball centered at q with radius r intersects L.

If q falls into $N_r(L)$, partitions comprising both halves of the space defined by L have to be searched. If q does not fall into $N_r(L)$, only the partition on the side of L ,where q falls into, need to be searched. The partition on the other side can be pruned.

Average query performance can be improved by reducing the probability of queries falling into r-neighborhood. Assume queries have the same distribution as the data points in the database, the total number of data points in the r-neighborhood, or r-neighborhood size, is critical to query performance. That is, the smaller the r-neighborhood size, the better the query performance.

R-neighborhood is also useful when redundancy is introduced in the index tree. To guarantee that only one partition will be further considered at each level of the index tree, one can duplicate the r-neighborhood of the partition boundaries [14].

4.2   Minimal Width of the r-Neighborhood

The r-neighborhood size, a dominant factor of query performance, is jointly determined by the width of the r-neighborhood and the density of data along the width. Since density is determined by data distribution, it is impossible to know it when design the index. Therefore, in this subsection, we find out the minimal width of r-neighborhood among all rotations of ball partition and GH partition.

As discussed in Section 3, ball partition and GH partition boundaries are straight lines in 2-d pivot spaces. Let line $L: y = ax$ be an arbitrary partition line rotated from ball partition or GH partition (without losing generality, the constant intercept is omitted). We name such type of partitioning **linear partition** in the pivot space. The shape of r-neighborhood of linear partition in the pivot space is given by Theorem 1.

Theorem 1. The r-neighborhood of line $L: y = ax$, $0 \leq a$ is $N_r(L): |y\text{-}ax| \leq R_L(r)$, where $R_L(r)$ is a real valued function of r and L.

In other words, Theorem 1 states the fact that the r-neighborhood of linear partition is delimited by two lines, which are parallel and with the same distance to the linear partition line. The proof is straightforward and is omitted.

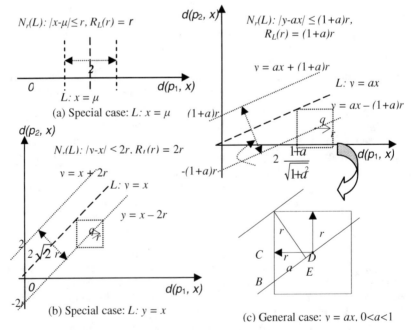

Fig. 7. R-neighborhood of hyper-plane partitions

In the following, we discuss the value of $R_L(r)$, and the width of $N_r(L)$. We start from two special cases, and followed by the general cases.

Case 1: L: $x = \mu$

This is the ball partition. From simple analysis (Fig. 7 (a)), the r-neighborhood of L: $x = \mu$ is: $N_r(L) : |x-\mu| \leq r$, $R_L(r) = r$, and the width of $N_r(L)$ is $2r$.

Case 2: L: $y = x$

This is the GH partition. According to the pivot space model, the query ball in metric space is covered by a cube in the pivot space. It can be observed (Fig. 7 (b)) that the r-neighborhood of L: $y = x$ is:

$N_r(L)$: $|y-x| \leq 2r$, $R_L(r) = 2r$, and the width of $N_r(L)$ is $2\sqrt{2}\,r$.

Case 3: General case: $y = ax$

Because of the symmetry, we only need to consider the case that $0<a<1$. The result is stated in Theorem 2.

Theorem 2. Let L: $y = ax$, $0<a<1$, then $R_L(r) = (1+a)r$, and the width of $N_r(L)$ is $2\dfrac{1+a}{\sqrt{1+a^2}}\,r$.

Theorem 3 is illustrated in Fig. 7(c). It can be proved with basic geometry.

The following corollary is correct due to the fact that $0 \leq a$.

Corollary 1. The width of $N_r(L)$ is minimized to $2r$ when $a = 0$ or ∞, i.e. the partition line is parallel to the axes, or ball partition. The width of $N_r(L)$ is maximized to $2\sqrt{2}\,r$, when $a=1$, i.e. the partition line has $45°$ angle to the axes, or GH partition.

According to Corollary 1, ball partition is optimal among its rotations with respect to the width of r-neighborhood. Although the r-neighborhood size is determined by both the width and the density, a smaller width is still an important sign of fewer points in the r-neighborhood, thus better query performance. To further investigate the r-neighborhood size, next, we mathematically calculate it for a particular data distribution, 2-dimensional normal distribution, in the pivot space.

4.3 r-Neighborhood Size of 2-d Normal Distribution in Pivot Space

Let us consider the case where data is normally distributed in a 2-d pivot space (Fig. 8). For simplicity, let the joint distribution of d_1 and d_2 to be $N(0, 1, 0, 1, -\rho)$. That is, the marginal distribution of both d_1 and d_2 is the standard normal distribution, and correlation coefficient is $-\rho$, $0 \leq \rho \leq 1$.

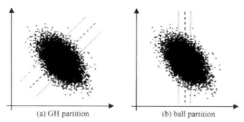

(a) GH partition   (b) ball partition

Fig. 8. Example: normal distribution in 2-d pivot space

In the following, we calculate the numbers of points in the r-neighborhoods of ball and GH partitions, denoted by $/N_{Lb}(r)/$ and $/N_{LG}(r)/$, respectively. Let $P_{Lb}(r)$ and $P_{LG}(r)$ be the probability that the data falls into the r-neighborhoods of ball and GH partitions. It is sufficient to compare $P_{Lb}(r)$ with $P_{LG}(r)$.

Theorem 3. $P_{Lb}(r) \leq P_{LG}(r)$, and they equal when $\rho = 1$.

Proof: In statistics, it have been proved that for normal distribution $N(0, 1, 0, 1, -\rho)$, the projected distribution on the first principle component (perpendicular to the GH partition line in Fig.8(a)) is $N(0,1+\rho)$, normal distribution with variance $1+\rho$ [20].

From Theorem 2, the widths of r-neighborhoods of ball and GH partitions are $2r$ and $2\sqrt{2}\,r$, respectively. Therefore:

$P_{Lb}(r) = P(/x/ \leq r \mid x \sim N(0,1))$, and $P_{LG}(r) = P(/x/ \leq \sqrt{2}\,r, \mid x \sim N(0, 1+\rho))$
After normalization: $P_{LG}(r) = P(/x/ \leq \sqrt{2/(1+\rho)}\,r, \mid x \sim N(0, 1))$

Since $0 \leq \rho \leq 1$, we have $1 \leq \sqrt{2/(1+\rho)}$, and the equality holds when $\rho = 1$.

That is, $P_{Lb}(r) \leq P_{LG}(r)$, and the equality holds when $\rho = 1$.   □

In conclusion, for 2-d normal distribution in the pivot space, better query performance can be expected from ball partition than GH partition since there are fewer points in the r-neighborhood of ball partition boundary than those of GH partition.

5   Empirical Results

In this section, we present experimental results to indicate that ball partition outperforms GH partition. We first introduce a comprehensive test suite we use, then the r-neighborhood sizes of both partitions on the test suite, and finally the range query performance of both partition methods on the test suite.

Table 1. Summary of test suite

Workload	Db. size	Distance metric	Dom.dim.	radius
Uniform vector	1Million		2, 8	0.02/0.3
Texas	36463	L^2 norm	2	0.02
Hawaii	9290		2	0.02
Protein	100k	Weighted edit distance	18	2
DNA	100k	Hamming distance	6	2
Image	10,221	L-norms	66	0.02
English dictionary	69069	Edit distance	N/A	1
NASA image	40150	L^2 norm	20	0.03

5.1   Test Setup

Our test data consists of the MoBIoS test suite [27] and the SISAP test suite [31] (Table 1). The MoBIoS test suite consists of synthetic vector data (dimensions 2 and

8), biological data, real world vector data and an image dataset. Different dimensions of the synthetic vector data have independent identical uniform distributions. Two types of biological data are considered: (1) amino-acid sequence fragments of the yeast proteome with weighted-edit distance based on the metric PAM substitution matrix [36]; (2) the DNA sequence fragments of the Arabidopsis genomes with Hamming distance. The real world vector data consists of the US cartographic boundary data of Texas and Hawaii. The images are represented by 66 dimensional feature vectors with a linear combination of L^1 and L^2 norms. Two datasets from the SISAP test suite [31] are involved: the English dictionary uses the edit distance. The NASA images are represented by 20-dimensional vectors with the Euclidean distance.

The sizes of the databases are all $100k$, except for uniform vector, which is 1 million, and those small workloads, where only limited amount of data is available. Although the data volume is not huge, it is large enough to clearly show the trends.

The number of pivots is 2 for all cases. Adding more pivots will increase the dimension of the pivot space. In Section 3, we have shown that ball partition and GH partition are rotations of each other. The only difference between them is the angle of rotation. Follow this idea, in the multiple-pivot scenario, one can still rotate the partition boundary of ball partition to get GH partition, or vice versa. This transformation is similar to the linear transformation of bases of vector space in linear algebra. Therefore, we limit our number of pivots to 2 for simplicity.

Two pivot selection heuristics are examined. One is *Farthest-First-Traversal* (FFT) [18], which selects corner of data as pivots. The other is a PCA-based heuristic, which has been shown to perform well generally [24].

For each pivot, data is partitioned into 3 parts. Thus, the total number of partitions is 9. Two data partition heuristics are examined. One is balanced partition [9], and the other is *Clusteringkmeans*, which derives the partition from k-means clustering [25]. Please note that GHT only adopts the balanced partition.

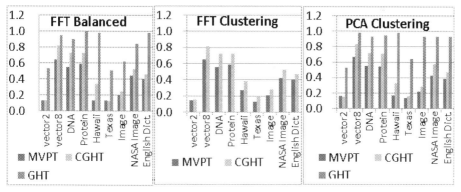

Fig. 9. Percentage of data to further examine at the index root

For each dataset, combination of pivot selection and data partition heuristics, statistics with a number of range query radii are collected. However, to save space, only the results of one representative radius are presented for each dataset and combination of heuristics. Those representative radii are listed in Table 1. For each dataset, 5000 data objects are chosen sequentially from the beginning of the dataset files as range query objects. The experimental results collected are averaged over queries.

We first compare the r-neighborhood size of MVPT, GHT and CGHT, then the range query performance.

5.2   Number of Points in r-Neighborhood

This subsection examines the index performance only at the tree root level, and the whole index is examined in the next subsection, based on range query performance.

In the presence of multiple pivots, a query in the r-neighborhood might lead to more than 2 partitions being further search. Therefore, instead of simply counting the number of points in r-neighborhood, we collect the average percentage of data to further examine, or the data that cannot be pruned, at the index root, as a more accurate indicator of the query performance.

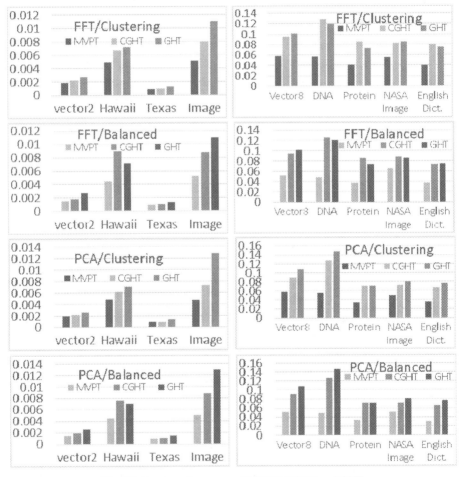

Fig. 10. Query performance of MVPT, CGHT and GHT

Fig. 9 shows the percentage of data cannot be pruned, or pruning power, at the index root level. Obviously, CGHT has more pruning power than GHT in all the cases, indicating a significant improvement of CGHT over GHT. Furthermore, MVPT has more pruning power than CGHT for almost all the cases. The only two exceptions are for Texas data (FFT, Balance) and 2-dimensional vector (PCA, Clustering), where the difference is neglectable.

5.3   Range Query Performance

Since distance evaluation in a metric space is usually costly, to focus on the algorithmic behavior, we use the number of distance calculations normalized over dataset sizes, which is implementation independent, as a performance measure.

The range query performance of MVPT, CGHT and GHT are listed in Fig. 10. It is clear to see that MVPT outperforms CGHT and GHT. The differences are significant in most cases. Moreover, the performance is more than doubled in several cases.

In the total 36 comparisons of CGHT and GHT, CGHT outperforms GHT in 28 cases. However, there are 8 of 36 cases that GHT outperforms CGHT, while the differences are significant in a few cases. We would like to point out that it does not prevent one from concluding that MVPT generally outperforms CGHT or GHT, or, in other words, ball partition generally outperforms GH partition.

6   Conclusions and Future Work

In this paper, we unify data partition methods of tree structure metric-space indexes under the pivot space model, and propose an approach to compare and predict query performance through r-neighborhood size. Theoretical and experimental results indicate that ball partition outperforms GH partition.

The contributions of our paper are as follows:

(1) Propose the Complete General Hyper-plane Tree which takes full use of the pivots and outperforms GHT.

(2) Unify ball partition and GH partition in the context of the pivot space model.

(3) Show that ball partition possesses the optimal rotation angle with respect to the width of r-neighborhood.

(4) Prove that ball partition has fewer points in r-neighborhood than GH partition for 2-d normal distribution in the pivot space.

(5) Show that ball partition has fewer points in r-neighborhood than GH partition, based on our test benchmark.

(6)   Show that MVPT outperforms CGHT by experiments, which consists with the theoretical analysis.

The methodology developed in this paper forms a basis for resolving the merits of the different classes of metric-space indexing algorithms.

The pivot space model establishes a bridge connecting metric-space indexing and high-dimensional indexing. It also lays a foundation for theoretical analysis. In the future work, we plan to take advantage of the fruitful results along the much longer research history in high-dimensional indexing to study metric-space indexing. We believe some of the dimension reduction and data partition approaches in high-dimensional indexing may benefit to metric-space indexing. Moreover, we have only considered linear partition so far. Non-linear partitioning deserves more attention. Actually, there are already some efforts on this topic, such as Li and Zhang's Haperplane tree [22].

Acknowledgment. This research was supported by the following grants: China 863: 2012AA010239; NSF-China: 61033009, 61170076, U1301252, 61202377, 61103001; Shenzhen Foundational Research Project: JCYJ20120613161449764.

References

1. Aizerman, M., Braverman, E., Rozonoer, L.: Theoretical foundations of the potential function method in pattern recognition learning. Automation and Remote Control 25, 821–837 (1964)
2. Bentley, J.L.: Multidimensional binary search trees used for associative searching. ACM Commun. 18(9), 509–517 (1975)
3. Beygelzimer, A., Kakade, S., Langford, J.: Cover trees for nearest neighbor. In: ICML, pp. 97–104 (2006)
4. Boser, B.E., Guyon, I.M., Vapnik, V.N.: A training algorithm for optimal margin classifiers. In: Haussler, D. (ed.) 5th Annual ACM Workshop on COLT, pp. 144–152. ACM Press, Pittsburgh (1992)
5. Bozkaya, T., Ozsoyoglu, M.: Indexing large metric spaces for similarity search queries. ACM Trans. Database Syst. 24(3), 361–404 (1999)
6. Brin, S.: Near Neighbor Search in Large Metric Spaces. In: The 21th International Conference on Very Large Data Bases (VLDB 1995). Morgan Kaufmann Publishers Inc. (1995)
7. Bustos, B., Navarro, G., Chavez, E.: Pivot selection techniquesfor proximity searching in metric spaces. Pattern Recogn. Lett. 24(14), 2357–2366 (2003)
8. Casella, G., Berger, R.L.: Statistical Inference. Duxbury Press (2001)
9. Chavez, E., Navarro, G., Baeza-Yates, R., Marroqu, J.: Searching in metric spaces. ACM Computing Surveys 33(3), 273–321 (2001)
10. Ciaccia, P., Patella, M.: Bulk loading the M-tree. In: 9th Australasian Database Conference, ADO 1998 (1998)
11. Ciaccia, P., Patella, M.: Proceedings of the Third International Conference on Similarity Search and Applications, Istanbul, Turkey, September 18-19. ACM, New York (2010)
12. Ciaccia, P., Patella, M., Zezula, P.: M-tree: An Efficient Access Method for Similarity Search in Metric Spaces. Presented at the 23rd International Conference on Very Large Data Bases (VLDB 1997), Athens, Greece (1997)
13. Cortes, C., Vapnik, V.: Support-Vector Networks. Machine Learning 20 (1995)
14. Feng, Y.-C., Kui, C., Cao, Z.-S.: A Multidimensional Index Structure for Fast Similarity Retrieval. Journal of Software 13(8), 1678–1685 (2002)
15. Filho, R.F.S., Traina, A.J.M., Traina, C., Faloutsos, C.: Similaritysearch without tears: The OMNI family of all-purpose accessmethods. In: ICDE, pp. 623–630 (2001)
16. Fuchs, H., Kedem, Z.M., Naylor, B.F.: On visible surface generation by a priori tree structures. In: Proceedings of the 7th Annual Conference on Computer Graphics and Interactive Techniques (SIGGRAPH 1980), SIGGRAPH Computer Graphics, vol. 14(3), pp. 124–133. ACM, New York (1980)

17. Hjaltason, G.R., Samet, H.: Index-driven similarity search in metric spaces. ACM Transactions on Database Systems (TODS) 28(4), 517–580 (2003)
18. Hochbaum, D.S., Shmoys, D.B.: A best possible heuristic for the k-center problem. Mathematics of Operational Research 10(2), 180–184 (1985)
19. Jagadish, H.V., Ooi, B.C., Tan, K., Yu, C., Zhang, R.: iDistance: An adaptive B+-tree based indexing method for nearest neighbor search. ACM Trans. Database Syst. 30(2), 364–397 (2005)
20. Johnson, R.A., Wichern, D.W.: Applied Multivariate Statistical Analysis, 6th edn. Prentice Hall (2007)
21. Karger, D., Ruhl, M.: Finding nearest neighbors in growth restricted metrics. In: Proceedings of the 34th Annual ACM Symposium on Theory of Computing, pp. 741–750 (2002)
22. Li, J.-Z., Zhang, Z.-G.: Haperplane Tree: A Structure of Indexing Metric Spaces for Similarity Search Queries. Journal of Computer Research and Development 40(8), 1209–1215 (2003)
23. Lokoč, J., Skopal, T.: On applications of parameterized hyper-plane partitioning. Poster in the Proceedings of the Third International Conference on Similarity Search and Applications (SISAP2010), Istanbul, Turkey, September 18-19, pp. 131–132 (2010)
24. Mao, R., Miranker, W., Miranker, D.P.: Pivot Selection:Dimension Reduction for Distance-Based Indexing. Journal of Discrete Algorithms 13, 32–46 (2012)
25. Mao, R., Xu, W., Ramakrishnan, S., Nuckolls, G., Miranker, D.P.: On Optimizing Distance-Based Similarity Search for Biological Databases. In: The 2005 IEEE Computational Systems Bioinformatics Conference (2005)
26. Matousek, J.: Lectures on Discrete Geometry, p. 497. Springer-Verlag New York, Inc. (2002)
27. MoBIoS test suite, http://www.cs.utexas.edu/~mobios/
28. Navarro, G.: Searching in Metric Spaces by Spatial Approximation. In: Proceedings of the String Processing and Information Retrieval Symposium & International Workshop on Groupware. IEEE Computer Society (1999)
29. Samet, H.: Foundations of Multidimensional and Metric Data Structures. Morgan-Kaufmann (2006)
30. Shen, H.T., Ooi, B.C., Zhou, X.: Towards Effective Indexing for Very Large Video Sequence Database. In: SIGMOD Conference 2005, pp. 730–741 (2005)
31. SISAP test suite, http://sisap.org/Metric_Space_Library.html
32. Traina Jr., C., Traina, A.J.M., Seeger, B., Faloutsos, C.: Slim-trees: High performance metric trees minimizing overlap between nodes. In: Zaniolo, C., Grust, T., Scholl, M.H., Lockemann, P.C. (eds.) EDBT 2000. LNCS, vol. 1777, pp. 51–65. Springer, Heidelberg (2000)
33. Uhlmann, J.K.: Satisfying General Proximity/Similarity Queries with Metric Trees. Information Processing Letter 40(4), 175–179 (1991)
34. Venkateswaran, J., Kahveci, T., Jermaine, C.M., Lachwani, D.: Reference-based indexing for metric spaces with costly distance measures. VLDB J. 17(5), 1231–1251 (2008)
35. Xu, W., Miranker, D.P.: A Metric Model of Amino Acid Substitution. Bioinformatics 20(8), 1214–1221 (2004)
36. Yianilos, P.N.: Data structures and algorithms for nearest neighbor search in general metric spaces. In: The Fourth Annual ACM-SIAM Symposium on Discrete Algorithms.Society for Industrial and Applied Mathematics (1993)
37. Zezula, P., Amato, G., Dohnal, V., Batko, M.: Similarity Search: the Metric Space Approach. Springer, Heidelberg (2006)
38. Zhang, Z.-G., Li, J.-Z.: An Algorithm Based on RGH-Tree for Similarity Search Queries. Journal of software 13(10), 1969–1976 (2002)

Improving Performance of Graph Similarity Joins Using Selected Substructures

Xiang Zhao[1], Chuan Xiao[2], Wenjie Zhang[3], Xuemin Lin[3], and Jiuyang Tang[1]

[1] National University of Defense Technology, China
[2] Nagoya University, Japan
[3] The University of New South Wales, Australia
xiangzhao@nudt.edu.cn

Abstract. Similarity join of complex structures is an important operation in managing graph data. In this paper, we investigate the problem of
graph similarity join with edit distance constraints. Existing algorithms
extract substructures – either rooted trees or simple paths – as features,
and transform the edit distance constraint into a weaker count filtering
condition. However, the performance suffers from the heavy overlapping
or low selectivity of substructures. To resolve the issue, we first present
a general framework for substructure-based similarity join and a tighter
count filtering condition. It is observed under the framework that using either too few or too many substructures can result in poor filtering
performance. Thus, we devise an algorithm to select substructures for
filtering. The proposed techniques are integrated into the framework,
constituting a new algorithm, whose superiority is witnessed by experimental results.

1 Introduction

Graph is a universal data structure for describing complex structured data.
Due to the emergence of a wide spectrum of applications, including chem-
informatics and automatic pattern recognition, graph data management system
has received continuous attention lately. Effective and efficient techniques were
studied in many fundamental problems, especially subgraph pattern mining and
matching [3,11].

Due to the existence of errors from various sources, e.g., natural noise and
measurement limit, a recent trend is to study similarity matches to tackle the
issue. We investigate graph similarity joins in this paper. A similarity join finds
all pairs of similar graphs, respectively, from two collections of graphs. It attracts
interests from many domain applications [13]. For instance, a user may want to
join two datasets of unlabeled chemical structures to find the molecules that
belong to the same category of compounds. Among existing graph similarity
measures, graph edit distance (GED) stands out for its elegant properties, being
(1) a metric applicable to all types of graphs, and (2) able to identify structural
differences on both vertices and edges. Consequently, in this paper, we say two
graphs are *similar* if the edit distance between them is no larger than a given
threshold.

S.S. Bhowmick et al. (Eds.): DASFAA 2014, Part I, LNCS 8421, pp. 156–172, 2014.

Nevertheless, verifying GED is difficult (NP-hard [12]). Thus, existing solutions to graph similarity queries employ a *filter–verify* framework, utilizing either tree-based or path-based substructures for filtering. The basic filtering principle is that two graph must share a portion of substructures if they are dissimilar within a certain distance τ. In light of this, count filtering condition is established based on an upper bound of maximum number of substructures affected by τ edit operations; any pair of graph does not meet the condition will not be a result. For tree-based substructures, the notion of κ-AT [8] was proposed – a tree composed of a vertex of the graph and those that can be reached in κ-hops. κ-AT algorithm is associated with the drawback of usually loose lower bound of matching κ-ATs, which tends to be small or even negative. Path-based substructures [13] adopt fixed-length simple paths. While it is experimentally more advanced, the issue of loose lower bound is not solved. Hence, it resorts to other sophisticated filtering for good performance. Another line of studies [9,12] is based on star structure, which is the same structure as 1-hop κ-AT. Unlike count filtering condition, these methods convert the edit distance constraint to a matching distance constraint among the star structures. However, implicit parameters of the algorithms need to be tuned in order to achieve good performance, and the graph edit distance computation was not involved in its evaluation, resulting in unclear overall runtime performances of these methods.

Contributions. This paper tries to address the aforementioned issue.

(1) We review the existing solutions to graph similarity join in-depth, and discuss the underlying reasons behind the issue (Section 2). Given two graphs, the count filtering condition relies on not only the maximum number of substructures affected by τ edit operations, but also the total number of substructures. We identify that having a tight upper bound of the former boosts the filtering capacity; moreover, using all substructures does not necessarily guarantee the best performance.

(2) We put forward a generalized framework for substructure-based graph similarity join (Section 3). The framework encompasses all the substructure-based solutions to the problem. In addition, we re-design the index structure by adding embedding numbers, and explore two perspectives to improve filtering capacity.

(3) We propose a tighter count filtering condition based on an upper bound of the maximum number of substructures affected by τ edit operations (Section 4). We observe that edit operations at different locations of the graph affect different number of substructures. Hence, the filtering condition is strengthened in polynomial time.

(4) We devise a novel technique to further improve the filtering capacity by using selected substructures (Section 5). Due to different selectivities of substructures, the count filter may not work well under the entire substructure space. Thus, we propose an algorithm to choose selected substructures and optimize performance.

(5) Using public real data, we experimentally verify the effectiveness and efficiency of the proposed techniques (Section 6). We find that the proposed tech-

niques effectively tighten the count filtering condition, and improve the filtering capacity and overall performance, on top of the existing solutions.

Related Work. Similarity joins are of importance in numerous domain applications, e.g., set [7] and linked data [14]. We investigate similarity joins on graphs.

Graph similarity queries receive considerable attention lately. Graph edit distance was employed to measure the difference between graphs in [12]. A recent advance was to utilize κ-AT [8] for edit distance based similarity search. It builds an inverted index by decomposing graphs into κ-AT's, and performs filtering by comparing a count filtering based distance lower bound with the threshold. A recent effort SEGOS [9] proposed a novel indexing and query processing framework pertinent to similarity search. GSimJoin [13] is among the first to study graph similarity joins. It defines path-based q-grams to reduce the overlap among substructures. The contribution of this paper is orthogonal to the aforementioned work.

Subgraph similarity search is to retrieve graphs that approximately contain the query. Grafil [10] developed a feature based pruning technique for subgraph similarity search, and similarity is defined as the number of missing edges with respect to maximum common subgraph. GrafD-index [6] exploited effective pruning and validation rules to tackle the problem of connected subgraph similarity search. The problem was also recently studied on single large graph setting, e.g., [2,4].

Another line of related research focuses on *graph edit distance computation*. Currently, the fastest exact solution is an A*-based algorithm incorporating a bipartite heuristic [5]; there are also approximate methods that find suboptimal answers [1].

2   Preliminaries

2.1   Problem Definition

For ease of exposition, we focus on *simple undirected* graphs. A labeled graph g is represented as a triple (V_g, E_g, l_g), where V_g is a set of vertices, $E_g \subseteq V_g \times V_g$ is a set of edges, and l_g is a label function that assigns labels to vertices and edges. $|V_g|$ and $|E_g|$ are the numbers of vertices and edges in g, respectively.

A *graph edit operation* is an edit operation to transform one graph to another [12], including insertion/deletion of an isolated vertex, relabeling of a vertex, insertion of an edge between two disconnected vertices, and deletion/relabeling of an edge.

The *graph edit distance* between g and g', denoted by $\mathrm{GED}(g, g')$, is the *minimum* number of edit operations that transform g to g', or vice versa. It is shown that computing the edit distance between two graphs is NP-hard in general [12].

Example 1. Figure 1 shows the molecular structures of 1,4-dichlorobenzene (g_1) and 1-chloro-2-fluorobenzene (g_2) after omitting hydrogen atoms. For ease of illustration, we add subscripts to atoms of same symbols: for instance, $\mathtt{Cl}_1$ and

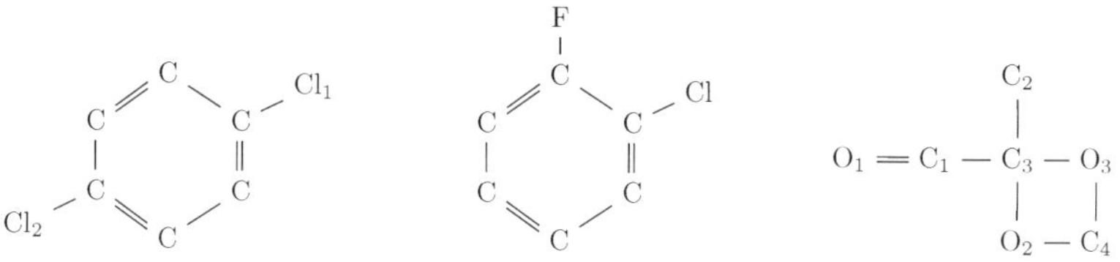

Fig. 1. Sample Graphs

Cl_2 correspond to the same label in real data. Single and double lines indicate different chemical bonds, represented in edge labels in real data. The graph edit distance between g_1 and g_2 is 4 – remove F and its incident edge, and insert Cl and an edge.

Problem 1. Given two collections of graphs R and S, a *graph similarity join* with edit distance threshold τ returns pairs of graphs from each collection such that their GED is no larger than τ; i.e., $\{(r, s) \mid \mathrm{GED}(r, s) \leq \tau, r \in R, s \in S\}$. Assuming there is a unique identifier *id* for each graph, this paper focuses on *self-join* case; i.e., $\{\langle g_i, g_j \rangle \mid \mathrm{GED}(g_i, g_j) \leq \tau \wedge i < j, g_i, g_j \in G\}$, where G is a graph collection.

2.2   Prior Work

Existing solutions to graph similarity queries using substructures were inspired by q-grams (substrings of length q) idea for string similarity queries. Given two similar strings, they must share a certain number of common q-grams, called *matching* q-grams. Tree-based q-gram, a.k.a. κ-ATs, was firstly defined on graphs [8]. For each vertex v in a graph, κ-AT is the breadth-first-search tree of depth κ rooted at v.

Example 2. Consider graph g_1 in Figure 1, and $\kappa = 1$. There are 8 1-ATs of g_1, as shown in Figure 2(a). The trailing numbers indicate the number of embeddings.

We say a κ-AT is *affected* by an edit operations if the edit operation changes the content of the κ-AT. The number of substructures affected by an edit operation is the *edit effect* of the operation. In graph g, the total number of κ-ATs is $|V_g|$, and the maximum number of κ-ATs affected by an edit operation is $D_{\kappa\text{-AT}}(g) = 1 + \gamma_g \cdot \frac{(\gamma_g - 1)^{\kappa} - 1}{\gamma_g - 2}$, where γ_g is the maximum vertex degree in g.

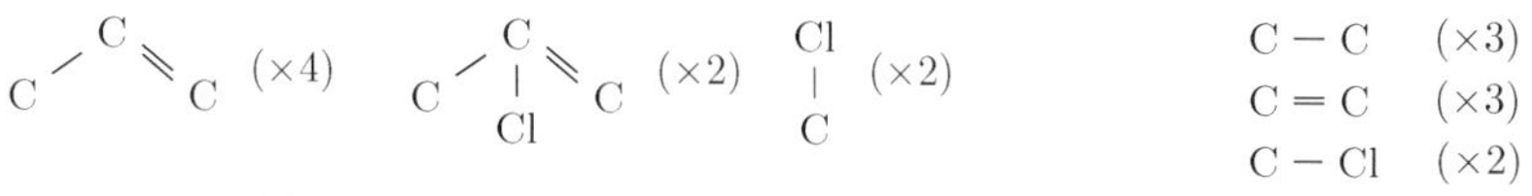

(a) Tree-based Substructures   (b) Path-based Substructures

Fig. 2. Substructures of graph g_1 in Figure 1

Hence, if any graph is within distance τ to g, it must share $LB_{\kappa\text{-AT}}(g)$ matching κ-ATs with g, where $LB_{\kappa\text{-AT}}(g) = |V_g| - \tau \cdot D_{\kappa\text{-AT}}(g)$. Given a pair of graphs g and g', the condition can be applied from g' too. Thus, they must satisfy the larger of the two lower bounds, i.e., $LB_{\kappa\text{-AT}} \triangleq \max(LB_{\kappa\text{-AT}}(g), LB_{\kappa\text{-AT}}(g'))$.

A pair of graphs satisfying the lower bound is called a *candidate* pair. Because it does not necessarily satisfy the edit distance constraint, graph edit distance calculation will be invoked for every candidate pair that survives this *count filtering*.

κ-AT algorithm is associated with the drawback of loose lower bound of matching κ-ATs; $LB_{\kappa\text{-AT}}$ tends to be small or even negative if there is a vertex with high degree or a large distance threshold. This is due to the heavy overlap among κ-ATs, and thus, one edit operation can affect many κ-ATs, rendering $D_{\kappa\text{-AT}}$ large.

κ-ATs are categorized to *tree*-based substructures. There are also path-based q-grams [13]. A path-based q-gram is a simple path of length q. To put it into our framework, we refer it as *path*-based substructure.

Example 3. Consider graph g_1 in Figure 1, and $q = 1$. There are 8 path-based q-grams of g_1, as shown in Figure 2(b).

Accordingly, a count filtering condition LB_{path} was derived on matching path-based q-grams, similar to that of κ-AT. It was experimentally showed that path-based q-grams present tighter count filtering lower bounds than κ-ATs in most cases but not always [13]. Thus, path-based q-gram does not tackle the issue directly. We also observe that the *exponential* number of possible paths in length q and the *low* selectivity of path features may impede the performance.

We are also aware of a star structure-based method for similarity search [9,12], which adopts a disparate filtering scheme based on bipartite matching. We will adapt it to similarity join, and compare with the proposed method in Section 6.

3   Substructure-Based Similarity Join Framework

This section develops a general framework encompassing all the existing substructure-based solutions for graph similarity join, i.e., tree-based and path-based.

By decomposing two graphs into two multisets (or bags) of substructures, a general observation that underlies the prior work is that if two graphs are within a small edit distance, large portion of their *substructure multisets* must overlap.

Formally, S_g denotes the substructure space of g comprising all the unique substructures in g, and there is a universe of substructures S constituted of all the unique substructures of graph g and g' such that $S = S_g \cup S_{g'}$. Accordingly, we use a vector $N_g = \{ n_g^i \}$ to record the numbers of embeddings in g for each substructure $s_i \in S$, $i \in [1, |S|]$. Let $T_S(g) \triangleq \sum_{i=1}^{|S|} n_g^i$ denote the total number of substructures in g, and $D_S(g)$ denote the maximum number of substructures in g that can be affected by *one* edit operations. $D_S(g)$ can be derived by analyzing the number of substructures affected by changing the label of the vertex of largest degree [13]. This is usually done in $O(|\mathcal{S}|)$ as a by-product while decomposing a graph.

Afterwards, for any graph which is within edit distance τ to g, the lower bound of the number of matching substructures between them is

$$LB_{\mathcal{S}}(g) = \begin{cases} T_{\mathcal{S}}(g) - \tau \cdot D_{\mathcal{S}}(g), & \text{if } T_{\mathcal{S}}(g) \geq \tau \cdot D_{\mathcal{S}}(g), \\ \text{invalid}, & \text{otherwise.} \end{cases}$$

Note we refine $LB_{\mathcal{S}}$ by enforcing *non-negativity*, since negative lower bounds are invalid, having no filtering capacity. We omit the subscript $\mathcal{S}$ when context is clear.

Lemma 1. *If the number of matching substructures between graphs g and g' is less than $LB(g)$, g and g' are determined to be not within edit distance threshold τ.*

Proof. (sketch) Assume all τ edit operations affect $D(g)$ (distinct embeddings of) substructures, respectively. The total number of substructures affected by τ operations is $\tau \cdot D(g)$, and hence, the number of substructures remaining unaffected is the total number of substructures excluding $\tau \cdot D(g)$, i.e., $T(g) - \tau \cdot D(g)$. Any graph similar to g must share at least this number of substructures; otherwise, it requires at least $\tau + 1$ operations to produce more mismatching substructures.

The lemma converts the distance constraint into a numerical condition on matching substructures. Since there exist multiple embeddings of a substructure corresponding to different parts of a graph, to obtain a tight condition, we enforce *one-to-one* match. For $s_i \in \mathcal{S}$, if there are n_g^i embeddings in g and $n_{g'}^i$ in g', respectively, the number of matching substructures at s_i is $\min(n_g^i, n_{g'}^i)$. Thus, the total number matching substructures between g and g' is $\Sigma_{i=1}^{|\mathcal{S}|} \min(n_g^i, n_{g'}^i)$.

To facilitate such filtering, we rely on an inverted index. Particularly, each index entry corresponds to a substructure in the universe $\mathcal{U}$, containing all possible substructures generated from the database G. For each index entry $s_i \in \mathcal{U}$, it has a postings list such that a posting is a tuple $\langle g_j, n_j^i \rangle$, where g_j is a graph in G having s_i as a substructure, and n_j^i is the number of embeddings of s_i in g_j.

Example 4. Consider graphs in Figure 1, and the path-based substructures in Figure 2(b) as s_1 to s_3 from top to bottom. Figure 3 depicts a partial inverted index; for instance, s_2 is contained in g_1 and g_2 with both 3 embeddings, respectively.

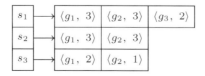

Fig. 3. Partial Index for Substructures in Example 3

Next, we describe the framework in Algorithm 1, which builds the index on-the-fly and executes self-joins. It takes as input a graph collection, and a distance threshold, and outputs the graph pairs conforming the constraint. For each graph g_j, we initialize a map M of size $O(|G|)$ to record the number of matching substructures for each g_i $(1 \leq i < j)$. Thus, we only join g_j with graphs having smaller identifiers. We construct the substructure space $\mathcal{S}_{g_j}$ of g_j, generate tuples $\langle s, n \rangle$ for each substructure in $\mathcal{S}_{g_j}$, and store them in SP (Line 4). Afterwards, for each $\langle s, n \rangle$ in SP, we iterate through the corresponding entry of s (Lines 7 – 10). For each $\langle g_i, m \rangle$ in I_s, we either conduct size filtering between g_j and g_i if $M[g_i]$ is uninitialized (Line 9), or increase $M[g_i]$ by the number of matching substructures $\min(m, n)$ (Line 10). The current tuple $\langle s, n \rangle$ is indexed for g_j in the end of the postings list of s (Line 11). In the end, we verify each g_i such that $M[g_i]$ satisfies the count filtering condition (Lines 12 – 13).

Algorithm 1. SubstructureBasedJoin (G, τ)

Input   : G is a collection of graphs; τ is a distance threshold.
Output : $A = \{ \langle g_i, g_j \rangle \mid \text{GED}(g_i, g_j) \leq \tau \}$, initialized as $\emptyset$.
1 $I_k \leftarrow \emptyset$ $(1 \leq k \leq |\mathcal{U}|)$;   /* inverted index */
2 **foreach** $g_j \in G$ $(j \in [1, |G|])$ **do**
3 $\quad$ $M \leftarrow$ empty map from identifier to **integer**;
4 $\quad$ $SP \leftarrow$ generate tuples $\{ \langle s, n \rangle \}$ for each substructure in $\mathcal{S}_{g_j}$;
5 $\quad$ **for** $k = 1$ to $|\mathcal{S}_{g_j}|$ **do**
6 $\quad\quad$ $s \leftarrow SP[k].s$;
7 $\quad\quad$ **foreach** $\langle g_i, m \rangle \in I_s$ **do**
8 $\quad\quad\quad$ **if** $M[g_i]$ is uninitialized **then**
9 $\quad\quad\quad\quad$ **if** $\text{abs}(|V_{g_i}| - |V_{g_j}|) + \text{abs}(|E_{g_i}| - |E_{g_j}|) > \tau$ **then** $M[g_i] \leftarrow -1$
10 $\quad\quad\quad$ **else** $M[g_i] \leftarrow M[g_i] + \min(m, n)$
11 $\quad\quad$ $I_s \leftarrow I_s \cup \{ \langle g_j, SP[k].n \rangle \}$;   /* index current $\langle s, n \rangle$ for g_j */
12 $\quad$ **foreach** $M[g_i]$ such that $M[g_i] \geq LB(g_i, g_j)$ **do**
13 $\quad\quad$ **if** Verify$(g_i, g_j) = $ **true then** $A \leftarrow A \cup \{ \langle g_i, g_j \rangle \}$

---

14 **return** A

Correctness and Complexity Analysis. Based on Lemma 1, it is immediate that Algorithm 1 correctly computes the results. Apart from verification, the time complexity of indexing and filtering is $O(L \cdot |\mathcal{S}_g| \cdot |G|)$, where L is the average length of $|\mathcal{U}|$ lists, and $|\mathcal{S}_g|$ is the average number of substructures in a graph.

The performance of Algorithm 1 largely relies on the power of count filters. Intuitively, we desire a tight filtering condition so that more candidates can be pruned. Observe that the filtering condition depends on not only the total number $T(g)$ but also the maximum edit effect $D(g)$. We put forward the following equation

$$\alpha(g) \triangleq \frac{LB(g)}{T(g)} = \frac{T(g) - \tau \cdot D(g)}{T(g)} = 1 - \frac{\tau \cdot D(g)}{T(g)}. \qquad (1)$$

to assist the analysis, when $LB(g)$ is not invalid. Particularly, the higher value of Equation (1), the more matching substructures required regarding the substructure multiset, the stronger constraint on the common structure, the tighter count filtering, and hence, the better filtering capacity, the better runtime performance.

Given a fixed substructure space, one way to increase the value of Equation (1) is to reduce $D(g)$. Recall that the lower bound employed in Lemma 1 is directly adapted from the count filtering condition for q-grams on strings, i.e., the difference between the total number of q-grams and τ times the maximum number of q-grams affected by one edit operation. Since the number of string q-grams affected by one edit operation – insertion, deletion and substitution (unit cost) – is always q, the lower bound on strings is straightforward. However, cases become intriguing on graphs, as different edit operations affect various number of substructures, and moreover, edit operations invoked at different locations of the graph have disparate effects. In light of this, we develop a tighter condition in Section 4.

Afterwards, one may conclude that, for a given choice of substructure, e.g., tree or path, the filtering performance could not be improved further, since $T(g)$ and $D(g)$ are fixed then. However, we have not explored the opportunity of composing a filter based on selected substructures, rather than include all of them. An interesting question is: *does a filter achieve good filtering performance if not all of the substructures are used simultaneously?* A seemingly attractive intuition is that, the more substructures are used, the greater pruning power is achieved. After all, we are utilizing more information provided by g_j to compare with g_i's. Unfortunately, though a bit counterintuitive, using all substructures together will not necessarily give the optimal filtering solution such that Equation (1) is maximized. Further, chances are that this may even impede the performance. In Section 5, we will investigate the principles behind this phenomenon, and then devise a solution.

4   Tightening Count Filtering Condition

We first look at an example to illustrate the idea.

Example 5. Consider graph g_3 in Figure 1 adopting path-based substructures with length 1, and $\tau = 2$. The total number of substructures is 7; and we derive $D(g_3)$ as 4, since relabeling vertex C_3 affects 4 path-based substructures. Consequently, the count filter becomes invalid, due to $T(g_3) \le \tau \cdot D(g_3)$. We observe that there is only 1 vertex, relabeling which affects 4 substructures. The operation affecting the second largest number of substructures is relabeling C_1

(or C_4, 0_2, 0_3), which affects 2 substructures. Thus, $\tau \cdot D(g_3)$ overestimates the number of affected substructures.

The example evidences that the assumption of *maximum* edit effect of all edit operations is unlikely to be achieved on graphs. Nevertheless, the existing solutions take this for granted, and have not developed the count filtering condition pertinent to graphs. We address the issue below.

4.1   A Tighter Lower Bound

Since the scenario that τ edit operations all incur the maximum edit effect is rare, adopting D for all τ edit operations results in a usually loose count filtering lower bound. Instead, we can enumerate τ edit operations with top-τ edit effects to tighten the lower bound, and filter out unpromising candidates.

Formally, $D^\tau(g)$ denotes the maximum number of substructures in g affected by τ edit operations. We discuss the computation of $D^\tau(g)$ shortly. For any graph within distance τ to g, the lower bound of the number of matching substructures is

$$LB^+(g) = \begin{cases} T(g) - D^\tau(g), & \text{if } T(g) \geq D^\tau(g), \\ \text{invalid}, & \text{otherwise.} \end{cases}$$

Lemma 2. *If the number of matching substructures between graphs g and g' is less than $LB^+(g)$, g and g' are determined to be not within edit distance threshold τ.*

Proof. The proof is by an argument similar to that of Lemma 1.

The major difference between Lemmas 1 and 2 is that we replace the original lower bound LB with LB^+. In other words, based on the observation of non-uniform edit effects on graphs, $\tau \cdot D$ is substituted by D^τ, and in this way, a tighter count filtering condition is obtained. Accordingly, Equation (1) is updated to

$$\alpha^+(g) \triangleq \frac{LB^+(g)}{T(g)} = \frac{T(g) - D_\tau(g)}{T(g)} = 1 - \frac{D_\tau(g)}{T(g)}. \tag{2}$$

Example 6. The 2 edit operations with top-2 edit effects are relabelling vertices C_3 and C_1 (or C_4, 0_2, 0_3), which affect 4 and 2 substructures, respectively. As a consequence, $LB^+(g_3) = 7 - (4+2) = 1$, and $\alpha^+ = 1/7$. Recall that in Example 5 $LB(g_3)$ and $\alpha(g)$ are both invalid, and the count filter is furtile.

4.2   Algorithm

Thus far, we have not discussed how to compute the new lower bound. There exists no closed-form formula for D^τ. Following first shows a result to simplify the computation of D^τ (and LB^+), based on which we present the algorithm.

Algorithm 2. ComputeLB$^+$ $(g, \tau, \mathcal{S}_g)$

    Input   : g is a graph; $\mathcal{S}_g$ is the substructure space of g; τ is a distance
              threshold.
    Output : $LB^+(g)$.
  1 get edit effects F_g of all vertex relabeling under $\mathcal{S}_g$;   /* done while
    extracting substructures */
  2 sort F_g in descending order of edit effects;
  3 $D^\tau(g) \leftarrow 0$;
  4 for $i = 1$ **to** τ **do**
  5 $\llcorner$ $D^\tau(g) \leftarrow D^\tau(g) + |F_g^i|$;
  6 return $|S_g| - D^\tau(g)$

Lemma 3. *Consider τ edit operations of vertex relabelling with top-τ edit effects. D^τ equals the number of substructures affected by these edit operations.*

Proof. (sketch) It is easy to verify that for a specific vertex in a graph, vertex relabelling is the single edit operation that affects the maximum number of substructures. Moreover, the edit effect of any other edit operations can be instantiated (or covered) by a certain vertex relabelling. Thus, the edit effect any τ edit operations can be instantiated (or covered) by vertex relabelling on τ vertices. To maximize this effect for an upper bound, we choose the vertices with top-τ edit effects of vertex relabelling. Hence, the correctness follows.

Based on Lemma 3, Algorithm 2 describes the computation of LB^+. It takes as input a graph, distance threshold, and a set of substructures, to produces $LB^+(g)$. Particularly, we first compute the edit effects of all vertex relabeling, and sort them in descending order of their cardinalities (Lines 1 − 2). Then, $D^\tau(g)$ is initialized as 0, and increased by the cardinalities of first τ edit effects in F_g (Lines 3 − 5).

Correctness and Complexity Analysis. The evaluation above never underestimates the real number of substructures affected by τ edit operations. This guarantees that any algorithm relies on the filtering principle will not miss results. The algorithm is of $O(\tau \cdot |V_g| \cdot \log|V_g|)$ time, since the total number of edit effects in F_g is $|V_g|$. Additionally, the complexity of the sort operations can be reduced to $O(\tau \cdot \log|V_g|)$, if we have employed a priority queue to record the edit effect sizes while extracting the substructures from a graph (Line 4 in Algorithm 1).

    LB^+ is tight in the sense that it is reachable when τ edit operations are invoked exactly on the locations of vertex relabelling occur, and hence, it cannot be reduced anymore in this case. Further, LB^+ is, if not tighter, as tight as LB; it deteriorates to LB only under the aforementioned scenario. To integrate Algorithm 2 into the framework, we replace Line 12 in Algorithm 1 with "**foreach** $M[g_i]$ such that $M[g_i] \geq LB^+(g_i, g_j)$ **do**", where $LB^+(g_i, g_j) \triangleq \max(LB^+(g_i), LB^+(g_j))$.

5   Improving Performance with Selected Substructures

In previous sections, both LB and LB^+ are applied on all the substructures of a graph, i.e., the entire substructure space. In this section, we investigate a novel idea to improve the filtering performance using a selected portion of the substructures in a graph, i.e., a substructure subspace. We start with a motivating example.

Example 7. Consider graphs in Figure 4, and $\tau = 1$. The count filtering lower bound on g_1 is $LB^+(g_1) = 7 - 4 = 3$. It is easy to verify that g_2 passes the filter, since they matches 4 substructures, i.e., C = O and C - H ($\times 3$). However, we can use the substructures excluding C - H (bounded by dashed lines), to prune g_2. Let the remaining subgraphs be g_1' and g_2', respectively. Apply the tight count filtering on the subgraphs, $LB^+(g_1', g_2') = 2$, while they only match 1 substructures. Thus, we are able to prune g_1 and g_2, without considering the excluded parts.

We discuss the correctness and the choice of selected substructure space below.

5.1   Count Filtering with Selected Substructure

We observe that vertices with large edit effects are those with large vertex degrees, and there are some frequent substructures (e.g., C - H in chemical data) associated with vertices of large degrees. Due to this, under the entire substructure space, top-τ edit effects are usually large. Nonetheless, those frequent substructures are unlikely to contribute mismatches. In order to not miss results, we cannot avoid such phenomenon. In contrast, if we can exclude these substructures when applying count filtering, we expect smaller edit effects under the subspace of selected substructures, and hence, tighter lower bound with greater filtering capacity.

Consider a subspace $\mathcal{S}'$ with respect to the entire substructure space $\mathcal{S}$. Let $T_{\mathcal{S}'}(g) = \sum_{i=1 \wedge s_i \in \mathcal{S}'}^{|\mathcal{S}|} n_g^i$ denote the total number of substructures in g such that the substructure is in $\mathcal{S}'$, and $D_{\mathcal{S}'}^\tau(g)$ be the maximum number of such substructures in g affected by τ edit operations. For any graph within edit distance τ to g, the lower bound of the number of matching substructures in the subspace $\mathcal{S}'$ is

$$LB_{\mathcal{S}'}^+(g) = \begin{cases} T_{\mathcal{S}'}(g) - D_{\mathcal{S}'}^\tau(g), & \text{if } T_{\mathcal{S}'}(g) \geq D_{\mathcal{S}'}^\tau(g), \\ \text{invalid}, & \text{otherwise.} \end{cases}$$

Lemma 4. *Consider a subspace $\mathcal{S}'$ with respect to the entire substructure space $\mathcal{S}$. If the number of matching substructures in $\mathcal{S}'$ between g and another graph g' is less than $LB_{\mathcal{S}'}^+(g)$, the edit distance between g and g' is not within the threshold τ.*

Proof. (sketch) We construct a graph sg from g such that only the vertices and edges appearing in the embeddings of substructures in $\mathcal{S}'$ are retained.

Immediate is that sg is a subgraph of g, denoted $sg \subseteq g$. Similarly, we have $sg' \subseteq g'$. Based on Lemmas 2 and 3, if the number of matching substructures between sg and sg' is less than $LB_{S'}^{+}(sg, sg')$, $\text{GED}(sg, sg') > \tau$. Afterwards, any substructure in $S \setminus S'$ will only introduce new edit errors; the best case is that $\forall s_i \in S \setminus S', n_g^i = n_{g'}^i$, which brings 0 edit error. Therefore, $\text{GED}(g, g')$ must be no less than $\text{GED}(sg, sg')$; the equality holds only when $g \setminus sg$ is graph isomorphic to $g' \setminus sg'$, where $g \setminus sg$ denotes the subgraph of g excluding sg. Hence, the lemma follows.

Note Lemma 2 is a special case of Lemma 4 when $S' = S$. Nonetheless, as demonstrated, using all substructures does not necessarily filter the most candidates. Next, we incorporate the idea of selected substructures to strengthen count filters.

5.2   Optimal Substructure Set

We formulate the substructure subspace selection as an optimization problem.

Problem 2. Consider a substructure space S. Find the substructures subspace S' for g such that Equation (2) is maximized.

By maximizing the quotient, the target is that under the framework, for each graph g_j, the least number of candidates with identifier smaller than g_j passing the filters. However, choosing the set of substructures for best performance is difficult.

Theorem 1. *Consider a graph g and a substructures space S. In the worst case, it takes $\Omega(2^{|S|})$ steps to compute the optimal set of substructures.*

Proof. We omit the proof in the interest of space.

In practice, we are interested in the heuristics that are good for a large number of graphs in question. Intuitively, we rely on the following definition to measure the filtering capacity of a substructure s for all graphs in the database.

Given a collection G, graph g, and substructure s, the *selectivity* of s is defined as

$$\delta_s(G, g) = \text{avg}(|n_{g'}^s - n_g^s|)/\max(deg(v)),\ g' \in G,\ v \in V_s \in V_g,$$

where $n_{g'}^s$ is the number of embeddings of s in g'. The larger value $\delta_s(G, g)$, the more selective the substructure. Before elaborating the algorithm, we first conceptualize some general principles that provide guidance on selection.

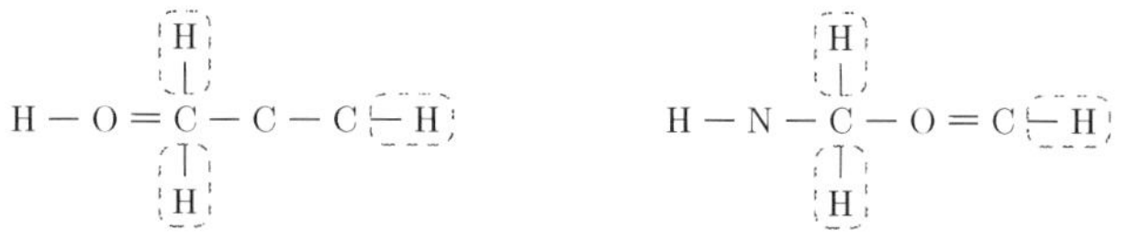

Fig. 4. Example of Leveraging Selected Substructures

- Select many but not too many substructure;
- Prefer substructures of high selectivity; and
- Ensure substructures cover g uniformly.

The first principle is necessary, since a too small substructure subspace implies little structural information for pruning. If too large, the maximum affected substructures by τ edit operations may become large. In that case, the filtering algorithm loses its pruning power. The second is more intuitive, since high selective substructures lead to fewer candidates. In particular, we prefer infrequent substructures with small vertex degrees. The third is more subtle than the previous. If most substructures cover several common vertices, using a few edit operations can affect these substructures. Additionally, this also enable a full reflection of structural characteristics of g. Unfortunately, these three criteria are not consistent with each other. For instance, if we pick only $\frac{T(g)}{2}$, roughly half of g is not represented by the substructures. Hence, the graphs having edit errors regarding g in the excluded part will not be identified by the selected substructures. On the other hand, we cannot use the most selective substructures alone, as we could have only a few of them in g, and some edit errors may not appear on these highly selective substructures.

5.3   Algorithm

We devise a simple heuristic algorithm in Algorithm 3 trying to capture the principles. It takes as input a graph, a substructures space, and the distance threshold, and outputs a selected subspace of substructures. Specifically, we first sort the substructures in ascending order of their selectivities (Line 1). Let LB_{max} record the maximum $\alpha^+(g)$ so far, initialized 0. Then, we go through the ordered substructures and pick a subset (Lines 3 – 6). Considering S as an ordered set, $S(i)$ denotes the first i substructures in S. During the sequential scan of S, we consider all choices conforming $S_g(i) = S_g(i-1) + S_g[i]$, and choose $S(i)$ of the best value via Equation (2), which is returned by S'_g eventually.

Algorithm 3. SelectSubstructure (g, S_g, τ)

> **Input**   : g is a graph; S_g is the set of substructures of g; τ is a distance threshold.
> **Output** : S'_g is a set of selected substructures, initialized as $\emptyset$.
> 1 sort S_g in ascending order of selectivity;
> 2 $\alpha^+_{max}(g) \leftarrow 0$;
> 3 **for** $i = 1$ to $|S_g|$ **do**
> 4 $\quad LB^+_{S(i)}(g) = $ ComputeLB$^+(g, \tau, S(i))$;
> 5 $\quad \alpha^+_{S(i)}(g) \leftarrow LB^+_{S(i)}(g)/T_{S(i)}(g)$;
> 6 $\quad$ **if** $\alpha^+_{S(i)}(g) \geq \alpha^+_{max}(g)$ **then** $\alpha^+_{max}(g) \leftarrow \alpha^+_{S(i)}(g)$, $S'_g \leftarrow S(i)$
> 7 **return** S'_g

Currentness and Complexity Analysis. It is immediate that Algorithm 3 computes a subspace from $\mathcal{S}$, and it has the highest value of Equation (2) among all choices of $\mathcal{S}_g(i)$, $i \in [1, |\mathcal{S}_g|]$. It is in $O(|\mathcal{S}_g| \cdot |V_g| \cdot \log |V_g|)$ time.

Composing substructure subspaces by adding elements not only reduces the search space heuristically but also partially reflects the first principle. As the substructures are sorted in ascending order of selectivity, front substructures have better pruning power in general than those behind. Since selectivity is factorized by frequency and degree, this heuristic also reflects the second and third principles.

6   Experimental Study

6.1   Experiment Setup

We conducted experiments on both real and synthetic datasets:

- **AIDS** is an antivirus screen compound dataset from the Developmental Therapeutics Program, containing 42,687 chemical compound structures.
- **PROT** comprises 600 protein structures from the Protein Data Bank. Vertices are types of secondary structure elements; edges are lengths in amino acids.

Table 1 lists the dataset statistics. PROT is denser with less unique labels than AIDS.5, 000 graphs were randomly sampled from AIDS to make up the database.

Experiments were run on a machine of Quad-Core AMD Opteron 8378@800MHz with 96G RAM under Ubuntu. All the algorithms were implemented in C++. We measured (1) index size; (2) number of candidates that need further verification; (3) running time, including indexing, candidate generation and GED computation.

Table 1. Dataset Statistics

Dataset	$|G|$	avg $|V|/|E|$	$|l_V|/|l_E|$
AIDS	5, 000	25.60 / 27.60	62 / 3
PROT	600	32.63 / 62.14	3 / 5

Table 2. Comparing Index Size (kB)

Dataset	κ-AT-Join	SEGOS-Join	Select-Tree
AIDS	183.36	527.75	326.17
PROT	74.85	158.28	107.63

6.2   Evaluating Proposed Techniques

We first verify the effectiveness of the proposed techniques. This set of experiments are demonstrated by path-based substructures, and path length was set to 4 on AIDS and 3 on PROTEIN, respectively. The basic algorithm under the framework in Algorithm 1 using path-based substructures is denoted as "Basic-Path" or "BP"; adopting the tight count filtering produces "Tight-Path" or "TP", and further utilizing selected substructures constitutes "Select-Path" or "SP".

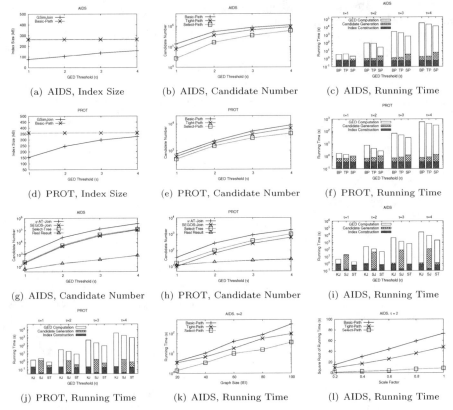

(a) AIDS, Index Size   (b) AIDS, Candidate Number   (c) AIDS, Running Time

(d) PROT, Index Size   (e) PROT, Candidate Number   (f) PROT, Running Time

(g) AIDS, Candidate Number   (h) PROT, Candidate Number   (i) AIDS, Running Time

(j) PROT, Running Time   (k) AIDS, Running Time   (l) AIDS, Running Time

Fig. 5. Experiment Results

We plot the results on AIDS in Figures 5(a) – 5(c). Since all the three algorithms rely on the same index, they have the same index size; thus, we put the index size of GSimJoin as reference, which also uses path-based substructures. We observe that the index size of Basic-Path is not influenced by the threshold, while that of GSimJoin grows according to τ due to the choice of minimum prefix length for indexing. As to candidates number, Tight-Path produces up to 36.2% less than Basic-Path, and Select-Path further improves it to 70.1%. Thus, the total running time of the algorithms becomes expectable, as in Figure 5(c). The three algorithms spend the same amount of time on index construction, but Select-Path needs slightly more time for candidate generation, compared with Basic-Path and Tight-Path. Nevertheless, this effort is rewarding in the verification phase, as GED computation is rather time-consuming. Thus, Select-Path provides the best overall performance.

Similar trends can be observed on PROT, as shown in Figures 5(d) – 5(f). We highlight some results below: (1) The index size of GSimJoin is more close to Basic-Path, as the minimum prefix length is less effective on dense graphs in PROT. (2) Select-Path has smaller reduction ratio of candidate number on PROT, compared with that on AIDS. (3) The GED computation per graph pair on PROT takes longer than that on AIDS. In the following subsection, we applied all the proposed techniques when comparing with others existing algorithms.

6.3 Comparing with Existing Algorithms

This set of experiments demonstrate the performance improvement over the existing solutions of tree-based substructures. Following algorithms were involved:

- **Select-Tree** , labeled "ST", is an algorithm with all proposed techniques under our framework using tree-based substructures.
- κ-**AT-Join** , labeled "KJ", is an algorithm using κ-ATs, essentially "Basic-Tree" under our framework. κ was set to 1 to achieve the best performance.
- **SEGOS-Join** , labeled "SJ", is adapted from SEGOS. In order to make SE-GOS support self-joins, we ran SEGOS in an index nested loops join mode. It iterates through the dataset, and selects each graph as a query with the corresponding database contains all the graphs with smaller identifiers than that of the query.

We first compare the index size in Table 2. κ-AT-Join constructs the smallest index, and **Select-Tree** is about twice larger. The advanced index by **SEGOS-Join** consumes the largest space. Figures 5(g) – 5(h) compare the candidate number on AIDS and PROT, respectively. On both datasets, **SEGOS-Join** produces the smallest candidate sets, followed by **Select-Tree** and κ-AT-Join. However, this is achieved by sacrificing the overall runtime performance. As seen in Figures 5(i) – 5(j), **Select-Tree** always consumes the least time; in comparison, **SEGOS-Join** builds the index efficiently, but takes longer for candidate generation and GED computation. Specifically, **SEGOS-Join** does not show advantage over the other two when $\tau = 1$, as it needs longer filtering time for candidate generation. For **Select-Tree**, we observe a speedup of as much as 3.5x and 4.7x over κ-AT-Join on AIDS and PROT, respectively, 2.0x and 2.4x over **SEGOS-Join**, respectively.

6.4 Evaluating Scalability

This set of experiments evaluate the scalability of the proposed techniques against graph size ($|E|$) and dataset cardinality ($|G|$) at $\tau = 2$. We first randomly selected five sets of 100 graphs from AIDS such that the average size of graphs is in $\{20, 40, 60, 80, 100\}$ and the variation is at most 5; e.g., the size of graphs in the first set is within $[15, 25]$. Then, we expand the datasets to 5000, respectively, by generating similar graphs via randomly applying $[0, 5]$ edit operations to the original graphs. The result is shown in Figure 5(k). We observe that the running time increases steadily towards large graphs. The proposed techniques scale well with graph size, with **Select-Path** being the best. The margin is more remarkable on larger graphs, as we can choose a substructure subspace from a wider spectrum.

The second experiment was conducted on portions of AIDS, with the scale factor ranging in $\{0.2, 0.4, 0.6, 0.8, 1\}$. We plot the results in Figure 5(l), where the y-axis is set as the square root of running time. It is observed that the running time grows quadratically with the increase of data set cardinality. In particular, Select-Path performs the best among the three, up to 10.4x faster than Basic-Path, and Tight-Path comes at the second place, up to 4.1x faster than Basic-Path.

7   Conclusion

In this paper, we have investigated graph similarity joins with edit distance constraints. By conceiving a general framework for substructure-based methods, we first tightened the count filtering condition, and further strengthened it using selected substructures. The performance of the algorithms was empirically verified.

Acknowledgement. This work is in part supported by the Research Fund for Doctoral Program of Higher Education of China No. 20114307110008 and NSFC No. 61302144.

References

1. Fankhauser, S., Riesen, K., Bunke, H.: Speeding up graph edit distance computation through fast bipartite matching. In: Jiang, X., Ferrer, M., Torsello, A. (eds.) GbRPR 2011. LNCS, vol. 6658, pp. 102–111. Springer, Heidelberg (2011)
2. Hung, H., Bhowmick, S., Truong, B., Choi, B., Zhou, S.: QUBLE: towards blending interactive visual subgraph search queries on large networks. The VLDB Journal, 1–26 (2013)
3. Kang, U., Tong, H., Sun, J., Lin, C.-Y., Faloutsos, C.: gbase: an efficient analysis platform for large graphs. VLDB J. 21(5), 637–650 (2012)
4. Khan, A., Wu, Y., Aggarwal, C.C., Yan, X.: NeMa: Fast graph search with label similarity. PVLDB 6(1), 181–192 (2013)
5. Riesen, K., Fankhauser, S., Bunke, H.: Speeding up graph edit distance computation with a bipartite heuristic. In: MLG (2007)
6. Shang, H., Lin, X., Zhang, Y., Yu, J.X., Wang, W.: Connected substructure similarity search. In: SIGMOD Conference, pp. 903–914 (2010)
7. Vernica, R., Carey, M.J., Li, C.: Efficient parallel set-similarity joins using mapreduce. In: SIGMOD Conference, pp. 495–506 (2010)
8. Wang, G., Wang, B., Yang, X., Yu, G.: Efficiently indexing large sparse graphs for similarity search. IEEE Trans. Knowl. Data Eng. 24(3), 440–451 (2012)
9. Wang, X., Ding, X., Tung, A.K.H., Ying, S., Jin, H.: An efficient graph indexing method. In: ICDE, pp. 210–221 (2012)
10. Yan, X., Yu, P.S., Han, J.: Substructure similarity search in graph databases. In: SIGMOD Conference, pp. 766–777 (2005)
11. Yuan, D., Mitra, P., Giles, C.L.: Mining and indexing graphs for supergraph search. PVLDB 6(10), 829–840 (2013)
12. Zeng, Z., Tung, A.K.H., Wang, J., Feng, J., Zhou, L.: Comparing stars: On approximating graph edit distance. PVLDB 2(1), 25–36 (2009)
13. Zhao, X., Xiao, C., Lin, X., Wang, W., Ishikawa, Y.: Efficient processing of graph similarity queries with edit distance constraints. The VLDB Journal, 1–26 (2013)
14. Zheng, W., Zou, L., Feng, Y., Chen, L., Zhao, D.: Efficient simrank-based similarity join over large graphs. PVLDB 6(7), 493–504 (2013)

Linear Path Skyline Computation
in Bicriteria Networks

Michael Shekelyan, Gregor Jossé, Matthias Schubert, and Hans-Peter Kriegel

Institute for Informatics, Ludwig-Maximilians-Universität München, Oettingenstr. 67,
D-80538 Munich, Germany
{shekelyan,josse,schubert,kriegel}@dbs.ifi.lmu.de

Abstract. A bicriteria network is an interlinked data set where edges
are labeled with two cost attributes. An example is a road network where
edges represent road segments being labeled with traversal time and energy consumption. To measure the proximity of two nodes in network
data, the common method is to compute a cost optimal path between
the nodes. In a bicriteria network, there often is no unique path being optimal w.r.t. both types of cost. Instead, a path skyline describes
the set of non-dominated paths that are optimal under varying preference functions. In this paper, we examine the subset of the path skyline
which is optimal under the most common type of preference function,
the weighted sum. We will examine characteristics of this more strict
domination relation. Furthermore, we introduce techniques to efficiently
maintain the set of linearly non-dominated paths. Finally, we will introduce a new algorithm to compute all linearly non-dominated paths
denoted as linear path skyline. In our experimental evaluation, we will
compare our new approach to other methods for computing the linear
skyline and efficient approaches to compute path skylines.

1 Introduction

In recent years graph data has gained much importance in numerous information management systems. For example, spatial databases no longer rely on free
movement and simple distance measures but more restrictive movement patterns, e.g. along streets or railways being modelled as a network. Other applications for graph data include social networks, computer networks or the world
wide web itself. In all these networks it is often of interest to determine cost-
optimal paths between nodes in order to determine connections or to simply
measure the distance. The cost for traversing an edge usually depends on the
application, e.g. the number of common friends in social networks or the energy consumption within a street network. Using a single cost criterion is often
strongly restrictive. For example, when planning a bicycle route, minimizing only
the distance is usually not sufficient. To determine the difficulty of a route, considering the ascension of a path is essential. In order to employ multiple criteria
when computing optimal paths, a straight forward way is to combine the cost
criteria into a single value and optimize the combined costs. However, finding a

S.S. Bhowmick et al. (Eds.): DASFAA 2014, Part I, LNCS 8421, pp. 173–187, 2014.
© Springer International Publishing Switzerland 2014

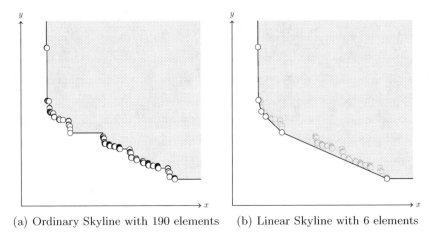

(a) Ordinary Skyline with 190 elements   (b) Linear Skyline with 6 elements

Fig. 1. The Linear Skyline consists of a subset of the elements of the Ordinary Skyline

meaningful cost combination means weighting the impact of each cost criterion. This usually depends on personal preferences, the semantics of the application as well as the scale of the cost criteria. Hence, finding an appropriate weighting is often very difficult. An alternative way to cope with multiple cost criteria is to determine all results that could potentially be optimal under an arbitrary cost combination. In the database community this type of query is known as skyline query [1]. The result set of a skyline query contains elements which are not dominated in the following sense: Object o is dominated by object q if for each cost criterion the cost of o is equal or larger than the cost of q, and for at least one criterion the cost of q is strictly smaller than the cost of o. This means, there is no non-negative cost combination where p would be preferable to q. The task of computing a skyline of paths in a network is called Route Skyline Computation [2] or Multiobjective Shortest Path Search [3].

Though posing a skyline query does not require the explicit specification of attribute weights, it still yields two practical problems limiting its usability. The number of non-dominated solutions typically grows superlinear in the number of criteria. In consequence, there often is an abundance of results having very similar costs. Furthermore, for more than two cost criteria, it is hard to determine the conditions and the combination functions which would lead to the optimality of a particular element of the skyline. In order to keep the result set easier to interpret and the amount of potentially optimal paths on a moderate level, we will focus on the basic case of optimizing two cost criteria. This case also exhibits mathematical benefits.

To reduce the amount of results to a more intuitive set, we propose the novel concept of linear skyline queries. A linear skyline consists of the subset of the conventional skyline which is optimal under all linear combination functions. This concept is introduced more formally in Section 3 and visualized in Figure 1. The linear skyline contains fewer and more diverse results. Also, the characteristics of both skyline notions differ greatly. In contrast to conventional domination, linear domination is defined through sets of objects. In the course of this paper, we will introduce and discuss the concept and the computation of linear skylines.

Though the task of computing a linear skyline itself is novel, there exist approaches which employ the concept as a preprocessing step towards computing conventional skylines. In contrast to existing approaches (cf. Section 2), we propose a two-step method which is based on the idea of label correction for computing the conventional skyline. Our novel method is mainly based on new techniques to efficiently manage linearly non-dominated solutions, allowing efficient updates and deletions. To summarize, the contributions of the paper are the following: the introduction of linear skyline computation as a novel type of query within a network, a discussion of the properties of linear skylines, and an efficient algorithm that exploits the more restrictive dominance relation.

The rest of the paper is organized as follows: In Section 2, we discuss related work. The general setting and formalizations are given in Section 3. Section 4 describes our new algorithm for computing the linear path skyline. In Section 5, we compare our new algorithm to methods that can be adapted to compute linear skylines. Section 6 summarizes the paper and presents some ideas for future work.

2   Related Work

In the database community, the skyline operator was introduced in [1]. Multiple approaches to compute skylines in database systems on sets of cost vectors followed [4–6]. Also, different notions of dominance have been presented. In [7] k-dominant skylines are proposed which generalize the dominance relationship by requiring that an object needs to improve any other object in at least k attributes. The approach in [8] is based on identifying subspaces in which the skyline contains a limited amount of solutions. [9] demonstrates that the selection of the k most representative skyline objects is a non-trivial task and proposes a dynamic programming algorithm for two cost criteria and a polynomial time algorithm for higher dimensionalities. In [10], the authors propose to group skyline points w.r.t. the subspace for which they are part of the skyline.

For network data, the following approaches have been introduced. In [11], the authors introduce a method for calculating a skyline of landmarks in a road network that are compared w.r.t. their network distance to several query objects. In [12] the authors propose in-route skyline processing in road networks. Assume that a user is moving along a predefined route to a known destination, the algorithm processes minimal detours to sets of landmarks being distributed along the path. In [13] the authors discuss continuous skyline queries in road networks, i.e. a moving user queries for a skyline of point of interest. The most similar setting to the approach being presented in this paper is computing route skylines [2]. In this task, we want to find all paths between two nodes in a multicost network where the costs of paths are not dominated by any other path between the same two nodes. In Operations Research the task of computing a set of non-dominated solutions is known as multiobjective optimization or pareto optimization [3]. In case there are only two cost criteria the task is named bicriteria shortest path problem. This setting is of special interest because the amount of results is usually less extensive than in higher dimensions, and the result is much easier to interpret by users. Furthermore, the multicriteria and bicriteria shortest path problems are known to be NP-hard [14]. Thus, the limitation helps to keep the computational effort at an applicable level. Finally, the limitation to a two-dimensional cost space allows several optimizations which cannot be transferred to higher dimensional cost spaces.

In [15] the authors describe a comparison of several state-of-the-art methods. The label correcting method [16] maintains a set of non-dominated solutions at each node. Another approach is the near-shortest path method [17] which is based on the idea of computing all paths having a length within a certain deviation from the length of the optimal path for one criterion. Finally, [18, 15] finds all non-dominated paths in a two-phase approach. The first phase computes the set of so-called supported solutions, followed by the second phase determining the remaining results. These supported solutions are equivalent to the linearly non-dominated solutions we aim to find in this paper. In [15], the authors compared several algorithms for determining the supported solutions. The most successful method for computing k supported solutions is based on $2 \cdot k - 1$ Dijkstra searches, each with a different linear cost combination as optimization criterion. We will compare two improved versions of this approach in our experiments.

Our method does not employ any precomputation step [19] and thus, it is applicable to dynamically changing network costs. To guide our search towards the target, our method generates optimal lower bound estimations as part of the query processing algorithm as proposed in [20].

The linear path skyline as described in this paper is part of of the convex hull over the cost vectors of all paths between start and target. However, the set of all paths is not available at query time. Thus, efficient methods to determine the convex hull cannot be applied to solve the problem being described in this paper.

3   Preliminaries

A bicriteria network is a directed graph $G = (V, E)$ where V denotes the set of vertices and $E \in V \times V$ denotes the set of directed edges. Furthermore, we assume the existence of two cost functions $c_1, c_2 : E \to \mathbb{R}_{\geq 0}$. In case G describes a road network the cost functions may represent height differences, travel distances or the number of traffic lights. Every edge $(u, v) \in E$ is labeled with a non-negative two-dimensional cost vector, $c((u, v)) := ((u, v)_1, (u, v)_2)$. A path or route P is a consecutive set of edges $((s, v), \ldots, (u, t))$ which does not visit any node twice. Likewise, any path has a cost vector $p = (p_1, p_2) = \sum_{(u,v) \in P} c((u, v))$. Throughout this paper, we shall denote paths with capital letters (e.g. P) and the corresponding cost vector with the same lowercase letter (e.g. p). For multiple paths, we either choose different capital letters or superscript their indices.

Given a start node $s \in V$ and a target node $t \in V$, our algorithm computes a subset of all the routes $\mathcal{R} := \mathcal{R}(s, t)$ starting at s and ending at t. Let us note that the following definitions and theoretical results are equally applicable to any set of vectors in $\mathbb{R}^d_{\geq 0}$ as well.

In order to distinguish between linear skylines and skylines in the ordinary sense, we first recall the definiton of conventional skylines and the conventional dominance relation they are based on. For reasons of generality, we do not restrict our definitions to the bicriteria case.

Definition 1. *Conventional Route Skyline*
Let $\mathcal{R}$ be a set of paths in a d-dimensional cost space. Then $P \in \mathcal{R}$ dominates $Q \in \mathcal{R}$, denoted as $p \prec q$ (or $P \prec Q$), iff

$$\exists\, 1 \leq i \leq d : p_i < q_i \wedge \nexists\, 1 \leq i \leq d : p_i > q_i.$$

The set of non-dominated routes, i.e. $\{P \in \mathcal{R} \mid \nexists Q \in \mathcal{R} : q \prec p\}$, is denoted conventional skyline.
Let § be a set of non-dominated paths and Q be a route. If there exists $P \in §$ such that $p \prec q$, we say Q is dominated by §. Conversely, we say Q is non-dominated by § if no such P exists.

After describing conventional dominance, we now present the definintion of linear dominance which is more strict.

Definition 2. *Linear Dominance*
Let $\mathcal{R}$ be a set of paths in a d-dimensional cost space. $P \in \mathcal{R}$ w-dominates $Q \in \mathcal{R}$, where $0 \neq w \in \mathbb{R}^d_{\geq 0}$, iff $w^T p < w^T q$. w is called a weight vector.

Definition 3. *Linear Skyline*
Let $\mathcal{R}$ be a set of paths in a d-dimensional cost space.
A subset $\mathcal{S}' = \{P^1, \ldots, P^K\} \subseteq \mathcal{R}$ linearly dominates a path $Q \in \mathcal{R}$, denoted as $\mathcal{S}' \prec_{lin} Q$ (or $\{p^1, \ldots, p^K\} \prec_{lin} q$) iff

$$\exists\, P \in \mathcal{S}' : P \prec Q) \vee (\forall\, w \in \mathbb{R}^d_{\geq 0}\, \exists P \in \mathcal{S}' : w^T p < w^T q.$$

The maximal set of linearly non-dominated paths (i.e. for all $\hat{\mathcal{S}} \supsetneq \mathcal{S}'$ exists $Q \in \hat{\mathcal{S}} \setminus \mathcal{S}'$ such that $\mathcal{S}' \prec_{lin} Q$) is referred to as linear skyline.

In contrast to conventional domination, testing for linear domination might require comparing an object to more than one other object. That is, unless an object is already conventionally dominated. The definition also implies that the set of linearly non-dominated paths is a subset of the conventionally non-dominated paths. Figure 1(a) and 1(b) illustrate both concepts on the same two-dimensional dataset.

The term linear dominance has a graphical intuition: Consider a two-dimensional cost space. One would expect K linearly non-dominated paths $\{P^1, \ldots, P^K\}$ to dominate another path Q if q lies "behind" any line that can be spanned by the elements of $\{p^1, \ldots, p^K\}$. Thus, the area being linearly dominated by $\{p^1, \ldots, p^K\}$ corresponds to the the part of the convex hull being directed towards the origin. Furthermore, the line connecting q to the origin has to intersect this hull.

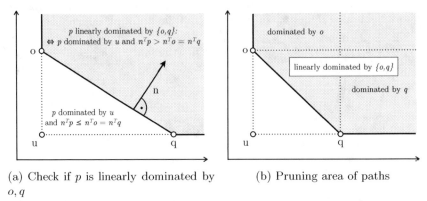

(a) Check if p is linearly dominated by o, q

(b) Pruning area of paths

Fig. 2. Linear Domination areas

We show that this intuition coincides with our definition for two-dimensional cost spaces. The proof for an arbitrary number of dimensions is not straight forward and therefore omitted here.

Theorem 1. *Let O and Q be two paths with $o_1 < q_1$ and $q_2 < o_2$ and let P be another path, for which we want to know if it is linearly dominated by $\{O, Q\}$. Let n be the normal vector of the line between O and Q, such that $n^T o = n^T q$. u shall be the component-wise minimum of o and q, such that $u_1 = min(o_1, q_1)$ and $u_2 = min(o_2, q_2)$. Then, it follows that:*

$$\{o, q\} \prec_{lin} p \quad \Leftrightarrow \quad u \prec p \wedge n^T p > n^T o = n^T q$$

This is illustrated in Figure 2.

Proof. If P is conventionally dominated by O or Q, P has to lie above the line between o and q such that $n^T p > n^T o = n^T q$ and p is also dominated by the component-wise minimum u of o and q. If p is not dominated by u, p has either a smaller x-value than both o and q, a smaller y-value than both, or dominates o or q. If p is not dominated by u, it is therefore not linearly dominated by $\{o, q\}$. In the remaining cases, p is not dominated by o or q but is dominated by u. From the statement $\{o, q\} \prec_{lin} p$ then follows that $\forall w \in \mathbb{R}_{\geq 0} : w^T o < w^T p \vee w^T q < w^T p$. This contradicts $n^T o = n^T q \geq n^T p$ for $w = n$ and therefore requires $n^T p > n^T o$. If p is dominated by u, p therefore lies above the line between o and q if it is linearly dominated by $\{o, q\}$. Accordingly if $n^T p \leq n^T o = n^T q$ there is a contradiction with $\{o, q\} \prec_{lin} p$. If p is dominated by u, p is therefore not linearly dominated by $\{o, q\}$ if it lies below the line or on the line between o and q.

Based on this comprehension of linear skylines, we now move on to introduce the algorithmic structure of our approach.

4   Linear Skyline Computation in Bicriterion Networks

In this section, we will describe our new method and its theoretical foundations. We will start with describing our solution to the most critical aspect of our algorithm, efficiently maintaining all linearly non-dominated paths between two nodes s and v. Afterwards, we will describe our complete algorithm and all its remaining steps in detail.

4.1   Managing Linear Skylines

As for conventional domination, it can be shown that for any linearly non-dominated path P from s to t all subpaths from s to any intermediate node $v \in P$ are non-dominated in the linear skyline $\mathcal{S}(s, v)$.

Lemma 1. *Let* $P = ((s, n_1).., (n_i, v), ..(n_j, t)) \in \mathcal{S}(s, t)$ *then any subpath* $Q = ((s, n_1), .., (n_i, v))$ *is linearly non-dominated in* $\mathcal{S}(s, v)$.

Proof. Given that the cost vector p of path P is linearly non-dominated, we know that $\exists \ w \in \mathbb{R}_{\geq 0}^d$ where $w^T p$ is optimal. If there would be a subpath $Q = ((s, n_1), .., (n_i, v))$ of P which is linearly dominated in $LS(s, v)$ then $w^T q$ cannot be optimal because that linear domination implies w-domination as well. Thus, exchanging Q with $\hat{Q}$ with $w^T \hat{q} < w^T q$ implies the existence of a path $\hat{P}$ with $w^T \hat{p} < w^T p$. This is a contradiction to $P \in \mathcal{S}(s, t)$.

Based on this lemma, it makes sense to only extend paths not being linearly dominated during traversal. Furthermore, after extending a path P from s to v, it has to be checked whether there are any other paths P^i, P^j linearly dominating P. If P is linearly non-dominated, it must be stored with v as part of its local linear skyline $\mathcal{S}(s, v)$. Furthermore, we have to delete any other $P^i \in \mathcal{S}(s, v)$ for which $\exists \ P^j \in \mathcal{S}(s, v) : \{p, p^j\} \prec_{\text{lin}} p^i$ holds.

This step is the most critical operation in our algorithm because it is performed frequently during each query. Moreover, the complexity increases with the number of elements in $\mathcal{S}(s, v)$. Thus, optimizing this operation is the key to efficiently computing linear skylines.

A naive solution to solve this problem is to compare the new path P to any pair $P^i, P^j \in \mathcal{S}(s, v), i \neq j,$. If a new path P is non-dominated, we also need to check whether for any pair $P^i, P^j \in \mathcal{S}(s, v), \{p, p^j\} \prec_{\text{lin}} p^i$ holds. All paths P^i for which this property holds must be deleted from $\mathcal{S}(s, v)$. Both steps have a quadratic complexity in the number of elements K being contained in $\mathcal{S}(s, v)$.

A characteristic of linear skylines in a two-dimensional cost space is that sorting the values of the first criterion in ascending order implies an descending order in the second criterion. Formally, an ordering of $\mathcal{S}(s, v)$ can be expressed as a tuple $(P^1, \ldots, P^K)$ where $\forall \ i < j \in \{1, .., K\}$ the following conditions hold: $p_1^i < p_1^j \wedge p_2^i > p_2^j$. This follows directly from the definition of conventional domination.

For a given path P and a linear skyline $\mathcal{S}(s, v)$, we refer to the elements $P^k, P^{k+1} \in \mathcal{S}(s, v)$ being closest to p w.r.t. the first cost criterion as neighbors. We refer to P^k with $p_1^k \leq p_1$ as left neighbor and denote P^{k+1} with $p_1^{k+1} > p_1$ as right neighbor. At least one of both neighbors must exist, unless $\mathcal{S}(s, v) = \emptyset$. For checking linear dominance, it is only necessary to compare to the potential neighbors of P in $\mathcal{S}$ which is shown in the following lemma.

Lemma 2. *Given a path $P = ((s, v), \ldots, (u, t))$ and the linear skyline $\mathcal{S}(s, v) = (P^1, .., P^K)$. Then the following statements hold:*
Case 1: If $p_1^1 > p_1$ then P is linearly non-dominated in $\mathcal{S}(s, t)$.
Case 2: If $p_1^K < p_1 \wedge p_2^K > p_2$ then P is linearly non-dominated.
Case 3: If $\exists\, k \in \{1, .., K-1\}$ with $p_1^k \leq p_1 \wedge p_1^{k+1} \geq p_1$ and $\exists O, Q \in \mathcal{S}(s, v)$ with $\{o, q\} \prec_{lin} p$ then $\{p^k, p^{k+1}\} \prec_{lin} p$ holds as well.

Proof. Case 1: $p_1 < p_1^1 \Rightarrow p_1 < p_1^i$ with $1 \leq i \leq K$. Thus, P w-dominates all P^i for $w = (1, 0)$. Case 2: $p_2 < p_2^K \Rightarrow p_2 < p_2^i$ with $1 < i < K$. Thus, P w-dominates all P^i for $w = (0, 1)$. If neither case 1 nor case 2 applies, P has two neighbors and case 3 applies. Case 3: There are only two valid orderings for the values $o_1, q_1, p_1^k, p_1, p_1^{k+1}$ where $\{o, q\} \prec_{lin} q$ and $p_1^k \leq p_1 \leq p_1^{k+1}$ can both hold.
Subcase 1: $o_1 \leq q_1 \leq p_1^k \leq p_1 \leq p_1^{k+1}$. From $q_1 \leq p_1 \wedge o_1 < p_1$ follows $q_2 < p_2$ Since $P^k \in \mathcal{S}(s, v) \wedge q_1 \leq p_1^k \Rightarrow q_2 > p_2^k$. Therefore, $p_1^k \leq p_1 \wedge p_2^k \leq p_2$ which implies $p^k \prec p \Rightarrow \{p^k, p^{k+1}\} \prec_{lin} p$.
Subcase 2: $o_1 \leq p_1^k \leq p_1 \leq p_1^{k+1} \leq q_1$ This case can be shown by contradiction. Assume $\{p^k, p^{k+1}\} \prec_{lin} p$ does not hold. Then either $\{o, q\} \prec_{lin} p^k$ or $\{o, q\} \prec_{lin} p^{k+1}$ or both would hold. This is a contradiction to $P^k, P^{k+1} \in \mathcal{S}(s, v)$.

Practically, this observation allows us to reduce the amount of domination checks to one single check for each insertion candidate. Furthermore, determining the neighbors of P in $\mathcal{S}(s, v)$ can be performed using a binary search w.r.t. the first criterion. Thus, the complexity of checking whether P has to be inserted into $\mathcal{S}(s, v)$ has a time complexity of $O(\log |\mathcal{S}(s, v)|)$.

As mentioned above, we have to check whether inserting P leads to the deletion of other elements from $\mathcal{S}(s, v)$. The general idea of how to efficiently determine elements which are now linearly dominated is the following: All dominated elements have to be either conventionally dominated by P or linearly dominated by P and some other element $P^i \in \mathcal{S}(s, v)$. To find all linearly dominated paths efficiently (rather than testing all pairs), we propose to start two linear searches. The first search beginning with the left neighbor and the second search beginning with the right neighbor of P. As was shown in Lemma 2 a path P^i is only dominated, if it is dominated by both of its neighbors. Thus, it is sufficient to check whether P^k is dominated by P and its other neighbor P^{k-1} when searching to the left side and P^{k+1} when searching to the right side. Let us note that the left neighbor P^{k-1} can already conventionally dominate a path P^k. Thus, we can improve the search in the left direction by checking this simpler property before testing for linear dominance. If P^k indeed is dominated, there might be more dominated objects in the same search direction and we would have to repeat the domination check with the new neighbor. On the other hand, if P^k is not linearly dominated, we can stop the search in this direction. Formally, this is formulated in the following lemma:

Lemma 3. *Given a linear skyline $\mathcal{S}(s, v) = (P^1, .., P^K)$ being ordered w.r.t. the first cost criterion and a new non-dominated Path P. For any $P^k \in \mathcal{S}(s, v)$ with $p_1^k < p_1$ it holds that if $\{p, p^{k-1}\} \prec_{lin} p^k$ does not hold then $\nexists\, i \in \{1, .., k\}:$*

$\{p, p^{i-1}\} \prec_{lin} p^i$. *Correspondingly, for any* $P^k \in \mathcal{S}(s, v)$ *with* $p_1^k > p_1$ *it holds that if* $\{p, p^{k+1}\} \prec_{lin} p^k$ *does not hold, then* $\nexists\, i \in \{k, .., K\} : \{p, p^{i+1}\} \prec_{lin} p^i$.

Proof. From Lemma 2 we know that if P^k is linearly dominated it has to be linearly dominated by its neighbors P^{k-1} and P^{k+1}. If neither P^{k-1} nor P^{k+1} have been deleted from $\mathcal{S}(s, v)$, the new path P cannot be a neighbor of P^k and since P^{k-1}, P^k, P^{k+1} were members in the original linear skyline $\mathcal{S}(s, v)$, P^k must remain linearly non-dominated.

To conclude, when determining the dominated elements of $\mathcal{S}(s, v)$, we have to repeatedly check the current neighbors of P. If a neighbor is not linearly dominated, we can terminate the search in this direction. Updating the linear skyline $\mathcal{S}(s, v)$ has a linear worst case complexity in $K = |\mathcal{S}(s, v)|$ because in the worst case we have to delete any former elements of $\mathcal{S}(s, v)$. Let us note that the results of the dominance checks are negative in the majority of cases. Typically, we only have to perform the domination test having a logarithmic time complexity.

4.2   Computing Bicriterion Linear Skylines

After describing the efficient management of linear skylines, we will now specify our complete algorithm for computing linear skylines in bicriteria networks. In general, our algorithm starts to construct linearly non-dominated paths beginning with the start node s by successively selecting nodes and expanding all linearly non-dominated paths being discovered so far. Furthermore, we mark paths which have been already expanded to avoid duplicate expansions.

Our method employs two pruning rules to exclude paths which cannot be extended into a linearly non-dominated result path. The first is that a path ending at node v has to be linearly non-dominated in $\mathcal{S}(s, v)$. To enforce this rule, we maintain a local linear skyline for each visited node v and delete all paths that are linearly dominated.

The second pruning method is based on the condition whether a path P might be extended into a linearly non-dominated path between s and t. In other words, any path $P = ((s, n_1), ..(n_i, v))$ having a cost vector being already linearly dominated in $\mathcal{S}(s, t)$ does not have to be extended because any extension from v to t would only add costs. Thus, the extension would be dominated in $\mathcal{S}(s, t)$ as well. This pruning method can be considerably improved by adding a lower bound of the cost values for each cost criterion. In [2] a reference point embedding is used to estimate the lower bound costs. This method has two drawbacks. The first is that the quality of the cost approximation strongly depends on the selected reference points. The second drawback is that a reference point embedding requires offline precomputation which limits the usablitiy of this solution under dynamically changing cost values. Recently, [20] proposed to generate optimal forward estimations during query processing by performing two reverse single-source all-target Dijkstra searches from the target t to the start s. The method starts by determining the shortest path $SP^1 = ((s, n_1), .., (n_i, t)$ with respect to criterion 1 und thus, derives an upper bound w.r.t. criterion 2 sp_2^1. Afterwards,

---

Algorithm 1. BLRSC

Require: start s and target t
Ensure: Linear Route Skyline between s and t
  Compute Minimal costs to t using backward Dijkstra search
  init queue Q of nodes with s
  while $\neg$ Q.isEmpty **do**
    $n := Q$.pop
    for all Path $P \in S(s, n)$ **do**
      if $\neg$P.isExtended **then**
        if $(p + n^{min})$ not linearly dominated in $S(s, t)$ /*forward estimation*/ **then**
          for all Outgoing Link (n, m) of Node n **do**
            O := extend(P,(n,m))
            Search P^k, P^{k+1} for O in $S(s, m)$ using binary search
            if $\neg(\{p^k, p^{k+1}\} \prec_{\mathrm{lin}} o)$ **then**
              $i = k$ // prune dominated elements left of o
              while $i > 2 \wedge \{p^{i-1}, o\} \prec_{\mathrm{lin}} p^i$ **do**
                Delete p^i from $S(s, m)$
                $i = i - 1$
              end while
              $i = k + 1$ // prune dominated elements right of o
              while $i < (|S(s, m)| - 1) \wedge \{o, p^{i+1}\} \prec_{\mathrm{lin}} p^i$ **do**
                Delete p^i in $S(s, m)$
                $i = i + 1$
              end while
              Q.insertOrUpdate(m)
            end if
          end for
        end if
        P.setExtended
      end if
    end for
  end while
  return S(s,t)

---

we compute the shortest path to $SP^2 = ((s, m_1), .., (m_j, t))$ w.r.t. criterion 2 and thus, receive an upper bound for the costs in criterion 1 sp_1^2. Now we continue the search w.r.t. criterion 1 until all nodes having a smaller cost than sp_1^2 are found. Finally, the search w.r.t. criterion 2 is continued until all nodes having a smaller distance than sp_2^1 are retrieved. This way, each visited node v can be labeled with a cost vector (v_1^{min}, v_2^{min}) representing the costs of the shortest paths between v and t. Though this precomputation step seems to be rather expensive, it offers a tight lower bound for any intermediate node which extremely increases the pruning power of the global pruning criterion testing forward estimations against already retrieved paths. Since the step constructing all linearly non-dominated paths is multiple times more expensive than the backsearch determining the lower bound costs, using this step still yields a dramatic runtime improvement.

Algorithm 1 describes our method in pseudocode. In the first step, we perform the precomputation step by performing the reverse single-source all-target Dijkstra searches starting with the target node t. In the next step, the forward traversal is started. We maintain a priority queue of nodes being ordered w.r.t. the first attribute in ascending order. The queue is initialized by adding the start node s. For each visited node v, we store a linear skyline which is managed as an ordered list. The priority of a node always corresponds to the smallest cost value of any linearly non-dominated path which has not been extended yet. Already extended paths are kept in the skyline to use them to dominate other paths, but are marked to prevent multiple extensions. In the main loop the algorithm pops the top node v. For each unprocessed path $P \in \mathcal{S}(s, v,)$, we check linear dominance in the current result skyline $\mathcal{S}(s, t)$ by adding the lower bound costs from v to t to the cost vector p. If p is dominated, it is flagged as already extended. If the forward estimation of p is not linearly dominated yet, the path is extended with all outgoing edges of node v. Each new path $\hat{P}$ ending at node n is checked for linear dominance in the current skyline $\mathcal{S}(s, n)$. If $\hat{P}$ has to be inserted into $\mathcal{S}(s, n)$, all linearly dominated results are removed from $\mathcal{S}(s, n)$ by successively searching the left and the right neighbors until a linearly non-dominated path is found in each direction. Furthermore, the node n is either inserted into the priority queue or its position in the queue is checked for a potential improvement. The algorithm terminates, if the queue is empty meaning that there is no path left which can be potentially extended into a result path.

5   Evaluation

All experiments are performed on a dedicated machine with an 3.0 Ghz Intel Xeon 5160 processor and 32 GB RAM. The algorithms were implemented in Java 1.7 and do not make use of multiple cores. Each task is consecutively solved by all compared algorithms and the execution order is randomized for each task. If an algorithm is not able to solve a task in less than 60 seconds the computation is aborted and it is counted as a timeout. Three separate runs of all tasks are performed and the average of these three runs is used. This experimental setup is intended to ensure similar evaluation conditions, eliminate possible evaluation order effects and smooth out runtime discrepancies.

There are four different graphs on which routing tasks are performed:

1. G_{munich} is a OpenStreet Map (OSM) road network of the area around Munich which covers $6\,992$ km^2 including the neighboring city Augsburg and consists of $782\,030$ nodes and $1\,595\,261$ edges. The 2450 routing tasks on this graph are between 50 city districts and municipalities in and around Munich.
2. G_{bavaria} is a OSM road network of Bavaria which covers $70\,549$ km^2 and consists of $4\,044\,556$ nodes and $8\,298\,017$ edges. The 90 routing tasks on this graph are between 10 major Bavarian cities.

Table 1. Overview of absolute runtime averages. In case of timeouts (>60s) the rounded number of timeouts is denoted next to the † symbol.

cost criteria	time & distance		time & crossings		rand.$_1$ & rand.$_2$	
graph	G_{munich}	G_{bavaria}	G_{munich}	G_{bavaria}	$G_{250\times250}$	$G_{50\times50\times50}$
# of tasks	2450	90	2450	90	4	8
BLRSC	0.2s	4s	0.36s	4.83s	17s	22.49s
Multi-A^*	0.21s	5.13s	0.37s	>8.47s †3	30.4s	>55.4s †4
ARSC	0.25s	>21.06s †18	0.4s	>11.12s †6	>60s †4	>60s †8
Multi-BD	0.61s	>33.05s †32	0.41s	>17.54s †13	30.85s	41.3s

3. $G_{250\times250}$ is a 2d grid network with 62 500 nodes and 249 000 edges. The 4 routing tasks on this graph are between opposite corners of the grid.
4. $G_{50\times50\times50}$ is a 3d grid network with 125 000 nodes and 735 000 links. The 8 routing tasks on this graph are between opposite corners of the lattice.

On G_{munich} and G_{bavaria}, the criteria time & distance and time & crossings are used for the routing tasks. On $G_{250\times250}$ and $G_{50\times50\times50}$, the criteria rand.$_1$ & rand.$_2$ are used which have independent pseudorandom cost values.

We compare our new method BLRSC to three other algorithms: The first is ARSC [2] to a have a comparision partner for computing conventional route skylines. We modified ARSC by the same precomputation step proposed in [20] to achieve a better comparability to BLRSC. Multi-BD represents the state-of-the-art for computing supported solutions as proposed in [15]. For Multi-BD, the single objective shortest paths are computed using Bidirectional Dijkstra. However, the lower bounds in [20] can be used in combination with the approach from [15] as well which allows us to employ A^*-search instead of Bidirectional Dijkstra. We will refer to this algorithm as Multi-A^* and compare to this appraoch to evaluate how much of the performance gain is caused by the optimal lower bounds.

As can be seen in Table 1 the proposed algorithm BLRSC has the lowest runtime average in all experimental conditions, and it is the only algorithm in the experiments which solves all tasks in less than 60 seconds. As can be seen in Figure 3 this is the case throughout different task difficulties which is assessed through the number of hops between end points. The only exception can be seen in Figure 4(a), where Multi-BD is faster than BLRSC for less than 300 hops. BLRSC is also clearly faster for grid networks as can be seen in Figure 4. BLRSC is on average significantly faster than the previous approach Multi-BD. As can be seen in Table 2, on G_{munich} with criteria time & distance it is about three times faster and on G_{bavaria} more than eight times faster, and on G_{bavaria} with criteria time & crossings it is more than three times faster.

ARSC performs very poorly on $G_{250\times250}$ and $G_{50\times50\times50}$, where it cannot solve a single task in less than 60 seconds which is also the worst performance out of all tested algorithms. ARSC is also more than five times slower for than BLRSC on G_{bavaria} with time & distance. Multi-A^* also performs poorly on $G_{250\times250}$ and

Table 2. Overview of runtime averages relative to runtime averages of our main contribution BLRSC. In case of timeouts (>60s) the rounded number of timeouts is denoted next to the † symbol.

cost criteria	time & distance		time & crossings		rand.$_1$ & rand.$_2$	
graph	G_{munich}	G_{bavaria}	G_{munich}	G_{bavaria}	$G_{250\times250}$	$G_{50\times50\times50}$
# of tasks	2450	90	2450	90	4	8
BLRSC	1	1	1	1	1	1
Multi-A^*	1.05	1.28	1.02	>1.75 †3	1.78	>2.46 †4
ARSC	1.22	>5.25 †18	1.09	>2.29 †6	>3.52 †4	>2.66 †8
Multi-BD	2.98	>8.24 †32	1.12	>3.62 †13	1.81	1.83

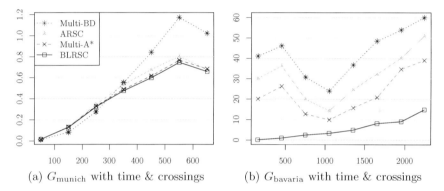

(a) G_{munich} with time & crossings   (b) G_{bavaria} with time & crossings

Fig. 3. Values on the x-axis are numbers of hops between end points and values on the y-axis are average runtimes in seconds

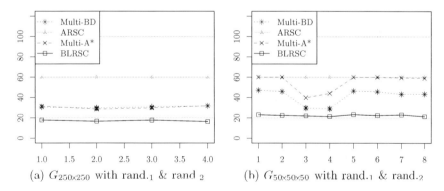

(a) $G_{250\times250}$ with rand.$_1$ & rand $_2$ (b) $G_{50\times50\times50}$ with rand.$_1$ & rand.$_2$

Fig. 4. Values on the x-axis are ids of the tasks and values on the y-axis are runtimes of three runs in seconds. If the value is 60s, it's a timeout and took longer than 60s.

$G_{50\times50\times50}$, where it is even slower than Multi-BD and much slower than BLRSC. It is also at least about twice as slow on G_{bavaria} using time & crossings than BLRSC. The experiments therefore show that BLRSC clearly outperforms all other tested approaches and that its performance gain is not only caused by the use the forward estimation. In other words, making use of the forward estimation with ARSC or using multiple A^* searches performs worse than BLRSC.

6 Conclusions

In this paper, linear skyline queries are introduced as a new query type in bi-criteria networks. We propose a two-step algorithm which is based on efficient techniques for managing and updating linear skylines. In a first precomputation step, we perform two network traversals to determine lower bounds for both cost criteria and all nodes being potentially visited during the search. In the next step, we use these lower bounds to perform a third graph traversal determining the linear skylines for all visited nodes. In our experimental evaluation, we compare our new algorithm to the state-of-the-art algorithm for computing supported solutions which is the same task from a technical point of view. It can be observed that our new approach outperforms the compared methods for linear and conventional path skyline queries.

For future work, we would like to develop new efficient methods for determining linear path skylines in general multicriteria datasets. Furthermore, we would like to investigate linear path skyline computation under time-dependency, i.e. at least one cost criterion depends on the time an edge is visited.

Acknowledgements. This research has been supported by the IKT II program in the Shared-E-Fleet project. They are funded by the German Federal Ministry of Economics and Technology under the grant number 01ME12107. The responsibility for this publication lies with the authors.

References

1. Borzsonyi, S., Kossmann, D., Stocker, K.: The skyline operator. In: Proceedings of the 17th International Conference on Data Engineering (ICDE), Heidelberg, Germany (2001)
2. Kriegel, H.P., Renz, M., Schubert, M.: Route skyline queries: a multi-preference path planning approach. In: Proceedings of the 26th International Conference on Data Engineering (ICDE), Long Beach, CA, pp. 261–272 (2010)
3. Ehrgott, M.: Multicriteria optimization. Springer (2005)
4. Tan, K.L., Eng, P.K., Ooi, B.C.: Efficient progressive skyline computation. In: Proceedings of the 27th International Conference on Very Large Data Bases (VLDB), Roma, Italy (2001)
5. Kossmann, D., Ramsak, F., Rost, S.: Shooting stars in the sky: an online algorithm for skyline queries. In: Proceedings of the 28th International Conference on Very Large Data Bases (VLDB), Hong Kong, China (2002)
6. Papadias, D., Tao, Y., Fu, G., Seeger, B.: An optimal and progressive algorithm for skyline queries. In: Proceedings of the ACM International Conference on Management of Data (SIGMOD), San Diego, CA (2003)

7. Chan, C.-Y., Jagadish, H.V., Tan, K.-L., Tung, A.K.H., Zhang, Z.: On high dimensional skylines. In: Ioannidis, Y., Scholl, M.H., Schmidt, J.W., Matthes, F., Hatzopoulos, M., Böhm, K., Kemper, A., Grust, T., Böhm, C. (eds.) EDBT 2006. LNCS, vol. 3896, pp. 478–495. Springer, Heidelberg (2006)

8. Zhang, Z., Guo, X., H. L., Tung, A.K.H., Wang, N.: Discovering strong skyline points in high dimensional spaces. In: CIKM 2005: Proceedings of the 14th ACM International Conference on Information and Knowledge Management, Bremen, Germany (2005)

9. Lin, X., Yuan, Y., Zhang, Q., Zhang, Y.: Selecting stars: the k most representitive skyline operator. In: Proceedings of the 23th International Conference on Data Engineering (ICDE), Istanbul, Turkey (2007)

10. Pei, J., Jin, W., Ester, M., Tao, Y.: Catching the best views of skyline: a semantic approach based on decisive subspaces. In: VLDB 2005: Proceedings of the 31st International Conference on Very Large Data Bases,Trondheim, Norway (2005)

11. Deng, K., Zhou, Y., Shen, H.T.: Multi-source query processing in road networks. In: Proceedings of the 23th International Conference on Data Engineering (ICDE), Istanbul, Turkey (2007)

12. Huang, X., Jensen, C.S.: In-route skyline querying for location-based services. In: Proc. of the Int. Workshop on Web and Wireless Geographical Information Systems (W2GIS), Goyang, Korea, pp. 120–135 (2004)

13. Jang, S.M., Yoo, J.S.: Processing continuous skyline queries in road networks. In: International Symposium on Computer Science and its Applications, CSA 2008 (2008)

14. Hansen, P.: Bicriterion path problems. In: Multiple Criteria Decision Making Theory and Application, pp. 109–127. Springer (1980)

15. Raith, A., Ehrgott, M.: A comparison of solution strategies for biobjective shortest path problems. Computers & Operations Research 36(4), 1299–1331 (2009)

16. Brumbaugh-Smith, J., Shier, D.: An empirical investigation of some bicriterion shortest path algorithms. European Journal of Operational Research 43(2), 216–224 (1989)

17. Matthew Carlyle, W., Kevin Wood, R.: Near-shortest and k-shortest simple paths. Networks 46(2), 98–109 (2005)

18. Mote, J., Murthy, I., Olson, D.L.: A parametric approach to solving bicriterion shortest path problems. European Journal of Operational Research 53(1), 81–92 (1991)

19. Delling, D., Sanders, P., Schultes, D., Wagner, D.: Engineering route planning algorithms. In: Lerner, J., Wagner, D., Zweig, K.A. (eds.) Algorithmics of Large and Complex Networks. LNCS, vol. 5515, pp. 117–139. Springer, Heidelberg (2009)

20. Machuca, E., Mandow, L.: Multiobjective route planning with precalculated heuristics. In: Proc. of the 15th Portuguese Conference on Artificial Intelligence (EPIA 2011), pp. 98–107(2011)

Label and Distance-Constraint Reachability Queries in Uncertain Graphs

Minghan Chen, Yu Gu, Yubin Bao, and Ge Yu

Northeastern University, China
cmhshirley@gmail.com, {guyu,baoyubin,yuge}@ise.neu.edu.cn

Abstract. A fundamental research problem concerning edge-labeled uncertain graphs is called label and distance-constraint reachability query (LDCR): Given two vertices u and v, query the probability that u can reach v through paths whose labels and lengths are constrained by a label set and a distance threshold separately. Considering LDCR is not tractable as a #P-complete problem, we aim to propose effective and efficient approximate solutions for it. We first introduce a subpath-based filtering strategy which combines divide-conquer algorithm and branch path pruning to compress the original graph and reduce the scale of DC-tree. Then to approximate LDCR, several estimators are presented based on different sampling mechanisms and a path/cut bound is proposed to prune large-deviation values. An extensive experimental evaluation on both real and synthetic datasets demonstrates that our approaches exhibit prominent performance in term of query time and accuracy.

1 Introduction

The growing popularity of graph databases has generated many interesting data management problems. One important type of queries over graphs is reachability queries which have been handled by a number of algorithms proposed in recent years [1, 2]. With the advent of Semantic Web and RDF, graphs contain more complex semantics where edge labels are introduced to represent the type of relationship between two nodes. Therefore, constrained reachability is proposed in response to the need of querying the existence of any path which satisfies certain constraints between nodes. The definition of constraints, e.g. number of hops, type of edge labels, or a regular expression etc., depends on the targeted problem. Constrained reachability queries find their importance in a variety of graph databases, such as bioinformatics, social networks and so on.

Nevertheless, most of these constrained reachability queries are performed on certain graph databases. Since complex networks, such as biological, social, and communication networks, often entail uncertainty which can be modeled as probabilistic graphs, the previous studies that do not consider the uncertain properties are unsuitable to current complex networks. For instance, the interactions between metabolic compounds in metabolic networks are often associated with uncertainties in that they are inevitably derived through noisy and error-prone lab experiments. In metabolic networks (Fig. 1), each vertex represents

S.S. Bhowmick et al. (Eds.): DASFAA 2014, Part I, LNCS 8421, pp. 188–202, 2014.

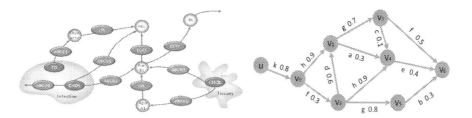

Fig. 1. Lipid Metabolism   **Fig. 2.** An Uncertain Graph

a compound, and a directed uncertain edge between two compounds represents the probability that one compound can be transformed into another through a chemical reaction, which is controlled by a certain enzyme represented as the edge's label. Chemists may be concerned about the following questions: whether there is an active transformation between two compounds using some enzymes, and what is the probability of the transformation. This can be described as a reachability query with constraints on the edge labels in uncertain graphs.

Different from the membership-based constraint proposed in [2], which just requires the path's edge labels to be in the label constraint set, we query paths whose labels must contain all the labels in the label constraint set. Furthermore, we add a distance threshold to refine our query criteria. Here, the distance refers to the number of hops between two nodes. This paper focuses on a large uncertain graph and study a new type of constrained reachability problem–LDCR. In order to address the queries efficiently, we make the following contributions:

1. We define the label and distance-constraint reachability (LDCR) problem over uncertain graphs for the first time and develop approximate algorithms with high efficiency and effectiveness.
2. Based on the concept of subpaths, we propose pruning methods to find all the satisfactory subpaths and then transform the original graph into a subpath graph. A divide-conquer approach optimized by a branch path filter algorithm is proposed for LDCR queries on the new subpath graph.
3. We investigate several sampling methods to approximate the reachability and an upper-lower bound to filter sampling results.

The rest of the paper is organized as follows. In Sect. 2, we give a broad level overview of the prior work in reachability queries and uncertain graphs. Before introducing our method, the problem definition and its background is presented in Sect. 3. Section 4 describes our subpath-based algorithm, while Sect. 5 puts forward approximate solutions with different sampling estimators. Experimental results are reported in Sect. 6. Finally, we conclude the paper in Sect. 7.

2   Related Work

2.1   Uncertain Graphs

Managing and mining uncertain graphs have attracted extensive attention in the data management and data mining communities [3–5]. Aggarwal systematically introduced the topic of incorporating and handling uncertainty in the database system [6]. With respect to uncertain graphs, Potamias et al. studied the k-nearest neighbor queries (kNN) over uncertain graphs [7], i.e., computing the k closest nodes to a query node. In [4], Yuan et al. searched shortest paths in uncertain graphs based on the internal structure and the interplay among all the uncertain paths. Zou et al. studied frequent subgraph mining over uncertain graph data under the probabilistic or expected semantics [3, 5].

Another line of related work is concerned with distance-constraint query in uncertain graphs. Jin et al. developed efficient approximate algorithms to study a novel s-t distance-constraint reachability problem concerning uncertain graphs [8]. They also presented effective sampling methods to estimate the reachability, which on average reduce both variance and running time significantly. However, labels are not involved in the query.

2.2   Reachability Queries

The prior work on reachability queries in certain graphs is divided into two main categories: simple reachability queries and constrained reachability queries.

Simple reachability queries just query if a source vertex has any path to the destination vertex. The recent works by Yildirim and Schaik et al. on simple reachability queries summarized an extensive overview of the previous literatures [9, 10].

Recently, substantial attention has been paid to constrained reachability queries. Fan et al. addressed the problem of adding regular expressions and patterns to the reachability queries [11]. Gubichev et al. implemented a technique of evaluating path queries over RDF graphs using purely database style indexing and efficient join processing techniques [12]. Jin et al. proposed a method for evaluating label-constraint reachability (LCR) queries [2], which is a special case of the problem in [11]. They employed a spanning tree and a partial transitive closure to compress the full transitive closure for labeled graphs. Then, Xu further improved the efficiency of LCR by transforming the edge-labeled directed graph into an augmented directed acyclic graph (DAG) [13].

However, the above solutions do not consider underlying uncertainty in graph data and thus fail to support our problem.

3   Preliminaries

For an uncertain labeled directed acyclic graph $\mathcal{G} = ((V, E, \Sigma), (\lambda, p, w))$, where V is the set of vertices, E is the set of edges, and Σ is the set of edge labels. λ is a function that assigns each edge $e \in E$ with a label $\lambda(e) \in \Sigma$, while $p : E \rightarrow (0, 1]$ is a function that assigns each edge with a probability that indicates the likelihood of the edge's existence, and $w : E \rightarrow (0, \infty)$ associates each edge with a weight (length). In this paper, for simplicity of description, we assume each edge has unit-weight (unit-length).

In the possible world model, an uncertain graph can derive a series of standard graphs $G = ((V_G, E_G, \Sigma_G), (\lambda_G, w_G))$, which are called possible graphs. According to [4], the possible graph G has the sampling probability of $Pr[G]$:

$$Pr[G] = \prod_{e \in E_G} p(e) \prod_{e \in E \setminus E_G} (1 - p(e)) \tag{1}$$

For each edge e, there are two cases: e exists in G or not. Thus, there are a total of 2^n possible graphs. In the example illustrated in Fig. 2, graph $\mathcal{G}$ has 2^{12} possible graphs.

Given two vertices, u and v in the graph, and a label set L, where $u, v \in V$. If there is a path p from vertex u to v whose path labels $L(p)$ contains L, i.e., $L \subseteq L(p)$, then we say u can reach v with label-constraint L, denoted as $(u \xrightarrow{L} v | G)$. The distance of an L-path is the sum of all the edges' weights (lengths) on the path, denoted as $dis(u \xrightarrow{L} v | G)$. Given a distance-constraint d, we say vertex v is LD-reachable from u if the distance of the L-path from u to v in G is smaller than or equal to d, namely $dis(u \xrightarrow{L} v | G) \leq d$. And we define the path as $(u \xrightarrow{LD} v | G)$.

Definition 1. *(**Label and Distance Constraint Reachability**) The problem of computing u-v label and distance constraint (LD-constraint) reachability in uncertain graph $\mathcal{G}$ is to compute the probability of the possible graphs G, in which vertex v is LD-path from u. Specifically, we have*

$$I_{u,v}^{ld}(G) = \begin{cases} 1, & if \ min(u \xrightarrow{L} v | G) \leq d \\ 0, & if \ min(u \xrightarrow{L} v | G) > d \end{cases} \tag{2}$$

Then, the u-v LD-constraint reachability in uncertain graph $\mathcal{G}$ with respect to parameters L and D is defined as

$$R_{u,v}^{ld}(\mathcal{G}) = \sum_{G \sqsubseteq \mathcal{G}} I_{u,v}^{ld}(G) \cdot Pr[G] \tag{3}$$

Theorem 1. *The label and distance-constraint reachability query on uncertain graphs is a #p-complete problem.*

The simple reachability problem is known to be #P-Complete [14]. As its generalization, the LDCR problem which is much harder than the simple reachability problem, is also #p-complete. Detailed proof of Theorem 1 is omitted due to space limitation.

4   A Subpath-Based Filtering Strategy

Due to the complexity of LDCR query, we can begin by searching a subpath starting and ending with vertices that have labels included in the label constraint set, and then from the end vertex we continue to search another subpath with labels in L on the path's head and tail vertices. Therefore, the evaluation of LDCR query can be broken down into the connectivity probability of many desirable subpaths. In this section, we first introduce the concept of subpaths which can facilitate fast computation of u-v LD-constraint reachability.

4.1   Subpaths

As the graph is of large scale and the edges are numerous, searching edges will incur high overhead. Here, we define the subpath to reduce the query size and storage cost.

Definition 2. *(Subpath) For arbitrary reachability queries, given two vertices u and v in the graph, a label set L, and a distance threshold d, a subpath is a path that meets the following requirements: the head and tail labels of the path are included in the label set L, while the remaining edges' labels are excluded to L, and the total length of the path is no larger than d.*

Paths are also regarded as subpaths when they begin with the source vertex and end with a label-constraint edge, or begin with a label-constraint edge and end with the destination vertex. We use the head and tail edges and their labels to represent a subpath, denoted as $sp = (e_i, e_j, k, e, d, p)$, where d is the path's length and p is the probability of the path. For instance, a subpath $\{u, v_0, v_2, v_4, v_6\}$ in Fig. 2 is $sp = (e_{u0}, e_{46}, k, e, 3, 0.108)$, where p and d is computed from node v_0 to node v_6 to avoid double counting, e.g. $p = 0.3 \times 0.9 \times 0.4 = 0.108$ and $d = 3$. However, the occurrence frequency of required labels will affect the filtering efficiency. Furthermore, we have the following theorem.

Theorem 2. *When an edge-labeled graph satisfies $m < \frac{1+\sqrt{1+8n}}{2}$ and $m < \frac{n}{d_a^d}$, the number of subpaths is smaller than that of edges. Where d is the distance threshold, n is the total edge number, m is edge numbers meeting the label requirement, and d_a is the average out-degree of edges.*

Proof. For m edges meeting the label requirement, we have at most C_m^2 subpaths (i.e., directed pair-wise edges). On the other hand, if the average out-degree of edges in the graph is d_a, then the maximum number of subpaths is $m \cdot d_a^d$ starting from the m edges. Therefore, the upper bound of subpaths is the minimum of C_m^2 and $m \cdot d_a^d$. To make sure $\min\{C_m^2, m \cdot d_a^d\}$ is smaller than the edge number n in the graph, we can get the above conclusion.

Theorem 2 can be used to judge whether the subpath-based filtering strategy is executable on various labeled graphs. Since most queries are constraint to a small size of label set, m meets Theorem 2 under normal circumstances. We consequently transform the problem into querying reachability over subpath graphs. Next, we propose a basic method using the breadth-first search with the help of a hash index of graph data to identify all the desired subpaths.

Algorithm 1. FindSP($\mathcal{G},u,v,L,d$)

---

1   **Generate** two queues $BlackNodes$, $QNodes$ and initialize them with u;
2   **while** $BlackNodes \neq \phi$ **do**
3       $startV = BlackNodes.poll()$;
4   **while** $QNodes \neq \phi$ /* Breadth-First Traverse */ **do**
5           $curV = QNodes.poll()$;
6   **if** $curV = v$ or $curV.dis > d$ **then**
7   break;
8   **end**
9   **for** each outgoing edge e of $curV$ **do**
10   **if** $e.tag \in L$ /* e.tag is the label of edge e */ **then**
11   get a sp;
12   **if** $e.node \notin BlackNodes$ /* e.node is the tail node of e */ **then**
13   add $e.node$ to $BlackNodes$;
14   **end**
15   **end**
16   **else**
17   add $e.node$ to $QNodes$;
18   **end**
19   **end**
20   **end**
21 **end**

Starting with an source node u as the root, Algorithm 1 sequentially searches each adjacent edges of node u until all the outgoing edges are visited (Line 9). If we reach node v_1 along certain edge whose label is included in the label set, we obtain a subpath and mark it as black (Line 10-15). In the breadth-first order, we search all the children of the non-black nodes until the distance from node u to the current node is larger than the threshold d (Line 6-8), or we reach the destination node. Repeat above operations for the generated tree rooted at black nodes, and search new subpaths until no black nodes remain (Line 2). In this way, we can obtain all the desired subpaths.

4.2   Pruning Techniques

When searching subpaths, we first prune subpaths based on the distance threshold. Therefore, besides storing each subpath's length, we also need to record the current total distance, i.e., the cumulative distance from the source node to the current node, which can help avoid searching useless subpaths and thus reducing the memory cost.

Note that some paths with different expressions can be merged. For another example in Fig. 3, when visiting the adjacent nodes of v_2, we will encounter node v_3 again. To avoid revisiting node v_3, we try to merge the two paths.

For paths $\{v_0, v_1, v_3\}$ and $\{v_0, v_4, v_9, v_3\}$ in Fig. 3, one v_3 is in the level of d_1, and the other is in d_2 (suppose $d_2 > d_1$). Similarly, we assume the probability of the common edges in the two paths is p_0, and the probabilities of different edges are p_1 and p_2 respectively. Then we add a new record $\{d_2, p_{new}\}$ to the former node, where p_{new} is the probability that the current distances of paths are smaller than d_2, i.e., $p_{new} = p_0[1 - (1 - p_1)(1 - p_2)]$. Pay attention to the descendants of node v_3. Once the probability at v_3 is changed, the probability

of its descendants has to be changed too. To avoid repeatedly walking on each
descendant paths, we utilize the following equation to add new record $\{d'_{child},$
$p'_{child}\}$ to its descendants.

$$d'_{child} = d_{child} + d_2 - d_1; \quad p'_{child} = \frac{p_{new}}{p_{old}} \cdot p_{child} \qquad (4)$$

where p_{old} is the original probability at node v_3.

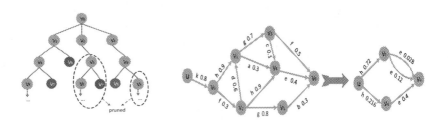

Fig. 3. Pruned by path merge   **Fig. 4.** Transformation of graph

In the example illustrated in Fig. 3, it can be observed that the two paths
bifurcates at node v_0, which is the least common ancestor of the duplicate node
v_3. Thus, we can update the probabilities quickly as long as we find the least
common ancestors (LCA) of the duplicate nodes.

An LCA algorithm is further proposed to search the least common ancestors
in a tree. For arbitrary two nodes, we select one node and search upwards to
generate an ancestor linked list, and then create a hash table to store nodes.
Go upwards from the other node and check each encountered node in the hash
table. If the node already exists in the hash table, then it is the least common
ancestor.

4.3   Fast Query on Subpath Graphs

In this subsection, we first transform the original graph into a subpath graph
which largely reduces memory costs, and then investigate a fast query on it
combining the divide-conquer framework with a novel branch path filter strategy.

For the example in Fig. 3, connecting subpaths by common edges will generate
a new graph as displayed in Fig. 4. In the subpath graph, the length of arbitrary
path starting from source node u is no larger than the distance threshold d,
which is the most important property of the subpath graph. Since we only need
to work on the subpath graph in the remainder of the paper, we simply use G
for the subpath graph. Note that using the subpath to transform graph may give
rise to a situation that more than one edge exists between two nodes because
there are two different subpaths between two nodes, as shown in Fig. 4. But it
will not influence the execution of our algorithm.

Due to the property of uncertainty, we can adopt the divide-conquer method to handle our problem. In this subsection, we give a new definition of prefix groups based on subpaths rather than edges as below:

Definition 3. (SP_1, SP_2)-*prefix group: The* (SP_1, SP_2)-*prefix group of possible graphs from uncertain graph* $\mathcal{G}$, *denoted as* $\mathcal{G}(SP_1, SP_2)$, *includes all the possible graphs of* $\mathcal{G}$ *which includes all subpaths in subpath set* $SP_1 \subseteq SP$ *and does not contain any subpath in subpath set* $SP_2 \subseteq SP$, *i.e.,*

$$\mathcal{G}(SP_1, SP_2) = \{G \sqsubseteq \mathcal{G} | SP_1 \subseteq SP_G \wedge SP_2 \cap SP_G = \emptyset\} \tag{5}$$

If SP_1 contains a LD-path from u to v, then, $R_{u,v}^{ld}(\mathcal{G}(SP_1, SP_2)) = 1$; if SP_2 contains a LD-cut between u and v, then, $R_{u,v}^{ld}(\mathcal{G}(SP_1, SP_2)) = 0$. Here, a LD-cut between u and v is an edge set C_d of $\mathcal{G}$ if $\mathcal{G} \backslash C_d$ does not have a LD-path, i.e., $min(u \overset{L}{\rightarrow} v | G) > d$. Also, the factorization has changed:

$$R_{u,v}^{ld}(\mathcal{G}(SP_1, SP_2)) = p(sp)R_{u,v}^{ld}(\mathcal{G}(SP_1 \cup \{sp\}, SP_2)) + \\ (1 - p(sp))R_{u,v}^{ld}(\mathcal{G}(SP_1, SP_2 \cup \{sp\})) \tag{6}$$

Figure 5 illustrates the divide-conquer procedure for $R_{u,v}^{ld}(\mathcal{G})$ based on the above factorization. The computational complexity is $O(2^m)$, where m is the average prefix length, i.e., the average numbers of subpaths $|SP_1 \cup SP_2|$ we have to select in order to determine whether node v is LD-reachable from u, which is much smaller compared with the average number of edges $|E_1 \cup E_2|$ in [8].

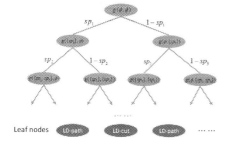
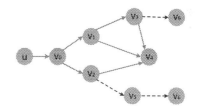

Fig. 5. Divide-Conquer Tree(DC-Tree)   **Fig. 6.** Branch Paths of Query (u, v_4)

Next, we focus on the problem in selecting the uncertain subpaths for each prefix group of the possible graph to minimize the average recursive depth. Even though we have reduced the average depth to some extent, experimental results reveal that it becomes ineffective when the scale of graph exceeds certain crucial value due to the exponential feature. It is observed that some branch paths that can not reach the destination are inevitably explored in the DFS search process. Obviously, enumerating the subpaths on the branch paths can neither form a L-path nor a L-cut, just leading to a high overhead on the contrary. Here we

---

Algorithm 2. BPFilter($\mathcal{G}, u, v, t$)

---

1 **if** $t = v$ /* t is the current node, u, v are source and destination nodes */ **then**
2 | **return** 1;
3 **end**
4 **if** t has no outgoing subpaths **then**
5 | **return** 0;
6 **end**
7 **for** each $t' \in N(t)$ /* t' is the neighbour node of t */ **do**
8 | **if** !BPFilter($\mathcal{G}, u, v, t'$) **then**
9 | | **delete** all sps between t, t';
10 | **end**
11 **end**

---

introduce a branch path pruning presented in Algorithm 2 that can significantly improve the efficiency of our algorithm by filtering branch paths.

For the example in Fig. 2, assuming the source and destination vertices are u and v_4, we can find five paths using the above traverse sequence. The paths are displayed in a tree form, as shown in Fig. 6. There are two branch paths that cannot reach v_4, marked as a dotted line. Obviously, enumerating the subpaths on the branch paths can neither form a L-path nor a L-cut. We delete these branch paths once they are recognized as having no connection with destination nodes. Algorithm 2 describes the pruning strategy in detail, which recursively deletes branch subpaths until reaching the split node that engenders the branch paths. Nevertheless, exponential run time can not be avoided, even we have removed a subpaths and reduced the complexity to $O(2^{m-a})$. Therefore, sampling techniques are further proposed in next section to give approximate solutions.

5   Sampling

We now focus on sampling mechanisms to gain quick and high precision approximations. The variance of an unbiased estimator proves to be the key criterion for evaluating the quality of an approximate approach based on an estimator. The smaller the variance, the more accurate the estimator.

A naive way is random walk sampling on possible graphs using the estimator, $\hat{R}_W = \frac{\sum_{i=1}^{n} I_{u,v}^{ld}(G_i)}{n}$. Since $\hat{R}_W$ is an unbiased estimator of the u-v LD-constraint reachability, its accuracy measured by variance can refer to $Var(\hat{R}_W) \approx \frac{1}{n}\hat{R}_W(1 - \hat{R}_W)$. However, this method can be rather computationally expensive due to the large size of graphs. In the following we will propose two more advanced sampling estimators which are more suitable to our problems.

5.1   DC-Tree Sampling

This subsection discusses how to estimate $R_{u,v}^{ld}$ using a more accurate approximation scheme. Instead of directly sampling on the graph, we propose to sample paths on the divide and conquer tree. Accordingly, we enumerate subpaths on the tree to find n paths (LD-paths or LD-cuts) and let $\hat{R}$ be the estimation result.

It is observed that $R^{ld}_{u,v}$ is only influenced by LD-paths as the values of LD-cuts are 0, thus contributing nothing to the reachability. Therefore, the more LD-paths we find, the more accurate the approximation is. On the other hand, when the depth of the tree becomes deeper, the sampling result $\hat{R}$ will be inaccurate if the sample size n is fixed. Next, we will give an effective estimation of $R^{ld}_{u,v}$ even for a fixed sample size n.

In the worst case, the DC-tree has at most 2^m leaf nodes, where m is the total number of subpaths. If we sample a path (either LD-path or LD-cut) in an improved tree with a relatively small average depth, the real number of paths we sample is 2^{m-k}, where k is the depth of path in the worst case. Then the total real number of paths we sample is $\sum_{i=1}^{n} 2^{m-k_i}$ when the sample size is n. If the n paths are uniformly sampled in the tree, we have the estimation $\hat{R}_T$:

$$\hat{R}_T = \frac{2^m \hat{R}}{2^{m-k_1} + 2^{m-k_2} + \cdots + 2^{m-k_n}} = \frac{\hat{R}}{\sum_{i=1}^{n} 2^{-k_i}} \qquad (7)$$

And the variance can be proved as:

$$Var(\hat{R}_T) = \frac{1}{n} \sum_{i=1}^{n} (\frac{\hat{R}}{2^{-k_i}} - R^{ld}_{u,v}(\mathcal{G}))^2 \qquad (8)$$

5.2 Yates-Grundy-Sen Estimator

In this subsection, we leverage the Yates-Grundy-Sen estimator to estimate $R^{ld}_{u,v}$. By sampling distinctive nodes, this estimator can improve the sample representativeness and the sampling efficiency. Assume we sample n distinctive leaf nodes in the divide and conquer tree, referred as the effective sample size. Let Pr_i be the weight associated with the i-th sampled leaf node, q_i be the leaf sampling probability, and the inclusion probability π_i be the probability to include leaf i in the sample, which is defined as $\pi_i = 1 - (1 - q_i)^n$. The Yates-Grundy-Sen estimator for $R^{ld}_{u,v}$ is:

$$\hat{R}_{YGS} = \sum_{i=1}^{n} \frac{Pr_i I^{ld}_{u,v}(\mathcal{G})}{\pi_i} \qquad (9)$$

The Yates-Grundy-Sen estimator is an unbiased estimator, and its variance can be derived as [15]:

$$Var(\hat{R}_{YGS}) = \sum_{i=1}^{n} \sum_{j>1}^{n} \left(\frac{\pi_i \pi_j - \pi_{ij}}{\pi_{ij}} \right) \left(\frac{Pr_i}{\pi_i} - \frac{Pr_j}{\pi_j} \right)^2 \qquad (10)$$

When n is fixed, Yates-Grundy-Sen estimator is more stable than other estimators (such as Hansen-Hurwitz estimator in [8]). The time complexity of this sampling estimator is $O(nm)$, where m is the average recursive depth from the root node to the leaf node in the enumeration tree.

To sum up, the random walk sampling is more costly than the other two for it has to invoke the corresponding algorithms computing $I_{u,v}^{ld}(G)$ in each sampled graph. With respect to effectiveness, DC-Tree sampling is more stable considering the sampling depth. Furthermore, to improve the estimator's accuracy Yates-Grundy-Sen estimator is introduced to avoid sampling the same nodes.

5.3   Probabilistic Filter of Sampling Results

For an uncertain graph $\mathcal{G}$, if there are a total m desired paths between nodes u and v, the reachability probability satisfies the following bounds based on Bonferroni and Chung-Erdos inequality.

$$\frac{\left(\sum\limits_{j=1}^{m} \Pr[P_k]\right)^2}{\left(\sum\limits_{k<j}^{m} \Pr[P_k \cap P_j] + \sum\limits_{j=1}^{m} \Pr[P_k]\right)} \leq \Pr\left[\bigcup_{k=1}^{m} P_k\right] \leq \sum_{k=1}^{m} \Pr[P_k] \qquad (11)$$

where P_k and P_j are arbitrary path-set events satisfying label and distance constraints. However, the above bounds need to find all the paths between nodes u, v. It become useless as we can't guarantee all the reachable paths would be found during the sampling process. We now give another upper-lower bound based on path/cut set presented in Theorem 3.

Theorem 3. *For arbitrary m desired paths of an uncertain graph, the reachability probability satisfies the following inequality.*

$$\prod_{i=1}^{n}\left(1 - \prod_{e \in C_i} \Pr(e)\right) \leq \Pr[P] \leq 1 - \prod_{i=1}^{m}\left(1 - \prod_{e \in P_i} \Pr(e)\right) \qquad (12)$$

Proof. Assume P_i is the path-set event, C_i is cut-set event, and e is the edge in the path/cut set. As path-set events are independent of each other when the paths are disjoint, we have $\Pr[P] = \Pr(P_1 \vee P_2 \vee ... \vee P_m) = 1 - \Pr(\bar{P}_1 \wedge \bar{P}_2 \wedge ... \wedge \bar{P}_m) = 1 - \Pr(\bar{A}_1)\Pr(\bar{A}_2) \wedge ... \wedge \Pr(\bar{A}_t)$, where $\bar{A}_k = \bar{P}_i \wedge ... \wedge \bar{P}_j$, and the paths between each $\bar{A}_k$ are disjoint. By proving $\Pr(\bar{A}_k) = \Pr(\bar{P}_i \wedge ... \wedge \bar{P}_j) \geq \Pr(\bar{P}_i) \wedge ... \wedge \Pr(\bar{P}_j)$, we can get the upper bound. Similarly, we get the lower bound based on cut-set.

Since it is hard to ensure that each sampling result is approximate to the real value, the above bounds can be used to prune large-deviation values. During each sampling process, we check if the sampling result falls in the bounds calculated by the current path and cut sets.

6   Experiments

In this section, comprehensive experimental studies are performed on both real and synthetic datasets. We are interested in the accuracy and computational efficiency of the estimators presented in this work: 1) random walk sampling on original graphs, $\hat{R}_{OW}$; 2) random walk sampling on subpath graphs, $\hat{R}_W$; 3) sampling on DC-Tree, $\hat{R}_T$; 4) Yates-Grundy-Sen estimator, $\hat{R}_{YGS}$. Experiments were performed on a 3.4GHz eight Core Intel Processor with 8GB of memory.

6.1   Synthetic Datasets

To study the accuracy and efficiency for answering LDCR queries, the synthetic datasets are generated in three steps. First, we generate random directed graphs based on the Scale-Free model. Second, we assign a random label associated with each edge in these generated random graphs, according to the power-law distribution. Last, the edge probability is randomly generated between 0 to 1 according to the uniform distribution.

The query time of each estimator is utilized to evaluate the computational efficiency. The effectiveness of these different sampling estimators can be mathematically captured by the mean squared error, $E(\hat{R} - R_{u,v}^{ld}(\mathcal{G}))^2$. Since all estimators studied in this paper are unbiased estimators, the mean squared error equals to $Var(\hat{R})$. Therefore, the variance of an unbiased estimator is the indicator of its accuracy. Specifically, the query time reported is the overall running time for each estimators to process a total of 1,000 random LDCR queries in seconds (s). The default parameters are setted as follows: $|V| = 50,000$, density $|E|/|V| = 4$, the total number of edge labels $|\Sigma| = 20$, the label constraint size $|L|/|\Sigma| = 20\%$, and the distance threshold $d = 15$.

Scalability: This experiment studies the scalability of different estimators by varying the graph size $|V|$ from 20,000 to 100,000. Figure 7(a) shows the query time of different sampling estimators. We can see that the computational costs of these estimators increase with the vertex number. The $\hat{R}_{YGS}$ and $\hat{R}_T$ estimators cost roughly the same time and are more effective than $\hat{R}_W$ and $\hat{R}_{OW}$ as we expected. Random walk sampling on original graphs is slower than on subpath graphs. In Fig. 7(d), compared with the baseline variance, the estimators $\hat{R}_W$, $\hat{R}_T$ and $\hat{R}_{YGS}$ achieves the variance reduction by 30%, 60% and 70% on average, respectively. Generally, the R_{YGS} and R_T estimators return more accurate answers for LDCR queries than random walk sampling.

Varying Label Constraint Size: In this experiment, we change the label constraint size from 10% to 90% of the total edge labels to compare how these approaches differ from each other. In Fig. 7(b), the estimators $\hat{R}_{YGS}$ and $\hat{R}_T$ have similar performance in that they both sample on DC-Tree and only vary in the computation of sampling. Note that the query time of $\hat{R}_W$ becomes close to that of $\hat{R}_{OW}$ since the subpath graph is almost equivalent to the original graph when $|L|/|\Sigma| = 90\%$. Figure 7(e) displays the corresponding variances of the four sampling estimators. When the label constraint size exceeds 50%, most queries become non-reachable. So the variances of all estimators keep stable with the increase of $|L|/|\Sigma|$.

Varying Distance Threshold: We change the distance threshold d from 10 to 30 to study how the distance constraint affects the estimation accuracy and performance. Figure 7(c) shows that the computational cost of $\hat{R}_W$ estimator is about double that of $\hat{R}_T$, while the cost of $\hat{R}_{OW}$ is even worse, about four times that of $\hat{R}_T$. On the other hand, the cost tendency for the four estimators at first increases slightly, then remains stable when the distance threshold exceeds the average diameter of graphs. Similarly, the variance efficiency of estimators $\hat{R}_W$, $\hat{R}_T$ and $\hat{R}_{YGS}$ remains unchanged when d increases, as shown in Fig. 7(f).

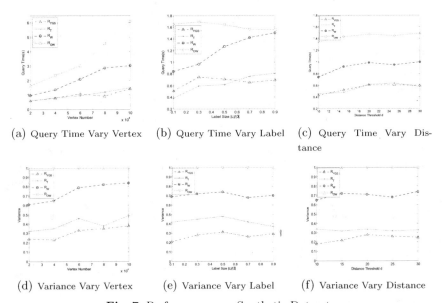

(a) Query Time Vary Vertex   (b) Query Time Vary Label   (c) Query Time Vary Distance

(d) Variance Vary Vertex   (e) Variance Vary Label   (f) Variance Vary Distance

Fig. 7. Performances on Synthetic Datasets

FindSP and BPFilter Performance: In this experiment, we evaluate the performance of our pruning methods: FindSP and BPFilter. With the vertex number varying from 20,000 to 100,000, Fig. 8(a) compares the size of original graphs and subpath graphs that directly affects the efficiency of DC-Tree. Based on the analysis of experimental results, we find that subpath number grows slightly with the vertex number and will keep the low increasing rate when the density and label constraint size are within the values of 8 and 50% respectively. BPfilter further decreases subpaths to at least one-tenth of the total subpaths, as shown in Fig. 8(b).

6.2   Real Datasets

We employ a real graph dataset in our experiments, i.e., the semantic knowledge network, Yago [16], which is extracted from Wikipedia, WordNet and GeoNames.

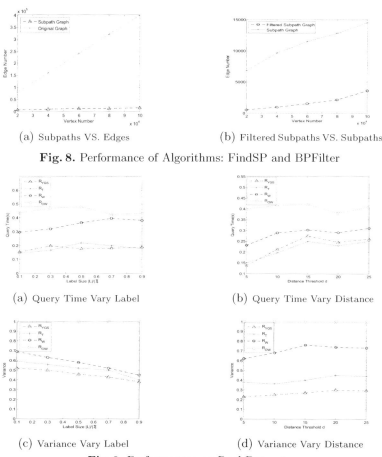

(a) Subpaths VS. Edges

(b) Filtered Subpaths VS. Subpaths

Fig. 8. Performance of Algorithms: FindSP and BPFilter

(a) Query Time Vary Label

(b) Query Time Vary Distance

(c) Variance Vary Label

(d) Variance Vary Distance

Fig. 9. Performances on Real Datasets

The Yago graph contains about 41,000 vertices with 80 labels. Since it is a standard graph, we generate each edge with a probability in the uniform distribution.

Figure 9 reports the running time and relative variance with respect to different approaches. Due to a relatively small graph size, the superiority of $\hat{R}_{YGS}$ and $\hat{R}_T$ is not conspicuous in Fig. 9(a) and 9(d). In addition, a larger label size leads to more unreachable sampled results, which can accordingly reduce the computation costs and the obtained variance. We also study the influence of varying distance thresholds on the effectiveness and efficiency of the four estimators. Again, the two sampling estimators $\hat{R}_{YGS}$ and $\hat{R}_T$ are the winners as they take only half the time of $\hat{R}_{OW}$. It is observed that both the query time and variances become stable when $d > 20$, as shown in Fig. 9(b) and 9(d).

7   Conclusions

In this paper, we address a new reachability problem with label and distance constraints in uncertain graphs. We introduce the subpath-based filtering framework by searching all the satisfactory subpaths to generated a subpath graph. Based on this graph, we utilize the divide-conquer approach and propose a branch path pruning strategy for LDCR queries. Different sampling methods and probabilistic bounds are proposed to approximate the reachability. Our experimental evaluation on both real and synthetic datasets validate that our pruning methods achieve high performance and the DC-Tree and Yates-Grundy-Sen sampling outperform the random walk sampling in both running time and accuracy.

Acknowledgments. This research is supported by the National Basic Research Program of China (973 Program) under Grant No. 2012CB316201, and the National Natural Science Foundation of China (61272179, 61033007, 61173028).

References

1. Jin, R., Xiang, Y., Ruan, N., Fuhry, D.: 3-hop: a high-compression indexing scheme for reachability query. In: SIGMOD Conference, pp. 813–826 (2009)
2. Jin, R., Hong, H., Wang, H., Ruan, N., Xiang, Y.: Computing label-constraint reachability in graph databases. In: SIGMOD Conference, pp. 123–134 (2010)
3. Zou, Z., Li, J., Gao, H., Zhang, S.: Finding top-k maximal cliques in an uncertain graph. In: ICDE, pp. 649–652 (2010)
4. Yuan, Y., Chen, L., Wang, G.: Efficiently answering probability threshold-based shortest path queries over uncertain graphs. In: Kitagawa, H., Ishikawa, Y., Li, Q., Watanabe, C. (eds.) DASFAA 2010. LNCS, vol. 5981, pp. 155–170. Springer, Heidelberg (2010)
5. Zou, Z., Gao, H., Li, J.: Discovering frequent subgraphs over uncertain graph databases under probabilistic semantics. In: KDD, pp. 633–642 (2010)
6. Aggarwal, C.C.: Managing and Mining Uncertain Data. Advances in Database Systems, vol. 35. Kluwer (2009)
7. Potamias, M., Bonchi, F., Gionis, A., Kollios, G.: k-nearest neighbors in uncertain graphs. PVLDB 3, 997–1008 (2010)
8. Jin, R., Liu, L., Ding, B., Wang, H.: Distance-constraint reachability computation in uncertain graphs. PVLDB 4, 551–562 (2011)
9. Yildirim, H., Chaoji, V., Zaki, M.J.: Grail: Scalable reachability index for large graphs. PVLDB 3, 276–284 (2010)
10. van Schaik, S.J., de Moor, O.: A memory efficient reachability data structure through bit vector compression. In: SIGMOD Conference, pp. 913–924 (2011)
11. Fan, W., Li, J., Ma, S., Tang, N., Wu, Y.: Adding regular expressions to graph reachability and pattern queries. Frontiers of Computer Science 6, 313–338 (2012)
12. Gubichev, A., Neumann, T.: Path query processing on very large rdf graphs. In: WebDB (2011)
13. Xu, K., Zou, L., Yu, J.X., Chen, L., Xiao, Y., Zhao, D.: Answering label-constraint reachability in large graphs. In: CIKM, pp. 1595–1600 (2011)
14. Garey, M.R., Johnson, D.S.: Computers and Intractability: A Guide to the Theory of NP-Completeness. W. H. Freeman (1979)
15. Yates, F., Grundy, P.: Selection without replacement from within strata with probability proportional to size. Journal of the Royal Statitical Society 15, 235–261 (1953)
16. Suchanek, F.M., Kasneci, G., Weikum, G.: Yago: a core of semantic knowledge. In: WWW, pp. 697–706 (2007)

Privacy-Preserving Reachability Query Services

Shuxiang Yin[1], Zhe Fan[2], Peipei Yi[2], Byron Choi[2],
Jianliang Xu[2], and Shuigeng Zhou[1]

[1] Shanghai Key Lab of Intelligent Information Processing, Fudan University, China
{sxyin,sgzhou}@fudan.edu.cn
[2] Computer Science Department, Hong Kong Baptist University, Hong Kong, China
{zfan,csppyi,choi,xujl}@comp.hkbu.edu.hk

Abstract. Due to the massive volume of graph data from a wide range of recent applications and resources required to process numerous queries at large scale, it is becoming economically appealing to outsource graph data to a third-party service provider ($\mathcal{SP}$), to provide query services. However, $\mathcal{SP}$ cannot always be trusted. Hence, data owners and query clients may prefer not to expose their data graphs and queries. This paper studies privacy-preserving query services for a fundamental query for graphs namely the reachability query where *both* clients' queries and the structural information of the owner's data are protected. We propose *privacy-preserving 2-hop labeling* (pp-2-hop) where the queries are computed in an encrypted domain and the input and output sizes of any queries are indistinguishable. We analyze the security of pp-2-hop with respect to cipher-text only and size based attacks. We verify the performance of pp-2-hop with an experimental study on both synthetic and real-world datasets.

1 Introduction

There is a wide range of emerging applications of graph-structured data, *e.g.*, bioinformatics, communication networks, social networks, web topology and semi-structured data. Many graph queries have been proposed to retrieve such graph data. However, as the volume of graph data is increasing at an unprecedented rate, hosting efficient query services has become a technically challenging task. The *owners* of graph data may not always be equipped with the expertise required to provide such services and therefore may employ *query service providers* ($\mathcal{SP}$s) to host *query services*, which are often supported by high performance computing. Security (such as the confidentiality of messages exchanged) has been stated as one of the attributes of Quality of Services (QoS) [23], as $\mathcal{SP}$ s cannot always be trusted. This attribute may influence the willingness of both data owners and query clients to use $\mathcal{SP}$'s services.

Take the reachability query — one of the most *fundamental and popular* graph queries [3, 4, 7, 8, 10, 16–18, 25, 26, 28–30, 32] — as an example: *Given two nodes u and v of a graph, the reachability query is used to test if v is reachable from u or not.* A query client may prefer not to expose his/her reachability query to an $\mathcal{SP}$. On the other hand, the data owner may prefer that the $\mathcal{SP}$ not be able to infer the structure of their graph data. Therefore, the query results must be protected as malicious $\mathcal{SP}$s can exploit the results from multiple queries to infer the graphs' structures.

Motivating Example. Consider a pharmaceutical company whose revenue depends mostly on the invention of Health Care Products, illustrated in Fig. 1. The company may

S.S. Bhowmick et al. (Eds.): DASFAA 2014, Part I, LNCS 8421, pp. 203–219, 2014.

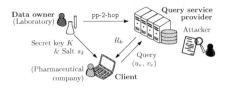

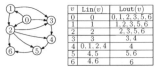

Fig. 1. Overview of the system model

Fig. 2. A (partial) schematic of a biological network (LHS) and its 2-hop labeling (RHS)

have discovered new compounds for a new product. To save laboratory experiments, it may often query the compounds from web-accessible biological pathway networks of massive size (such as [2]) to understand the compounds' characteristics, such as whether it is possible for the compounds to form other compounds via any chemical reactions (reachability in the network). However, on the one hand, the company does not want the $\mathcal{SP}$ to know about the queries (the compounds), as it may apply for patents for the synthesis. On the other hand, the owner of the pathway networks may not only lack the experience to host query services (which is best supported by IT companies, *e.g.*, [15]) but also be reluctant to release the networks to the public. Alternatively, the owner is willing to release a license only to paid users. Therefore, it is crucial to protect *both* the queries and the network from the $\mathcal{SP}$.

This paper studies the problem of *evaluating the reachability query at an $\mathcal{SP}$ without compromising the privacy of the reachability of the query nodes and the graph structure under cipertext-only and size-based attacks, in the paradigm of query services* (to be detailed with Fig. 1 in Sec. 3). To our knowledge, this has not been addressed before.

There have been many recent studies on efficient reachability query (*e.g.*, [3, 4, 6–8, 17, 18, 25, 26, 28, 29, 32]). Jin et al. [16] show that these studies can be roughly categorized into *transitive closure compressions*, *refined online search* and *hop labeling*. The first two categories suffer from high storage costs and require online searches, respectively (to be discussed in Sec. 2). In this paper, we propose our techniques based on hop labeling [10], in particular 2-hop labeling. The benefits of adopting 2-hop labeling are threefold. Firstly, the structures of 2-hop labels are simple, where each node is associated with two sets of nodes called Lin and Lout. Secondly, the query evaluation with 2-hop labeling is an intersection between an Lin and an Lout. Such simple structure and algorithm make privacy preservation plausible. Thirdly, 2-hop labeling is an active research topic. The recent works on large graph partitioning (*e.g.*, [8]), compression (*e.g.*, [7]) and maintenance (*e.g.*, [3, 25]) of 2-hop labeling can be readily adopted.

This paper proposes pp-2-hop (*privacy-preserving* 2-hop), which adopts 2-hop labeling. Firstly, the evaluation of each query on 2-hop labeling only involves an intersection between two sets of center nodes. Hence, we minimize the size of the maximum cardinality of the intersection results and add *minimal* artificial nodes (called surrogate nodes) to Lins and Louts such that the intersection results for all possible queries are of the *same* size. We unify each of the Lin and Lout labels such that the difference of the label set sizes are within a user-defined parameter. Secondly, we encrypt the 2-hop labels, after adding surrogate nodes and evaluate queries in the encrypted domain. We analyze the privacy of pp-2-hop.

The contributions of this paper are summarized as follows.

- We propose algorithms to unify the sizes of 2-hop labels and the query result sizes;
- We propose private query processing over the encrypted 2-hop labels;
- We propose a new heuristic 2-hop construction that yields 2-hop labeling that minimizes the intermediate results in private query processing; and
- We conduct an empirical study to confirm that our techniques are efficient.

The rest of this paper is organized as follows. We introduce some related work in Sec. 2. We then present the background, the problem and the overview of our solution in Sec. 3. We propose pp-2-hop, the index construction, optimization and query processing in Sec. 4. We conduct a privacy analysis in Sec. 5. We present an experimental evaluation in Sec. 6. We end this paper with the conclusion in Sec. 7.

2   Related Work

In this section, we present some related works on security of graph queries.

Authentication of Graph Queries. Query authentication is a security problem where the $\mathcal{SP}$ cannot be trusted. It requires the client to verify the correctness of the data graphs returned. Kundu et al. [21, 22] focus on the verification of the *authenticity* of a *given portion of data* (subtree/subgraph that users' have the right to access to) without leakage of extraneous information of the data (tree/graph/forest). They optimize the signature needed [22]. Zhe et al. [12] propose an efficient authenticated subgraph query services framework under the outsourced graph databases system. In comparison, the input of our problem is a reachability query not a subgraph for verification. Our problem focuses on *protecting both the query and its answer from $\mathcal{SP}$s*.

Privacy-Preserving Graph Query. He et al. [14] analyze the node reachability of the graph data, with the preservation of edge privacy. Unfortunately, the method reveals the reachability of the query nodes and partial structure of the graph data to the $\mathcal{SP}$. Gao et al. [13] propose neighborhood-privacy protected shortest distance in the paradigm of cloud computing. This aims to preserve all the neighborhood connections and the shortest distances between each pair of nodes in outsourced graph data. When this work is directly applied to reachability queries, some information about the graph structure and the reachability of the query nodes are still exposed to the $\mathcal{SP}$.

Mouratidis et al. [24] determine the shortest path of the query nodes with no information leakage by using the PIR [9] protocol. Firstly, the high computational cost of PIR has been well known. Secondly, the PIR approach requires the transfer of the same amount of data for every query, which can be large. Reachability queries are simple (yet fundamental) queries and do not require the use of the costly PIR method. Cao et al. [5] propose to support subgraph query over an encrypted database of small graphs. Their work protects the query privacy, index privacy and feature privacy. However, reachability queries cannot be expressed as subgraph queries. Karwa et al. [19] present an efficient algorithm for outsourcing useful statistics of graph data by protecting the edge *differential privacy* and they support counting queries but not reachability queries.

Approaches for Reachability Query. Numerous approaches for reachability query have been proposed recently in the literature. Jin et al. [16] recently discuss the approaches based on three main categories: *transitive closure compressions*, *refined online* *search* and *hop labeling*. (i) Transitive closure compressions (*e.g.*,[1, 17, 29]) offer the best query performance. It is known that their storage costs are the highest. (ii) Refined online search (*e.g.*, [26, 28, 32]) relies on online searches (*e.g.*, DFS and BFS) by definition leaks graph structure information and it is not clear how they can be adopted here. (iii) The seminal 2-hop labeling is proposed by Cohen et al. [10] and further optimized by many studies [7, 8, 25]. Due to the simple structure of the 2-hop labeling approach and the simple query evaluation (as motivated in Sec. 1), in this paper, we propose our techniques based on the 2-hop labeling approach.

3   Problem Formulation and Overview

This section formulates the problem studied and gives an overview of our solution.

Data Model. We consider *directed node-labeled graphs*. A graph is denoted as G, and $V(G)$ and $E(G)$ are the node and edge sets of G, respectively. Since the reachability information of nodes in a strongly connected component is identical, we assume directed acyclic graphs (DAG), for presentation brevity. A reachability query takes two nodes u and v as input, denoted as $u \rightsquigarrow v$, and returns true *iff* v is reachable from u.

System Model. We follow the system model that is commonly used in the literature of database outsourcing, presented in Fig. 1. The model consists of three parties:

- *Data owner*: An owner owns the graph data and computes the privacy-preserving 2-hop labeling offline *once*. It outsources them to the service provider and delivers query clients a salt s_2 to encrypt queries and the secret key K to decrypt results;
- *Service provider* ($\mathcal{SP}$): The $\mathcal{SP}$ has high computational utility such as cloud computing. The $\mathcal{SP}$ handles massive query requests over the encrypted data on behalf of the data owner and returns the encrypted results to clients; and
- *Client*: A client encrypts a query using s_2, sends the encrypted query to the $\mathcal{SP}$ and decrypts the result with K. We assume that the clients and $\mathcal{SP}$ do not collude.

Privacy Target. Our privacy target is required to keep the following two pieces of information private from *attackers* using the two attacks defined in the attack model:

- *Reachability of the query nodes* a.k.a the query result. In particular, given a reachability query $u \rightsquigarrow v$, attackers cannot infer whether u can reach to v; and
- *Graph structure* a.k.a. the topology of the data graph, *e.g.*, the existence of an edge.

Attack Model. As other outsourcing work, we assume the $\mathcal{SP}$s are *honest-but-curious* [20]. The attackers may be the $\mathcal{SP}$ or another adversary hacking the $\mathcal{SP}$. For simplicity, we *term the attackers as the* $\mathcal{SP}$. The $\mathcal{SP}$ can adopt the following popular attacks.[1]

- *Ciphertext only attack*, where the $\mathcal{SP}$ can access only the ciphertext (encrypted graph data) and does not know what their original graph is; and
- *Size based attack*, where the $\mathcal{SP}$ attempts to infer the two pieces of private information from the sizes of the data and the query results.

---

[1] There are admittedly many other attacks in the literature. We do not claim them in this paper since their theoretical privacy guarantees have yet to be established.

2-hop Labeling. As motivated in Sec 1, we undertake the 2-hop approach to address the problem. In 2-hop labeling, each node $u \in V(G)$ is associated with two sets of nodes, denoted as $\text{Lout}(u)$ and $\text{Lin}(u)$, called 2-hop *labels*. Nodes in $\text{Lout}(u)$ (respectively, $\text{Lin}(u)$) can be reachable from u (respectively, can reach u), which are also called *center nodes*. Given two nodes u and v, $u \rightsquigarrow v$ *iff* $\text{Lout}(u) \cap \text{Lin}(v) \neq \emptyset$.

Example 1. Consider the simplified biological pathway network shown in the LHS of Fig. 2. The node ID represents the chemical ID and the edge denotes a chemical reaction. The nodes' labels of the graph data are omitted for simplicity of presentation. The original 2-hop labeling (RHS) of the graph is shown in the RHS of Fig. 2. Consider two nodes 1 and 5. Node 1 can reach Node 5, *i.e.*, $1 \rightsquigarrow 5$. $\text{Lout}(1) \cap \text{Lin}(5) = \{5\}$. However, Node 0 is not reachable from Node 6 as $\text{Lout}(6) \cap \text{Lin}(0) = \emptyset$.

2-hop Construction. To ensure that the 2-hop labels of $V(G)$ contain all the reachability (a.k.a connectivity) information of G, the labels must cover all the elements in the transitive closure $T(G)$ of G. 2-hop labels (covers) are known to be costly to construct. In practice, they are constructed offline. In addition, there are various works that significantly optimize the construction time (*e.g.*, [8]).

The majority of previous 2-hop constructions focus on minimizing the size of 2-hop labeling, defined as $\sum_{u \in V(G)}(|\text{Lout}(u)| + |\text{Lin}(u)|)$. As we shall present our heuristic in Sec 4.1, we briefly outline the heuristic construction approach [10] for 2-hop construction. Initially, a variable T' is defined to represent the uncovered elements of $T(G)$, *i.e.*, $T' = T(G)$. Elements of $T(G)$ are iteratively covered and removed from T'. For each node $w \in G$, an undirected bipartite graph (*a.k.a* center graph) $G_w(L_w, R_w, E_w)$ is built, where L_w are nodes that can reach w and R_w are those w can reach. $(u, v) \in E_w$ *iff* (u, v) is in T'. The heuristic algorithm *selects* the center w whose induced subgraph $G_i(L_i, R_i, E_i)$ of G_w has the largest ratio $\texttt{maxDensCover}$ defined below.

$$\texttt{maxDensCover} = \frac{|E_i \cap T'|}{|L_i \cup R_i|} \qquad (1)$$

In each iteration, the node w with the largest $\texttt{maxDensCover}$ is selected as a center. For all $u \in L_i$ and $v \in R_i$, the algorithm adds w into $\text{Lout}(u)$ and $\text{Lin}(v)$, and removes (u, w), (w, v) and (u, v) from T'. The iterations terminate when T' is empty.

4   Privacy-Preserving 2-hop Labeling

This section presents the details of privacy-preserving 2-hop labeling. Sec. 4.1 presents a heuristic, to replace $\texttt{maxDensCover}$, that minimizes *the maximum cardinalities of the intersection results* $I_{\max}$ of 2-hop labels. In Sec 4.2, we propose a greedy algorithm that introduces a minimal number of surrogate nodes to Lins and Louts of the 2-hop labels derived from Sec. 4.1. In Sec. 4.3, we encrypt the 2-hop labels. In Sec. 4.4, we present their private query processing.

4.1   Imax-Aware 2hop Construction

In practice, $I_{\max}$ constructed by using $\texttt{maxDensCover}$ (outlined in Sec. 3) are very often large. In our experiments, the average $I_{\max}$ of the 2-hop labels using $\texttt{maxDensCover}$ of our real datasets is 378 (Table 7). Furthermore, large I_{max}s lead to large surrogate labels and thus high query costs.

In this section, we propose a heuristic to minimize I_{max} as 2-hop labels are constructed. The idea is that the heuristic includes the intersection information in 2-hop construction. Specifically, the objectives are to minimize the following two quantities:

1. the center node w that covers the most uncovered elements in T': As in previous work, such a heuristic leads to few center nodes and small 2-hop labels; and
2. the maximum cardinality of the intersection between $\text{Lout}(u)$ and $\text{Lin}(v)$, *i.e.*, $max(|\text{Lout}(u) \cap \text{Lin}(v)|)$.

To achieve the objectives, we propose the heuristic ratio maxISCover as follows:

$$\text{maxISCover} = \frac{|E_w \cap T'|}{max(|\text{Lout}(u) \cap \text{Lin}(v)|)}, \quad (2)$$

where $u \in L_w$, $v \in R_w$, T' is the uncovered elements in $T(G)$. Eqn. 2 contains two main parts. The details are as follows:

1. $E_w \cap T'$ is the uncovered elements in $T(G)$ covered by selecting w; and
2. $max(|\text{Lout}(u) \cap \text{Lin}(v)|)$ is the largest $|\text{Lout}(u) \cap \text{Lin}(v)|$ of all $u, v \in V(G)$.

Putting these together, we use the 2-hop construction presented at the end of Sec. 3, replacing maxDensCover with maxISCover.

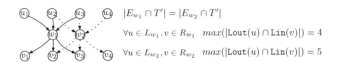

$|E_{w_1} \cap T'| = |E_{w_2} \cap T'|$

$\forall u \in L_{w_1}, v \in R_{w_1} \quad max(|\text{Lout}(u) \cap \text{Lin}(v)|) = 4$

$\forall u \in L_{w_2}, v \in R_{w_2} \quad max(|\text{Lout}(u) \cap \text{Lin}(v)|) = 5$

Fig. 3. Illustration of selection of center node by maxISCover

Example 2. We compare maxDensCover and maxISCover with a schematic shown in Fig. 3. The graph data is on the LHS. Suppose there are only two center nodes (w_1 and w_2) checked in an iteration respectively. $L_{w_1} = \{u_1, u_2, u_3\}$, $R_{w_1} = \{v_1, v_2, v_3\}$, and $L_{w_2} = \{u_2, u_3, u_4\}$, $R_{w_2} = \{v_2, v_3, v_4\}$. We assume $|E_{w_1} \cap T'| = |E_{w_2} \cap T'|$. maxDensCover may select either w_1 or w_2. Further assume that $\forall u \in L_{w_1}, v \in R_{w_1}$, $max(|\text{Lout}(u) \cap \text{Lin}(v)|) = 4$ and $\forall u \in L_{w_2}, v \in R_{w_2}, max(|\text{Lout}(u) \cap \text{Lin}(v)|) = 5$. maxISCover selects w_1 as it results in smaller intersections.

4.2   Addition of Surrogate Nodes

Next, we add surrogate nodes to 2-hop labeling to unify the sizes of intersection results and the Lins and Louts.

Unification of Intersection Results. The main idea of adding surrogate nodes into 2-hop labels (Lins and Louts) is to achieve that for each u and v, the intersection between surrogate labels $\text{Lout}^s(u)$ and $\text{Lin}^s(v)$ always equals to I_{max}.[2] Our objective

---

[2] For brevity, we assume the data graph does not contain highly disconnected nodes where its I_{max} equals 0 or 1. To cater for such special cases, we may further set a minimum I_{max} (*e.g.*, 4) so that the $\mathcal{SP}$ cannot infer whether the graph is highly disconnected or not.

Algorithm 1. Unify Intersection Sizes unifyIS(Lout, Lin)

Input: 2-hop labeling Lout and Lin
Output: Lout^s and Lin^s

1: Initialize surrogate node set $D_w = [\ ]$

2: create a priority queue Q_v for $V(G)$, where the rank of $v \in V(G)$ is defined by:
   $\quad max_{u \in V}(I_{\max} - |\text{Lout}(u) \cap \text{Lin}(v)|)$

3: **while** $v \neq$ null, where $v \leftarrow Q_v.\text{getNext}()$ //scan Lins

4:     $D_w.\text{movetofront}()$ //move to the front of D_w

5:     **while** $d_i \leftarrow D_w.\text{getNext}()$

6:         **if** $d_i =$ null and $\exists u', v'\ |\text{Lout}(u') \cap \text{Lin}(v')| < I_{\max}$
               $d_i \leftarrow$ **new** node() and $D_w.\text{push}(d_i)$
           else break

7:         **for each** $u \in V(G)$, where $|\text{Lout}(u) \cap \text{Lin}(v)| < I_{\max}$ //scan Louts
           //if the *intersection constraint* is satisfied.

8:         **if** $\psi(\text{2-hop}, u, v, d_i)$ is true

9:             $\text{Lout}(u) \leftarrow \text{Lout}(u) \cup \{d_i\}; \text{Lin}(v) \leftarrow \text{Lin}(v) \cup \{d_i\}$

10: **return** (Lout, Lin) **as** (Lout^s, Lin^s)

---

is thus to introduce the smallest number of surrogate nodes (*i.e.*, the 2-hop label size is minimized). As the cardinality for each intersection is the same, the $\mathcal{SP}$ gains no connectivity knowledge about specific information of each pair of nodes.[3] We call the problem the *minimum addition of surrogate nodes* (MASN), defined below.

Definition 1. *The problem of minimum addition of surrogate nodes (MASN) is that given the* 2-hop *labels of a graph G, introduce surrogate nodes in both* Lins *and* Louts *to obtain* Lin^{s}s *and* Lout^{s}s, *such that*

- $\forall u, v \in V(G), |\text{Lout}^s(u) \cap \text{Lin}^s(v)| = I_{\max};$
- *the largest* Lin^s *and the largest* Lout^s *are minimized; and*
- $\sum_{u \in V(G)}(|\text{Lin}^s(u)| + |\text{Lout}^s(u)|)$ *is minimized.*

Proposition 1. *The problem of* MASN *is NP-hard.*

It is not surprising that the problem of MASN is NP-hard, as presented in Prop. 1. Due to space constraints, we provide its proof in [27]. We propose a greedy algorithm called unifyIS (shown in Algo. 1) to solve the problem of MASN.

The input of unifyIS is 2-hop labels (*i.e.*, Louts and Lins). The output is the 2-hop labels with surrogate nodes, *i.e.*, Lout^s and Lin^s. The main idea is that for any surrogate node d, we add d to as many Lins and Louts as possible, in $O(|V|)$, and hence more entries in $T(G)$, in $C(|V|^2)$, can be covered. Specifically, we maintain a list of surrogate nodes D_w as we add them to the 2-hop labels (Line 1). unifyIS processes the Lins whose intersection with some other Louts has the largest gap between $I_{\max}$ earlier (Lines 2-3).[4] As unifyIS may use new surrogate nodes to reduce such a gap, the surrogate nodes can be used to reduce the gaps between the Lin and many other Louts. In Lines 5-9, unifyIS processes the Louts iteratively. Line 6 checks if there is a Lout to be processed (*i.e.*, $|\text{Lout}(u) \cap \text{Lin}(v)| < I_{\max}$). If so, and all existing surrogate

---

[3] In 2-hop labeling, the sizes of the intersection results of a query indicate the reachability of nodes. Only unreachable query nodes have a value of 0.

[4] Given Lins and Louts, the value of $I_{\max}$ can be easily computed.

v	$\mathrm{Lin}^s(v)$	$\mathrm{Lout}^s(v)$
0	$0,7,8,9$	$0,1,2,3,5,6,7,8$
1	$1,7,8,10$	$1,2,3,5,6,7,8,9$
2	$2,7,8,11$	$2,3,5,6,7,8,9,10$
3	$3,7,8,12$	$3,4,7,8,9,10,11$
4	$0,1,2,4,9,10,13$	$4,7,8,9,10,11,12$
5	$4,5,7,8,14$	$5,6,7,8,9,10,11,12,13$
6	$4,6,7,8$	$6,7,8,9,10,11,12,13,14$

v	$\mathrm{Lin}^s(v)$	$\mathrm{Lout}^s(v)$
:	:	:
4	$0,1,2,4,9,10,13$	$4,7,7_1,7_3,.,8,9,10,11,12$
5	$4,5,7_1,7_2,8,14$	$5,6,7,7_2,8,9,10,11,12,13$
6	$4,6,7_2,7_3,8$	$6,7,7_2,8,9,10,11,12,13,14$

Encryption of $\mathrm{Lin}^s(6)$:
$\mathrm{Lin}^e(h_{s_2}(6)) = \{(h_{s_1}(4), E(0)), \cdots, (h_{s_1}(8), E(1))\}$

Fig. 4. The 2-hop labels of the network in Figure 2 after the addition of surrogate nodes

Fig. 5. One iteration of split on 2-hop labels shown in Figure 4 and the encrypted $\mathrm{Lin}^s(6)$

nodes have been used in previous iterations, then a new surrogate node is created (Line 6). Otherwise, unifyIS terminates. The surrogate node is added to both $\mathrm{Lout}(u)$ and $\mathrm{Lin}(v)$ (Line 9), if the *intersection constraint* ψ (defined below) in Line 8 is satisfied:

$$\psi(2\text{-hop}, u, v, d_i) : \forall v', |(\mathrm{Lout}(u) \cup \{d_i\}) \cap \mathrm{Lin}(v')| \leq I_{\max} \wedge \\ \forall u', |\mathrm{Lout}(u') \cap (\mathrm{Lin}(v) \cup \{d_i\})| \leq I_{\max}, \quad (3)$$

where $u', v' \in V(G)$. In other words, there is no $|\mathrm{Lout}(u) \cap \mathrm{Lin}(v)| > I_{\max}$ after adding d_i to both $\mathrm{Lout}(u)$ and $\mathrm{Lin}(v)$. Thus, if ψ is true, we can add d_i into both $\mathrm{Lout}(u)$ and $\mathrm{Lin}(v)$. The algorithm terminates when for all pairs of nodes u and v, $|\mathrm{Lout}(u) \cap \mathrm{Lin}(v)| = I_{\max}$.

Discussion. The number of distinct surrogate nodes used by unifyIS is $I_{\max}|V|$ in the worst case and $I_{\max}$ in the best case. Therefore, the total number of surrogate nodes added by unifyIS is $I_{\max}|V|(1 + |V|)$ in the worst case and $2I_{\max}|V|$ in the best case, where $|V|$ is the number of nodes in the graph.

Example 3. Figure 4 shows the 2-hop labeling after the addition of surrogate nodes to that in Figure 2. The IDs of the real centers are highlighted in bold in Lin^ss and Lout^ss. Other IDs represent surrogate nodes. The $I_{\max}$ of the 2-hop labels is 3.

Consider the first iteration of unifyIS. Since $I_{\max} - |\mathrm{Lin}(0)| \cap |\mathrm{Lout}(1)| = I_{\max}$, Node 0 is one of the nodes with the highest rank in Q_u (Lines 2 and 3). There are obviously some intersection results of Lin and Lout smaller than $I_{\max}$. Moreover, d_i is null as D_w is empty. A new surrogate (Node 7) is created (Line 6). Node 7 is introduced to both $\mathrm{Lin}(0)$ and $\mathrm{Lout}(1)$ as this does not violate ψ (Lines 8-9). Next, Node 7 is added to Louts of Nodes 0 and 2-6 as the addition does not violate ψ (Lines 7-9). In the next few iterations (Lines 5-9), unifyIS adds new surrogate nodes 8 and 9 to $\mathrm{Lin}(0)$ and the corresponding Louts until all intersection results of $\mathrm{Lin}(0)$ and other Louts are of the size $I_{\max}$.

Then, unifyIS processes the next node in Q_u (Line 3), for example, Node 1. unifyIS uses Node 7 (Lines 4-5) and adds it to $\mathrm{Lin}(1)$ as it does not violate ψ (Lines 7-9). Similarly, Node 8 is added to $\mathrm{Lin}(1)$. However, Node 9 cannot be added to $\mathrm{Lin}(1)$ as $|\mathrm{Lin}(1) \cap \mathrm{Lout}(1)| = |\{\mathbf{1}, 7, 8\}| = I_{\max}$.

Consider the surrogate node 7 from Figure 4. It has been added 13 places. However, it covers 6×7 elements in $T(G)$. For any query u and v, the intersection size is the same. Following the same queries of Example. 1, $\mathrm{Lout}^s(1) \cap \mathrm{Lin}^s(5) = \{5, 7, 8\}$ and $\mathrm{Lout}^s(6) \cap \mathrm{Lin}^s(0) = \{7, 8, 9\}$. The result size is always 3.

Unification of Labeling Sizes. In order to unify the sizes of both Lin^ss and Lout^ss, we introduce a (postprocessing) unifyLin operation of surrogate nodes. Since the task

Algorithm 2. Unify $\texttt{Lin}^s$ $\texttt{unifyLin}(\texttt{Lout}^s, \texttt{Lin}^s, \delta)$

Input: 2-hop labeling $\texttt{Lout}^s$ and $\texttt{Lin}^s$ after addition of surrogate nodes, and the user-specified allowable differences δ in the sizes of $\texttt{Lin}^s$ and $\texttt{Lout}^s$.

Output: $\texttt{Lout}^s$ and $\texttt{Lin}^s$

1: **for each** u, $|\texttt{Lin}^s(u)| + \delta \leq \texttt{Lin}_{\max}$
2:   choose a surrogate node w from $\texttt{Lin}^s(u)$
3:   $W = \{w_1, \cdots, w_n, w_{n+1}\}, n \leq \texttt{Lin}_{\max} - |\texttt{Lin}^s(u)| - 1$
4:   $\texttt{Lin}^s(u) \leftarrow (\texttt{Lin}^s(u)/\{w\}) \cup \{w_1, \cdots, w_{n+1}\}$
5:   **for each** u', where $|\texttt{Lin}^s(u')| < \texttt{Lin}_{\max} \wedge w \in \texttt{Lin}^s(u')$
6:     $\texttt{Lin}^s(u') \leftarrow (\texttt{Lin}^s(u')/\{w\}) \cup \{w_{n+1}, w_{n+2}\}$
7:   **for each** v, where $|\texttt{Lout}^s(v)| = \texttt{Lout}_{\max} \wedge w \in \texttt{Lout}^s(v)$
8:     $\texttt{Lout}^s(v) \leftarrow \texttt{Lout}^s(v) \cup \{w_{n+1}\}$
9:   **for each** v, where $|\texttt{Lout}^s(v)| < \texttt{Lout}_{\max} \wedge w \in \texttt{Lout}^s(v)$
10:     $\texttt{Lout}^s(v) \leftarrow \texttt{Lout}^s(v) \cup \{w_i, w_{n+2}\}$,
        where $i \in [1,n]$ and each i is added to some $\texttt{Lout}^s$ at least once.

---

of unifying $\texttt{Lin}^s$ and $\texttt{Lout}^s$ are symmetric, we only discuss $\texttt{Lin}^s$ for a concise presentation. We first denote the size of the largest $\texttt{Lin}^s$ (respectively, $\texttt{Lout}^s$) as $\texttt{Lin}_{\max}$ (respectively, $\texttt{Lout}_{\max}$). The intuitions of our approach can be described as follows: (1) for each node u in the graph where $|\texttt{Lin}^s(u)| < \texttt{Lin}_{\max}$, we split its surrogate nodes in $\texttt{Lin}^s(u)$ such that the size of $\texttt{Lin}^s(u)$ is closer to $\texttt{Lin}_{\max}$; and (2) we never increase the $I_{\max}$ during unification. We formally define the problem, namely *unification of the labeling size* (ULS) as follows.

Definition 2. *The problem of the unification of the labeling size (ULS) is that given* $\texttt{Lin}^s s$ *and* $\texttt{Lout}^s s$*, unify both of their sizes, s.t.,* $\forall u, v$,

$$\frac{||\texttt{Lin}^s(u)| - \texttt{Lin}_{\max}|}{\texttt{Lin}_{\max}} \leq \delta \text{ and } \frac{||\texttt{Lout}^s(u)| - \texttt{Lout}_{\max}|}{\texttt{Lout}_{\max}} \leq \delta,$$

where δ *is a user-specified parameter for the allowable difference in the label sizes.*

We propose $\texttt{unifyLin}$ algorithm presented in Algo. 2 to solve the problem of ULS. Suppose that we have sorted the $\texttt{Lin}^s$s by their sizes in descending order and we apply $\texttt{unifyLin}$ to each $\texttt{Lin}^s$ accordingly. We first choose a surrogate node w from $\texttt{Lin}^s(u)$ (Lines 1-2). We declare new surrogate nodes $\{w_1, \cdots, w_n, w_{n+1}\}$, where $n \leq \texttt{Lin}_{\max}$ - $|\texttt{Lin}^s(u)|$ - 1 (Line 3). We replace w by $\{w_1, \cdots, w_{n+1}\}$ in $\texttt{Lin}^s(u)$ (Line 4). For each u' where $w \in \texttt{Lin}^s(u')$ and $|\texttt{Lin}^s(u')| < \texttt{Lin}_{\max}$, we replace w with w_{n+1} and w_{n+2} in $\texttt{Lin}^s(u')$ (Lines 5-6). For each v where $|\texttt{Lout}^s(v)| = \texttt{Lout}_{\max}$ and $w \in \texttt{Lout}^s(v)$, we add w_{n+1} (Lines 7-8). For each v where $|\texttt{Lout}^s(v)| < \texttt{Lout}_{\max}$ and $w \in \texttt{Lout}^s(v)$, we add $\{w_i, w_{n+2}\}$, where $i \in [1,n]$ (Lines 9-10).

Example 4. We illustrate Algo. 2 with reference to the 2-hop labels shown in Figure 4. The result after one $\texttt{unifyLin}$ operation is shown in Figure 5. Suppose we choose Node 7 from $\texttt{Lin}^s(5)$ (Line 2). $\texttt{Lin}_{\max} = 7$ and $|\texttt{Lin}^s(5)| = 5$. Therefore, $n+1$ can be 2. We replace Node 7 with $\{7_1, 7_2\}$ in $\texttt{Lin}^s(5)$ (Line 4). Since Node 7 appears in some other $\texttt{Lin}^s$, we add $\{7_2, 7_3\}$ to $\texttt{Lin}^s(i)$, $i \in \{0,1,2,3,6\}$ (Lines 5-6). $\texttt{Lout}_{\max} = |\texttt{Lout}^s(5)| = |\texttt{Lout}^s(6)|$. We add Node 7_2 into $\texttt{Lout}^s(5)$ and $\texttt{Lout}^s(6)$ (Lines 7-8). We add Node 7_3 into $\texttt{Lout}^s(j)$ (Lines 9-10), where $j = \{0,1,2,3,4\}$.

One may verify that the sizes of intersection results do not change with a simple case analysis. In the meantime, all the surrogate nodes due to unifyLin appear in some intersection results. The increase in the sizes of $\mathtt{Lin}^s$ and $\mathtt{Lout}^s$ due to a unifyLin operation can be listed as follows: (i) the size of $\mathtt{Lin}^s(u)$ is increased by $n+1$ (Line 4); (ii) the size of $\mathtt{Lin}^s$ that contains w is increase by two (Lines 5-6); (iii) the sizes of $\mathtt{Lout}^s$ with w whose sizes are $\mathtt{Lout}_{max}$ are increased by one (Lines 7-8); and (iv) the $\mathtt{Lout}^s$s that contains w whose sizes are smaller than $\mathtt{Lout}_{max}$ increase by two (Lines 9-10). We then alternately apply unifyLin operations on $\mathtt{Lin}^s$ and $\mathtt{Lout}^s$ until the sizes do not differ from the largest labels by δ (as stated in Def. 2).

4.3   Index Encryption

After adding the surrogate nodes, the remaining task is to encrypt the labels . In order to distinguish the real nodes and surrogate nodes, we implement the nodes with flag values (see Def. 3). The flag values of real nodes are 0, and 1 otherwise. We will present how to use the flags to encode the query result in Sec. 4.4. Below is the new definition of *center*s for $\mathtt{Lout}^s$s or $\mathtt{Lin}^s$s.

Definition 3. *Each* center *of* $\mathtt{Lout}^s(u)$ *or* $\mathtt{Lin}^s(v)$ *is a binary tuple* (w, f), *where* $f = 0$ *if* w *is a real center, and* 1 *otherwise.*

Based on Def. 3, we encrypt the surrogate labels in order to protect both the reachability of the query nodes and the graph structure. (i) To hide any association between the nodes and the center nodes, we hash the w in (w, f) and the u in $\mathtt{Lout}^s(u)$ and $\mathtt{Lin}^s(u)$ with a *one-way collision-resistant hash function* with different salts, denoted as $h_{s_1}(w)$ and $h_{s_2}(u)$, to hash them respectively. Recall that that $w = u$ does not imply $h_{s_1}(w) = h_{s_2}(u)$, where $s_1 \neq s_2$. (ii) Regarding the encryption of the flag value, we use Elgamal $\mathtt{E}(\cdot)$ [11], which is a *multiplicative homomorphic encryption* method. The benefits of Elgamal are twofold: (1) since the flag has binary values, Elgamal ensures randomness in the encrypted flags; (2) Elgamal allows one decryption at the client side. To sum up, the definition of the privacy-preserving 2-hop labeling is given as follows.

Definition 4. *Each* encrypted center *is a binary tuple* (w_e, f_e), *where* $w_e = h_{s_1}(w)$ *and* $f_e = \mathtt{E}(f)$. *The* privacy-preserving 2-hop (pp-2-hop) *is a* 2-hop *labeling where each encrypted node* u_e, *where* $u_e = h_{s_2}(u)$, *is associated with two sets of encrypted* $\mathtt{Lout}^s(u)$ *and* $\mathtt{Lin}^s(u)$, *denoted as* $\mathtt{Lout}^e(u)$ *and* $\mathtt{Lin}^e(u)$.

Example 5. Fig. 5 illustrates an example of the encryption of $\mathtt{Lin}^s(6)$ for node 6. The encryption of $\mathtt{Lin}^s(6)$ is denoted as $\mathtt{Lin}^e(h_{s_2}(6))$, where $h_{s_2}(6)$ is the encryption of node 6. For example, the first center of $\mathtt{Lin}^s(4)$, $(4, 0)$, is encrypted as $(h_{s_1}(4), \mathtt{E}(0))$.

4.4   Private Query Processing

Based on the encryption of the pp-2-hop labeling in Def. 4, we present its query processing without decryption. There are three main steps: (1) The client encrypts the query — the query $u \rightsquigarrow v$ is hashed to $u_e \rightsquigarrow v_e$; (2) The $\mathcal{SP}$ intersects $\mathtt{Lout}^e(u_e)$

and $\text{Lin}^e(v_e)$ and returns the encrypted result R_e to the client; and (3) The client uses the secret key K and an Elgamal decryption to decrypt the result decryption.

Naïve Solution. The naïve solution for processing a query $u \rightsquigarrow v$ is to perform an intersection on the centers in $\text{Lout}^e(u_e)$ and $\text{Lin}^e(v_e)$ and transmit the encrypted flag of the centers in the intersection results to clients. The client decrypts each of the encrypted flag and checks if there is at least one flag that signifies a real center. However, this solution requires $I_{\max}$ decryptions.

Multiplicative Homomorphic Query Processing. It is known that decryption is costly, especially when the client is not equipped with powerful hardware. Therefore, we propose a query processing that requires *one* decryption at the client side. We define the intersection result of u_e and v_e as $R(u_e,v_e)$, or simply R, where

$$R = \{(w_e, f_e) \mid (w_e, f_e) \in \text{Lout}^e(u_e) \text{ and } (w_e, f'_e) \in \text{Lin}^e(v_e)\}.$$

The encrypted result R_e, defined as $\prod_{(w_e, f_e) \in R} f_e$,[5] is transmitted to the client. At the client side, the client decrypts R_e by using the secret key K. If the decrypted message is 0, then u can reach v. Otherwise, u cannot reach v. Note that R_e is a product of flag values. The product is 0 *iff* there is a real node (whose flag is 0) in the intersection result. That is, if all centers in the results are surrogates, the product R_e is 1.

Example 6. Consider the private query processing of the query $1 \rightsquigarrow 5$, following Example 1 for clarity. The query processing on pp-2-hop in Example 3 is similar: (1) the client hashes the query nodes as $h_{s_2}(1)$ and $h_{s_2}(5)$ by using the salt s_2 from the data owner, and issues to the $\mathcal{SP}$; (2) the $\mathcal{SP}$ performs $\text{Lout}^e(h_{s_2}(1)) \cap \text{Lin}^e(h_{s_2}(5))$ and obtains $\{(h_{s_1}(5), \text{E}(0)), (h_{s_1}(7), \text{E}(1)), (h_{s_1}(8), \text{E}(1))\}$. Based on the result, the $\mathcal{SP}$ computes the result $R_e = \text{E}(0) \times \text{E}(1) \times \text{E}(1) = \text{E}(0)$ and returns it to the client; and (3) the client decrypts the R_e, which is 0, and obtains that Node 1 can reach Node 5.

5   Analysis of Privacy

In this section, we provide an analysis of the privacy under the assumptions of our attack model, *i.e.*, the size based attack and ciphertext only attack (stated in Sec. 3).

Privacy against Ciphertext only Attack. We prove that the reachability of the query nodes and the topology of the graph have been protected from the $\mathcal{SP}$ under the *ciphertext only attack*.

Proposition 2. *The $\mathcal{SP}$ breaks the reachability of query nodes only if the $\mathcal{SP}$ breaks either the one-way collision-resistant hash function or the Elgamal encryption.*

Proof. (Sketch) **Case 1:** (i) Suppose the $\mathcal{SP}$ can break the Elgamal encryption. The $\mathcal{SP}$ can determine whether the flag of a center signifies a real center or not. During query processing, the $\mathcal{SP}$ can analyze the intersection result R. The $\mathcal{SP}$ identifies the reachability of a pair of query nodes by checking if there is a real center R.

(ii) If the $\mathcal{SP}$ can break the hash function (*e.g.*, SHA-1), it can determine the center identities, *i.e.*, the center IDs in Lin^s or Lout^s. Then, the $\mathcal{SP}$ can check if a center is real by checking if it has corresponding Lin^s and Lout^s in pp-2-hop.

---

[5] We use $\prod$ and $\times$ to denote the modular multiplications in the Elgamal encryption scheme.

Case 2: Suppose the $\mathcal{SP}$ cannot break the one-way collision-resistant hash function (*e.g.*, SHA-1) and the Elgamal encryption. We analyze step by step the information the $\mathcal{SP}$ obtains during query processing. Given a query u_e and v_e, the $\mathcal{SP}$ retrieves $\text{Lout}^s(u_e)$ and $\text{Lin}^s(v)$. The $\mathcal{SP}$ computes R_e under the Elgamal encryption.

Since the $\mathcal{SP}$ cannot break the one-way collision-resistant hash function, it cannot determine either the nodes of the query (u_e and v_e) or the centers in $\text{Lout}^s(u_e)$ and $\text{Lin}^s(v_e)$. Moreover, since we assume that the $\mathcal{SP}$ cannot break the Elgamal encryption, it cannot determine the flags of the centers in $\text{Lout}^s(u_e)$ and $\text{Lin}^s(v_e)$. Due the homomorphic multiplication supported by the Elgamal encryption, the $\mathcal{SP}$ cannot determine the plaintext of R_e. Thus, the $\mathcal{SP}$ does not know the reachability of query nodes.

By exploiting the preservation of the reachability of any two nodes, we prove that pp-2-hop protects the graph structure from the $\mathcal{SP}$. It is straightforward to argue that it is not possible to determine the *existence of an edge* in a graph under pp-2-hop. Hence, it is not possible to infer the topology of the graph structure.

Proposition 3. *The $\mathcal{SP}$ can determine the existence of an edge only if it breaks either the one-way collision-resistant hash function or the Elgamal encryption.*

Proof. We establish the proposition via proof by contradiction. Suppose the $\mathcal{SP}$ can determine the existence of one edge (u,v). The $\mathcal{SP}$ has broken the reachability of at least one query $u \rightsquigarrow v$. By Prop. 2, this is possible only if the $\mathcal{SP}$ breaks either the one-way collision-resistant hash function or the Elgamal encryption.

Privacy against Size-Based Attack. In addition to the analysis of privacy against ciphertext only attack, we prove privacy under *size-based attack*.

Proposition 4. *When δ is set to 0, the reachability of the query nodes is perfectly protected against size-based attack.*

Proof. We prove the proposition via proof of contradiction. Suppose the $\mathcal{SP}$ can determine the reachability of the query nodes, $u_e \rightsquigarrow v_e$, under size-based attack. The $\mathcal{SP}$ can thus infer the reachability from (1) the size of $\text{Lout}^e(u_e) \cap \text{Lin}^e(v_e)$; and (2) the size of both $\text{Lout}^e(u)$ and $\text{Lin}^e(v_e)$. However, the size of $\text{Lout}^e(u_e) \cap \text{Lin}^e(v_e)$ always exactly equals $I_{\max}$, and $|\text{Lout}^e(u_e)| = \text{Lout}_{\max}$, $|\text{Lin}^e(v_e)| = \text{Lin}_{\max}$. Therefore, the $\mathcal{SP}$ gains zero information content from the sizes.

Proposition 5. *When δ is set to 0, the graph structure is perfectly protected against size based attack.*

Proof. The proof is similar to that of Prop. 4.

In practice, δ may not necessarily be set to 0 as the sizes of Lin^s and Lout^s do not directly represent the connectivity of a node after surrogates are added to Lin^s and Lout. However, a non-zero value of δ requires non-trivial privacy analysis to quantify the information leakage. Hence, we omit its analysis.

6   Experimental Evaluation

In this section, we present the experimental evaluation that verifies the performance of our proposed techniques and the effectiveness of our optimization.

6.1   Experimental Setup

Running Platform. We conducted all experiments using a machine with Intel Core i3-2310 2.10GHz CPU and 4G RAM running Windows 7 OS. All algorithms were implemented using C++ based on the implementation of 2-hop labelings provided by R.Bramandia et al. [3]. The hash function (h_{s_1} and h_{s_2}) was 160-bit SHA-1 using two different salts. The encryption E was 1024-bit Elgamal [11].

Datasets. We used three synthetic datasets (denoted as SYN) and four real-world datasets. Some of their characteristics are shown in Tables 1 and 2. The synthetic datasets were all scale-free graphs, which are popular in experimentation. The generator used was provided by Choi et al. [33]. We controlled the sizes and densities of the graphs by setting $\alpha = 0.27$ and $\beta = 10$. The real-world datasets are all publicly available.[6]

Table 1. Synthetic datasets

Synthetic graph G	$|V(G)|$	$|E(G)|$	$|E(G)|/|V(G)|$
SYN-1	3073	37615	12.24
SYN-2	5651	15968	2.83
SYN-3	4880	27946	5.73

Table 2. Real-world datasets

Real graph G	$|V(G)|$	$|E(G)|$	$|E(G)|/|V(G)|$
YEAST	2361	7182	3.04
ODLIS	2909	18419	6.33
ERDOS	6927	11850	1.71
ROGET	1022	5075	4.97

Query Sets. For each of the synthetic and real-world datasets, we generated 1000 random queries, 50% of which were positive queries, generated from the transitive closure, and 50% were negative queries.

Heuristics. We have implemented the classical 2-hop heuristic maxDensCover [10], the heuristic maxSetCover $= |E_w \cap T'|$, also proposed by Cheng et al. [8], and our heuristic maxISCover. These heuristics are plugged into 2-hop construction (Sec. 3). δ is set to 0 by default.

6.2   Experiments on Synthetic Datasets

Effectiveness of maxISCover. Table 3 reports the comparison on I_{max} of the above three heuristics. We can see that maxISCover *always* produced the smallest I_{max} when compared to maxDensCover and maxSetCover, as maxISCover considers the intersection size for each iteration of the 2-hop construction. For instance, I_{max} with maxISCover heuristic was 88 and 2.5 times smaller than that with maxDensCover and maxSetCover heuristics on average, respectively. We note that while maxSetCover was not proposed to minimize I_{max}, maxSetCover greedily determines centers that

---

[6] YEAST: http://vlado.fmf.uni-lj.si/pub/networks/data/bio/Yeast/Yeast.htm
ODLTS: http://vlado.fmf.uni-lj.si/pub/networks/data/dic/odlis/Odlis.htm
ERDOS: http://vlado.fmf.uni-lj.si/pub/networks/data/Erdos/Erdos02.net
ROGET: http://vlado.fmf.uni-lj.si/pub/networks/data/dic/roget/Roget.htm

most cover the uncovered $T(G)$. Our experiment showed that this sometimes led to small $I_{\max}$s.

Effectiveness of `unifyIS`. Next, we tested the algorithms for unification of the intersection results. In particular, Table 4 reports the comparison of the number of distinct surrogate nodes introduced by `unifyIS` (Algo. 1) and a baseline algorithm `Naive`. `Naive` chooses to add unused surrogate nodes into the index rather than checking if the surrogate nodes from previous iterations can be reused. The number of added surrogate nodes in `MASN` was almost always at least three times fewer than that of `Naive` under all other heuristics. Moreover, as the $I_{\max}$ was the smallest under `maxISCover` (Table 3), such an $I_{\max}$ leads to the smallest distinct number of surrogate nodes, except in SYN-1.

Table 3. The maximum intersection size $I_{\max}$

Graph	$I_{\max}$		
	maxDensCover	maxSetCover	maxISCover
SYN-1	2558	22	15
SYN-2	17	7	3
SYN-3	1169	48	13

Table 4. # of distinct added surrogate nodes

Graph	Naive vs. MASN		
	maxDensCover	maxSetCover	maxISCover
SYN-1	7.86M vs. 12.11k	67.61k vs. 17.30k	46.10k vs. 13.73k
SYN-2	96.07k vs. 8.75k	39.56k vs. 6.24k	16.95k vs. 6.08k
SYN-3	5.70M vs. 23.03k	0.23M vs. 18.92k	63.44k vs. 11.11k

Table 5. Query time at $\mathcal{SP}$ and client

Graph	$\mathcal{SP}$ (ms) vs. Client (ms)		
	maxDensCover	maxSetCover	maxISCover
SYN-1	106.54 vs. 0.52	2.55 vs. 0.43	2.02 vs. 0.46
SYN-2	2.01 vs. 0.56	1.37 vs. 0.67	1.79 vs. 0.47
SYN-3	52.35 vs. 0.54	4.44 vs. 0.52	2.15 vs. 0.52

Table 6. Throughput at $\mathcal{SP}$

Graph	$\mathcal{SP}$ (query per second)		
	maxDensCover	maxSetCover	maxISCover
SYN-1	9	392	495
SYN-2	495	730	559
SYN-3	19	225	465

Query Performance and Throughput of `pp-2-hop`. Table 5 presents the query time at both the $\mathcal{SP}$ and the client side. Each of the reported times is the average of 1000 queries. For the query time at the $\mathcal{SP}$ side, as the $I_{\max}$s due to `maxISCover` were small, the times of multiplications on the flags were small. Therefore, the query times of `maxISCover` were the best in all cases. For SYN-1 and SYN-3, `maxISCover` is more than an order of magnitude faster than `maxDensCover`; for SYN-3, `maxISCover` is more than twice as fast as `maxDensCover`. At the client side, the client only needs to perform *one* decryption of R_e for every query and the decryption algorithm essentially did the same amount of computation. Thus, the times at the client side were roughly the same and very small.

Based on the query performances, we calculate the corresponding throughput of the $\mathcal{SP}$ in Table 6. The results showed that with a commodity machine, the $\mathcal{SP}$ using `maxISCover` consistently offers a throughput around 500 queries per second. In comparison, `maxDensCover` is the least efficient. While `maxSetCover` sometimes has comparable throughputs, it is more sensitive to the datasets used.

Table 7. The maximum intersection size $I_{\max}$

Graph	$I_{\max}$		
	maxDensCover	maxSetCover	maxISCover
YEAST	237	6	4
ODLIS	274	3	3
ERDOS	250	5	3
ROGET	752	6	4

Table 8. # of distinct added surrogate nodes

Graph	Naive vs. MASN		
	maxDensCover	maxSetCover	maxISCover
YEAST	0.56M vs. 7.72K	14.17K vs. 2.81K	9.44K vs. 2.72K
ODLIS	0.80M vs. 8.39K	8.73K vs. 2.98K	8.73K vs. 2.98K
ERDOS	1.73M vs. 8.86K	34.64K vs. 7.03K	20.78K vs. 7.01K
ROGET	0.77M vs. 3.75K	6.13K vs. 1.40	4.09K vs. 1.28K

Table 9. Query time at $\mathcal{SP}$ and client

Graph	$\mathcal{SP}$ (ms) vs. Client (ms)		
	maxDensCover	maxSetCover	maxISCover
YEAST	12.38 vs. 0.58	0.89 vs. 0.50	0.73 vs. 0.47
ODLIS	13.33 vs. 0.68	0.59 vs. 0.50	0.65 vs. 0.53
ERDOS	11.98 vs. 0.55	1.34 vs. 0.59	0.97 vs. 0.52
ROGET	31.91 vs. 0.64	0.57 vs. 0.59	0.29 vs. 0.64

6.3   Experiments on Real-World Datasets

Finally, we conducted a similar evaluation on four publicly available real-world datasets. Since the results were similar to those obtained from synthetic datasets, we only highlight some major results here.

Table 7 shows the performances of maxISCover. Our proposed maxISCover heuristic consistently produced the smallest I_{max} when compared to the other two heuristics. Since the I_{max}s due to maxISCover were the smallest, the number of distinct added surrogate nodes by unifyIS were also the smallest as shown in Table 8. The query time at both the $\mathcal{SP}$ side and the client side are shown in Table 9. The query time of maxISCover at the $\mathcal{SP}$ side was almost always at least an order of magnitude faster than that of maxDensCover.

7   Conclusion

In this paper, we investigated privacy-preserving reachability query services. We proposed heuristic algorithms to determine a 2-hop labeling called pp-2-hop. We proposed and analyzed its private query processing over pp-2-hop. We conducted experiments to show the performance of our techniques. In the future, we plan to (i) integrate the large body of optimizations for 2-hop labeling into pp-2-hop (*e.g.*, [31]) and (ii) implement the shortest distance queries which are supported by the original 2-hop labeling [10].

Acknowledgement. Zhe Fan, Peipei Yi and Byron Choi were partially supported by GRF 210510. Shuxiang Yin and Shuigeng Zhou were supported by the Research Innovation Program of Shanghai Municipal Education Commission under grant No.13ZZ003.

References

1. Agrawal, R., Borgida, A., Jagadish, H.V.: Efficient management of transitive relationships in large data and knowledge bases. SIGMOD (1989)
2. Bader, G.D., Cary, M.P., Sander, C.: Pathguide: a pathway resource list. Nucleic Acids Research 34(suppl. 1), D504–D506 (2006)
3. Bramandia, R., Choi, B., Ng, W.K.: Incremental maintenance of 2-hop labeling of large graphs. TKDE 22(5), 682–698 (2010)
4. Cai, J., Poon, C.K.: Path-hop: efficiently indexing large graphs for reachability queries. CIKM, pp. 119–128 (2010)

5. Cao, N., Yang, Z., Wang, C., Ren, K., Lou, W.: Privacy-preserving query over encrypted graph-structured data in cloud computing. In: ICDCS, pp. 393–402 (2011)
6. Cheng, J., Huang, S., Wu, H., Fu, A.: Tf-label: a topological-folding labeling scheme for reachability querying in a large graph. SIGMOD, pp. 193–204 (2013)
7. Cheng, J., Yu, J.X., Lin, X., Wang, H., Yu, P.S.: Fast computation of reachability labeling for large graphs. In: Ioannidis, Y., Scholl, M.H., Schmidt, J.W., Matthes, F., Hatzopoulos, M., Böhm, K., Kemper, A., Grust, T., Böhm, C. (eds.) EDBT 2006. LNCS, vol. 3896, pp. 961–979. Springer, Heidelberg (2006)
8. Cheng, J., Yu, J.X., Lin, X., Wang, H., Yu, P.S.: Fast computing reachability labelings for large graphs with high compression rate. In: EDBT, pp. 193–204 (2008)
9. Chor, B., et al.: Private information retrieval. J. ACM 45, 965–981 (1998)
10. Cohen, E., et al.: Reachability and distance queries via 2-hop labels. In: SODA (2002)
11. El Gamal, T.: A public key cryptosystem and a signature scheme based on discrete logarithms. In: Blakely, G.R., Chaum, D. (eds.) CRYPTO 1984. LNCS, vol. 196, pp. 10–18. Springer, Heidelberg (1985)
12. Fan, Z., Peng, Y., Choi, B., Xu, J., Bhowmick, S.S.: Towards efficient authenticated subgraph query service in outsourced graph databases. IEEE Transactions on Services Computing 99 (2013)
13. Gao, J., Yu, J.X., Jin, R., Zhou, J., Wang, T., Yang, D.: Neighborhood-privacy protected shortest distance computing in cloud. In: SIGMOD, pp. 409–420 (2011)
14. He, X., Vaidya, J., Shafiq, B., Adam, N., Lin, X.: Reachability analysis in privacy-preserving perturbed graphs. In: WI-IAT, pp. 691–694 (2010)
15. Informatics Outsourcing. Outsourcing Solution Service, http://www.informaticsoutsourcing.com/
16. Jin, R., Ruan, N., Dey, S., Xu, J.Y.: Scarab: scaling reachability computation on large graphs. In: SIGMOD, pp. 169–180 (2012)
17. Jin, R., Ruan, N., Xiang, Y., Wang, H.: Path-tree: An efficient reachability indexing scheme for large directed graphs. TODS, 7:1–7:44 (2011)
18. Jin, R., Xiang, Y., Ruan, N., Fuhry, D.: 3-hop: a high-compression indexing scheme for reachability query. In: SIGMOD, pp. 813–826 (2009)
19. Karwa, V., Raskhodnikova, S., Smith, A., Yaroslavtsev, G.: Private analysis of graph structure. In: VLDB, pp. 1146–1157 (2011)
20. Katz, J., Lindell, Y.: Introduction to Modern Cryptography. Chapman & Hall/CRC (2007)
21. Kundu, A., et al.: How to authenticate graphs without leaking. In: EDBT, pp. 609–620 (2010)
22. Kundu, A., et al.: Efficient leakage-free authentication of trees, graphs and forests. IACR Cryptology ePrint Archive, p. 36 (2012)
23. Menascé, D.A.: Qos issues in web services. Internet Computing 6(6), 72–75 (2002)
24. Mouratidis et al., K.: Shortest path computation with no information leakage. PVLDB (2012)
25. Schenkel, R., Theobald, A., Weikum, G.: HOPI: An efficient connection index for complex XML document collections. In: Bertino, E., Christodoulakis, S., Plexousakis, D., Christophides, V., Koubarakis, M., Böhm, K. (eds.) EDBT 2004. LNCS, vol. 2992, pp. 237–255. Springer, Heidelberg (2004)
26. Seufert, S., Anand, A., Bedathur, S., Weikum, G.: Ferrari: Flexible and efficient reachability range assignment for graph indexing. ICDE, pp. 1009–1020 (2013)
27. Yin, S., Fan, Z., Yi, P., Choi, B., Xu, J., Zhou, S.: Privacy-Preserving Reachability Query Services, http://www.comp.hkbu.edu.hk/~zfan/TR20140201.pdf
28. Trissl, S., Leser, U.: Fast and practical indexing and querying of very large graphs. In: SIGMOD, pp. 845–856 (2007)

29. van Schaik, S.J., de Moor, O.: A memory efficient reachability data structure through bit vector compression. In: SIGMOD, pp. 913–924 (2011)
30. Xu, K., Zou, L., Yu, J.X., Chen, L., Xiao, Y., Zhao, D.: Answering label-constraint reachability in large graphs. CIKM, pp. 1595–1600 (2011)
31. Yi, P., Fan, Z., Yin, S.: Privacy-preserving reachability query services for sparse graphs. In: GDM (2014)
32. Yildirim, H., Chaoji, V., Zaki, M.J.: Grail: scalable reachability index for large graphs. PVLDB 3(1-2), 276–284 (2010)
33. Zhu, L., Choi, B., He, B., Yu, J.X., Ng, W.K.: A uniform framework for ad-hoc indexes to answer reachability queries on large graphs. In: Zhou, X., Yokota, H., Deng, K., Liu, Q. (eds.) DASFAA 2009. LNCS, vol. 5463, pp. 138–152. Springer, Heidelberg (2009)

SKY R-tree: An Index Structure
for Distance-Based Top-k Query

Yuya Sasaki[1], Wang-Chien Lee[2], Takahiro Hara[1], and Shojiro Nishio[1]

[1] Osaka University, Osaka, Japan
{sasaki.yuya,hara,nishio}@ist.osaka-u.ac.jp
[2] The Pennsylvania State University, PA, USA
wlee@cse.psu.edu

Abstract. Searches for objects associated with location information and non-spatial attributes have increased significantly over the years. To address this need, a top-k query may be issued by taking into account both the location information and non-spatial attributes. This paper focuses on a distance-based top-k query which retrieves the best objects based on distance from candidate objects to a query point as well as other non-spatial attributes. In this paper, we propose a new index structure and query processing algorithms for distance-based top-k queries. This new index, called *SKY R-tree*, drives on the strengths of R-tree and Skyline algorithm to efficiently prune the search space by exploring both the spatial proximity and non-spatial attributes. Moreover, we propose a variant of SKY R-tree, called *S2KY R-tree* which incorporates a similarity measure of non-spatial attributes. We demonstrate, through extensive experimentation, that our proposals perform very well in terms of I/O costs and CPU time.

Keywords: Top-k query, Spatial database, R-tree, Skyline, Location-based service.

1   Introduction

With the development of positioning technology and wide availability of mobile devices, efficient location-based search becomes essential for many applications, where both the location information and non-spatial attributes are considered. For example, Yelp and Foursquare[1], two well known location-based social networking services (LBSNs), may employ location-based search to assist their users to find restaurants based on specified location and user preference. Consider Figure 1, where 9 objects $o_1, o_2, \cdots, o_9$ are located in spatial area (illustrated on the left of the figure), and each of them is associated with 2 non-spatial attributes, i.e., price and rating (as illustrated by the table on the right). Given a query point q, a user wants to find the best restaurant o based on both the distance between q and o as well as the user's preference in terms of price and rating of

---

[1] http://www.yelp.com/,https://foursquare.com/

S.S. Bhowmick et al. (Eds.): DASFAA 2014, Part I, LNCS 8421, pp. 220–235, 2014.
© Springer International Publishing Switzerland 2014

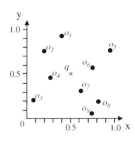

	price	rating
o_1	0.1	1.0
o_2	0.8	0.1
o_3	0.3	0.8
o_4	0.3	0.4
o_5	0.9	0.3
o_6	0.2	0.7
o_7	0.3	0.3
o_8	0.6	0.2
o_9	0.6	0.6

Fig. 1. A spatial area containing objects and a query point

the restaurants.[2] The best object is determined based on the user's preference and the query point.

Several top-k queries for spatial database have been proposed recently [3, 10]. In [10], a *distance-based top-k query* which takes into account both the location and other non-spatial attributes has been proposed. In this query, a user designates the number of requested objects k, a query point q, and a weight α that controls the importance between the roles of distance and other non-spatial attributes in the search. In other words, the distance-based top-k query returns k objects ranked based on "the distance between the objects and q" and "the user preference over the non-spatial attributes". Distance-based top-k queries in [10] retrieve the best k objects based on the branch-and-bound algorithm [9]. However, this branch-and-bound algorithm must construct the tree structure for each distance-based top-k query in the system. Since the paper [10] focuses on a continuous query, it does not care about the cost that constructs the tree structures, but it is absolutely inefficient for a snapshot query. To the best knowledge of the authors, efficient indexing techniques for distance-based top-k queries have not been reported in the literature.

In this paper, we propose a new indexing framework for processing distance-based top-k queries. This framework exploits the ideas behind R-tree [5] and skyline algorithm [2], and thus called *SKY R-tree*. Notice that R-tree is a very efficient index for supporting spatial search while skyline algorithms are very effective for pruning the search space to find the best objects (i.e., skyline points) in terms of non-spatial preference. Thus, SKY R-tree aims to prune the search space by exploring the location and non-spatial attributes. Each node of SKY R-tree records a summary of location information and skyline points of objects in the sub-tree rooted at the node. Moreover, to avoid unnecessarily increasing the number of skyline points, each node employs *Abstract skyline* [11] to limit the number of skyline points stored. Utilizing the location information to estimate the spatial distance of candidate objects to a query point, and the skyline points to estimate the values of their non-spatial attributes, the proposed query processing algorithm is able to efficiently answer the distance-based top-k query.

---

[2] In this example, smaller values of non-spatial attributes are better.

To incorporate a similarity measure of non-spatial attributes, we also propose a variant of SKY R-tree, called *S2KY R-tree*. While SKY R-tree mainly considers location information in its construct, S2KY R-tree considers both the location information and non-spatial attributes, and thus is able to process distance-based top-k queries efficiently.

The main contributions of this paper are summarized as follows.

- We propose a new indexing technique which incorporates R-tree and skyline to support distance-based top-k queries.
- We propose an efficient algorithm for processing distance-based top-k queries.
- We demonstrate, through extensive experimentation, that our proposal performs very well in terms of both the I/O costs and CPU time.

The remainder of this paper is organized as follows. Section 2 describes the problem formulation. Section 3 presents the proposed index structure. Section 4 presents the enhanced approach. Section 5 summarizes the results obtained in the simulation experiments. Section 6 introduces related work. Section 7 summarizes the paper.

2   Preliminaries

2.1   Problem Formulation

Given an object dataset O in which each object o has location information $o.loc$ and non-spatial attribute $o.att$. We assume $o.loc$ is in a two dimensional geographical space composed of latitude and longitude, and $o.att$ is a d-dimensional vector where $o.att_i \in [0, 1](i = 1, \cdots, d)$. We further assume that smaller values of these non-spatial attributes, e.g., price, are preferable, without loss of generality. Each score of object is calculated based on a query q which is represented by a query point $q.loc$ and a query weight $q.w$, where $q.w$ is a vector of weights such that $q.w_i \geq 0(i = 1, \cdots, d)$ and $\sum_{i=1}^{d} q.w_i = 1$. Accordingly, the ranking score of an object o is calculated by the following equation.

$$score(q, o) = \alpha \frac{D(q.loc, o.loc)}{maxD}$$

$$+ (1 - \alpha) \frac{\sum_{i=1}^{d} q.w_i \cdot o.att_i}{maxA} \tag{1}$$

where $\alpha \in [0, 1]$ is a parameter for balancing spatial proximity and non-spatial attributes, $D(q.loc, o.loc)$ is the Euclidian distance between q and o, and $maxD$ and $maxA$ are maximal distance and dissimilarity for normalization, respectively. Based on the above scoring function, the distance-based top-k query is defined as follows.

Definition 1 (Distance-Based Top-k Query). *Given a query q and the number of objects of interest k, the result of the distanced-based top-k query q includes a set of objects $TOP_k(q)$ such that $TOP_k(q) \subset O$, $|TOP_k(q)| = k$ and $\forall o_i, o_j : o_i \in TOP_k(q), o_j \in O - TOP_k(q)$, it holds that $score(q, o_i) \leq score(q, o_j)$.*

Notice that, in this paper, if some objects have the same kth score, the result retrieves one (or some) of the objects that meet the requirements of the $TOP_k(q)$.

Example: Recall the example in Figure 1. Let the query be q ($q.loc=\{0.5, 0.5\}$, $q.w=\{0.5, 0.5\}$), α be 0.5, and $k = 2$. By calculating the score of each object by Eq. (1), the top-1 and top-2 objects are o_7 and o_4, respectively.

2.2 Background

Here, we introduce Skyline and Max aR-tree which inspire our proposal of integrating R-tree and Skyline.

Skyline. In distance-based top-k queries, the scores of objects factor in both the location and non-spatial attributes. When focusing on only the non-spatial attributes, we can find the best object from skyline set because the top-k query is defined by linearly weighting the non-spatial attributes.

In the following we define the skyline set and discuss its relation to top-k queries.

Definition 2 (Skyline). *An object $o_i \in O$ is said to dominate another object $o_j \in O$, if for each attribute att_x $o_i.att_x \leq o_j.att_x$ and at least one attribute att_y $o_i.att_y < o_j.att_y$. The skyline is a set of points $SKY \subseteq O$ which are not dominated by any other points. The points in SKY are called skyline points.*

Observation 1. *The top-1 object obtained for a query that employs an increasingly monotone function on the non-spatial attributes belongs to the skyline set [12].*

From the skyline points, we can find the best objects (without considering the factor of location). Thus, skyline is useful to prune the search space from the aspect of non-spatial attributes.

Example: Consider the right graph in Figure 2 which plots the non-spatial attributes of the objects in Figure 1 in a two-dimensional space. As o_4 is dominated by o_7, o_4 is not a skyline point. In this figure, skyline points are o_1, o_2, o_6, o_7, and o_8.

Max aR-tree. *Max aR-tree* proposed in [7] can be utilized for distance-based top-k queries. In Max aR-tree, each node has a summary of location information and non-spatial attributes of descendant nodes. First, as location information, each node records a Minimum Bounding Rectangle (*MBR*) which is a rectangle contains all objects in the sub-tree rooted at the node. Additionally, each node in Max aR-tree records the best non-spatial attribute of each dimension in the sub-tree rooted at the node. Max aR-tree can prune the search space by non-spatial attribute, but it is not efficient because the best non-spatial attribute recorded in a node is quite different from an actual best score of object.

Example: Figure 3 illustrates a Max aR-tree indexing for the nine objects in Figure 1. Notice that, while Figure 2 shows the MBR recoded in each node and

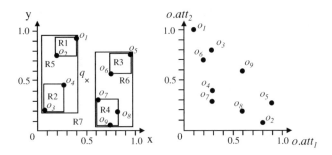

Fig. 2. Minimum bounding rectangles and non-spatial attributes

non-spatial attributes of each object, Table 1 shows the best values recorded in an internal node and non-spatial attributes of objects recorded in a leaf node. In this example, a leaf node R1 records $(o_1:0.1, 1.0)$ and $(o_2:0.8, 0.1)$, and an internal node R5 records $(R1:0.1, 0.1)$ and $(R2:0.3, 0.4)$. From R5's view, R1 seems to record the object with the best non-spatial attribute, however, the actual non-spatial attributes in the objects recorded in R1 are quite different.

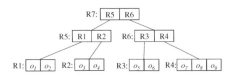

Fig. 3. Index structure in corresponding R-tree

Table 1. Best values which are recorded in each node in Max aR-tree

Node	Best value
R7	(R5:0.1,0.1), (R6:0.2,0.2)
R6	(R3:0.2,0.3), (R4:0.6,0.2)
R5	(R1:0.1,0.1), (R2:0.3,0.4)
R4	(o_7:0.3,0.3), (o_8:0.6,0.2), (o_9:0.6,0.6)
R3	(o_5:0.9,0.3), (o_6:0.2,0.7)
R2	(o_3:0.3,0.8), (o_4:0.3,0.4)
R1	(o_1:0.1,1.0), (o_2:0.8,0.1)

3   Hybrid Indexing for R-tree and Skyline

We present a new indexing framework that exploits the strengths of R-tree and skyline, called *SKY R-tree*. Moreover, we develop an algorithm for processing distance-based top-k queries.

3.1   SKY R-tree

SKY R-tree is essentially an R-tree, where each index node records skyline points of all objects in the sub-tree rooted at the node.

The design of SKY R-tree aims at pruning the search space by exploring non-spatial attribute. Ultimately, each node records all the non-spatial attributes of objects in a sub-tree rooted at the node in R-tree. To reduce the I/O cost in

R-tree, the number of child nodes is limited by a size of information recorded in one index node. Hence, if each node contains too large a volume of information, the efficiency decreases. Due to this constraint, it is important to select the information needed for efficiently pruning the search space. Our design considers two goals: (1) to guarantee an exact result, and (2) to represent a large amount of information in a compact form. In SKY R-tree, we choose skyline points because they elegantly meet these two goals.

When processing distance-based top-k queries, skyline points are very helpful to process the non-spatial attributes of candidate objects. However, the number of skyline points depends on the distribution of non-spatial attributes. If there are a large number of skyline points, the size of some nodes may become very large, resulting in deterioration of the performance. Therefore, SKY R-tree reduces the number of stored skyline points, while guaranteeing the correctness of query processing (and thus the result). If the number of skyline points exceeds a preset maximum number of skyline point $|\overline{SKY}|$, the skyline points are aggregated by *Abstract skyline* [11], which picks $|\overline{SKY}|$ *representative* skyline points from the entire skyline points.

In SKY R-tree, each leaf node contains a number of entries of the form *(o, o.loc, o.att)*. Meanwhile, an internal (non-leaf) node contains a number of entries of the form *(np, rectangle, sp)*, where *np* is a pointer to one of its child node, *rectangle* is the MBR, and *sp* is the skyline points of the child nodes.

Example: Figure 3 illustrates the SKY R-tree for the nine objects in Figure 1. Note that, while the index is structured as a MAX aR-tree, the information associated with the tree nodes is different. Let the maximum number of skyline points be 2. Table 2 shows the skyline points associated with an internal node as well as non-spatial attributes associated with objects in a leaf node. For example, R5 stores 3 skyline points (i.e., <0.1,1.0>, <0.8,0.1> for R1 and <0.3,0.4> for R2). As R7 can record only 2 skyline points for R5, we merge <0.1,1.0> and <0.3,0.4> into <0.1,0.4> and records it along with <0.8,0.1> for R5.

Table 2. Skyline points recorded in each node

Node	Skyline point
R7	(R5:<0.1,0.4>,<0.8,0.1>), (R6:<0.2,0.7>,<0.3,0.2>)
R6	(R3:<0.9,0.3>,<0.2,0.7>), (R4:<0.3,0.3>,<0.6,0.2>)
R5	(R1:<0.1,1.0>,<0.8,0.1>), (R2:<0.3,0.4>)
R4	$(o_7{:}0.3,0.3)$, $(o_8{:}0.6,0.2)$, $(o_9{:}0.6,0.6)$
R3	$(o_5{:}0.9,0.3)$, $(o_6{:}0.2,0.7)$
R2	$(o_3{:}0.3,0.8)$, $(o_4{:}0.3,0.4)$
R1	$(o_1{:}0.1,1.0)$, $(o_2{:}0.8,0.1)$

We next present how to construct SKY R-tree. Similar to the R-tree algorithm [5], it repeats the Insert function in Algorithm 1, which uses a ChooseLeaf to select a leaf node with the smallest increase in MBR of the node in order to insert an object into the node as an entry. If the insertion exceeds the maximum node capacity, a Split is incurred to divide the node into two MBRs distant from each other.

Algorithm 1. Insert algorithm

1: $N \leftarrow$ ChooseLeaf($object$)
2: Add $object$ to node N.
3: **if** exceed maximum number of child nodes of N **then**
4:     $\{N_1, N_2\} \leftarrow N$.Split()
5:     **if** N is root **then**
6:         Initialize new root NR and add N_1 and N_2 to NR
7:     **else**
8:         Ascend from N to root, and update MBR and skyline points
9:     **end if**
10: **else if** $N \neq$root **then**
11:     Update the MBR and skyline points of the ancestor node of N
12: **end if**

What makes SKY R-tree essentially different from the conventional R-tree is the update of the skyline points in ancestor node (line 11).

3.2 Distance-Based Top-k Query Processing

To process distance-based top-k queries, we exploit the best-first traversal. The best-first traversal algorithm searches the entry with the smallest score in a heap. The score for an individual object has been defined previously in Eq. (1). Therefore, we define the score for an index node N here as follows.

$$score(q, N) = \alpha \frac{MinD(q.loc, N.MBR)}{MaxD}$$
$$+ (1 - \alpha)\text{argmin}_{\forall p \in N.sp} \sum_{i=1}^{d} q.w_i \cdot p.att_i \tag{2}$$

where $MinD(q.loc, N.MBR)$ denotes a minimum Euclidian distance between the MBR of N and $q.loc$. Eq. (2) provides a possible minimum score (i.e., lower bound) corresponding to the descendant objects of the node N.

In Algorithm 2, MINHEAP is a heap which contains index nodes and objects sorted in ascending order of their scores. Hence, if the first entry in MINHEAP is an object, it is the best object in MINHEAP (because the other objects definitely have larger scores than the first one). When the number of object in $TOP_k(q)$ exceeds k, the processing terminates (lines 6–7). Here, since objects and nodes in MINHEAP are sorted in ascending order, objects and nodes with larger scores than the top-k objects in MINHEAP are discarded (line 11). While the number of objects and nodes in MINHEAP becomes small, the computational overhead decreases because the sorting time becomes small.

Algorithm 2. Search algorithm

Input: k, q and index structure
Output: $TOP_k(q)$
 1: $MINHEAP$.insert($root$, 0)
 2: **while** $MINHEAP$.size() $\neq 0$ **do**
 3:     $N \leftarrow MINHEAP$.first()
 4:     **if** N is object **then**
 5:         $TOP_k(q)$.insert(N)
 6:         **if** $|TOP_k(q)| \geq k$ **then**
 7:             return $TOP_k(q)$
 8:         **end if**
 9:     **else**
10:         **for** $\forall n_i \in N.entry$ **do**
11:             **if** Number of objects with smaller score in $MINHEAP$ than $score(q, n_i) <$ $(k - |TOP_k(q)|)$ **then**
12:                 **if** N is a leaf node **then**
13:                     $MINEHEAP$.insert($Object$, $score(q, n_i)$)
14:                 **else**
15:                     $MINHEAP$.insert($Node$, $score(q, n_i)$)
                        //If $score$ is a same value, a smaller MBR is better.
16:                 **end if**
17:             **end if**
18:         **end for**
19:     **end if**
20: **end while**

We prove that $TOP_k(q)$ is an exact result by the following theorem.

Theorem 1. *The score to an index node N is smaller than the scores of its descendant object o for any query q.*

Proof. The score factors in (i) location proximity and (ii) non-spatial attributes. First, the MBR of N includes all descendant objects, i.e., $\forall o \in$ descendant objects of N; $MinD(q.loc, N.MBR) \leq D(q.loc, o.loc)$. Second, the skyline points of N dominate or equal to all descendant objects, i.e., $\text{argmin}_{\forall p \in N.sp} \sum_{i=1}^{d} q.w_i \cdot p.att_i \leq \sum_{i=1}^{d} q.w_i \cdot o.att_i$. These inequalities imply that $score(q, N) \leq score(q, o)$. □

From this theorem, it follows that if the score of an object is smaller than the score of an index node, the score of the object is smaller than that of all descendant objects of the index node, i.e., the first object in MINHEAP is the best object.

4   Enhanced SKY R-tree

SKY R-tree is constructed based on locations of objects. However, the non-spatial attributes also play a role in the distance-based top-k queries. Thus, we argue that the locality of non-spatial attributes should also be considered in construction of the index such that the cost of storage access can be reduced. Therefore, by considering a similarity measure of non-spatial attributes of objects, we propose a variant of SKY R-tree, called *S2KY R-tree* (similarity skyline R-tree).

4.1   S2KY R-tree

The S2KY R-tree inherits a similar idea from SKY R-tree, i.e., each index node records skyline points of its subtree. However, its clustering strategy is different. Thus, its ChooseLeaf and Split are different from that of SKY R-tree in Algorithm 1 by recording different child nodes at each internal node. Notice that SKY R-tree takes into account only the areas of MBRs under examination. On the other hand, S2KY R-tree takes into account both the areas of MBRs as well as the similarity of non-spatial attributes. To choose an appropriate insertion path for a new object, S2KY R-tree selects a child node as described below.

Let $E_1, \cdots, E_p$ be the entries in the current node, and let o be the object to be inserted. The corresponding R-tree calculates an increased area of MBR by the following equation.

$$AreaIncrease(o, E_x) = Area(o + E_x) - Area(E_x) \tag{3}$$

where $Area(x)$ is an area of the MBR including x, and E_x ($1 \leq x \leq p$) is an entry (child node) in the current node. A small $AreaIncrease$ means the new object is close to the descendant objects in the child node.

On the other hand, S2KY R-tree also considers the similarity of non-spatial attributes which is defined as follows.

Definition 3 (Similarity measure of non-spatial attributes). *Similarity is a distance between non-spatial attributes of new object and skyline points in entry E.*

$$Similarity(o, E_x) = \min_{\forall p \in E_x.sp} dist(o.att, p) \tag{4}$$

where $dist(x, y)$ denotes the distance between non-spatial attribute of x and y.

Finally, we calculate $Increase$ by summing of $AreaIncrease$ and $Similarity$.

$$Increase(o, E_x) = \beta \frac{AreaIncrease(o, E_x)}{MaxArea}$$
$$+ (1 - \beta) \frac{Similarity(o, E_x)}{MaxSim} \tag{5}$$

where $MaxArea$ and $MaxSim$ are the size of entire area and the maximum similarity used for normalization, and β is a weighting parameter. The node with the smallest $Increase$ is chosen as the next node because a small $Increase$ means that the distance between E and o is a small (and the non-spatial attributes are similar). This calculation is repeated until progressing to a leaf node in Algorithm 3.

Algorithm 3. ChooseLeaf algorithm

1: $N \leftarrow$ the root
2: **while** 1 **do**
3:　　**if** N is a leaf node **then**
4:　　　　return N
5:　　**end if**
6:　　$N \leftarrow np$ in the entry in N with the smallest *increase*
7: **end while**

If β is 1 in Eq. (5), S2KY R-tree is the same as SKY R-tree. The best β is empirically determined by queries and distribution of non-spatial attribute.

Moreover, we define Split function and the difference between skyline sets in two child nodes by the following equation.

$$Sdist(SKY_A, SKY_B) = \frac{\sum_{p \in SKY_A} min_{\forall q \in SKY_B} dist(p,q)}{|SKY_A| + |SKY_B|}$$
$$+ \frac{\sum_{q \in SKY_B} min_{\forall p \in SKY_A} dist(q,p)}{|SKY_A| + |SKY_B|}. \tag{6}$$

This equation calculates the average minimum distance between skyline points. A small $Sdist(x, y)$ means that skyline sets x and y in the two nodes is closer to each other.

Algorithm 4. Split algorithm

1: **for** $\forall$ pair of E_i and E_j in entries **do**
2:　　$d_{ij} \leftarrow area(E_i + E_j) - area(E_i) - area(E_j)$.
3:　　$sim_{ij} \leftarrow Sdist(SKY_{E_i}, SKY_{E_j})$;
4:　　$dif_{ij} = \beta d_{ij} + (1 - \beta)sim_{ij}$
5: **end for**
6: Choose the pair with the largest dif value to be the first elements of the two group

7: **while not** all entries in N belong a group **do**
8:　　**if** one group needs to include all remaining entries **then**
9:　　　　Assign all remaining entries to it and **break**
10:　　**end if**
11:　　Calculate *Increase* for all remaining entries to two groups
12:　　Choose the entry with the largest *Increase* and add it to the other group
13: **end while**

In Algorithm 4, the first two entries are selected as the first elements (lines 1–6). Other entries are assigned to either groups. To avoid assigning an entry with a large *Increase* to a group, the entry with the largest *Increase* is assigned to the other group in order (each entry has two *Increases* due to two groups) (lines 11–12). Moreover, to avoid a bias of the number of child nodes in two groups, the minimum number of child nodes are determined in advance (basically a half of the maximum number of child nodes) (lines 8–9).

4.2   Query Processing

We use Algorithm 2 for processing of distance-based top-k queries on S2KY R-tree. In S2KY R-tree, each leaf node are more likely to record similar objects. Since k objects with better scores are inserted to MINHEAP earlier than SKY R-tree, the number of objects and nodes in MINHEAP becomes smaller. As a result, the computational overhead may be reduced.

5   Performance Evaluation

5.1   Simulation Model

We evaluate SKY R-tree and S2KY R-tree to validate our idea in processing the distance-based top-k query, by using Max aR-tree as the baseline for comparison.

Datasets. We use real location sets for objects, where the real location set is a set of points of interests located in Tokyo (10km × 10km) extracted from Foursquare. The dataset includes 45,129 objects. As the location set extracted from Foursquare has no non-spatial attribute, we added synthetic attributes to those points of interests. We use 2 types of synthetic datasets for non-spatial attributes: uniform and anti-correlated.

Setup. All index structures are in memory, and we assume the page size is 4KB. The size of an index node is $(16 + 4d \cdot |\overline{SKY}|)$B, i.e., each node has $\lceil 4,096/(16 + 4d \cdot |\overline{SKY}|) \rceil$ child nodes (In Max aR-tree, the size of an index is the same when $|\overline{SKY}| = 1$). For SKY R-tree and S2KY R-tree, the default values of $|\overline{SKY}|$ and β are set at 5 and 0.8, respectively. All algorithms implemented in c++, and experimented on a server with Intel(R) Core(TM) i7-3770 CPU @ 3.40GHz with 8.00 GB RAM.

For each experiment, we generate 100 queries with randomly generated locations and weights. We use the average measures of "the number of I/O" (a number of visited nodes) and "CPU time" (query processing time) as the performance metrics. Table 3 shows parameters and values used in our experiments (the values in bold are the default values).

5.2   Experimental Result

To evaluate the performance of algorithms for the distance-based top-k queries, we vary two query parameters, k and α, as well as the dimension of non-spatial attributes, d, with 2 types of synthetic datasets for non-spatial attributes.

Uniform Non-spatial Attribute
Varying k. With k increasing, the search space increases. Figures 4a and 4d show the result in the uniform non-spatial attribute. We can see that SKY R-tree and S2KY R-tree outperform Max aR-tree for all values of k in terms of both the number of I/O and CPU time. From this result, we validate our idea of using skyline points to efficiently prune the search space. SKY R-tree works better than

Table 3. Setting

Parameter	Value
Requested # of object k	1, 5, 10, **20**, 50, 100
Alpha	0, 0.2, **0.3**, 0.4, ,0.6, 0.8, 1.0
Dimension d	1, 2, **3** , 4
Attribute set	uniform, anti-correlated

S2KY R-tree in terms of the number of I/O for a small value of k, while S2KY R-tree works better for a large value of k. The MBR of an index node in SKY R-tree is smaller than that in S2KY R-tree, and thus the search space becomes smaller. In S2KY R-tree, as k increases, the probability for the top-k objects to have the same ancestor nodes increases, resulting in decrease in the number of I/O. Moreover, S2KY R-tree significantly outperforms SKY R-tree in terms of CPU time. Since S2KY R-tree likely finds objects with good score at one traversal, the number of nodes and objects in MINHEAP decreases, resulting in decrease of CPU time.

Varying α. Parameter α in Eq. (1) represents importance of location for a user. With α increasing, the location proximity contributes more than the non-spatial attributes to the ranking score. Figures 4b and 4e show the result in the uniform non-spatial attribute. When α is small, the performance S2KY R-tree is better than the other two methods. When α is large, the performance is worse than the others because S2KY R-tree takes into account the similarity measure of non-spatial attributes. In SKY R-tree and Max aR-tree, although the search space can be pruned by non-spatial attributes, the MBR of each node likely increases. Hence, with α increases, the number of I/O gradually increases. SKY R-tree outperforms Max aR-tree for all values of α because SKY R-tree can prune the search space more efficiently than Max aR-tree.

Varying d. With the number of dimensions increasing, the diversity of non-spatial attributes increases. Due to the curse of dimensionality, it becomes difficult to prune the search space by non-spatial attributes. Figures 4c and 4f show the result in the uniform non-spatial attribute. Our proposal outperforms Max aR-tree for all values of d, except for 1. Note that Max aR-tree works better than our proposal in the one dimension space because default setting for the number of skyline points is 5, but there is only one skyline point in one dimensional space. As a result, our setting wastes some redundant information in this case. With d increasing, a similarity of non-spatial attributes efficiently works, resulting in increase in the performance of S2KY R-tree more than SKY R-tree.

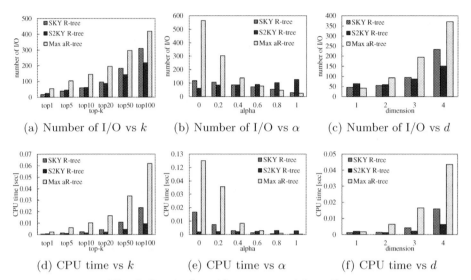

(a) Number of I/O vs k (b) Number of I/O vs α (c) Number of I/O vs d

(d) CPU time vs k (e) CPU time vs α (f) CPU time vs d

Fig. 4. Results in uniform non-spatial attribute

Anti-correlated Attribute

Varying k. Figures 5a and 5d show the result with the anti-correlated non-spatial attributes. The performance in S2KY R-tree does not change much compared with the uniform attribute set. On the other hand, SKY R-tree and Max aR-tree on this dataset work less efficiently than the uniform attribute set. In the anti-correlated set, SKY R-tree and Max aR-tree cannot efficiently prune the search space upon the non-spatial attributes, while S2KY R-tree keeps high efficiency by taking into account the similarity of non-spatial attributes. Specially, in Max aR-tree, almost all nodes record the non-spatial attributes near the best one (i.e., non-spatial attributes of each dimension are almost zero), and thus the performance is significantly worse. From this result, we can see that S2KY R-tree is robust for any non-spatial attributes by considering the similarity of non-spatial attributes.

Varying α. Figures 5b and 5e show the result with anti-correlated non-spatial attribute. The trend of results in our proposal is similar to Figure 4b. On the other hand, when α is small, Max aR-tree works worse because pruning the search space by non-spatial attribute does not work well in the anti-correlated attributes.

Varying d. Figures 5c and 5f show the result in the anti-correlated non-spatial attribute. S2KY R-tree outperforms the other methods for all values of d except for 1. The variation of non-spatial attributes in anti-correlated is smaller than the uniform, and thus the similarity of non-spatial attributes works better.

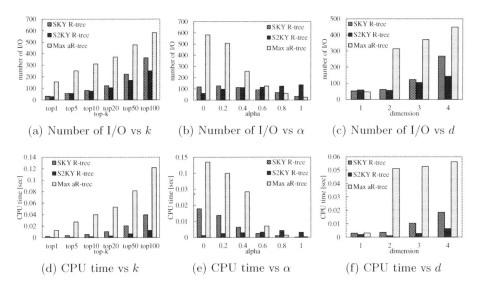

(a) Number of I/O vs k (b) Number of I/O vs α (c) Number of I/O vs d

(d) CPU time vs k (e) CPU time vs α (f) CPU time vs d

Fig. 5. Results in anti-correlated non-spatial attribute

6   Related Work

Top-k Query. Conventional top-k query processing has been extensively studied [6]. Fagin et al. [4] propose a class of algorithms known as threshold algorithms for computing the top-k query from multiple lists. Threshold algorithm accesses tuples from the database in a certain order, and maintains an upper bound as the maximum score for the unseen objects. If the upper bound is lower than the score of top-k objects, the algorithm terminates. Various threshold algorithms [1] have been proposed to improve some of their limitation and to study in other application areas. Tao et al. [9] propose a branch-and-bound algorithm for top-k queries. This branch-and-bound algorithm, based on R-tree [5], visits nodes which record the potential object with the best score.

These algorithms can be used for processing distance-based top-k queries, but they are inefficient. For example, by finding the top-1 object based on non-spatial attributes, and then adding the score of location proximity, we can repeatedly find the top-k objects. However, these algorithms probably search redundant objects whose locations are further from a query points.

Nearest Neighbor Query. k nearest neighbor (kNN) query processing in spatial databases, which retrieves the k objects closest to a query points, is also related to our work. kNN query processing proposed by Roussopoulos et al. [8] maintains potential nearest neighbors in a queue based on R-tree [5], and traverses the tree to find the best objects. For example, it finds the nearest neighbor, and then adds the score of non-spatial attributes. However, kNN queries processing based on R-tree are worse than a MAX aR-tree approach.

7   Conclusion

In this paper, we propose a new index structure for distance-based top-k query. The index structure, named *SKY R-tree*, incorporates R-tree and skyline to prune the search space by exploring both the spatial proximity and non-spatial attributes. Accordingly, we propose an algorithm for processing distance-based top-k queries. The index structure and processing algorithm, forming a new framework, is very efficient to retrieve the best objects. Moreover, we consider the similarity measure of non-spatial attributes of objects in its construction in a new index, named *S2KY R-tree*. Through extensive experimentation, we demonstrate that our proposal efficiently processes distance-based top-k queries in terms of both the I/O costs and CPU time. The difference between performance of SKY R-tree and S2KY R-tree depends on the situations under different importance of location proximity and the distribution of locations of objects and non-spatial attributes.

This work can be extended in several directions for future work. First, it would cooperate with keyword search, e.g., the "Japanese" restaurant close to a query point with high rating. Second, continuous distance-based top-k query processing for mobile users would be interesting. Third, in this paper we calculate the distance between objects and a query point in Euclidian distance, while a road-based distance is also an important challenge.

Acknowledgment. This research is partially supported by the Grant-in-Aid for Scientific Research (S)(21220002), (B)(24300037), and JSPS Fellows (24-293) of the Ministry of Education, Culture, Sports, Science and Technology, Japan.

References

[1] Akbarinia, R., Pacitti, E., Valduriez, P.: Best position algorithms for top-k queries. In: VLDB, pp. 495–506 (2007)

[2] Borzsony, S., Kossmann, D., Stocker, K.: The skyline operator. In: ICDE, pp. 421–430 (2001)

[3] Cong, G., Jensen, C.S., Wu, D.: Efficient retrieval of the top-k most relevant spatial web objects. VLDB Journal 2(1), 337–348 (2009)

[4] Fagin, R., Lotem, A., Naor, M.: Optimal aggregation algorithms for middleware. Journal of Computer and System Sciences 66(4), 614–656 (2003)

[5] Guttman, A.: R-trees: a dynamic index structure for spatial searching. SIGMOD 14, 47–57 (1984)

[6] Ilyas, I.F., Beskales, G., Soliman, M.A.: A survey of top-k query processing techniques in relational database systems. ACM CSUR 40(4), 11 (2008)

[7] Lazaridis, I., Mehrotra, S.: Progressive approximate aggregate queries with a multi-resolution tree structure. ACM SIGMOD Record 30(2), 401–412 (2001)

[8] Roussopoulos, N., Kelley, S., Vincent, F.: Nearest neighbor queries. ACM SIGMOD Record 24(2), 71–79 (1995)

[9] Tao, Y., Hristidis, V., Papadias, D., Papakonstantinou, Y.: Branch-and-bound processing of ranked queries. Information Systems 32(3), 424–445 (2007)

[10] Vlachou, A., Doulkeridis, C., Nørvåg, K.: Monitoring reverse top-k queries over mobile devices. In: MobiDE, pp. 17–24 (2011)

[11] Vlachou, A., Doulkeridis, C., Nørvåg, K.: Distributed top-k query processing by exploiting skyline summaries. IEEE Tran. on Distributed and Parallel Databases 30(3-4), 239–271 (2012)

[12] Vlachou, A., Doulkeridis, C., Nørvåg, K., Vazirgiannis, M.: On efficient top-k query processing in highly distributed environments. In: ACM SIGMOD, pp. 753–764 (2008)

Causal Structure Discovery
for Spatio-temporal Data

Victor W. Chu[1], Raymond K. Wong[1], Wei Liu[2], and Fang Chen[2]

[1] School of Computer Science and Engineering,
University of New South Wales, Australia
[2] National ICT Australia

Abstract. Numerous causal structure discovery methods have been proposed recently but none of them has taken possible time-varying structure into consideration. In this paper, we introduce a notion of causal time-varying dynamic Bayesian network (CTV-DBN) and define a causal boundary to govern cross time information sharing. Although spatio-temporal data have been investigated by multiple disciplines; by reducing structure discovery into a set of optimization problems, CTV-DBN is a scalable solution targeting large datasets. CTV-DBN is constructed using asymmetric kernels to address sample scarcity and to adhere to causal principles; while maintaining good variance and bias trade-off. We explore trajectory data collected from mobile devices which are known to exhibit heterogeneous patterns, data sparseness and distribution skewness. Contrary to a naïve method to divide space by grids, we capture the moving objects' view of space by using density-based clustering to overcome the problems. In our experiments, CTV-DBN is used to reveal the evolution of time-varying region macro structure in a ring road system based on trajectories, and to obtain a local time-varying road junction dependency structure based on static traffic flow sensor data.

1 Introduction

Structure discovery for spatio-temporal data has been a cross disciplinary subject among databases [8] [15] [21], data mining [5] [30], artificial intelligence [20], etc., though they may have different focus and scalability requirements. While there is not much previous work done in analyzing the relationships among network conditions across time, Song *et al.* [27] outline a time-varying dynamic Bayesian networks (TV-DBN) to model gene-to-gene interaction networks. In this paper, we extend TV-DBN to model time-varying network causal relationships. By discovering their causal dependency network structure, we will be able to detect region-to-region and/or junction-to-junction interactions.

Most spatio-temporal observations from mobile devices exhibit unfavorable statistical properties: heterogeneous patterns, data sparseness and distribution skewness [19] [31]. By using a time-varying model, the network structure discovery process reveals the evolution of region connections to mitigate the issue of heterogeneous patterns. To address data sparseness and distribution skewness, we use density-based clustering methods to identify regions for a particular

S.S. Bhowmick et al. (Eds.): DASFAA 2014, Part I, LNCS 8421, pp. 236–250, 2014.
© Springer International Publishing Switzerland 2014

space-time interval from trajectories directly. Density-based clustering methods [8] [16] are argued to be the best solution for spatio-temporal clustering [24].

We introduce a notion of causal time-varying dynamic Bayesian networks (CTV-DBN) based on a relaxed version of TV-DBN. In CTV-DBN, causal Markov assumption [28] [29] is satisfied by considering causal boundary. It is achieved by asymmetric kernels [22] to make the information sharing across time in better adherence to causal principles; but still, ensure sharing among suitable neighbors to address data scarcity while maintaining similar level of variance and bias trade-off. Asymmetric kernels also provide a solution to rectify boundary problem created by real-world data where they mirror a causal function [22]. Causal inference can be made based on manipulation rule once a Bayesian network is made isomorphic with a causal model [18] [28] [29].

In our experiments, we apply CTV-DBN to a macro ring road system and also to a local road system which serves as a gateway to a major highway. Although there exists some investigations into the economical impact [6] and spatio relationship impact to travelers' route choice [14] of ring road systems, little has been done to identify their region causal relationships. On the other hand, a better understanding of the junction-to-junction dependence of an area which serves as a gateway to a major highway will certainly help to achieve optimal throughput and to improve the efficiency and safety of this system.

To summarise, our primary contributions are:

1. A notion of causal time-varying dynamic Bayesian networks (CTV-DBN) is introduced that satisfies causal Markov assumption.
2. Asymmetric kernel is used to make the information sharing across time to better adhere to causal principles while maintaining a good level of variance and bias trade-off.
3. Upon satisfying causal Markov assumption, we base on manipulation rule to establish a structure for causal inference.
4. CTV-DBN network discovery is applied to both macro and micro level road systems to reveal the evolution of their causal time-varying structures.

The reminder of this paper is organized as follows. Section 2 presents related work. Section 3 introduces our region discovery method, describe a relaxed version of TV-DBN and then presents our proposed causal Markov assumption compliant CTV-DBN. Section 4 presents our experiment results and discuss our findings. Finally, Section 5 summarizes this paper and briefly outlines our future work.

2   Related Work

Yang *et al.* [31] introduce spatio-temporal hidden Markov models to model correlations among different traffic time-series. Each time-series have its own state set to address unfavorable data characteristics: sparse, dependent and heterogenous. Moreover, Lu *et al.* [21] explore spatio-temporal relationships among trajectories to extract semantic regions where users are likely to have some kinds of activities. Sequential clustering approach is used to develop a shared-nearest-neighbor-based clustering algorithm to discover frequent semantic regions but possible time-varying structure is not considered. In most literature, density-based clustering methods are suggested to be the best solution for trajectory clustering because: 1) they are able to handle clusters with no predefined shape, 2) they are able to cope with noises in the data, and 3) one can base on the parameters to fine tune the methods to fit a particular problem [24]. DBSCAN [8] is one of the popular methods in this class where it groups related objects by using density threshold, whilst OPTICS [1] is an alternative method which can handle data with varying densities. A more recent variant to DBSCAN is ST-DBSCAN [4].

In this paper, we investigate how to better model spatio-temporal data and their relationships by dynamic Bayesian networks (DBN). DBN [23] have been used to model sequences of variables and regarded as a method to overcome the expressive power limitation in hidden Markov models and Kalman filter models, in which state-space is represented in factored form. Nevertheless, DBN is in fact a time-invariant model, in which the structure of the network is fixed but is capable to model dynamic systems [27]. Non-stationary dynamic Bayesian networks (NS-DBN) are introduced by Robinson and Hartemink [25] [26] in recent time to model time-varying network structures. Markov chain Monte Carlo (MCMC) sampling method is used; however, Song et al. [27] point out that such an approach is unlikely to be scalable, and it is also prone to over-fitting.

In parallel, Grzegorczyk and Husmeier [11] [12] also develop an alternative approach. Their assumption of a fixed network structure is deemed to be too restrictive [7], even though the interaction parameters of the model can vary with time to cater for non-stationary systems. Song et al. propose TV-DBN [27] to overcome those weaknesses. More recently, Dondelinger et al. [7] propose a more complex non-homogeneous dynamic Bayesian networks for inferring gene regulatory networks with gradually time-varying structure. Although both Dondelinger and Song's proposals can be traced back to a common root of Robinson and Hartemink [25] [26], Dondelinger et al. work with continuous time data whilst the method proposed by Song et al. [27] is suitable to discretization data. From a different direction, Liu et al. [20] recently attempt to address the problem of temporal causal graph discovery assuming relational time-series exist.

3   Casual Structure Discovery

In this section, we first introduce our region discovery method (Section 3.1) and explain a relaxed version of time-varying dynamic Bayesian networks (Section 3.2). We next discuss how we make use of asymmetric kernel to formulate causal time-varying dynamic Bayesian networks (Section 3.3). Causal boundary is also defined by considering causal Markov assumption to eliminate contradiction; and therefore, manipulation rule can be applied to conduct causal inference.

3.1   Region Discovery

In spatio-temporal analysis, a spatio-temporal object is defined as a tuple $\langle id, timestamp, latitude, longitude \rangle$, whilst a trajectory is described as a spatio-temporal object traveling through space-time. An object could have both static and dynamic attributes. For example, $a_1 \in \mathcal{A}_1$ could be a static attribute, e.g., shape, which is invariant with time; whilst, $a_2 \in \mathcal{A}_2$ could represent the velocity of an object, which is a derived attribute from its original trajectory. The other derived attributes could be direction, time of stop, place of stop, etc.

There are known interesting spatio-temporal features in the real-world. For example, in a typical city like Melbourne or Beijing, the traffic of the central business districts (CBD) is normally very dense but the traffic of the outer suburb is much less. However, the outer suburbs' traffic might be in the middle between CBD and major infrastructure, e.g. airports; hence, they could give material impact to a society. In addition, it is common that people rush to work from home in the early morning and vice versa in late afternoon, and also the major goods and services distribution networks of industries may vary during a day. They all contribute to a periodic alternation of region macro-structure.

Instead of using traditional grid-based partitioning approach, we select density-based clustering method as it fits well to trajectory data analysis [24]. The region structure is found varying throughout a day from our experiments but tends to be stable within a specific time period, e.g., evening, morning rush hours, afternoon rush hours, etc. Hence, we propose to model time specific shape of regions by density-based clustering on historical trajectories of mobile sensors. The physical region areas can be mapped based on Voronoi diagram [2] on cluster centroids as shown in Figure 1, where the black dots are the centroids and the regions are divided by black lines. For static sensors, their regions are collapsed into points for the time they exist.

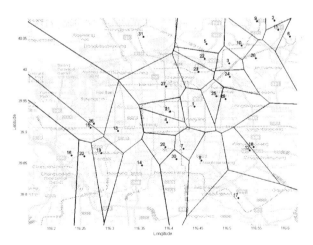

Fig. 1. Regions drawn by Voronoi diagrams at peak hour (8am) based on one week (Monday-Friday) of Beijing taxi trajectories

3.2   Time-Varying Dynamic Bayesian Networks (TV-DBN)

Time-varying dynamic Bayesian networks (TV-DBN) [27] is built based on dynamic Bayesian networks (DBN) [23]. Let $\mathbf{X}^t = (X_1^t, \ldots, X_r^t)^\top \in \mathbb{R}^r$ represents a random vector of feature values from r regions at time t, a dynamic process of such time dependant condition can be modeled by a first-order Markov model $p(\mathbf{X}^t | \mathbf{X}^{t-1})$. The probability of observing a scenario from these regions over a period $t \in \{1 \ldots T\}$ can be expressed by:

$$p(\mathbf{X}^1, \ldots, \mathbf{X}^T) = p(\mathbf{X}^1) \prod_{t=2}^{T} p(\mathbf{X}^t | \mathbf{X}^{t-1}). \tag{1}$$

Presume that the structure is specified by a set of regulatory relations $\mathbf{X}_{\pi_i}^{t-1} = \{X_j^{t-1} : X_j^{t-1} \text{ regulates } X_i^t\}$, where $i, j \in \{1 \ldots r\}$ and $\pi_i \subseteq \{1 \ldots r\}$, we can factorize the transition model $p(\mathbf{X}^t | \mathbf{X}^{t-1})$ over individual regions:

$$p(\mathbf{X}^1, \ldots, \mathbf{X}^T) = p(\mathbf{X}^1) \prod_{t=2}^{T} \prod_{i=1}^{r} p(X_i^t | \mathbf{X}_{\pi_i}^{t-1}). \tag{2}$$

Let graph $\mathcal{G}^t = (\mathcal{V}_\mathcal{G}, \mathcal{E}_\mathcal{G}^t)$ represents the conditional dependence between random vectors $\mathbf{X}^{t-1}$ and $\mathbf{X}^t$. Each vertex in $\mathcal{V}_\mathcal{G}$ corresponds to a sequence of variables $X_i^1, \ldots, X_i^T$, and the edge set $\mathcal{E}_\mathcal{G}^t \subseteq \mathcal{V}_\mathcal{G} \times \mathcal{V}_\mathcal{G}$ contains directed edges from components of $\mathbf{X}^{t-1}$ to components of $\mathbf{X}^t$. The time dependent transition model $p^t = (\mathbf{X}^t | \mathbf{X}^{t-1})$ is expressed by DBN in auto-regressive form $\mathbf{X}^t = \mathbf{A}^t \mathbf{X}^{t-1} + \epsilon$, where $\mathbf{A}^t \in \mathbb{R}^{r \times r}$ is a matrix of coefficients relating the variables at time $t-1$ from all regions to the variables of the regions in the next time point t and $\epsilon \sim \mathcal{N}(\mathbf{0}, \sigma^2 \mathbf{I})$ is an error term. The region time-varying structure is represented by the non-zero entries (connections) and zero entries (disconnections) in the estimated matrices $\hat{\mathbf{A}}^t$ at time t.

In this model, the strength of dependencies are estimated by minimizing a set of squared loss functions with regularization, one for each vertex at each time point $t^* \in \{1 \ldots T\}$:

$$\hat{\mathbf{A}}_{i.}^{t^*} = \underset{\mathbf{A}_{i.}^{t^*} \in \mathbb{R}^{1 \times r}}{\operatorname{argmin}} \frac{1}{T} \sum_{t=1}^{T} w^{t^*}(t)(\mathbf{A}_{i.}^{t^*} \mathbf{x}^{t-1} - x_i^t)^2 + \lambda \|\mathbf{A}_{i.}^{t^*}\|_1 \tag{3}$$

where λ is a regularization parameter which controls the sparsity of the networks, and $w^{t^*}(t)$ is the weighting of an observation from time t defined as $w^{t^*}(t) = \frac{K_h(t-t^*)}{\sum_{t=1}^{T} K_h(t-t^*)}$ in which $K_h(\cdot)$ is a symmetric and non-negative kernel function and h is the kernel bandwidth. The end product of TV-DBN estimation is a set of $\hat{\mathbf{A}}_{i.}^t$ (one per region) which can be combined to form an estimate of $\mathbf{A}^t$ where $t \in \{1 \ldots T\}$. The non-zero and zero entries in the matrices $\hat{\mathbf{A}}^t$ represent the time-varying connections and disconnections between regions:

$$\hat{\mathcal{E}}_\mathcal{G}^t = \{(i, j) \in \mathcal{V}_\mathcal{G} \times \mathcal{V}_\mathcal{G} | \hat{\mathbf{A}}_{ij}^t \neq 0\}. \tag{4}$$

Please note that we have removed the restriction that the regions between two adjacent time points must be different as specified in [27].

3.3   Causal Time-Varying Dynamic Bayesian Networks (CTV-DBN)

Although the structure of Bayesian networks (BN) may be directed, the directions of arrows do not define causal effects as the influence can flow both ways except a collider (v-structure) is hit. Therefore, all of the definitions in BN refer only to probabilistic properties, such as conditional independence [17]. Song *et al.* [27] recognise this problem – BN does not necessarily imply causality, but suggest that dynamic Bayesian networks (DBN) bears a natural causal implication in which TV-DBN is part of this family. Each edge in a DBN only points from time $t - 1$ to t contributing to a natural causal implication. However, the network discovery method established in TV-DBN ignores causal relationships allowing the sharing of information across the whole time period.

The difference between causal models and probabilistic models arises when we care about interventions in a model [17]. We aim to establish causal relationships between regions for causal inference by assigning manipulated probability density to a region of interest [28]. The condition of causal Markov assumption (CMA) is invoked to make a BN isomorphic with a causal model [18], where the condition is defined as: given a causal graph $\mathbb{G} = \langle \mathcal{V}_\mathbb{G}, \mathcal{E}_\mathbb{G}, \mathcal{P}_\mathbb{G} \rangle$, where $\mathcal{V}_\mathbb{G}$ is a set of vertices and $\mathcal{E}_\mathbb{G}$ is a set of edges between vertices in $\mathcal{V}_\mathbb{G}$ and $\mathcal{P}_\mathbb{G}$ is a probability distribution over the vertices in $\mathcal{V}_\mathbb{G}$. $\mathbb{G}$ satisfies CMA if and only if for every $v \in \mathcal{V}_\mathbb{G}$, v is independent of $\{\mathcal{V}_\mathbb{G} \setminus (Descendants(v) \cup Parents(v))\}$ given $Parents(v)$, where $Parents(v)$ is the set of parents of v in $\mathbb{G}$ and $Descendants(v)$ is the set of descendants of v in $\mathbb{G}$ [29].

A probability density $\mathcal{P}_\mathbb{G}(\mathcal{V}_\mathbb{G})$ can be factorized according to $\mathbb{G}$ if and only if

$$\mathcal{P}_\mathbb{G}(\mathcal{V}_\mathbb{G}) = \prod_{v \in \mathcal{V}_\mathbb{G}} \mathcal{P}_\mathbb{G}(v|Parents(v)) \quad [28]. \tag{5}$$

Assuming $\mathbf{n} \subset \mathcal{V}_\mathbb{G}$ with only non-descendants of m, a manipulation of $m \in \mathcal{V}_\mathbb{G}$ to $\mathcal{P}_\mathbb{G}'(m|\mathbf{n})$ can be achieved by replacing $\mathcal{P}_\mathbb{G}(m|Parents(m))$ in Equation (5) by a manipulated density $\mathcal{P}_\mathbb{G}'(m|\mathbf{n})$ to form a manipulation rule:

$$\mathcal{P}_\mathbb{G}(\mathcal{V}_\mathbb{G}||\mathcal{P}_\mathbb{G}'(m|\mathbf{n})) = \mathcal{P}_\mathbb{G}'(m|\mathbf{n}) \prod_{v \in \mathcal{V}_\mathbb{G} \setminus \{m\}} \mathcal{P}_\mathbb{G}(v|Parents(v)), \tag{6}$$

where the double bar indicates an assignment of probability and $\mathcal{P}_\mathbb{G}'$ is a new probability density. Hence, based on Equation (2), the manipulation rule at time $t = \zeta$ (where $\zeta \in \{2 \ldots T\}$) and region $i = m$ is:

$$p(\mathbf{X}^1, \ldots, \mathbf{X}^T || p(X_m^\zeta | \mathbf{X}_{\pi_m}^{\zeta-1})) = p(X_m^\zeta | \mathbf{X}_{\pi_m}^{\zeta-1}) \left(p(\mathbf{X}^1) \prod_{\substack{t=2\ldots T \\ t \neq \zeta}} \prod_{\substack{i=1\ldots r \\ i \neq m}} p(X_i^t | \mathbf{X}_{\pi_i}^{t-1}) \right). \tag{7}$$

If $\mathcal{V}_\mathbb{G}$ represents region variables from all time points, the network structure estimated by optimization as defined by Equation (3) does not satisfy CMA. It is because the weighting function $w^{t^*}(t)$ considers the time points in $\{\mathcal{V}_{\mathbb{G}} \setminus (Descendants(v) \cup Parents(v))\}$, e.g., $\hat{\mathbf{A}}_{i\cdot}^{t^*}$ at time t^* is not only determined by $\mathbf{x}_i^{t^*-1}$ but also $\mathbf{x}^{t^*-z}$ where $z > 1$. In order to comply with CMA, the weighting function should only gather evidence from $\mathcal{S} = \{Descendants(v) \cup Parents(v)\}$. We define causal boundary $\mathcal{B}$ as a set of points in the closure of $\mathcal{S}$ but not belonging to the interior of $\mathcal{S}$. We therefore propose to adopt a causal weighting function $w_c^{t^*}(t)$ to fulfil the requirement, such that $w_c^{t^*}(t) = \frac{K_h^a(t-t^*)}{\sum_{t=1}^{T} K_h^a(t-t^*)}$ where $K_h^a(\cdot)$ is an asymmetric and non-negative kernel function satisfying CMA. Hence, we rewrite Equation (3) to:

$$\hat{\mathbf{A}}_{i\cdot}^{t^*} = \underset{\mathbf{A}_{i\cdot}^{t^*} \in \mathbb{R}^{1 \times r}}{\operatorname{argmin}} \frac{1}{T} \sum_{t=1}^{T} w_c^{t^*}(t)(\mathbf{A}_{i\cdot}^{t^*} \mathbf{x}^{t-1} - x_i^t)^2 + \lambda \|\mathbf{A}_{i\cdot}^{t^*}\|_1. \tag{8}$$

The use of asymmetric kernel for non-parametric regression can be found in economic literature [9] [10] but rarely discussed in structure discovery. Mackenzie and Tieu [22] discuss an application of asymmetric kernel regression to radial-basis neural networks with an opinion that the available real-life data reproduce a causal function; and therefore, are naturally bounded by an interval. Hence, a truncation of a symmetric kernel at the boundary makes the model to suffer from material bias error. Although there are several attempts to resolve boundary problem [13] [32], most of them cannot correct bias without increasing noise level and/or variance error [22]. Mackenzie and Tieu propose to correct boundary error by replacing a symmetric kernel with an asymmetric one. Apart from the favorable boundary property [22] – maintaining similar level of variance and bias trade-off, asymmetric kernel can also provide a weighting function which is within causal boundary $\mathcal{B}$. We define $K_h^a(\cdot)$ by using gamma function [3].

For the case of using kernel regression to estimate functional relationship, e.g., $y_i = y(t_i) + \epsilon$, where $i \in \{1 \dots N\}$, $0 \leq t_i \leq T$ and ϵ is random noise [22]. By using symmetric Gaussian kernel ($K_{\mathcal{G}}$) with boundary, we obtain significant bias at the boundary as the odd moments of $K_{\mathcal{G}}$ are no longer zero due to truncation [22]:

$$bias[\hat{y}(\eta)] = \left\{ y(\eta) \int_0^\infty K_{\mathcal{G}}(t-\eta)dt + y'(\eta) \int_0^\infty (t-\eta)K_{\mathcal{G}}(t-\eta)dt \right.$$
$$\left. + \frac{1}{2}y''(\eta) \int_0^\infty (t-\eta)^2 K_{\mathcal{G}}(t-\eta)dt + \dots \right\} - y(\eta), \tag{9}$$

where $\hat{y}(\eta)$ is a Priestley-Chao estimator [22] of $y(\eta)$; versus the scenario of no boundary:

$$bias[\hat{y}(\eta)] = \int_{-\infty}^{\infty} y(t_i)K_{\mathcal{G}}(\eta-t_i)dt_i - y(\eta) \cong \frac{\sigma^2}{2}y''(\eta) + \frac{\sigma^4}{8}y''''(\eta). \tag{10}$$

However, the boundary error term is vanished by replacing symmetric kernel $K_{\mathcal{G}}$ by an asymmetric Gamma kernel (K_Γ) in the case of kernel regression with boundary:

$$bias[\hat{y}(\eta)] = \frac{\sigma^2}{2}y''(\eta) + \frac{\sigma^4}{3\eta}y'''(\eta) + \frac{\sigma^4}{8}\left\{1 + 2\left(\frac{\sigma}{\eta}\right)\right\}y''''(\eta) + \dots \tag{11}$$

As a result, we transform TV-DBN to CTV-DBN by adopting Equation (8) to be our new objective function, while Equations (1), (2) and (4) remain unchanged. The CTV-DBN structure discovery algorithm is summarized in Algorithm 1. In this implementation, the weighting function $w_c^{t^*}(t)$ is pushed into the square loss function by rescaling x_i^t and $\mathbf{x}^{t-1}$ such that it becomes a standard ℓ_1-regularized least-squares problem.

Algorithm 1. CTV-DBN Structure Discovery

1: Initialize $\hat{\mathbf{A}}^0$ randomly
2: **for** $i \in \{1 \ldots r\}$ **do**
3:     **for** $t^* \in \{1 \ldots T\}$ **do**
4:         $\hat{\mathbf{A}}_{i\cdot}^{t^*} \leftarrow \hat{\mathbf{A}}_{i\cdot}^{t^*-1}$ *# Warm-start initialization*
5:         $\mathring{x}_i^t \leftarrow \sqrt{w_c^{t^*}(t)}x_i^t$; $\mathring{\mathbf{x}}^{t-1} \leftarrow \sqrt{w_c^{t^*}(t)}\mathbf{x}^{t-1}, \forall t \in \{1 \ldots T\}$ *# Rescaling*
6:         **while** $\hat{\mathbf{A}}_{i\cdot}^{t^*}$ not converges **do**
7:             *# Search for the argument of the minimum based on Equation (8)*
8:             **for** $j \in \{1 \ldots r\}$ **do**
9:                 $S_j \leftarrow \frac{2}{T}\sum_{t=1}^T \left(\sum_{k \neq j} \mathbf{A}_{ik}^{t^*}\mathring{x}_k^{t-1} - \mathring{x}_i^t\right)\mathring{x}_j^{t-1}$; $b_j \leftarrow \frac{2}{T}\sum_{t=1}^T \mathring{x}_j^{t-1}\mathring{x}_j^{t-1}$
10:                 $\mathbf{A}_{ij}^{t^*} \leftarrow (\text{sign}\,(S_j - \lambda)\,\lambda - S_j)\,/b_j$, if $|S_j| > \lambda$; 0 otherwise
11:             **end for**
12:         **end while**
13:     **end for**
14: **end for**
15: **return** $\left\{\hat{\mathbf{A}}^1, \ldots, \hat{\mathbf{A}}^T\right\}$

---

4 Experiments

4.1 Structure Discovery for Ring Road System

We apply CTV-DBN on Beijing taxi trajectories from Complex Engineered Systems Lab, Tsinghua University, China[1], where Beijing is a densely populated city with special ring topology road structure. The dataset consists of one month of trajectories of 28,000 taxis in Beijing captured in May 2009, where each record includes the following information: 1) a taxi identifier, 2) a time-stamp in UTC of the time when the location was taken, and 3) latitude and 4) longitude specifying the position of the taxi. The trajectories are firstly passed through a spatial filter with a boundary[2] of Beijing city centred at the Forbidden City (city centre) and extended to its three international airport terminals (top right hand corner). We then apply density-based clustering method DBSCAN on one week (Monday-Friday) of trajectories at 8am to obtain a driver's view of regions

---

[1] http://sensor.ee.tsinghua.edu.cn/datasets.php
[2] A rectangle formed by latitude and longitude pairs (40.08200, 116.16054) and (39.75030, 116.62000)

during morning rush hours as shown in Figure 1. Please note that top right hand corner is the location of the airports (regions 2 and 6), and we can observe an expected region structure complexity from the city centre to the airports. The trajectory average speeds within clusters are calculated.

We obtain estimated network structures by using the structure discovery process of relaxed TV-DBN and CTV-DBN. The structures at 8:20am and 8:30am are shown in Figure 2 as examples, where the cells filled with black colour represent connections and blank otherwise. Because of the truncation at causal boundary $\mathcal{B}$, the method of CTV-DBN with truncated $K_{\mathcal{G}}$ is expected to suffer from higher bias as well as information loss (Figures 2(c) and 2(d)). This can be observed by comparing it with the results from relaxed TV-DBN with $K_{\mathcal{G}}$ (Figures 2(a) and 2(b)), although a subset of the structure can be recognised. So far, we can identify regions 3, 6, 11, 16, 22, 23, 24 and 28 (region list "A") are the ones heavily depending on nearly all regions in the city. Apart from region 16, all the other regions are between the city centre and the airports.

Finally, the method of CTV-DBN with K_{Γ} (Figures 2(e) and 2(f)) does not only come with a theoretical strength of low bias at the causal boundary $\mathcal{B}$ and satisfying CMA, it also reveals more details of causal relationships among regions. Based on the same level of regularization, regions with insufficient causal connections are eliminated in CTV-DBN (versus relaxed TV-DBN) and additional connections are added based on the evidence within causal boundary $\mathcal{B}$. The structural differences between CTV-DBN with K_{Γ} and relaxed TV-DBN with $K_{\mathcal{G}}$ are mainly from the enforced causal relationship in the former.

Out of all the regions in region list "A" above, only regions 3, 24 and 28 are the top 3 regions causally impacted by most of the regions and they are all located along the Beijing Airport Expressway (S12)[3] between the city and the airports, in which traffic jam is common[4]. Since the three regions are located just at or before ring roads[5] which diverge traffic to all major districts in the city, any congestion in the other regions would have ultimate impact to these three regions (major artery between the city and the airports). These findings have also been confirmed by [33] and from numerous published facts, such as a report by China Central Television[6] about the relationship between ring roads and other districts in the Beijing city.

4.2   Structure Discovery for Road Junction System

A proprietary traffic flow dataset from the Victoria government traffic authority (VicRoads) of Australia is selected in this paper for illustration purpose, due to the fact that the data is known to have high correlation, and that they are collected from road junctions which have distinct spatio structures (Figure 6). The data

---

[3] http://en.wikipedia.org/wiki/Airport_Expressway_(Beijing)
[4] http://www.bjjtgl.gov.cn/publish/portal1/, and
http://wikitravel.org/en/Beijing
[5] http://en.wikipedia.org/wiki/Ring_roads_of_Beijing
[6] http://www.cctv.com/lm/124/41/90128.html

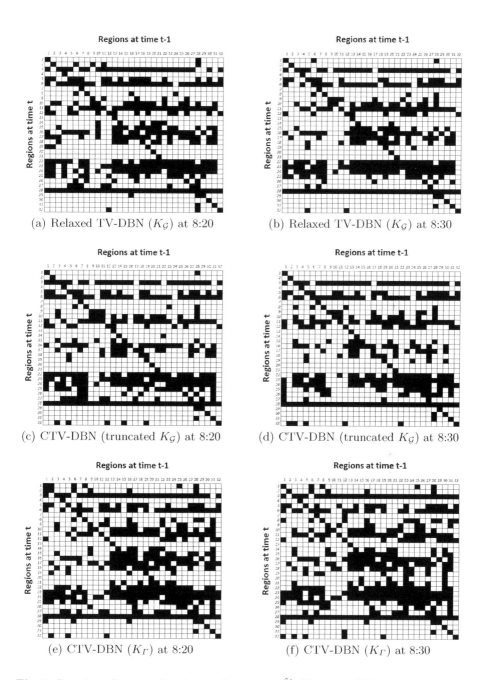

Fig. 2. Structure discovery for ring road system – $\hat{\mathcal{E}}_{\mathcal{G}}^t$ (Equation (4)) representing the directed edges from components of $\mathbf{X}^{t-1}$ (columns) to components of $\mathbf{X}^t$ (rows) based on $\hat{\mathbf{A}}^t$ with $\lambda = 0.4$

consist of traffic volumes by site and detector at static locations between 1 January 2011 to 31 December 2012 from an eastern suburb of Melbourne, Australia as shown in Figure 3, where each record includes the following information: 1) site number, 2) date of the reading is taken, 3) detector number, and 4) 96 15-minute intervals traffic volumes of the date starting from 00:00. In this experiment, we investigate junction-to-junction dependence of an area which serves as a gateway to a major highway (Monash Freeway) in Melbourne, Australia.

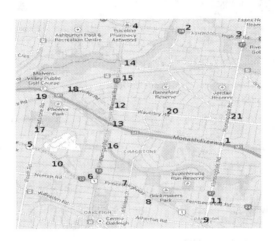

Fig. 3. An eastern suburb area of Melbourne, Australia with monitored junctions labeled from 1 to 21 based on the order of the site number in the dataset, where Melbourne CBD is located 18km north-west from this area

Different from trajectories obtained from mobile sensors, we define the regions by sites (same site number) in the dataset representing the road junctions numbered in Figure 3. The 96 15-minute interval weekdays' average utilisations of each junction is calculated by using the maximum average utilisation of the junctions as its estimated maximum capacity. We discovery the estimated network structures by using CTV-DBN with K_Γ, where the network structures at 8am and 6pm are shown in Figure 4 for comparison. Due to the expected higher correlation between junctions within a relatively small area, a higher regularization parameter $\lambda = 0.6$ is used to obtain a similar level of sparsity; versus $\lambda = 0.4$ used in Section 4.1. Based on visual comparison between Figures 2 and 4, and the inspection of unshown $\hat{\mathcal{E}}_\mathcal{G}^t$ in this experiment with t close to each others, the behaviors of our structure discovery mechanism are similar between a macro structure (Section 4.1) and a micro structure (this section); although an adjustment of regularization parameter is required as their level of correlations are different.

We capture the dependency of junctions 13 and 1 at t on the other junctions at $t - 1$ in Figure 5 by arrows. Each arrow represents the dependency of a

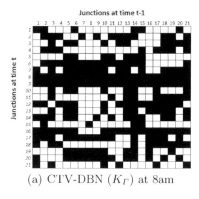
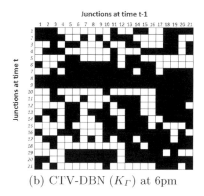

(a) CTV-DBN (K_Γ) at 8am   (b) CTV-DBN (K_Γ) at 6pm

Fig. 4. Structure discovery for road junction system – $\hat{\mathcal{E}}_{\mathcal{G}}^t$ (Equation (4)) representing the directed edges from components of $\mathbf{X}^{t-1}$ (columns) to components of $\mathbf{X}^t$ (rows) based on $\hat{\mathbf{A}}^t$ with $\lambda = 0.6$

junction at the arrow head on the junction at the arrow tail. Both junctions 13 and 1 are the only accesses of Monash freeway in the area. It is important to note that junction 13 provides both east-bound and west-bound, and both have entry and exit accesses to traffic, whilst Junction 1 only provides east-bound entry and west-bound exit. Moreover, there is no access to Monash freeway at junctions 19 and 18 where they are the monitored junctions along a local road (Waverley Road). The site layouts of junctions 13 and 1 are shown in Figure 6. The traffic to the Melbourne CBD (west-bound) via Monash Freeway have to go through junction 13, but the traffic to further out of the city (east-bound) can have an option to access the freeway via junctions 13 or 1.

In the morning (8am), both junctions 13 and 1 are mostly depending on the junctions in their regions, i.e., junction 13 relates to the junctions located to the left of the area, whilst junction 1 relates to the junctions located to the right of the area. The exceptions are a) junction 3 and 21 in Figure 5(a) which we can interpret that the the traffics have to travel to junction 13 to access Monash Freeway to the city where they cannot go via junction 1, b) junction 6, and junction 14 and 15 to a lesser extent in Figure 5(b) which we can interpret that some of the traffic prefer to travel further to junction 1 as east-bound traffic to access Monash Freeway. In the afternoon (6pm), the junction dependencies are centered to junction 13 where nearly all the junctions along Princes Highway and the other surrounding junctions are contributing to its utilisation as observed from Figure 5(c). However, the patterns of junction 1 between the morning (Figure 5(b)) and the afternoon (Figure 5(d)) are more or less the same; except there is a far dependence from junction 19. It can be interpreted that some of the east-bound traffic along Waverley Road access Monash Freeway despite they can access the freeway via junction 1.

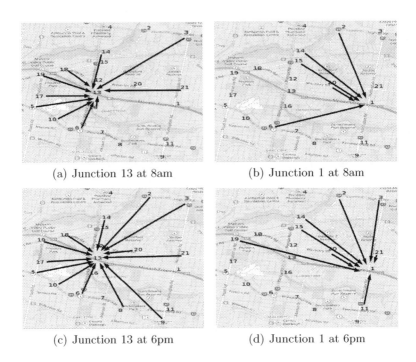

(a) Junction 13 at 8am   (b) Junction 1 at 8am

(c) Junction 13 at 6pm   (d) Junction 1 at 6pm

Fig. 5. Causal dependency summary

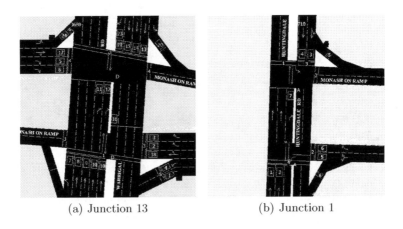

(a) Junction 13   (b) Junction 1

Fig. 6. Junction site layouts

5   Conclusion

This paper presents a notion of causal time-varying dynamic Bayesian networks (CTV-DBN), as well as defining causal boundary for cross time information sharing. CTV-DBN is then applied to spatio-temporal data by combining the model with density-based clustering and Voronoi diagram for region discovery.

By revealing the time-varying region structures using moving objects' view of territories, causal relationships among regions are captured and available for causal inference. The findings learnt from real-life trajectory dataset using CTV-DBN structure discovery method are consistent with the known facts [33]. The insights out of analysing the region-to-region and junction-to-junction relationships provide valuable information for road system improvement. Future work involves the formulation of mechanisms to fine tune the parameters of asymmetric kernels to obtain a best fit to causal boundary and automatic nomination of regularization parameter.

References

1. Ankerst, M., Breunig, M.M., Kriegel, H.-P., Sander, J.: OPTICS: Ordering points to identify the clustering structure. In: SIGMOD Conference, pp. 49–60 (1999)
2. Aurenhammer, F., Klein, R.: Voronoi Diagrams, ch. 5, pp. 201–290. Elsevier Science, Amsterdam (2000)
3. Bain, L.J., Engelhardt, M.: Introduction to probability and mathematical statistics, vol. 4. Duxbury Press, Belmont (1992)
4. Birant, D., Kut, A.: ST-DBSCAN: An algorithm for clustering spatial-temporal data. Data Knowl. Eng. 60(1), 208–221 (2007)
5. Chawla, S., Zheng, Y., Hu, J.: Inferring the root cause in road traffic anomalies. In: ICDM, pp. 141–150 (2012)
6. Clarke, H., Hawkins, A.: Economic framework for melbourne traffic planning. Agenda 13(1), 63–80 (2006)
7. Dondelinger, F., Lèbre, S., Husmeier, D.: Non-homogeneous dynamic bayesian networks with Bayesian regularization for inferring gene regulatory networks with gradually time-varying structure. Machine Learning 90(2), 191–230 (2013)
8. Ester, M., Peter Kriegel, H., J.S, Xu, X.: A density-based algorithm for discovering clusters in large spatial databases with noise, pp. 226–231. AAAI Press (1996)
9. Gospodinov, N., Hirukawa, M.: Time series nonparametric regression using asymmetric kernels with an application to estimation of scalar diffusion processes (2008)
10. Gospodinov, N., Hirukawa, M.: Nonparametric estimation of scalar diffusion models of interest rates using asymmetric kernels. Journal of Empirical Finance (2012)
11. Grzegorczyk, M., Husmeier, D.: Non-stationary continuous dynamic Bayesian networks. In: NIPS, pp. 682–690 (2009)
12. Grzegorczyk, M., Husmeier, D.: Non-homogeneous dynamic Bayesian networks for continuous data. Machine Learning 83(3), 355–419 (2011)
13. Hall, P., Wehrly, T.E.: A geometrical method for removing edge effects from kernel-type nonparametric regression estimators. Journal of the American Statistical Association 86(415), 665–672 (1991)
14. Hou, X., Zhang, J., Du, C., Zhang, L.: Research the influence of the ring road factor on route choice. In: 2009 17th International Conference on Geoinformatics, pp. 1–6. IEEE (2009)
15. Jeung, H., Yiu, M.L., Zhou, X., Jensen, C.S., Shen, H.T.: Discovery of convoys in trajectory databases. Proceedings of the VLDB Endowment 1(1), 1068–1080 (2008)
16. Kisilevich, S., Mansmann, F., Nanni, M., Rinzivillo, S.: Spatio-temporal clustering. In: Data Mining and Knowledge Discovery Handbook, pp. 855–874 (2010)

17. Koller, D., Friedman, N.: Probabilistic Graphical Models: Principles and Techniques. MIT Press (2009)
18. Lemmer, J.F.: The causal Markov condition, fact or artifact? SIGART Bulletin 7(3), 7–3 (1996)
19. Liu, W., Zheng, Y., Chawla, S., Yuan, J., Xing, X.: Discovering spatio-temporal causal interactions in traffic data streams. In: KDD, pp. 1010–1018 (2011)
20. Liu, Y., Niculescu-Mizil, A., Lozano, A.C., Lu, Y.: Learning temporal causal graphs for relational time-series analysis. In: Proceedings of the 27th International Conference on Machine Learning (ICML 2010), pp. 687–694 (2010)
21. Lu, C.-T., Lei, P.-R., Peng, W.-C., Su, I.-J.: A framework of mining semantic regions from trajectories. In: Yu, J.X., Kim, M.H., Unland, R. (eds.) DASFAA 2011, Part I. LNCS, vol. 6587, pp. 193–207. Springer, Heidelberg (2011)
22. Mackenzie, M., Tieu, A.K.: Asymmetric kernel regression. IEEE Transactions on Neural Networks 15(2), 276–282 (2004)
23. Murphy, K.: Dynamic Bayesian Networks: Representation, Inference and Learning. PhD thesis, UC Berkeley, Computer Science Division (2002)
24. Nanni, M., Pedreschi, D.: Time-focused clustering of trajectories of moving objects. J. Intell. Inf. Syst. 27(3), 267–289 (2006)
25. Robinson, J.W., Hartemink, A.J.: Non-stationary dynamic Bayesian networks. In: NIPS, pp. 1369–1376 (2008)
26. Robinson, J.W., Hartemink, A.J.: Learning non-stationary dynamic Bayesian networks. Journal of Machine Learning Research 11, 3647–3680 (2010)
27. Song, L., Kolar, M., Xing, E.P.: Time-varying dynamic bayesian networks. In: NIPS, pp. 1732–1740 (2009)
28. Spirtes, P.: Introduction to causal inference. Journal of Machine Learning Research 11, 1643–1662 (2010)
29. Spirtes, P., Glymour, C., Scheines, R.: Causation, Prediction, and Search, 2nd edn. MIT Press (2000)
30. Wang, M., Wang, A., Li, A.: Mining spatial-temporal clusters from geo-databases. In: Li, X., Zaïane, O.R., Li, Z.-h. (eds.) ADMA 2006. LNCS (LNAI), vol. 4093, pp. 263–270. Springer, Heidelberg (2006)
31. Yang, B., Guo, C., Jensen, C.S.: Travel cost inference from sparse, spatio-temporally correlated time series using markov models. Proceedings of the VLDB Endowment, 6 (2013)
32. Zhang, S., Karunamuni, R., Jones, M.: An improved estimator of the density function at the boundary. Journal of the American Statistical Association 94(448), 1231–1240 (1999)
33. Zhao, P., Lu, B., de Roo, G.: The impact of urban growth on commuting patterns in a restructuring city: Evidence from Beijing. Papers in Regional Science 90(4), 735–754 (2011)

Efficient Processing of Which-Edge Questions on Shortest Path Queries[⋆]

Petrie Wong[1], Duncan Yung[1], Ming Hay Luk[1], Eric Lo[1], Man Lung Yiu[1], and Kenny Q. Zhu[2]

[1] Hong Kong Polytechnic University
{cskfwong,cskwyung,csmhluk,ericlo,csmlyiu}@comp.polyu.edu.hk
[2] Shanghai Jiao Tong University
kzhu@cs.sjtu.edu.cn

Abstract. In this paper, we formulate a novel problem called *Which-Edge* question on shortest path queries. Specifically, this problem aims to find k edges that minimize the total distance for a given set of shortest path queries on a graph. This problem has important applications in logistics, urban planning, and network planning. We show the NP-hardness of the problem, as well as present efficient algorithms that compute highly accurate results in practice. Experimental evaluations are carried out on real datasets and results show that our algorithms are scalable and return high quality solutions.

1 Introduction

Shortest path queries have a wide range of applications in logistics, urban planning, and network planning. This paper introduces a novel problem called *Which-Edge* question on shortest path queries. The objective is to find k edges that minimize the total distance for a given set Q of shortest path queries on a weighted graph $G(V, E, W)$. Let *spd* denote the total distance of queries in Q on the graph. Specifically, there are two forms of *Which-Edge* questions.

- **Which-k-Edges-Insert question**: Given an edge set P s.t. $P \cap E = \emptyset$, which k edges in P would *minimize* the *spd* (if those k edges are *inserted* into G)?
- **Which-k-Edges-Delete question**: Given an edge set $\overline{P} \subseteq E$, which k edges in $\overline{P}$ would *minimize* the *spd* (if they are *removed* from G)?

The potential applications of *Which-Edge* question could be illustrated using a few examples. First, consider an express mail company's delivery network: a graph with nodes representing locations (e.g., cities, warehouses) and edges representing connections (e.g., flights between cities). Subject to rapidly changing business environments, the company may revise its resource allocation policy regularly—with additional resources, the management may want to reduce its overall/average delivery time (increasing its competitive advantage) by establishing a new connection between two indirectly connected locations. In the example above, a *Which-Edge* question natural arises: "*(Q1): which two locations should be connected by the new connection?*". In this situation, answers like "among all the possible choices, the maximum reduction

---

[⋆] This work is supported by the Research Grants Council of Hong Kong (GRF PolyU 520413), and Hong Kong Polytechnic University (ICRG grant A-PL99).

Table 1. Summary of Frequently Used Symbols

Symbol	Meaning
G	the input graph
$z_i(u_i, v_i)$	a bridge to be added to G to connect nodes u_i and v_i
$\|z_i\|$	capacity (resp. length) of a bridge/edge for shortest path
G^{+z_i} / G^{-e_i}	the graph G with bridge z_i added (with edge e_i removed)
$P / \bar{P}$	the set of candidate bridges to add to G (to remove from G)
K	the answer set of a Which-Edge question, containing k bridges/edges
B^{z_i} / D^{e_i}	the benefit (damage) of adding bridge z_i (removing edge e_i)
Z	an arbitrary set of bridges/edges
$sp^G_{(x,y)}$	the *shortest path distance* from x to y on graph G

in overall delivery time (35%) is achieved if a new connection between X and Y is established." will be very helpful in aiding the decision-making process. After said decision is made operational, a new connection (edge) is added to the delivery network (graph) reflecting the impact of the decision. Alternatively, during tough economic times, the management may ask *"(Q2): which k connections, if cut, have the minimal impact on overall delivery time?"* — a decision that is eventually reflected by some existing connections (edges) being removed from the delivery network (graph). Applications of *Which-Edge* questions abound in other domains, too. In urban planning where the road network is modeled as a graph, one may ask *"(Q3): which stretch of roads (edges) should we expand such that average travel times can be significantly reduced?"*. In network planning, one may ask *"(Q4): which optic fibre, if broken (e.g., due to a natural disaster), would have the largest negative impact on the average network communication time?"*;

Answering *Which-Edge* questions on graphs is an interesting yet challenging research issue. First, we have two types of *Which-Edge* questions, e.g., while Q1 and Q3 are related to *edge insertion*, Q2 and Q4 are related to *edge deletion*. Thus, it is necessary to develop general solutions that are applicable to both questions. Second, the evaluation of *Which-Edge* questions is computationally expensive because the solution space of *Which-Edge* questions can be very large. For example, in Q2, there are many possible k-combinations of connections that could be considered. That huge search space renders straightforward exhaustive algorithms impractical. In this paper, we make the following contributions:

- Introduce the concept of *Which-Edge* questions and present its specification.
- Establish the problem hardness of *Which-Edge* questions.
- Develop evaluation algorithms and efficiency optimizations for answering *Which-Edge* questions.

We will investigate the Which-k-Edges-Insert question and the Which-k-Edges-Delete question in Sections 2 and 3 respectively. In Section 4, we evaluate the solutions' quality and efficiency on real graph datasets. We discuss related work in Section 5 and conclude our paper in Section 6. Table 1 shows a summary of frequently used symbols.

2   Which-k-Edges-Insert Question

In this section, we first formulate the *Which-k-Edges-Insert* question and establish its hardness. Then, we propose our heuristic solutions and efficiency optimizations to compute highly accurate results in practice.

2.1   Problem Formulation

Given a weighted graph $G = (V, E, W)$, there are potentially many non-adjacent node-pairs in G that could be considered to be connected by some new edges. In practice, however, the number of non-adjacent node-pairs to be considered is usually domain-specific. For example in Q1, only flight connections offered by airlines should be considered. Thus, we model the set of non-adjacent node-pairs, P, as an input parameter (generated by other softwares or prepared manually). Specifically, each non-adjacent node-pair (u_i, v_i) in P is associated with an implied *bridge* z_i, which is the edge considered to be added into G for connecting u_i and v_i. The bridge z_i has a length $\|z_i\|$, and a *cost* c^{z_i}, which models the real-world cost of connecting u_i to v_i by z_i in G (e.g., the cost of a bridge can be a function of the bridge's length). In what follows, we use the term "non-adjacent node-pairs" and "bridge" interchangeably.

The *Which-k-Edges-Insert* question aims to find out which k edges in a given bridge-set P, if inserted into G, reduces the sum of shortest path distances of a query workload Q the most (optimization goal). Here we may have a *workload* Q of shortest-path queries with different sources and destinations on G. Each query $q_j(s_j, t_j) \in Q$ is a *distinct* shortest-path query with source s_j and destination t_j. If the user does not specify a workload, we consider Q to contain all-pairs shortest-path queries.

Each query $q_j \in Q$ is associated with an importance factor m_{q_j}—using deliver planning as an example, assuming that only one delivery makes a 100 mile trip from s_1 to t_1, and 50 deliveries make a 5 mile trip from s_2 to t_2, we may model them as two shortest-path queries $q_1(s_1, t_1)$ and $q_2(s_2, t_2)$ and set $m_{q_1} = 1$ and $m_{q_2} = 50$. Thus, query importance can model the number of beneficiaries of a bridge. For instance, a bridge $z_1(u_1, v_1)$, which reduces the shortest-path distance of q_1 from 100 to 40 miles, is not as beneficial as a bridge $z_2(u_2, v_2)$, which reduces the shortest-path distance of q_2 from 5 to 1 mile, because only one delivery makes the trip q_1. More precisely, let $sp_{q_j}^G$ be the shortest-path distance of query q_j on the graph G. Then the benefit of connecting u_i and v_i by z_i on a query $q_j(s_j, t_j)$, or simply the *benefit of bridge* z_i on query q_j, $b_{q_j}^{z_i}$, is the reduction in shortest-path distance of q_j in G^{+z_i} versus G, accounting for the query's importance factor m_{q_j}:

$$b_{q_j}^{z_i} = m_{q_j} \times (sp_{q_j}^G - sp_{q_j}^{G+z_i}) \tag{1}$$

In the example above, the bridge z_1, which shortens q_1 from 100 to 40 miles, has a benefit $b_{q_1}^{z_1} = 1 \times (100 - 40) = 60$; whereas the bridge z_2, which shortens q_2 from 5 miles to 1 mile, has a benefit $b_{q_2}^{z_2} = 50 \times (5 - 1) = 200$.

The above definitions can be extended to a subset K of bridges from P. The *benefit of a set K of bridges on query q_j* is defined as:

$$b_{q_j}^K = m_{q_j} \times (sp_{q_j}^G - sp_{q_j}^{G+K}) \tag{2}$$

The total benefit of K on a query workload Q is:

$$B^K = \sum_{q_j \in Q} b_{q_j}^K \tag{3}$$

PROBLEM 1 (WHICH-k-EDGES-INSERT QUESTION). *Given a graph G, a set P of non-adjacent node pairs (u_i, v_i), their associated bridges z_i and cost c^{z_i}, and a workload of shortest-path queries Q; find a subset $K \subseteq P$ of k bridges so that if they are added to G, they have the maximum benefit to workload Q, accounting of the cost C^K of adding K to G, i.e., $\arg\max_{K \subseteq P, |K| = k}(B^K - C^K)$, where $C^K = \sum_{z_i \in K} c^{z_i}$.*

In this formulation, we assume the cost c^{z_i} of a bridge z_i (e.g., 1000 USD) has been normalized to match the unit of benefit, as in any ranking function in database query processing. In fact, the relationship between the total benefit B^K and the cost C^K is flexible; in some applications we can consider a different formulation, for example, using another function $\arg\max_{K \subseteq P, |K| = k}(B^K / C^K)$.

2.2   Problem Hardness

We prove that this problem is $\mathcal{NP}$-hard.

Theorem 1. *The Which-k-Edges-Insert problem is $\mathcal{NP}$-hard.*

Proof. We present a *reduction scheme* that converts any given instance of the *Set-Cover problem* [1] into an instance $\langle k, G(V, E), P, Q \rangle$ of our Which-k-Edges-Insert problem. Let $\langle k, CS = \{S_i\}, U \rangle$ be an instance of Set-Cover, where k is an integer, U is a domain set of items, CS is a collection of subsets $S_i \subseteq U$. This problem asks whether there exists a sized-k collection $CS' \subseteq CS$ such that the size of its subset union $|\cup_{S_i \in CS'} S_i|$ equals to $|U|$. The reduction scheme is as follows:

– for each item $j \in U$, we insert a query $q_j(s_j, t_j)$ into Q, and insert the vertices s_j, t_j into V;
– for each subset $S_i \in CS$, we insert a directed bridge $z_i(u_i, v_i)$ with length 0 into P and insert the vertices u_i, v_i into V;
– for each query $q_j(s_j, t_j)$ of Q, we insert a directed edge (s_j, t_j) with length 1 into E;
– for each item $j \in U$ in a subset $S_i \in CS$, we insert directed edges (s_j, u_i) and (v_i, t_j) with length 0 into E.

An example reduction is illustrated in Figure 1. The bridges in P are shown as dashed lines. Observe that the size of the constructed instance $\langle k, G(V, E), P, Q \rangle$ is polynomial to the size of the given instance $\langle k, CS = \{S_i\}, U \rangle$. Also, the construction process takes polynomial time. The intuition behind this reduction scheme is that, if a bridge z_i is selected in a solution of Which-k-Edges-Insert, then the queries q_j benefit from it correspond to the items j covered by a chosen set S_i in a solution of Set-Cover.

Now, we only consider the subclass $\mathcal{C}$ of problem instances of Which-k-Edges-Insert that conform with the conditions in the above reduction scheme, specifically:

– all vertices in P and Q are unique; they are the only vertices in the graph G;
– each bridge $z_i(u_i, v_i) \in P$ has length 0;
– for each query $q_j(s_j, t_j) \in Q$, there must be an edge (s_j, t_j) with length 1 in the graph;
– in addition, the graph contains only the following edges: an edge (s_j, u_i) exists if and only if an edge (v_i, t_j) exists (for some z_i, q_j); such edges (if exist) must have length 0;

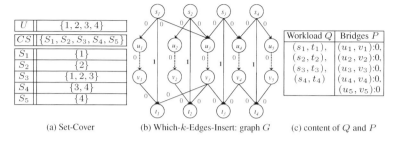

U	$\{1,2,3,4\}$
CS	$\{S_1,S_2,S_3,S_4,S_5\}$
S_1	$\{1\}$
S_2	$\{2\}$
S_3	$\{1,2,3\}$
S_4	$\{3,4\}$
S_5	$\{4\}$

Workload Q	Bridges P
(s_1,t_1),	(u_1,v_1):0,
(s_2,t_2),	(u_2,v_2):0,
(s_3,t_3),	(u_3,v_3):0,
(s_4,t_4)	(u_4,v_4):0,
	(u_5,v_5):0

(a) Set-Cover   (b) Which-k-Edges-Insert: graph G (c) content of Q and P

Fig. 1. Reduction: Set Cover to Which-k-Edges-Insert

Now, we show that a solution of Set-Cover $\langle k, CS = \{S_i\}, U\rangle$ corresponds to a solution of the $\mathcal{C}$ subclass of Which-k-Edges-Insert problem $\langle k, G(V,E), P, Q\rangle$ with benefit equal to $|Q|$, and vice versa.

Let CS' be a solution of the Set-Cover. Let K' be a solution of the Which-k-Edges-Insert problem. Let $M = |U| = |Q|$.

We first convert a given solution CS' to a corresponding solution K' and then derive the benefit value of K'. Let $U' = \cup_{S_i \in CS'} S_i$ be the union set of items covered by CS'. Recall that the size of CS' is k, and the size of U' is M. WLOG, we rename CS' as $\{S_{x_1}, S_{x_2}, \cdots, S_{x_k}\}$ and rename U' as $\{y_1, y_2, \cdots, y_M\}$. We claim that the corresponding solution of the Which-k-Edges-Insert problem is $K' = \{z_{x_1}, z_{x_2}, \cdots, z_{x_k}\}$, and the set of benefited queries is $Q' = \{q_{y_1}, q_{y_2}, \cdots, q_{y_M}\}$. Since CS' is a solution of the Set-Cover, each item in $y_j \in U'$ must be contained in S_{x_i} for some i. According to our reduction scheme, the corresponding edges (s_j, u_i) and (v_i, t_j) belong to the constructed graph. Thus, for each query $(s_j, t_j) \in Q'$, there exists a path s_j, u_i, v_i, t_j with distance 0, causing its benefit to be 1. Summing the benefit over all queries of Q', we obtain the benefit M.

We then convert a given solution K' to a corresponding solution CS' and then derive the union size of CS'. Since the size of CS' is k, we rename K' as $K' = \{z_{x_1}, z_{x_2}, \cdots, z_{x_k}\}$. The benefit of each query is either 0 or 1. Since the total benefit of K' on all queries is M, there must be a set of M benefited queries, say, $Q' = \{q_{y_1}, q_{y_2}, \cdots, q_{y_M}\}$. We claim that the corresponding solution of the Set-Cover problem is $CS' = \{S_{x_1}, S_{x_2}, \cdots, S_{x_k}\}$, with the union set of items covered as $U' = \{y_1, y_2, \cdots, y_M\}$. Since K' is a solution of the Which-k-Edges-Insert problem, each query in $q_{y_j} \in Q'$ must be benefited by the bridge z_{x_i} for some i. According to our reduction scheme, the corresponding item in $y_j \in U'$ must be contained in S_{x_i} for some i. Thus, all items in U' are covered by CS' and the union size is M.

Since the Set-Cover problem is $\mathcal{NP}$-hard [1], the above implies that Which-k-Edges-Insert problem is also $\mathcal{NP}$-hard.   □

Observe that Set-Cover instances correspond to only a subclass of problem instances of Which-k-Edges-Insert. Thus, the approximation algorithms (and their approximation ratio) for Set-Cover cannot be applied to all problem instances of Which-k-Edges-Insert.

Brute-Force Solution. We proceed to describe a brute-force algorithm (BF) for computing the exact result for the Which-k-Edges-Insert problem. It enumerates every possible subset K with k bridges from P. For each subset K, it temporarily inserts bridges in K to the graph G (denote that as G^{+K}) and uses an incremental shortest path algorithm (e.g., [2]) to compute the new shortest path. Finally, the subset K with the highest benefit B^K is reported as the result. Figure 2a shows an example[1] with $k=2$, a source s, a sink t, and with shortest path distance originally as 1. The running steps of BF with $P = \{(b, g), (g, c), (d, h), (j, d), (i, e)\}$ are illustrated in Figure 2b. There are $\binom{|P|}{k} = \binom{5}{2}$ subsets to be considered. The subset $K = \{(b, g), (g, c)\}$ is reported as the result because the new shortest path distance is 14 after adding them to G, which yields the highest benefit $B^K = 19 - 14 = 5$ among other subsets.

The time complexity of BF is $\mathcal{O}(\binom{|P|}{k} ISP(G))$, where $ISP(G)$ denotes the time complexity of an incremental shortest path algorithm. Note that an incremental shortest path algorithm [2] takes $\mathcal{O}(\log |V|)$ time. Furthermore, observe that the time complexity $\mathcal{O}(\binom{|P|}{k} ISP(G))$ is exponential to k. Therefore, the term $\binom{|P|}{k}$ renders BF only feasible for a tiny $|P|$.

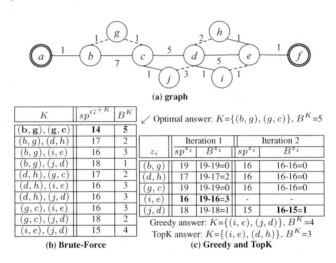

(a) graph

K	sp^{G+K}	B^K
$(b, g), (g, c)$	**14**	**5**
$(b, g), (d, h)$	17	2
$(b, g), (i, e)$	16	3
$(b, g), (j, d)$	18	1
$(d, h), (g, c)$	17	2
$(d, h), (i, e)$	16	3
$(d, h), (j, d)$	16	3
$(g, c), (i, e)$	16	3
$(g, c), (j, d)$	18	2
$(i, e), (j, d)$	15	4

(b) Brute-Force

✓ Optimal answer: $K=\{(b, g), (g, c)\}$, $B^K = 5$

z_i	Iteration 1		Iteration 2	
	sp^{z_i}	B^{z_i}	sp^{z_i}	B^{z_i}
(b, g)	19	19-19=0	16	16-16=0
(d, h)	17	19-17=2	16	16-16=0
(g, c)	19	19-19=0	16	16-16=0
(i, e)	**16**	**19-16=3**	-	-
(j, d)	18	19-18=1	15	**16-15=1**

Greedy answer: $K=\{(i, e), (j, d)\}$, $B^K = 4$
TopK answer: $K=\{(i, e), (d, h)\}$, $B^K = 3$

(c) Greedy and TopK

Fig. 2. Which-k-Edges-Insert Question: Running Steps of Algorithms ($k=2$)

2.3   Heuristics Solutions

We now present two polynomial-time heuristics algorithms that return a subset $K \subseteq P$ of k bridges whose benefit is an heuristics of the optimal benefit. Empirical results show that the algorithms can return high quality solutions using a very reasonable amount of time.

Both algorithms require a function called `CalSPBenefit`. Its goal is to compute the benefit B^{z_i} of every *single* (note: not combination of) bridge $z_i \in P$. A naive implementation of the function `CalSPBenefit` can be based on a nested-loop:

SP-NL-Benefit has a complexity of $\mathcal{O}(|Q||P|ISP(G))$, where $ISP(G)$ is the time complexity of an incremental shortest path algorithm `IncSP` (e.g,. $\mathcal{O}(\log |V|)$ [2]).

---

[1] For ease of illustration, we assume bridge costs being 0 in the example. In fact, our solutions can deal with arbitrary values of c^{z_i}.

Algorithm 1. Algorithm SP-NL-Benefit

1: **function** SP-NL-BENEFIT(G, Q, P) *implements* CalSPBenefit
2:     **for each** bridge z_i of P **do**   ▷ outer loop
3:         **for each** $q_j \in Q$ **do**   ▷ inner loop
4:             IncSP(q_j, G^{+z_i})
5:             Calculate the benefit $b_{q_j}^{z_i}$ of z_i
6:         $B^{z_i} = \sum_{q_j \in Q} b_{q_j}^{z_i}$ ▷ Equation 3

So, SP-NL-Benefit is not efficient enough because it invokes IncSP $|Q| \times |P|$ times. In Section 2.4, we will present more efficient implementations for CalSPBenefit. Now, we present the two heuristics algorithms first.

a) Greedy Algorithm. Our first heuristics algorithm is based on greedy heuristics. Algorithm 2 shows the pseudo-code of our Greedy algorithm. It invokes the function CalSPBenefit in k iterations. In each iteration, it greedily selects the bridge $z^* \in P$ with the highest benefit B^{z^*}, removes it from P, and then inserts it into G. The time complexity of Greedy is $\mathcal{O}(k \cdot CB(G, P))$, where $CB(G, P)$ denotes the time complexity of an implementation of function CalSPBenefit. Even using the (slow) naive implementation of CalSPBenefit (see Algorithm 1), Greedy is still a polynomial time algorithm.

Algorithm 2. Heuristics Algorithm (Greedy)

1: **function** GREEDY(G, s, t, P, k)
2:     $K = \emptyset$ ▷ result set
3:     **while** $|K| < k$ **do**
4:         CalSPBenefit(G, s, t, P)
5:         let $z^* \in P$ be the bridge with the highest benefit B^{z^*}
6:         remove z^* from P
7:         insert z^* into K ▷ z^* is one of the selected bridges
8:         $G = G \cup z^*$ ▷ update G to include z^* as well
9:     **return** the result set K

The running steps of Greedy on the example in Figure 2a are illustrated in Figure 2c. For $k = 2$, Greedy has two iterations. In the first iteration, Greedy first invokes the function CalSPBenefit to find the best bridge (i, e), which has a benefit of 3. So it is removed from P and gets inserted into the final result set K and into the graph G. In the second iteration, function CalSPBenefit is invoked the second time. Note that the benefit of some bridges (e.g., (d, h)) may change after the graph has become $G^{+(i,e)}$. Again, the bridge with the highest benefit in the second iteration (i.e., (j, d)) is inserted into the final result set K. Since $k = 2$, Greedy stops after this iteration. Greedy finds the optimal result $\{(i, e), (j, d)\}$ in this example.

b) TopK Algorithm. Our second heuristics algorithm, called TopK (Algorithm 3), attempts to further trade the result quality for better efficiency. It simply executes the function `CalSPBenefit` once and then returns the top-k most beneficial bridges. Its time complexity is $\mathcal{O}(CB(G, P))$, i.e., the complexity of function `CalSPBenefit`.

Algorithm 3. Heuristics Algorithm (TopK)

1: **function** TopK(G, s, t, P, k)
2:     `CalSPBenefit`(G, s, t, P)
3:     $K =$ the k most beneficial bridges   ▷ result set
4:     **return** the result set K

The running steps of TopK on the example in Figure 2a is the same as the first iteration of Greedy, as illustrated in Figure 2c. However, TopK has only one iteration. It triggers function `CalBenefit` once, and returns the top-2 bridges (i, e) and (d, h) as the result set. The total benefit of (i, e) and (d, h) is $19 - 16 = 3$.

2.4   Efficiency Optimization

(a) Pruning Candidate Bridges. The trick is very simple: if the length $\|z_i\|$ of a bridge z_i is longer than all queries' shortest-path distances, such bridge can be pruned right away because it yields no benefit. Furthermore, we skip a query q_j's `IncSP` call if its original shortest-path distance is smaller than the lengths of all the given candidate bridges.

(b) Input-Adaptive Benefit Calculation. The following lemmas enable us to determine the benefits of all bridges simply by (i) invoking `IncSP` (using each u_i or v_i as the source) $|P|$ times or (ii) invoking `IncSP` (using each s_j *and* each t_j as the sources) $2|Q|$ times. Therefore, our proposed algorithm, namely, SP-Fast-Benefit, implements `CalSPBenefit`, by doing (i) when $|P| < 2|Q|$ or doing (ii) otherwise. Briefly, in Lemma 1, by executing `IncSP` from a node u_i, we can obtain the shortest-path tree of u_i, which embeds the shortest-path distance of all nodes from u_i. For the path s_j to t_j via u_i to be shorter than $sp_{q_j}^G$, either the path s_j to u_i, or the path u_i to t_j must contain the node v_i and hence the bridge (u_i, v_i). So, we will have enough distance information to calculate the updated shortest path distances of all queries in Q.

Lemma 1. *Given a bridge $z_i(u_i, v_i)$, a graph G, and a query workload Q, the benefit B^{z_i} of bridge z_i can be obtained by executing* `IncSP` *only once, using either u_i or v_i as the source, and stops when all nodes in Q are seen.*

Proof. Assume z_i is inserted to G, so the graph becomes G^{+z_i}. If the shortest path of a query $q_j(s_j, t_j) \in Q$ on G^{+z_i} is shorter than its original shortest path on G, the new shortest path of q_j on G^{+z_i} must pass through z_i and its corresponding (new and shorter) shortest path distance $sp_{q_j}^{G^{+z_i}}$ would be:

$$sp_{q_j(s_j, t_j)}^{G^{+z_i}} = sp_{(s_j, u_i)}^{G^{+z_i}} + sp_{(u_i, t_j)}^{G^{+z_i}} \tag{4}$$

where $sp^{G+z_i}_{(s_j,u_i)}$ and $sp^{G+z_i}_{(u_i,t_j)}$ are the shortest path distances from u_i to s_j and t_j, respectively.

Given $sp^{G+z_i}_{q_j(s_j,t_j)}$, we derive the benefit $b^{z_i}_{q_j}$ of z_i on query q_j as:

$$b^{z_i}_{q_j} = m_{q_j} \times (sp^G_{q_j} - \min(sp^G_{q_j}, sp^{G+z_i}_{q_j(s_j,t_j)})) \tag{5}$$

As m_{q_j} and $sp^G_{q_j}$ are given, from Equations 4 and 5, we can see that as long as we have the values of $sp^{G+z_i}_{(s_j,u_i)}$ and $sp^{G+z_i}_{(u_i,t_j)}$, we are able to compute $b^{z_i}_{q_j}$. Obviously, given a graph G^{+z_i} with bridge $z_i(u_i, v_i)$ inserted, we can obtain the shortest path distances from u_i to all nodes using one Dijkstra's execution. After that, shortest path distances like $sp^{G+z_i}_{(s_j,u_i)}$ and $sp^{G+z_i}_{(u_i,t_j)}$ are readily available if we make sure the shortest-path execution stops only when all source nodes s_j and destination nodes t_j in the workload Q are seen. By repeatedly applying Equation 5 using the information above, all benefits $b^{z_i}_{q_j}$ can be obtained and the overall benefit B^{z_i} of bridge z_i on workload Q can be derived using Equation 3. The proof is the same if we use v_i as the source instead of u_i.   □

Lemma 2. *Given a query $q_j(s_j, t_j)$, a graph G, a set of bridges P, the benefit $b^{z_i}_{q_j}$ for all bridges $z_i(u_i, v_i)$ in P with respect to a query q_j in Q can be obtained by executing* IncSP *twice, once using s_j as the source, once using t_j as the source, and stop when all nodes in P are seen.*

Proof. If the shortest path of a query $q_j(s_j, t_j) \in Q$ on G^{+z_i} is shorter than its original shortest path on G, the new shortest path of q_j on G^{+z_i} must pass through $z_i(u_i, v_i)$ and its corresponding (new and shorter) shortest path distance $sp^{G+z_i}_{q_j}$ would be:

$$sp^{G+z_i}_{q_j(s_j,t_j)} = \min(\ sp^{G+z_i}_{(s_j,u_i)} + ||(u_i, v_i)|| + sp^{G+z_i}_{(v_i,t_j)},$$
$$sp^{G+z_i}_{(s_j,v_i)} + ||(u_i, v_i)|| + sp^{G+z_i}_{(u_i,t_j)}) \tag{6}$$

where $||(u_i, v_i)||$ denote the length of edge (u_i, v_i).

Given $sp^{G+z_i}_{q_j(s_j,t_j)}$, we can derive the benefit $b^{z_i}_{q_j}$ of z_i on query q_j using Equation 5. Since $||(u_i, v_i)||$, m_{q_j}, and $sp^G_{q_j}$ are given, from Equations 5 and 6, by using the values of $sp^{G+z_i}_{(s_j,u_i)}$, $sp^{G+z_i}_{(s_j,v_i)}$, $sp^{G+z_i}_{(v_i,t_j)}$, and $sp^{G+z_i}_{(u_i,t_j)}$, we are able to compute $b^{z_i}_{q_j}$. Obviously, given a graph G, we can obtain the shortest path distances from s_j and t_j to all nodes using two shortest-path executions. After that, shortest path distances like $sp^{G+z_i}_{(s_j,u_i)}$, $sp^{G+z_i}_{(s_j,v_i)}$, $sp^{G+z_i}_{(v_i,t_j)}$, and $sp^{G+z_i}_{(u_i,t_j)}$ are readily available if we make sure the shortest-path executions stop only when all nodes u_i and v_i in the P are seen. By repeatedly applying Equation 6 using the information above, the benefits $b^{z_i}_{q_j}$ of all bridges on a query $q_j(s_j, t_j)$ can be obtained.   □

3   Which-k-Edges-Delete Question

Next, we formulate the *Which-k-Edges-Delete* question and establish its hardness. Then, we adapt our heuristic solutions from the previous section to answer the *Which-k-Edges-Delete* question.

3.1   Problem Formulation

This problem aims to find out which k edges in a given set $\bar{P} \subseteq E$, if deleted, have the least impact on the overall shortest-path distances (e.g., Q2). In this case, the "damage" $D_{q_j}^{e_i}$ of deleting an edge e_i with respect to a query workload Q can be defined as:

$$D^{e_i} = \sum_{q_j \in Q} m_{q_j} (sp_{q_j}^{G-e_i} - sp_{q_j}^G) \tag{7}$$

where $sp_{q_j}^{G-e_i}$ denotes the shortest path of q_j on a graph G without edge e_i.

3.2   Problem Hardness and Solutions

Observe that an instance $\langle k, G, P, Q \rangle$ of the Which-k-Edges-Insert problem is equivalent to an instance $\langle |\bar{P}| - k, G^{+\bar{P}}, \bar{P}, Q \rangle$ of the Which-k-Edges-Delete problem, where $\bar{P} = P$. Since the Which-k-Edges-Insert problem is $\mathcal{NP}$-hard, the Which-k-Edges-Delete problem is also $\mathcal{NP}$-hard.

All our solutions (BF, Greedy, and TopK) are applicable here. First, we can also have a naive nested-loop implementation method for a function, CalSPDamage, that calculates the damage of deleting each edge in $\bar{P}$ (the set of candidate edges to be removed from G) on the query workload Q by calling a shortest incremental algorithm. Then for Greedy, we greedily choose the edge e with the least damage, delete e from G, and repeat k iterations. For TopK, we simply choose the k-lowest damage edges in the first iteration.

3.3   Efficiency Optimization

The problem of CalSPDamage is: Given a subset $\bar{P}$ of edges in E, calculate the damage D^{e_i} of each edge $e_i \in \bar{P}$ with respect to a query workload Q. A nested-loop algorithm, SP-NL-Damage, that invokes IncSP $|Q| \times |P|$ times is also applicable here.

To develop a more efficient algorithm. The question is how to obtain the item $sp_{q_j}^{G-e_i}$ for all queries $q_j \in Q$ in Equation 7 efficiently. Observe that the deletion of an edge e_i from G is equivalent to the insertion of the edges $\bar{P} - \{e_i\}$ into the graph $G^{-\bar{P}}$, i.e., $G^{-e_i} = G^{-\bar{P}+(\bar{P}-e_i)}$. Note that $sp_{q_j}^{G-e_i}$ means the shortest path distance of query q_j on the graph G without e_i. In other words, it is the shortest path distance of query q_j on the graph $G^{-\bar{P}+(\bar{P}-e_i)}$.

Suppose that $\bar{P} = \{e_1(u_1, v_1), e_2(u_2, v_2), e_3(u_3, v_3)\}$. In this example, the graph without e_1, i.e., G^{-e_1}, is the same as the graph without $\bar{P}$, but with edges e_2 and e_3 inserted back, i.e., $G^{-e_1} = G^{-\bar{P}+\{e_2,e_3\}}$. Therefore, if we want to compute the changes of shortest path values (indirectly, the damage) of deleting e_1 from G, we can consider the changes of shortest path values (indirectly, the benefit) of *inserting* e_2 and e_3 to $G^{-\bar{P}}$.

From the discussion above, our idea is to solve the deletion problem as an insertion problem on the graph $G^{-\bar{P}}$. Instead of using the entire $G^{-\bar{P}}$, we observe that the following shortest paths on $G^{-\bar{P}}$ are essential for solving the insertion problem on the graph $G^{-\bar{P}}$. Specifically, the following paths are necessary for computing shortest paths $sp_{q_j}^{G_{e_i}}$ for any edge e_i and query q_j.

- for each query $q_j(s_j, t_j)$, the shortest path $s_j \rightsquigarrow t_j$;
- for each combination of edges $e_i(u_i, v_i)$ and $e_h(u_h, v_h)$, the shortest paths $u_i \rightsquigarrow u_h$, $u_i \rightsquigarrow v_h$, $v_i \rightsquigarrow u_h$, $v_i \rightsquigarrow v_h$;
- for each combination of edge $e_i(u_i, v_i)$ and query $q_j(s_j, t_j)$, the shortest paths $u_i \rightsquigarrow s_j$, $u_i \rightsquigarrow t_j$, $v_i \rightsquigarrow s_j$, $v_i \rightsquigarrow t_j$.

Based on the above observation, we develop an efficient algorithm, SP-Fast-Damage, to create a *reduced graph* $\mathcal{G}$ that contains only nodes and edges involved in the shortest paths above. The computation is efficient because it works on the concise $\mathcal{G}$.

4   Experiments

In this section, we present experiment results based on real graph data. All experiments were run on a 2.5 GHz Intel PC running Ubuntu with 8 GB of RAM. We evaluated the algorithms by running experiments on four real undirected graphs of different sizes and types (Figure 3). The queries in Q are selected randomly. The bridge set P (for insertion) and the edge set $\bar{P}$ (for deletion) are selected randomly. Here, we also present results of both edge insertion case and edge deletion case only. The default bridge cost and query importance are 0 and 1, respectively. Experimental results using other values are largely similar, so we do not present them here.

Undirected Graphs (for shortest-path)	node Count	Edge Count	Avg Degree
Argentina Road Network (ARG) http://www.maproom.psu.edu/dcw	85,287	88,357	2.07
San Francisco Road Network (SF) http://www.maproom.psu.edu/dcw	174,956	223,001	2.54
CAIDA Internet Router Topology (LINKS) http://www.caida.org/tools/measurement/skitter	190,914	607,609	6.36

Fig. 3. Real Graph Data Used in Experiments

4.1   Which-k-Edges-Insert Question

The first row of Figures 4(a)–(c) shows the actual approximation ratio (A.A.R.) of BF, Greedy, and TopK on four real graphs. In this experiment, we limit the experiment setting to $|P| = |Q| = 25$ to let BF compute the optimal solution in a reasonable time. BF is an exact but exponential algorithm, it always has a ratio of 1. On datasets SF and LINKS, Greedy returns the optimal solution. On dataset ARG, the A.A.R. of Greedy degrades a little bit when $k = 4$, but it still returns 95% of the optimal benefit. The solutions of TopK are quite data dependent. On ARG, its A.A.R decreases when k increases. On SF, its A.A.R. is from 0.8 to 1. On LINKS, its A.A.R. drops initially when k increases but it rises again later.

The second row of Figures 4(a)-(c) shows the running times of BF, Greedy, and TopK. As a baseline, BF takes up to about a day when k is four on large graphs like SF and LINKS, even $|P|$ and $|Q|$ are so small ($= 25$) in this experiment. TopK and Greedy are orders of magnitude faster than BF and Greedy is about k times slower than TopK.

As the performances of Greedy and TopK mainly depend on the implementations of the `CalSPBenefit`, we evaluate the performance of the two implementations of

`CalSPBenefit` function: SP-NL-Benefit and SP-Fast-Benefit, by varying $|P|$ and $|Q|$. Figure 4(d) shows their running times when we vary the size of workload ($|Q| = 50$ to 5000; default $|P| = 500$) and the size of bridge sets ($|P| = 50$ to 5000; default $|Q| = 500$) on the ARG dataset. The results on the other three datasets are largely similar to the results on ARG, so we do not present them here. From the results we can see that SP-NL-Benefit is about two orders of magnitude slower than SP-Fast-Benefit. When fixing $|P|$ at 500 and varying $|Q|$ (Figure 4(d); top), the running time of SP-Fast-Benefit goes flat at $|Q| = 500$ because it decides to invoke `IncSP` $|P|$ times at that point. Similarly, when fixing $|Q|$ at 500 and varying $|P|$ (Figure 4(d); bottom), the running time of SP-Fast-Benefit goes flat at $|P| = 1000$ because it decides to invoke `IncSP` $2|Q|$ times at that point.

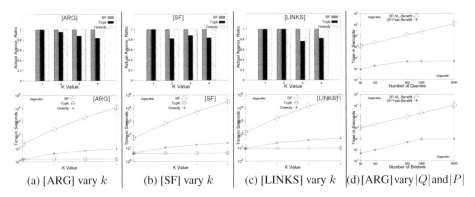

(a) [ARG] vary k (b) [SF] vary k (c) [LINKS] vary k (d) [ARG] vary $|Q|$ and $|P|$

Fig. 4. Which-k-Edges-Insert Question

4.2 Which-k-Edges-Delete Question

Figures 5(a)–(c) show the A.A.R. and the performance of BF, Greedy, and TopK, for the case of edge deletion ($|P| = |Q| = 25$). The 1.0 A.A.R of BF is put there as reference. Both Greedy and TopK found the same optimal solutions as BF and LINKS. On ARG and SF, the worst A.A.R of Greedy is 1.2 (meaning the damage of Greedy is 20% more than the optimal), which is a good result. The worst A.A.R 1.9 of TopK (at SF, $k=4$) is also low. TopK runs almost k times faster than Greedy. Both of them are orders of magnitude faster than BF.

Figure 5(d) shows the running times of SP-NL-Damage and SP-Fast-Damage, two implementations of `CalSPDamage`, under different workload size ($|Q| = 50$ to 5000) and edge set size ($|P| = 50$ to 5000), on ARG. The results on the other three datasets are largely similar to the results here, so we do not present them here. From the results we can see that both SP-NL-Damage and SP-Fast-Damage scale well but SP-NL-Damage is an order of magnitude slower than SP-Fast-Damage. We have also plotted a scalability graph showing the running times of SP-NL-Damage and SP-Fast-Damage on the four real graphs ($|Q| = |P| = 500$). Results show that they are both scalable to graphs of different sizes.

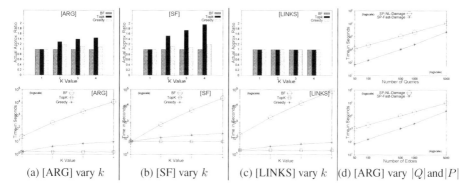

(a) [ARG] vary k (b) [SF] vary k (c) [LINKS] vary k (d) [ARG] vary $|Q|$ and $|P|$

Fig. 5. Which-k-Edges-Delete Question

4.3   Case Study

Here we present the findings of a case study of applying a Which-k-Edges-Insert question on the San Francisco Bay area road network. The goal is to determine the best locations for constructing a new bridge spanning the San Francisco Bay such that travel distances can be reduced the most (i.e., Q3 in Section 1). In the study, we concerned only with the construction of new bridges spanning the San Francisco Bay. We identified a number of possible locations for constructing new bridges, which resulted in 348 possible bridges to consider constructing. For all 348 candidate bridges, we assumed that their construction costs are directly related to the bridge's span length; thus, longer bridges are more costly to construct, but may shorten travel distances more, when compared with shorter bridges. To determine important travel destinations, we used San Francisco Bay Area commuter statistics supplied by the Metropolitan Transportation Commission. From the Metropolitan Transportation Commission, we mapped county to county commuter statistics back onto San Francisco Bay Area road network graph. After remapping the county to county commuter statistics, we identified 377 queries to characterize the commuter data supplied from the Metropolitan Transportation Commission. For each of these 377 commuter queries, their relative importance to San Francisco Bay Area residents were computed from census data[2], which were collected by the Metropolitan Transportation Commission.

Figure 6 shows the original San Francisco Bay Area road network (in black color). Using Greedy, the four most beneficial bridges to add into the San Francisco Bay Area road network are shown in Figure 6a (red colors): the single most beneficial bridge is shown just south of the Dumbarton Bridge (annotated with (1) in the figure). This bridge would help commuters traveling between the extreme ends of the Silicon Valley at the southern end of Alameda County and the southern end of San Mateo County. The second most beneficial bridge to construct is annotated with (2) in the figure. This bridge connects south San Francisco to the mid southern end of Alameda County. The third most beneficial bridge (annotated with (3)), connects central San Francisco to central Alameda County. Finally, the fourth most beneficial bridge (annotated with (4)) to construct connects Alameda County to San Mateo County. Figure 6b shows the four

---

[2] http://www.mtc.ca.gov/maps_and_data/datamart/
census/county2county/table1coco.html

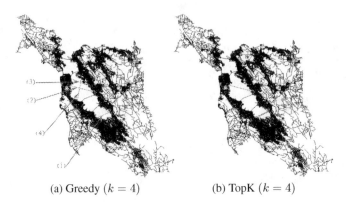

(a) Greedy ($k = 4$)   (b) TopK ($k = 4$)

Fig. 6. Case Study [***please view this figure from <u>color PDF file</u> or <u>color print copy</u>***]

bridges suggested by TopK. We can see that TopK selects to construct 3 bridges all south of the Dumbarton Edge in close proximity to each other. The running time of TopK algorithm is about 10 seconds, where Greedy is about 40 seconds, 4 times slower than TopK. Both running times are highly reasonable. However, the bridges suggested by Greedy are more insightful than bridges suggested TopK in this case study.

5   Related Work

Decision support and data mining on graphs (e.g., [3,4,5,6,7]) are related to this work. However, they have different focuses (e.g., summarizing, OLAP, or mining graphs) with us.

The broad applications of *Which-Edge* query makes it relevant to a number of other areas: spatial database, operations research, dynamic graph maintenance, and detection of high risk network links.

Optimal-location queries [8,9,10] are a class of spatial decision-support queries where users look for the best location, l, for a new facility such that the greatest *benefit* is obtained. For example, [9] considers the benefit of a location as the total weight of its reverse nearest neighbors (i.e., the total weight of objects that are closer to l than to any other data point in the dataset). Optimal-location queries are helpful in finding ideal locations for a new shop to attract the largest number of customers. However, *Which-Edge* queries (e.g., Q1–Q4) that work on graphs are more diverse than optimal-location queries, which only ask *"where to add a new point?"*

The reverse optimization problems [11] in operations research are relevant to us. An inverse optimization problem takes a *feasible solution* x as input and then tunes the problem's parameters, with as low of a cost as possible, such that x becomes the optimal solution. In reverse optimization problems, a *target value* v is specified, and then the problem's parameters are tuned, with as low cost as possible, such that v becomes either the optimal value or an upper bound of the optimal value for the problem. For example, in *reverse shortest-path problems* [11], an input of the desired shortest-path distance d to a shortest-path query q yields an output of a set of edge weight adjustments that make the shortest-path distance of q shorter than d. These existing operations research problems require the user to *explicitly state* the target to be tuned (e.g., the desired shortest-path distance d). In contrast, *Which-Edge* queries *tell* the user the target's identity and its optimized value.

If the set of graph elements that will be changed is explicitly known, then it is related to the dynamic graph maintenance (e.g., [12,13,2]) problems, whose goals are to efficiently update the graph measures (a.k.a. incremental update). However, those works do not tell users which set of graph elements is worthwhile to get updated, which is the objective of this work. Most network maintenance works (e.g., [14]) aim to localize faulty links *after* some links break. *Which-Edge* queries, like Q4, can be used to determine critical links in a network such that effective preventive maintenance measures can be implemented *before* any link breaks.

6   Conclusion

In this paper, we formulate a novel *Which-Edge* question on shortest path queries. It has important applications in logistics, urban planning, and network planning. We show the NP-hardness of the problem, as well as present efficient algorithms with optimizations for computing highly accurate results in practice. In future, we will investigate extensions of *Which-Edge* question for other graph queries (e.g., reachability queries).

References

1. Cormen, T.H., Leiserson, C.E., Rivest, R.L., Stein, C.: Introduction to Algorithms. MIT Press (2001)
2. Buriol, L.S., Resende, M.G.C., Thorup, M.: Speeding up dynamic shortest-path algorithms. INFORMS Journal on Computing 20(2), 191–204 (2008)
3. Chen, C., Lin, C.X., Fredrikson, M., Christodorescu, M., Yan, X., Han, J.: Mining graph patterns efficiently via randomized summaries. PVLDB 2(1), 742–753 (2009)
4. Chen, C., Yan, X., Zhu, F., Han, J., Yu, P.S.: Graph OLAP: Towards Online Analytical Processing on Graphs. In: ICDM, pp. 103–112 (2008)
5. Khan, A., Yan, X., Wu, K.L.: Towards proximity pattern mining in large graphs. In: SIGMOD, pp. 867–878 (2010)
6. Tian, Y., Hankins, R.A., Patel, J.M.: Efficient aggregation for graph summarization. In: SIGMOD, pp. 567–580 (2008)
7. Zhang, N., Tian, Y., Patel, J.M.: Discovery-driven graph summarization. In: ICDE, pp. 880–891 (2010)
8. Du, Y., Zhang, D., Xia, T.: The optimal-location query. In: Medeiros, C.B., Egenhofer, M., Bertino, E. (eds.) SSTD 2005. LNCS, vol. 3633, pp. 163–180. Springer, Heidelberg (2005)
9. Gao, Y., Zheng, B., Chen, G., Li, Q.: Optimal-location-selection query processing in spatial databases. TKDE 21, 1162–1177 (2009)
10. Xiao, X., Yao, B., Li, F.: Optimal location queries in road network databases. In: ICDE, pp. 804–815 (2011)
11. Zhang, J., Lin, Y.: Computation of reverse shortest-path problem. Journal of Global Optimization 25, 243–261 (2003)
12. Demetrescu, C., Italiano, G.F.: Algorithmic techniques for maintaining shortest routes in dynamic networks. Electron. Notes Theor. Comput. Sci. 171, 3–15 (April 2007)
13. Chan, E., Lim, H.: Optimization and evaluation of shortest path queries. The VLDB Journal 16, 343–369 (2007)
14. Pal, A., Paul, A., Mukherjee, A., Naskar, M., Nasipuri, M.: Fault detection and localization scheme for multiple failures in optical network. In: Distributed Computing and Networking, pp. 464–470 (2008)

Reconciling Multiple Categorical Preferences with Double Pareto-Based Aggregation

Nikos Bikakis[1,2,*], Karim Benouaret[3], and Dimitris Sacharidis[2]

[1] National Technical University of Athens, Greece
[2] IMIS, "Athena" Research Center, Greece
[3] Inria Nancy – Grand Est, France

Abstract. Given a set of objects and a set of user preferences, both defined over a set of categorical attributes, the *Multiple Categorical Preferences* (MCP) problem is to determine the objects that are considered preferable by all users. In a naïve interpretation of MCP, matching degrees between objects and users are aggregated into a single score which ranks objects. Such an approach, though, obscures and blurs individual preferences, and can be unfair, favoring users with precise preferences and objects with detailed descriptions. Instead, we propose an objective and fair interpretation of the MCP problem, based on two Pareto-based aggregations. We introduce an efficient approach that is based on a transformation of the categorical attribute values and an index structure. Moreover, we propose an extension for controlling the number of returned objects. An experimental study on real and synthetic data finds that our index-based technique is an order of magnitude faster than a baseline approach, scaling up to millions of objects.

Keywords: Group preferences, Categorical attributes, Rank aggregation, Skyline queries, Collective dominance.

1 Introduction

Given a collection of *objects* and a user's *preference*, both defined on a set of *attributes*, the *general recommendation problem* is to identify those objects that are most aligned to the user's preference. Several instances of this generic problem have appeared over the past few years in the Information Retrieval and Database communities; e.g., see [7,26]. This paper deals with an instance of the above class, termed the *Multiple Categorical Preferences* (MCP) problem. MCP has three characteristics. (1) Objects are described by a set of categorical attributes. (2) User preferences are defined on a subset of the attributes. (3) There are multiple users with distinct, possibly conflicting, preferences. The MCP problem may appear in several scenarios; for instance, colleagues arranging for a dinner at a restaurant, friends selecting a vacation plan for a holiday break.

To illustrate MCP, consider the following example. Assume that a three-member family is looking to buy a new car. Assume a list of available cars, where each is characterized by two categorical attributes, Body and Engine. Figure 1 depicts the hierarchies for these two attributes; Body is a three-level, and Engine is a four-level hierarchy. Table 1 shows the attribute values of four cars, and the family members' preferences. For

---

* This work is partially supported by the EU/Greece funded KRIPIS: MEDA Project & the FP7 project DIACHRON.

S.S. Bhowmick et al. (Eds.): DASFAA 2014, Part I, LNCS 8421, pp. 266–281, 2014.
© Springer International Publishing Switzerland 2014

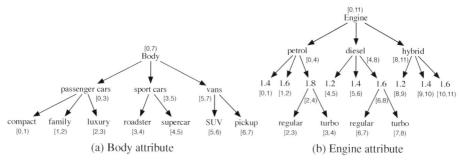

Fig. 1. Attribute hierarchies

Table 1. Objects, Users & Matching vectors

Car	Body	Engine
o_1	family	hybrid 1.4
o_2	roadster	petrol 1.8 turbo
o_3	SUV	diesel 1.6
o_4	compact	petrol 1.4

(a) Objects

User	Preference
u_1	{passenger cars, petrol}
u_2	{sport cars}
u_3	{petrol 1.8}

(b) Users

		User	
Car	u_1	u_2	u_3
o_1	$\langle 1/3, 0 \rangle$	$\langle 0, 0 \rangle$	$\langle 0, 0 \rangle$
o_2	$\langle 0, 1/4 \rangle$	$\langle 1/2, 0 \rangle$	$\langle 0, 1/2 \rangle$
o_3	$\langle 0, 0 \rangle$	$\langle 0, 0 \rangle$	$\langle 0, 0 \rangle$
o_4	$\langle 1/3, 1/4 \rangle$	$\langle 0, 0 \rangle$	$\langle 0, 0 \rangle$

(c) Matching vectors

instance, member u_1 prefers passenger cars with petrol engines, while u_2 likes sport cars but states no preference on the engine type.

Observe that if we look at a particular family member, it is straightforward to determine her/his ideal car based on existing methods. For instance, u_1 clearly prefers o_4, which is a passenger car with a petrol engine, while u_2 clearly favors o_2, which is a sport car. These conclusions are reached using the following reasoning. Each preference attribute value $u_j.A_k$ is *matched* with the corresponding object attribute value $o_i.A_k$ using, e.g., the Jaccard coefficient $\frac{|o_i.A_k \cap u_j.A_k|}{|o_i.A_k \cup u_j.A_k|}$, and a matching degree per preference attribute is derived. Given these degrees, the next step is to "compose" them into an overall matching degree between a user u_j and an object o_i. Note that several techniques are proposed for "composing" matching degrees; e.g., see [16,26]. The simplest option is to compute a linear combination, e.g., the sum, of the individual degrees. Returning to our example, the matching degrees of user u_1 are: $\langle 1/3, 0 \rangle$ for car o_1, $\langle 0, 1/4 \rangle$ for car o_2, $\langle 0, 0 \rangle$ for car o_3, and $\langle 1/3, 1/4 \rangle$ for car o_4. Note that independently of the "composition" method employed, o_4 is the most favorable car for user u_1. Using similar reasoning, car o_2, is ideal for both users u_2, u_3.

When all users are taken into consideration, as required by the MCP problem, several questions arise. Which is the best car that satisfies the entire family? And more importantly, *what does it mean to be the best car*? A simple answer to the latter, would be the car that has the highest "composite" degree of match to all users. Using a similar method as before, one can define a collective matching degree that "composes" the overall matching degrees for each user. This interpretation, however, enforces an additional level of "composition", the first being across attributes, and the second across users. These compositions obscure and blur the individual preferences per attribute of each user.

To some extent, the problem at the first "composition" level can be mitigated by requiring each user to manually define an importance weight among her/his specified attribute preferences. On the other hand, it is not easy, if possible at all, to assign weights to users, so that the assignment is fair. There are two reasons for this. First, *users may specify different sets of preference attributes*, e.g., u_1 specifies Body and Engine, while u_2 only Body. Second, even when considering a particular preference attribute, e.g., Engine, *users may specify values at different levels of the hierarchy*, e.g., u_1 specifies a petrol Engine, while u_3 a petrol 1.8 Engine, which is one level below. Similarly, objects can also have attribute values defined at different levels. Therefore, any "composition" is bound to be *unfair*, as it may favor users with specific preferences and objects with detailed descriptions, and disfavor users with broader preferences and objects with coarser descriptions. This is an inherent difficulty of the MCP problem.

In this work, we introduce the *double Pareto-based aggregation*, which provides an objective and fair interpretation to the MCP problem without "compositing" across preference attributes and users. Under this concept, the matching between a user and an object forms a *matching vector*. Each coordinate of this vector corresponds to an attribute and takes the value of the corresponding matching degree. The first Pareto-based aggregation is defined over attributes and induces a partial order on these vectors. Intuitively, *for a particular user*, the first partial order objectively establishes that an object is *better*, i.e., more preferable, than another if it is better on all attributes.

Then, the second Pareto-based aggregation, defined across users, induces the second partial order on objects. According to this order, an object is better than another, if it is more preferable according to *all users*. The answer to the MCP problem is the set of *maximal objects* under the second partial order. Note that since this order is only partial, i.e., two objects may not be comparable, there may exist multiple objects that are maximal; recall, that an object is maximal if there exists no other object that succeeds it in the order considered. In essence, it is the fact that this order is partial that guarantees objectiveness.

There exists a plethora of main-memory algorithms for finding the maximal elements according to some partial order, e.g., [17]. More recently, since [8], the problem has received great attention in the data management community, rechristened as the *skyline query*, for which several secondary memory algorithms have been proposed. Therefore, it is possible to adapt an existing algorithm to solve the MCP problem, as we discuss in Section 2.2. However, such an approach faces two performance limitations. First, it needs to compute the matching degrees and form the matching vectors for all objects, before actually executing the algorithm. Second, it makes little sense to apply index-based methods, which are known to be the most efficient, e.g., the state-of-the-art method of [23]. The reason is that the entries of the index depend on the specific MCP instance, and need to be rebuilt from scratch when the user preferences change, even though the description of objects persists.

To address these limitations, we introduce a novel index-based approach for solving the MCP problem. The key idea is to index the set of objects that, unlike the set of matching vectors, remains constant across MCP instances, and defer expensive computation of matching degrees. To achieve this, we apply a simple transformation of the categorical attribute values to intervals, so that each object translates to a rectangle in the

Euclidean space. Then, we can employ a space partitioning index, e.g., an R^*-Tree, to hierarchically group the objects. We emphasize that this transformation and index construction is a one-time process, whose cost is amortized across MCP instances, since the index requires no maintenance, as long as the collection of objects persists. Based on the transformation and the hierarchical grouping, it is possible to efficiently compute upper bounds for the matching degrees for *groups* of objects. We then introduce an algorithm that uses these bounds to guide the search towards objects that are more likely to belong to the answer set, and at the same time avoid computing unnecessary matching degrees.

There exists a potential issue of our MCP interpretation when the number of users becomes very large. In this case, the number of conflicting preferences also increases, which makes it harder to differentiate among objects. As a result, the cardinality of the result set increases, which is the cost to pay for being objective. To address this phenomenon, we relax our requirement for unanimity in the second Pareto-based aggregation, and require only a percentage $p\%$ of users to agree. We refer to this extension as the p-MCP problem; naturally, for $p = 100\%$, it reduces to the regular MCP.

2 Problem Statement

2.1 Definitions

Table 2 shows the most important symbols and their definition. Consider a set of d categorical *attributes* $\mathcal{A} = \{A_1, \ldots, A_d\}$. The domain of each attribute A_k is a *hierarchy* $\mathcal{H}(A_k)$. A hierarchy $\mathcal{H}(A_k)$ defines a tree, where a leaf corresponds to a lowest-level value, and an internal node corresponds to a category, i.e., a set, comprising all values within the subtree rooted at this node. The root of a hier-

Table 2. Notation

Symbol	Definition
$\mathcal{A}, d$	Set of attributes, number of attributes ($|\mathcal{A}|$)
$A_k, |A_k|$	Attribute, number of distinct values in A_k
$\mathcal{H}(A_k), |\mathcal{H}(A_k)|$	Hierarchy of A_k, number of hierarchy nodes
$\mathcal{O}, o_i$	Set of objects, an object
$\mathcal{U}, u_j$	Set of users, a user
$o_i.A_k, u_j.A_k$	Value of attribute A_k in object o_i, user u_j
$o_i.I_k, u_j.I_k$	Interval representation of the value of A_k in o_i, u_j
m_i^j	Matching vector of object o_i to user u_j
$m_i^j.A_k$	Matching degree of o_i to user u_j on attribute A_k
$o_a \succ o_b$	Object o_a is collectively preferred over o_b
$\mathcal{T}$	The R^*-Tree that indexes the set of objects
N_i, e_i	R^*-Tree node, the entry for N_i in its parent node
$e_i.ptr, e_i.mbr$	The pointer to node N_i, the MBR of N_i
M_i^j	Maximum matching vector of entry e_i to user u_j
$M_i^j.A_k$	Maximum matching degree of e_i to user u_j on A_k

archy represents the category covering all lowest-level values. We use the symbol $|A_k|$ (resp. $|\mathcal{H}(A_k)|$) to denote the number of leaf (resp. all hierarchy) nodes. With reference to Figure 1a, consider the "Body" attribute. The node "sport cars" is a category and is essentially a shorthand for the set {"roadster", "supercar"}, since it contains the two leaves, "roadster" and "supercar".

Assume a set of *objects* $\mathcal{O}$. An object $o_i \in \mathcal{O}$ is defined over *all* attributes, and the value of attribute $o_i.A_k$ is one of the nodes of the hierarchy $\mathcal{H}(A_k)$. For instance, in Table 1, the value of the "Body" attribute of object o_1, is the leaf "family" in the hierarchy of Figure 1a.

Further, assume a set of *users* $\mathcal{U}$. A user $u_i \in \mathcal{U}$ is defined over a *subset* of the attributes, and for each *specified* attribute $u_i.A_j$, its value in one of the hierarchy $\mathcal{H}(A_j)$

nodes. For all unspecified attributes, we say that user u_i is *indifferent* to them. Note that, a user may specify multiple values for each attribute (see [6] for details).

Given an object o_i, a user u_j, and a specified attribute A_k, the *matching degree* of o_i to u_j with respect to A_k, denoted as $m_i^j.A_k$, is specified by a *matching function* $\mathbf{M}: dom(A_k) \times dom(A_k) \to [0,1]$. The matching function defines the *relation* between the user's preferences and the objects attribute values. For an indifferent attribute A_k of a user u_j, we define $m_i^j.A_k = 1$.

Note that, different matching functions can be defined per attribute and user; for ease of presentation, we assume a single matching function. Moreover, note that this function can be *any user defined function* operating on the cardinalities of intersections and unions of hierarchy attributes. For example, it can be the Jaccard coefficient, i.e., $m_i^j.A_k = \frac{|o_i.A_k \cap u_j.A_k|}{|o_i.A_k \cup u_j.A_k|}$. The numerator counts the number of leaves in the intersection, while the denominator counts the number of leaves in the union, of the categories $o_i.A_k$ and $u_j.A_k$. Other popular choices are the Overlap coefficient: $\frac{|o_i.A_k \cap u_j.A_k|}{\min(|o_i.A_k|,|u_j.A_k|)}$, and the Dice coefficient: $2\frac{|o_i.A_k \cap u_j.A_k|}{|o_i.A_k|+|u_j.A_k|}$.

In our running example, we assume the Jaccard coefficient. Hence, the matching degree of car o_1 to user u_1 w.r.t. the "Body" attribute is $\frac{|\text{"family"} \cap \text{"passenger cars"}|}{|\text{"family"} \cup \text{"passenger cars"}|} = \frac{|\{\text{"family"}\}|}{|\{\text{"compact"},\text{"family"},\text{"luxury"}\}|} = \frac{1}{3}$, where we substituted "passenger cars" with the set $\{\text{"compact"},\text{"family"},\text{"luxury"}\}$.

Given an object o_i and a user u_j, the *matching vector* of o_i to u_j, denoted as m_i^j, is a d-dimensional point in $[0,1]^d$, where its k-th coordinate is the matching degree with respect to attribute A_k. Furthermore, we define the norm of the matching vector to be $\|m_i^j\| = \sum_{A_k \in \mathcal{A}} m_i^j.A_k$. In our example, the matching vector of car o_1 to user u_1 is $\langle 1/3, 0 \rangle$. All matching vectors of this example are shown in Table 1c.

In the following, we consider a particular user u_j and examine the matching vectors. The *first Pareto-based aggregation* across the attributes of the matching vectors, induces the following partial and strict partial "preferred" orders on objects. An object o_a is *preferred* over o_b, for user u_j, denoted as $o_a \succeq^j o_b$ iff for every specified attribute A_k of the user it holds that $m_a^j.A_k \geq m_b^j.A_k$. Moreover, object o_a is *strictly preferred* over o_b, for user u_j, denoted as $o_a \succ^j o_b$ iff o_a is preferred over o_b and additionally there exists a specified attribute A_k such that $m_a^j.A_k > m_b^j.A_k$. Returning to our example, consider user u_1 and its matching vector $\langle 1/3, 1/4 \rangle$ for o_4, and $\langle 1/3, 0 \rangle$ for o_1. Observe that o_4 is strictly preferred over o_1.

We now consider all users in $\mathcal{U}$. The *second Pareto-based aggregation* across users, induces the following strict partial "collectively preferred" order on objects. An object o_a is *collectively preferred* over o_b, if o_a is preferred over o_b for all users, and there exists a user u_j for which o_a is strictly preferred over o_b. From Table 1c, it is easy to see that car o_4 is collectively preferred over o_1 and o_3, because o_4 is preferred by all three users, and strictly preferred by user u_1.

The *collectively maximal objects* in $\mathcal{O}$ with respect to users $\mathcal{U}$, is defined as the set of objects for which there exists no other object that is collectively preferred over them. In our example, o_4 is collectively preferred over o_1, and all other objects are collectively preferred over o_3. There exists no object which is collectively preferred over o_2 and o_4, and thus are the collectively maximal objects.

We next formally define the MCP problem.

Problem 1. [MCP] Given a set of objects $\mathcal{O}$ and a set of users $\mathcal{U}$ defined over a set of categorical attributes $\mathcal{A}$, the *Multiple Categorical Preference* (MCP) problem is to find the collectively maximal objects of $\mathcal{O}$ with respect to $\mathcal{U}$.

2.2   Baseline MCP Algorithm

The MCP problem can be transformed to a maximal elements problem, or a skyline query, where the input elements are the matching vectors. Note, however, that the MCP problem is different than computing the conventional skyline, i.e., over the object's attribute values.

The *Baseline* (BSL) method, whose pseudocode is depicted in Algorithm 1, takes advantage of this observation. The basic idea of BSL is for each object o_i (loop in line 1) and for all users (loop in line 2), to compute the matching vectors m_i^j (line 3). Subsequently, BSL constructs a $|\mathcal{U}|$-dimensional tuple r_i (line 4), so that its j-th entry is a composite value equal to the matching vector m_i^j of object o_i to user u_j. When all users are examined, tuple r_i is inserted in the set $\mathcal{R}$ (line 5).

Algorithm 1. BSL

Input: objects $\mathcal{O}$, users $\mathcal{U}$
Output: CM the collectively maximal
Variables: $\mathcal{R}$ set of intermediate records

1 **foreach** $o_i \in \mathcal{O}$ **do**
2   **foreach** $u_j \in \mathcal{U}$ **do**
3   compute m_i^j
4     $r_i[j] \leftarrow m_i^j$
5   insert r_i into $\mathcal{R}$
6 $CM \leftarrow$ POSkylineAlgo$(\mathcal{R})$

The next step is to find the maximal elements, i.e., compute the skyline over the records in $\mathcal{R}$. It is easy to prove that tuple r_i is in the skyline of $\mathcal{R}$ iff object o_i is a collectively maximally preferred object of $\mathcal{O}$ w.r.t. $\mathcal{U}$. Notice, however, that due to the two Pareto-based aggregations, each attribute of a record $r_i \in \mathcal{R}$ is also a record that corresponds to a matching vector, and thus is partially ordered according to the preferred orders defined in Section 2.1. Therefore, in order to compute the skyline of $\mathcal{R}$, we need to apply a skyline algorithm for partially ordered attributes (line 6), such as [9,28,25,30].

The computational cost of BSL is the sum of two parts. The first is computing the matching degrees, which takes $O(|\mathcal{O}| \cdot |\mathcal{U}|)$ time. The second is computing the skyline, which requires $O(|\mathcal{O}|^2 \cdot |\mathcal{U}| \cdot d)$ comparisons, assuming a quadratic time skyline algorithms is used. Therefore, BSL takes $O(|\mathcal{O}|^2 \cdot |\mathcal{U}| \cdot d)$ time.

3   Index-Based MCP Algorithm

3.1   Hierarchy Transformation

This section presents a simple method to transform the hierarchical domain of a categorical attribute into a numerical domain. The rationale is that numerical domains can be ordered, and thus tuples can be stored in multidimensional index structures. The index-based algorithm of Section 3.2 takes advantage of this transformation.

Consider an attribute A and its hierarchy $\mathcal{H}(A)$, which forms a tree. We assume that any internal node has at least two children; if a node has only one child, then this node and its child are treated as a single node. Furthermore, we assume that there exists an ordering, e.g., the lexicographic, among the children of any node. Observe that such an ordering, totally orders all leaf nodes.

The hierarchy transformation assigns an interval to each node, similar to labeling schemes such as [1]. The i-th leaf of the hierarchy (according to the ordering) is assigned the interval $[i-1, i)$. Then, each internal node is assigned the smallest interval that covers the intervals of its children. Figure 1 depicts the assigned intervals for all nodes in the two car hierarchies.

Following this transformation, the value on the A_k attribute of an object o_i becomes an interval $o_i.I_k = [o_i.I_k^-, o_i.I_k^+)$. The same holds for a user u_j. Therefore, the transformation translates the hierarchy $H(A_k)$ into the numerical domain $[0, |A_k|]$.

An important property of the transformation is that it becomes easy to compute matching degrees for metrics that are functions on the cardinalities of intersections or unions of hierarchy attributes. This is due to the following properties, which use the following notation: for a closed-open interval $I = [\alpha, \beta)$, define $\|I\| = \beta - \alpha$. Note that the proofs can be found in the full version of the paper [6].

Proposition 1. For objects/users x, y, and an attribute A_k, let $x.I_k$, $y.I_k$ denote the intervals associated with the value of x, y on A_k. Then the following hold:

(1) $|x.A_k| = \|x.I_k\|$
(2) $|x.A_k \cap y.A_k| = \|x.I_k \cap y.I_k\|$
(3) $|x.A_k \cup y.A_k| = \|x.I_k\| + \|y.I_k\| - \|x.I_k \cap y.I_k\|$

3.2   Algorithm Description

This section introduces the *Index-based* MCP (IND) algorithm. The key ideas of IND are: (1) apply the hierarchy transformation, previously described, and index the resulting intervals, and (2) define upper bounds for the matching degrees of a group of objects, so as to guide the search and quickly prune unpromising objects.

We assume that the set of objects $\mathcal{O}$ and the set of users $\mathcal{U}$ are transformed so that each attribute A_k value is an interval I_k. Therefore, each object (and user) defines a (hyper-)rectangle on the d-dimensional cartesian product of the numerical domains, i.e., $[0, |A_1|) \times \cdots \times [0, |A_d|)$.

Figure 2 depicts the transformation of the objects and users shown in Table 1 for the hierarchies in Figure 1. For instance, object o_1 is represented as the rectangle $[9, 10) \times [1, 2)$ in the "Engine"×"Body" plane. Similarly, user u_1 is represented as two intervals,

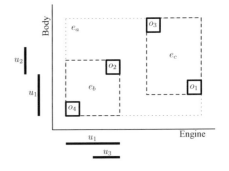

Fig. 2. Transformed objects and users

$[0, 4)$, $[0, 3)$, on the transformed "Engine", "Body" axes, respectively.

The IND algorithm indexes the set of objects in this d-dimensional numerical space. In particular, IND employs an R*-Tree $\mathcal{T}$ [5], which is well suited to index rectangles. Each $\mathcal{T}$ node corresponds to a disk page, and contains a number of entries. Each entry e_i comprises (1) a pointer $e_i.ptr$, and (2) a Minimum Bounding Rectangle (MBR) $e_i.mbr$.

A leaf entry e_i corresponds to an object o_i, its pointer $o_i.ptr$ is *null*, and $e_i.mbr$ is the rectangle defined by the intervals of o_i. A non-leaf entry e_i corresponds to a child node N_i, its pointer $e_i.ptr$ contains the address of N_i, and $e_i.mbr$ is the MBR of (i.e., the tightest rectangle that encloses) the MBRs of the entries within N_i.

Due to the enclosing property of MBRs, the following holds. The MBR of an entry e_i encloses all objects that are stored at the leaf nodes within the $\mathcal{T}$ subtree rooted at node N_i. It is often helpful to associate an entry e_i with all the objects it encloses, and thus treat e_i as a group of objects.

Consider a $\mathcal{T}$ entry e_i and a user $u_j \in \mathcal{U}$. Given only the information within entry e_i, i.e., its MBR, and not the contents, i.e., its enclosing objects, at the subtree rooted at N_i, it is impossible to compute the matching vectors for the objects within this subtree. However, it is possible to derive an *upper bound* for the matching degrees of any of these objects.

We define the *maximum matching degree* $M_i^j.A_k$ of entry e_i on user u_j w.r.t. specified attribute A_k as the highest attainable matching degree of any object that may reside within $e_i.mbr$. To do this we first need a way to compute lower and upper bounds on unions and intersections of a user interval with an MBR.

Proposition 2. Fix an attribute A_k. Consider an object/user x, and let I_x, denote the interval associated with its value on A_k. Also, consider another object/user y whose interval I_y on A_k is contained within a range R_y. Given an interval I, $\delta(I)$ returns 0 if I is empty, and 1 otherwise. Then the following hold:

(1) $1 \le |y.A_k| \le \|R_y\|$

(2) $\delta(I_x \cap R_y) \le |x.A_k \cap y.A_k| \le \|I_x \cap R_y\|$

(3) $\|I_x\| + 1 - \delta(I_x \cap R_y) \le |x.A_k \cup y.A_k| \le \|I_x\| + \|R_y\| - \delta(I_x \cap R_y)$

Then, defining the maximum matching degree reduces to appropriately selecting the lower/upper bounds for the specific matching function used. For example, consider the case of the Jaccard coefficient, $\frac{|o_i.A_k \cap u_j.A_k|}{|o_i.A_k \cup u_j.A_k|}$. Assume e_i is a non-leaf entry, and let $e_i.R_k$ denote the range of the MBR on the A_k attribute. We also assume that $u_j.I_k$ and $e_i.R_k$ overlap. Then, we define $M_i^j.A_k = \frac{\|e_i.R_k \cap u_j.I_k\|}{\|u_j.I_k\|}$, where we have used the upper bound for the intersection in the enumerator and the lower bound for the union in the denominator, according to Proposition 2. For an indifferent to the user attribute A_k, we define $M_i^j.A_k = 1$. Now, assume that e_i is a leaf entry, that corresponds to object o_i. Then the maximum matching degree $M_i^j.A_k$ is equal to the matching degree $m_i^j.A_k$ of o_i to u_j w.r.t. A_k.

Computing maximum matching degrees for other metrics is straightforward. In any case, the next proposition shows that an appropriately defined maximum matching degree is an upper bound to the matching degrees of all objects enclosed in entry e_i.

Proposition 3. The maximum matching degree $M_i^j.A_k$ of entry e_i on user u_j w.r.t. specified attribute A_k is an upper bound to the highest matching degree among all objects in the group that e_i defines.

In analogy to the matching vector, the *maximum matching vector* M_i^j of entry e_i on user u_j is defined as a d-dimensional vector whose k-th coordinate is the maximum matching degree $M_i^j.A_k$. Moreover, the norm of the maximum matching vector is $\|M_i^j\| = \sum_{A_k \in \mathcal{A}} M_i^j.A_k$.

Next, consider a $\mathcal{T}$ entry e_i and the entire set of users $\mathcal{U}$. We define the *score* of an entry e_i as $score(e_i) = \sum_{u_j \in \mathcal{U}} \| M_i^j \|$. This score quantifies how well the enclosed objects of e_i match against all users' preferences. Clearly, the highest the score, the more likely that e_i contains objects that are good matches to users.

Algorithm 2 presents the pseudocode for IND. The algorithm maintains two data structures: a heap H which stores $\mathcal{T}$ entries sorted by their score, and a list CM of collectively maximal objects discovered so far. Initially the list CM is empty (line 1), and the root node of the R*-Tree is read (line 2). The score of each root entry is computed and all entries are inserted in H (line 3). Then, the following process (loop in line 4) is repeated as long as H has entries.

The H entry with the highest score, say e_x, is popped (line 5). If e_x is a non-leaf entry (line 6), it is *expanded*, which means that the node N_x identified by $e_x.ptr$ is read (line 7). For each child entry e_i of N_x (line 8), its maximum matching degree M_i^j with respect to every user $u_j \in \mathcal{U}$ is computed (lines 10–11). Then, the list CM is scanned (loop in line 12). If there exists an object o_a in CM such that (1) for each user u_j, the matching vector m_a^j of o_a is better than M_i^j, and (2) there exists a user u_k so that the matching vector m_a^k of o_a is strictly better than M_i^k, then entry e_i is discarded (lines 13–14). It is straightforward to see (from Proposition 3) that if this condition holds, e_i cannot contain any object that is in the collectively maximal objects, which guarantees IND' correctness. When the condition described does not hold (line 15), the score of e_i is computed and e_i is inserted in H (line 16).

Algorithm 2. IND

Input: R*-Tree $\mathcal{T}$, users $\mathcal{U}$
Output: CM the collectively maximal
Variables: H a heap with $\mathcal{T}$ entries sorted by $score()$

```
1   CM ← ∅
2   read T root node
3   insert in H the root entries
4   while H is not empty do
5       e_x ← pop H
6       if e_x is non-leaf then
7           N_x ← read node e_x.ptr
8           foreach e_i ∈ N_x do
9               pruned ← false
10              foreach u_j ∈ U do
11                  compute M_i^j
12              foreach o_a ∈ CM do
13                  if ∀A_j : m_a^j ≽ M_i^j ∧ ∃A_k : m_a^k ≻ M_i^k
                    then
14                      pruned ← true
15              if not pruned then
16                  insert e_i in H
17      else
18          o_x ← e_x
19          result ← true
20          foreach o_a ∈ CM do
21              if o_a ≻ o_x then
22                  result ← false
23          if result then
24              insert o_x in CM
```

Now, consider the case that e_x is a leaf entry (line 17), corresponding to object o_x (line 18). The list CM is scanned (loop in line 20). If there exists an object that is collectively preferred over o_x (line 22), it is discarded. Otherwise (line 23–24), o_x is inserted in CM.

The algorithm terminates when H is empty (loop in line 4), at which time the list CM contains the collectively maximal objects.

IND performs in the worst case $O(|\mathcal{O}|^2 \cdot |\mathcal{U}| \cdot d)$ comparisons, and computes matching degrees on the fly at a cost of $O(|\mathcal{O}| \cdot |\mathcal{U}|)$. Overall, IND takes $O(|\mathcal{O}|^2 \cdot |\mathcal{U}| \cdot d)$ time, the same as BSL. However, in practice IND is more than an order of magnitude faster than BSL (see Section 6).

4   The p-MCP Problem

As the number of users increases, it becomes more likely that the users express very different and conflicting preferences. Hence, it becomes difficult to find a pair of objects such that the users unanimously agree that one is worst than the other. Ultimately, the number of maximally preferred objects increases. This means that the answer to an MCP problem with a large set of users becomes less meaningful.

The root cause of this problem is that we require unanimity in deciding whether an object is collectively preferred by the set of users. The following definition relaxes this requirement. An object o_a is *p-collectively preferred* over o_b, denoted as $o_a \succ_p o_b$, iff there exist a subset $\mathcal{U}_p \subseteq \mathcal{U}$ of at least $\frac{p}{100} \cdot |\mathcal{U}|$ users such that for each user $u_i \in \mathcal{U}_p$ o_a is preferred over o_b, and there exists a user $u_j \in \mathcal{U}_p$ for which o_a is strictly preferred over o_b. In other words, we require only $p\%$ of the users votes to decide whether an object is universally preferred. Similarly, the *p-collectively maximal objects* of $\mathcal{O}$ with respect to users $\mathcal{U}$, is defined as the set of objects in $\mathcal{O}$ for which there exists no other object that is p-collectively preferred over them. The above definitions give rise to the p-MCP problem.

Problem 2. [p-MCP] Given a set of objects $\mathcal{O}$ and a set of users $\mathcal{U}$ defined over a set of categorical attributes $\mathcal{A}$, the *p-Multiple Categorical Preference* (p-MCP) problem is to find the p-collectively maximal objects of $\mathcal{O}$ with respect to $\mathcal{U}$.

If an object o_1 is collectively preferred over o_2 is also p-collectively preferred for any p. Any object that is p-collectively maximal is also collectively maximal for any p. Therefore, the answere to the p-MCP problem is a subset of the answer to the corresponding MCP. Furthermore, the following transitivity relation holds for three objects o_1, o_2, o_3: if $o_1 \succ o_2$ and $o_2 \succ_p o_3$, then $o_1 \succ_p o_3$. This implies that if an object is not p-collectively maximal, then there must exist a collectively maximal object that is p-collectively preferred over it. These observations are similar to those for the k-dominance notion [10].

Based on this observation, we propose a baseline algorithm for the p-MCP problem, based on BSL (Section 2.2). The p-BSL algorithm first computes the collectively maximal objects applying BSL. Then among the results, it determines the objects for which there exists no other object that is p-collectively preferred over them, and reports them. This refinement is implemented using a block-nested loop procedure. Additionally, we propose an extension of the IND algorithm (Section 3.2) for the p-MCP problem, termed p-IND. Pseudocodes and details can be found in [6].

5   Related Work

Recommendation Systems. There exist several techniques to specify preferences on objects. The *quantitative* preferences, e.g., [2], assign a numeric score to attribute values, signifying importance. There also exist *qualitative* preferences, e.g., [16], which are relatively specified using binary relationships. This work assumes the case of boolean quantitative preferences, where some attribute values are preferred, while others are indifferent.

The general goal of *recommendation systems* [7,26] is to identify those objects that are most aligned to a user's preferences. Typically, these systems provide a *ranking* of the objects by *aggregating* user preferences; e.g., [18,2,16,26].

Recently, several methods for *group recommendations* are proposed [15]. These methods, recommend items to a group of users, trying to satisfy all the group members [24,21]. The existing methods are classified into two approaches. In the first, the preferences of each group member are combined to create a virtual user; the recommendations to the group are proposed w.r.t. to the virtual user. In the second, individual recommendations for each member is computed; the recommendations of all members are merged into a single recommendation.

Several methods to combine different ranked lists are presented in the IR literature. There the data fusion problem is defined. Given a set of ranked lists of documents returned by different search engines, construct a single ranked list combining the individual rankings; e.g., [3,13].

Pareto-Based Aggregation. The work of [8] rekindled interest in the problem of finding the maximal objects and re-introduces it as the skyline operator. An object is dominated if there exists another object before it according to the partial order enforced by the Pareto-based aggregation. The maximal objects are referred to as the skyline. The authors propose several external memory algorithms. The most well-known method is Block Nested Loops (BNL) [8], which checks each point for dominance against the entire dataset. Sort-based skyline algorithms (i.e., SFS [12], LESS [14], and SaLSa [4]) attempt to reduce the number of dominance checks by sorting the input data first. In other approaches, multidimensional indexes are used to guide the search and prune large parts of the space. The most well-known algorithm in this category is BBS [23] which uses an R-Tree. Specific algorithms are proposed to efficiently compute the skyline over partially ordered domains [9,28,25,30], metric spaces [11], or non-metric spaces [22].

Several lines of research attempt to address the issue that the size of skyline cannot be controlled, by introducing new concepts and/or ranking the skyline (see [20] for a survey). [29] ranks tuples based on the number of records they dominate, [10] relaxes the notion of dominance, and [19,27] find the k most representative skylines.

6   Experimental Analysis

6.1   Setting

Datasets. We use two datasets in our experimental evaluation. The first is *Synthetic*, where objects and users are synthetically generated. All attributes have the same hierarchy, a binary tree of height $\log |A|$, and thus all attributes have the same number of leaf hierarchy nodes $|A|$. To obtain the set of objects, we fix a level, ℓ_o (where $\ell_o = 1$ corresponds to the leaves), in all attribute hierarchies. Then, we randomly select nodes from this level to obtain the objects' attribute value. The number of objects is denoted as $|\mathcal{O}|$, while the number of attributes for each object is denoted as d. Similarly, to obtain the set of users, we fix a level, ℓ_u, in all hierarchies. We further make the assumption that a user specifies preferences for only one of the attributes. Thus, for each user we randomly select an attribute, and set its value to a randomly picked hierarchy node at level ℓ_u. The number of users is denoted as $|\mathcal{U}|$.

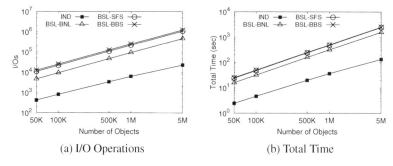

(a) I/O Operations   (b) Total Time

Fig. 3. Synthetic Data: Varying $|\mathcal{O}|$

The second dataset is *Cars*, where we use as objects $|\mathcal{O}| = 4261$ car descriptions retrieved from the Web[1]. We consider two categorical attributes, Body and Engine, having 3 and 4 levels, and 10 and 35 leaf hierarchy nodes, respectively. The user preferences are selected based on car rankings[2]. In particular, for the Body attribute, we use the most popular cars list to derive a subset of desirable nodes at the leaf level. For the Engine attribute, we use the most fuel efficient cars list to derive a subset of desirable nodes at the second hierarchy level. As in Synthetic, a user specifies preference on a single attribute, and thus randomly selects an attribute, and picks its value among the desirable subsets.

Synthetic Data Parameters. Table 3 lists the parameters that we vary and the range of values examined for Synthetic. To segregate the effect of each parameter, we perform six experiments, and in each we vary a single parameter from the table, while we set the remaining ones to their default values.

Table 3. Parameters (Synthetic)

Symbol	Values [Default]
$|\mathcal{O}|$	50K, 100K, [500K], 1M, 5M
d	2, 3, [4], 5, 6
$|\mathcal{U}|$	2, 4, [8], 16, 32
$\log|A|$	4, 6, [8], 10, 12
ℓ_o	[1], 2, 3, 4, 5
ℓ_u	[2], 3, 4, 5, 6

Methods. For the MCP problem, we implement IND (Section 3) and three flavors of the BSL algorithm (Section 2.2), denoted BSL-BNL, BSL-SFS, and BSL-BBS, which use the skyline algorithms BNL [8], SFS [12], BBS [23], respectively. Similarly, for the p-MCP problem (Section 4), we implement the respective extensions of all algorithms (IND and BSL variants), distinguished by a p prefix. All algorithms were written in C++, compiled with gcc, and experiments were performed on a 2GHz CPU.

Evaluation Metrics. To guage efficiency of all algorithms, we measure: (1) the number of disk I/O operations; (2) the number of dominance checks; and (3) the total execution time, measured in secs. Due to the space limitation, in most cases we present only the total time graphs; all the graphs can be found in [6].

---

[1] http://www.carlik.com
[2] http://www.edmunds.com/car-reviews/

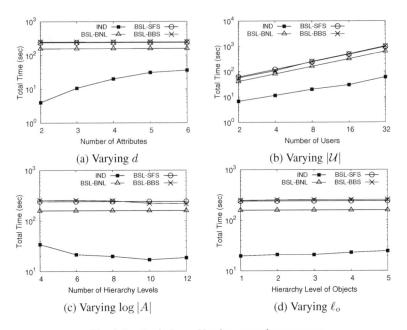

(a) Varying d (b) Varying $|\mathcal{U}|$

(c) Varying $\log|A|$ (d) Varying ℓ_o

Fig. 4. Synthetic Data: Varying several parameters

6.2   Efficiency of MCP Algorithms

Results on Synthetic Data

Varying the Number of Objects. In the first experiment, we study performance with respect to the objects' set cardinality $|\mathcal{O}|$. Particularly, we vary the number of objects from 50K up to 5M and measure the number of I/Os, the number of dominance checks, and the total processing time.

When the number of objects increases, the performance of all methods deteriorates. The number of I/Os (Figure 3a) performed by IND is much less than the BSL variants, the reason being BSL needs to construct a file containing matching degrees. Moreover, the SFS and BBS variants have to preprocess this file, to sort it and build the R-Tree, respectively. Hence, BSL-BNL requires the fewest I/Os among the BSL variants.

All methods require roughly the same number of dominance checks [6]. IND performs fewer checks, while BSL-BNL the most. Compared to the other BSL variants, BSL-BNL performs more checks because, unlike the others, computes the skyline over an unsorted file. IND performs as well as BSL-SFS and BSL-BBS, which have the easiest task. Overall, however, Figure 3b shows that IND is more than an order of magnitude faster than the BSL variants.

Varying the Number of Attributes. Figure 4a investigates the effect as we increase the number of attributes d from 2 up to 6. Figure 4a shows that the total time of IND increases with d, but it is still significantly smaller (more than 4 times) than the BSL methods even for $d = 6$.

Varying the Number of Users. In the next experiment, we vary the users' set cardinality $|\mathcal{U}|$ from 2 up to 32; results are depicted in Figure 4b. The performance of all methods deteriorates with $|\mathcal{U}|$. Figure 4b shows that IND is more than an order of magnitude faster than all the BSL variants, among which BSL-BNL is the fastest.

Varying the Hierarchy Height. In this experiment, we vary the hierarchy height $\log |A|$ from 4 up to 12 levels. Figure 4c illustrates the results. All methods are largely unaffected by this parameter. Overall, IND is more than an order of magnitude faster than all BSL variants.

Varying the Objects Level. Figure 4d depicts the results of varying the level ℓ_o from which we draw the objects' values. The performance of all methods is not significantly affected by ℓ_o.

Varying the Users Level. In this experiment, we vary he level ℓ_u from which we draw the users' preference values. The total time of IND takes its highest value of $\ell_u = 6$, as the number of required dominance checks increases sharply for this setting. Still IND is around 3 times faster than BSL-BNL [6].

Results on Real Data

In this experiment, we have $|\mathcal{O}| = 4261$ cars and we vary the number of users $|\mathcal{U}|$ from 2 up to 32. Figure 5 presents the results. As the number of users increase, the performance of the BSL variants worsens, while that of IND is slightly affected. This pattern is in analogy to the case of the Synthetic dataset (Figure 4). For more than 4 users, IND outperforms the BSL methods by at least an order of magnitude.

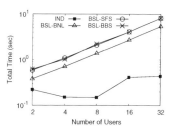

Fig. 5. Cars: varying $|\mathcal{U}|$

6.3 Efficiency of p-MCP Algorithms

In this section, we evaluate the p-MCP algorithms using the Car dataset; as before, we measure the number of I/Os, dominance checks and the total time.

Varying the Number of Users. Figure 6 shows the effect of varying the number of users from 8 up to 4096, while $p = 30\%$. The required number of I/Os operations increases with $|\mathcal{U}|$ for both methods, as Figure 6a shows; the rate of increase for p-BSL-BNL is much higher. Overall, Figure 6b shows that p-IND constantly outperforms p-BSL-BNL and is up to 1 order of magnitude faster.

Varying Parameter p. We increase the parameter p from 10% up to 50%. The performance of all methods (in terms of I/Os and total time) remains unaffected by p, and thus we omit the relevant figures in the interest of space.

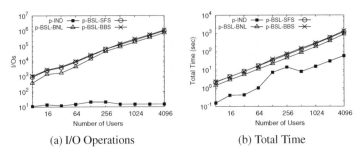

(a) I/O Operations   (b) Total Time

Fig. 6. p-MCP, Cars: varying $|\mathcal{U}|$

References

1. Agrawal, R., Borgida, A., Jagadish, H.V.: Efficient management of transitive relationships in large data and knowledge bases. In: SIGMOD (1989)
2. Agrawal, R., Wimmers, E.L.: A framework for expressing and combining preferences. In: SIGMOD (2000)
3. Aslam, J.A., Montague, M.H.: Models for metasearch. In: SIGIR (2001)
4. Bartolini, I., Ciaccia, P., Patella, M.: Efficient Sort-based Skyline Evaluation. TODS 33(4) (2008)
5. Beckmann, N., Kriegel, H.-P., Schneider, R., Seeger, B.: The r*-tree: An efficient and robust access method for points and rectangles. In: SIGMOD (1990)
6. Bikakis, N., Benouaret, K., Sacharidis, D.: Reconciling multiple categorical preferences with double pareto-based aggregation. Technical Report (2013),
   http://www.dblab.ntua.gr/~bikakis/MCP.pdf
7. Bobadilla, J., Ortega, F., Hernando, A., Gutiérrez, A.: Recommender systems survey. Knowl.-Based Syst. 46 (2013)
8. Börzsönyi, S., Kossmann, D., Stocker, K.: The skyline operator. In: ICDE (2001)
9. Chan, C.Y., Eng, P.-K., Tan, K.-L.: Stratified computation of skylines with partiallyordered domains. In: SIGMOD (2005)
10. Chan, C.Y., Jagadish, H.V., Tan, K.-L., Tung, A.K.H., Zhang, Z.: Finding k-dominant skylines in high dimensional space. In: SIGMOD (2006)
11. Chen, L., Lian, X.: Efficient processing of metric skyline queries. TKDE 21(3) (2009)
12. Chomicki, J., Godfrey, P., Gryz, J., Liang, D.: Skyline with presorting. In: ICDE (2003)
13. Dwork, C., Kumar, R., Naor, M., Sivakumar, D.: Rank aggregation methods for the web. In: WWW (2001)
14. Godfrey, P., Shipley, R., Gryz, J.: Algorithms and analyses for maximal vector computation. VLDBJ 16(1) (2007)
15. Jameson, A., Smyth, B.: Recommendation to groups. In: Brusilovsky, P., Kobsa, A., Nejdl, W. (eds.) Adaptive Web 2007. LNCS, vol. 4321, pp. 596–627. Springer, Heidelberg (2007)
16. Kießling, W.: Foundations of preferences in database systems. In: VLDB (2002)
17. Kung, H.T., Luccio, F., Preparata, F.P.: On finding the maxima of a set of vectors. Journal of the ACM 22(4) (1975)
18. Lacroix, M., Lavency, P.: Preferences:Putting more knowledge into queries. In: VLDB (1987)
19. Lin, X., Yuan, Y., Zhang, Q., Zhang, Y.: Selecting stars: The k most representative skyline operator. In: ICDE (2007)

20. Lofi, C., Balke, W.-T.: On skyline queries and how to choose from pareto sets. In: Catania, B., Jain, L.C. (eds.) Advanced Query Processing. ISRL, vol. 36, pp. 15–36. Springer, Heidelberg (2012)

21. Ntoutsi, E., Stefanidis, K., Nørvåg, K., Kriegel, H.-P.: Fast group recommendations by applying user clustering. In: Atzeni, P., Cheung, D., Ram, S. (eds.) ER 2012. LNCS, vol. 7532, pp. 126–140. Springer, Heidelberg (2012)

22. P, D., Deshpande, P.M., Majumdar, D., Krishnapuram, R.: Efficient skyline retrieval with arbitrary similarity measures. In: EDBT (2009)

23. Papadias, D., Tao, Y., Fu, G., Seeger, B.: Progressive skyline computation in database systems. TODS 30(1) (2005)

24. Roy, S.B., Amer-Yahia, S., Chawla, A., Das, G., Yu, C.: Space efficiency in group recommendation. VLDB J. 19(6) (2010)

25. Sacharidis, D., Papadopoulos, S., Papadias, D.: Topologically sorted skylines for partially ordered domains. In: ICDE (2009)

26. Stefanidis, K., Koutrika, G., Pitoura, E.: A survey on representation, composition and application of preferences in database systems. TODS 36(3) (2011)

27. Tao, Y., Ding, L., Lin, X., Pei, J.: Distance-based representative skyline. In: ICDE (2009)

28. Wong, R.C.-W., Fu, A.W.-C., Pei, J., Ho, Y.S., Wong, T., Liu, Y.: Efficient skyline querying with variable user preferences on nominal attributes. VLDB 1(1) (2008)

29. Yiu, M.L., Mamoulis, N.: Efficient processing of top-k dominating queries on multidimensional data. In: VLDB (2007)

30. Zhang, S., Mamoulis, N., Kao, B., Cheung, D.W.-L.: Efficient skyline evaluation over partially ordered domains. VLDB 3(1) (2010)

CARIC-DA: Core Affinity with a Range Index for Cache-Conscious Data Access in a Multicore Environment

Fang Xi[1], Takeshi Mishima[2], and Haruo Yokota[1]

[1] Department of Computer Science, Tokyo Institute of Technology, Japan
[2] Software Innovation Center, NTT Japan
xifang@de.cs.titech.ac.jp, mishima.takeshi@lab.ntt.co.jp,
yokota@cs.titech.ac.jp

Abstract. In recent years,the number of cores on a chip has been growing exponentially, enabling an ever-increasing number of processes to execute in parallel. Having been developed originally for single-core processors, database (DB) management systems (DBMSs) running on multicore processors suffer from cache conflicts as the number of concurrently executing DB processes (DBPs) increases. In this paper, we propose CARIC-DA, middleware for achieving higher performance in DBMSs on multicore processors by reducing cache misses with a new cache-conscious dispatcher for concurrent queries. CARIC-DA logically range-partitions the data set into multiple subsets. This enables different processor cores to access different subsets by ensuring that different DBPs are pinned to different cores and by dispatching queries to DBPs according to the data partitioning information. In this way, CARIC-DA is expected to achieve better performance via a higher cache hit rate for each core's private cache. It can also balance the loads between cores by changing the range of each subset. Note that CARIC-DA is *pure* middleware, which avoids any modification to existing operating systems (OSs) and DBMSs, thereby making it more practical. We implemented a prototype that uses unmodified existing Linux and PostgreSQL environments. The performance evaluation against benchmarks revealed that CARIC-DA achieved improved cache hit rates and higher performance.

Keywords: multicore, OLTP, middleware.

1 Introduction

In the computer industry, multicore processing is a growing trend as single-core processors rapidly reach the physical limits of possible complexity and speed [1]. Nowadays, multicore processors are widely utilized by many applications and are becoming the standard computing platform. However, these processors are far from realizing their potential performance when dealing with data-intensive applications such as database (DB) management systems (DBMSs). This is because advances in the speed of commodity multicore processors far outpace advances in

S.S. Bhowmick et al. (Eds.): DASFAA 2014, Part I, LNCS 8421, pp. 282–296, 2014.

memory latency [2], leading to processors wasting much time waiting for required items of data.

The cache level therefore becomes critical for overcoming the "memory wall" in DBMS applications. Some experimental researches indicate that the adverse memory-access pattern in DB workloads results in poor cache locality and imply that data placement should focus on the last-level cache (LLC) [3,4]. However, as increasing numbers of cores are becoming integrated into a single chip, researchers' attention is being diverted from the LLC to other cache levels. For modern multicore processors, it is usual to provide at least two levels of cache for the private use of each processor core in addition to an LLC for data sharing between several cores. For example, the AMD Opteron 6174 [5], which has 12 physical processor cores, provides two levels of private cache for each core, namely a 64-KB level-one (L1) instruction cache, a 64-KB L1 data cache, and a 512-KB level-two (L2) cache. Recent research indicates that increasing the number of cores that share an LLC does not cause an inordinate number of additional cache misses, with different workloads exhibiting significant data sharing between cores [6]. As the cache levels become more complex, access to the LLC involves more clock cycles because LLC access latency has increased greatly during recent decades. These changes in cache levels indicate that it is increasingly important to bring data beyond the LLC and closer to L1. In this paper, we first analyze how various scheduling strategies for concurrent DB processes (DBPs) on different processor cores affect the performance of private-cache levels, which are closer to the execution unit than is the LLC. We then propose a middleware-based solution to provide efficient data access to the private-cache levels for concurrent OLTP-style transactions on a multicore platform.

1.1   Private-Cache Contentions

For multicore systems, all concurrent DBPs dealing with the various queries will be scheduled to run concurrently on different processor cores. However, a different DBP-schedule decision will lead to different cache performance.

For example, Figure 1 shows three queries of Q1, Q2 and Q3 need to be dispatched to run on the two processor cores Core1 and Core2. In Plan1 of Figure 1, Q1 and Q2 are dispatched to co-run on Core1. Q1 is executed first and the data in the range [1–50] are loaded into the private cache of Core1. After the execution of Q1, the context switches to Q2. The data needed for Q2, namely [1–40], are already cached at that cache level and the resulting cache hit is as desired. However, the situation for Plan2 of Figure 1 is different. If Q1 and Q3 are dispatched for the same core, consider the situation when the execution of Q1 has finished and Q3 starts to run. The existing data, in the range [1–50] are not used at all by Q3. The cache has therefore to reload a new data set [50–100].

For OLTP applications which have big database, the whole data set is not uniformly accessed, and there is access skew for different subsets. If the queries which accessing data in the same subset co-run on one core, cache hit rate becomes better. Furthermore, the appropriate co-running of multiple DBPs on the same core (Plan1) can restrict the data access for each core's private cache

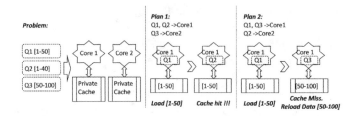

Fig. 1. Different query schedule strategies will cause different cache utilizations for private caches

to within a specific subset of the data in the whole DB, with the probability of cache hits thereby being improved.

The objective of our work is to provide a good co-running solution for concurrent queries to achieve higher private-cache utilization. The DBMS contains the data needs information for all its DBPs, and it has the ability to provide a good co-running strategy for the various DBPs. However, the scheduling of DBPs on processor cores is decided by the OS. The OS has no information about the co-running strategy provided by the DBMS.Therefore, in our approach, we achieve better private-cache utilization by importing OS functions into existing DBMSs.

1.2   The CARIC-DA Framework and Our Contribution

In this paper, we propose a framework—Core Affinity with Range Index for Cache-conscious Data Access (CARIC-DA)—that ensures that queries accessing the data set in a specific value range always run on the same processor core.

We use the core-affinity setting function provided by the OS to ensure that each DBP will always run on a specific processor core. The predefined value-range-based execute-core selection strategy serves as a range index for the CARIC-DA framework. CARIC-DA provides a function that enables the dispatching of all queries accessing the data in the same value range for execution by a specific DBP. The load balance between the processor cores is managed by a dynamic load-balancing process in CARIC-DA that adjusts the value range in the range index, according to the skew.

A major feature of our proposed solution is that all the functions provided by the CARIC-DA framework can be implemented as middleware over the existing OS and DBMS. Furthermore, CARIC-DA is *pure* middleware that does not require any modification to existing OSs and DBMSs, which is important because software for an existing DBMS is usually very large and complex, and any modification would be a time-consuming challenge.

The CARIC-DA is designed to optimize the performance for online transaction processing (OLTP) systems. A typical OLTP workload consists of a large number of concurrent, short-lived transactions, each accessing a small fraction (ones or tens of records) of a large dataset. Furthermore, the smaller cache footprints of these transactions make the data sharing between sequences of transactions possible in the private cache levels which are relatively small. In contrast, the OLAP applications with queries involving aggregation and join operations on large amounts of data cannot benefit from our middleware. Although this may appear a limitation, it is not so in the context of database applications as there is no solution optimal for all applications [7].

There are different access skews in both of the transactions and data subsets. Some transactions occur much more frequently than other transactions. The frequent occurring transactions only access specific parts of some tables instead of all tables. Therefore, even the whole database for the real world OLTP applications might be in different sizes, the primary working set is not changed too much. The relatively modest primary working set can be captured by modern LLC [6]. As more cache resources are becoming integrated into a single chip, there is much possibility to cache the modest primary working set in the higher cache levels (private caches). To our knowledge, CARIC-DA is the first multicore-optimized middleware addressing the private-cache levels for concurrent queries. Furthermore, our paper gives several insights about how individual cache levels contribute towards improving performance through detailed experiments. Especially we highlight the role of the L2 cache. Experiments show that our proposal can reduce the L2 cache miss rate by 56% and reduce the L1 cache miss rate by 6%–10%. In addition, CARIC-DA can increase the throughput by up to 25% for a TPC-C workload [8].

2   Related Work

In recent years, enlarged LLC caches have been used in an attempt to achieve better performance by capturing larger working sets. Unfortunately, many DB operations have relatively modest primary working sets and cannot benefit from larger LLCs. Furthermore, larger LLC caches require more time to service a hit. Hardavellas *et al.* [6] noticed this problem and pointed out that DB systems must optimize for locality in high-level caches (such as L1 cache) instead of LLC. STEPS [9] improves the L1 instruction-cache performance for OLTP workloads, but improvement in data locality remains a problem for high cache levels. Our CARIC-DA focuses on both of the L1 and L2 caches (private-cache levels).

The DORA [10] is a typical work which focus on the optimization for OLTP workloads on multicore platforms. Furthermore it shares similarities with CARIC-DA at the idea of range partitioning. Rather than the CARIC-DA which optimizes the cache performance, DORA is designed to reduce the lock contentions. DORA decomposes each transaction to smaller actions and assigns actions to different threads. It may be very challenging for existing DBMSs to benefit from the DORA, as making the conventional thread-to-transaction DBMS to support multiple threads for one transaction is rather complex. On the other hand, our CARIC-DA does not touch any existing functions of existing DBMSs.

The MCC-DB [4] introduces functions from the OS to improve cache utilization for DBMSs. However, they introduce a cache-partitioning function, which is not supported by general purpose OSs. In contrast, we rely on the CPU-affinity function, which is well supported by most modern OSs.

The CPU-affinity function is already used for performance improvement in a variety of applications. Foong et al. [11] present a full experiment-based analysis of network-protocol performance under various affinity modes on SMP servers and report good gains in performance by changing only the affinity modes. However, their experiences and results are only limited to network applications and cannot be directly adopted by DBMS applications, which are much more complex and have intensive data accesses.

3   The CARIC-DA Framework

To provide good private-cache utilization for each processor core, we should consider dispatching those queries that access the same data set to enable execution on the same core. One straightforward approach would be to change the process-scheduling strategy in the existing OS. However, considering the complexity of existing OSs, this would be impractical. Our CARIC-DA offers a more practical approach which can be easily implemented as *pure* middleware over existing software, with no modification being needed in either the DBMS or the OS.

3.1   Design Overview

Different queries have different data needs, and this determines how much a query can benefit from accessing accumulated private-cache data. Therefore, we should co-run different queries according to their data needs. In principle, the following two points are critical to achieving good private-cache utilization. First, we make queries that access data in the same value range co-run on the same processor core. This strategy can enable a query to reuse the cached data loaded by a forerunner. Second, we make queries that access data in different value ranges run on different processor cores. This is because queries with the same data needs that co-run with a query with different data needs will cause a cache-pollution problem in which the frequently accessed data are replaced by one-time-accessed data. Our CARIC-DA framework broadly follows these two principles and achieves its goal by the two-step strategy described below.

1. Data Set and DB-process Binding. The first step is for CARIC-DA to associate each DBP with a disjoint subset of the DB and to ensure that queries that access data in the same subset are executed by the same DBP. As an example, consider a table with 3k lines. Suppose that there are three DBPs, namely DBP1, DBP2, and DBP3, in our system. Each DBP can access a disjoint subset of 1k lines, such as DBP1 accessing the lines numbered 1–1k. All queries that access lines numbered 1–1k will be dispatched for execution by DBP1, which can access the data in this subset, but queries with a data need for line number 1500 will be assigned to DBP2.

2. DB-process and Processor-core Binding. Second, we aim to force each DBP to run only on a specific processor core by setting the CPU affinity for each DBP. The core affinity setting is achieved via a function provided by the

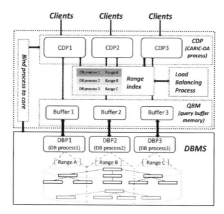

Fig. 2. Extensions of the CARIC-DA framework to conventional DBMSs

Linux OS called the CPU affinity. For example, if a CPU affinity of 1 is set for DBP1, this DBP will always run on processor core 1 and will never be scheduled for any other processor core by the OS.

In the first step, we create a binding between a data set and a DBP. In the second step, we create a binding between a DBP and a processor core. In this way, we reach the goal, namely bindings between data sets and cores. For the example, all queries that access lines numbered 1–1k are dispatched to run on processor core 1. In addition, all queries accessing data in different subsets are dispatched to run on other processor cores. This satisfies the two critical points needed to achieve the good private-cache utilization described above.

3.2   Core Components of CARIC-DA

Figure 2 shows an overview of the CARIC-DA framework and its main components, which extend an existing DBMS and OS by introducing a middleware subsystem comprising several subcomponents.

The binding between data set and DBP is achieved by a logical partitioning of the whole DB. A horizontal partition is adopted for each table. We first choose a suitable field for a table, which serves as its partition key. A mapping strategy for coupling the different partition key ranges with different DBPs is then formulated and stored in our CARIC-DA as the range index (RI). If there are multiple tables, the various tables are partitioned or overlapped, and mapped separately to all DBPs. That is, each process is mapped to several subsets for the different tables.

Theoretically any field of a table can be served as a partition key. But in practice the fields of the primary key of the table or only a subset of them can achieve good results as we have seen in our experiments. For multi-tables, it's better to choose the keys which can be commonly used to well partition most of the main tables (comparatively big and frequently accessed table). For example, the primary key of the Customers table of the TPC-C database consists of the

Warehouse id, the District id and the Customer id. The partition key fields may be the Warehouse id and the District id. The Warehouse id and the District id can also well be used to divide the other main tables. Furthermore, it is not necessarily to divide all of the tables in the database, as it is a trade-off between cache hit gain and query transfer cost.

The CARIC-DA processes (CDPs) use the mapping information provided by the RI to dispatch queries to DBPs with different data needs into an appropriate buffer in the query buffer memory (QBM). There is no physical partitioning of the DB or changes to the original DBMS, because the mapping work is done by the CDPs in terms of decomposing and dispatching the various queries according to the RI while considering access skews. Instead of obtaining queries directly from clients, each DBP only deals with queries in a specific buffer where the queries access only a specific data set.

3.3   Query Processing in CARIC-DA

In a traditional DBMS, all queries coming from the same client connection are always executed by the same DBP. In our CARIC-DA framework, CDPs communicate with the clients and dispatch the queries to different DBPs. For example, in Figure 2, CDP2 receives a "select" query from a client. CDP2 first checks the RI to identify the appropriate mapping between the data set related to this select query and the DBPs. Suppose DBP1 can access the data set this query requires. The query will then be put into Buffer1 in the QBM for checking by DBP1. Whenever DBP1 detects there is a query request, it picks up the query from Buffer1. The select query is executed and the answer is put back into Buffer1 by DBP1. CDP2 then transfers the answer in Buffer1 back to the relevant client.

When dealing with a range query that accesses a large data range, the query request can be divided into several subqueries that are transferred to several different buffers. If queries require data that are mainly mapped to one DBP, but that have a small portion (several lines of a table) being mapped to another DBP, we may prefer to dispatch this kind of query to one DBP instead of two different DBPs. This is because the query-and-answer transfer is an additional overhead for our platform, compared with traditional DBMS implementations. At the system level, if the query-dispatching function provided by the CARIC-DA framework can bring sufficient improvement to the whole system, the query transfer will be worth it. We can therefore suppose that a small imbalance will not affect the overall performance, because the imbalance will relate to only a few lines of the table and might not justify the extra query-transfer cost.

How to deal with cross-partition transaction is a challenge to our system as not all of the data sets can be well partitioned in many applications. For transactions which have lower isolation requirements, our middleware can decompose one transaction into several sub-transactions and dispatch them to different partitions. However we avoid transactions which need very high isolation level to cross different partitions. For the TPC-C benchmark, we assign the New-order transaction to a specific partition according to the district it related leaving the Item table no partitioned.

3.4   Processor-Core Binding for DBPs

Our DBP and processor-core binding is achieved by the CPU-affinity function, which is the Linux-based ability to bind one or more processes to one or more processor cores [12]. Other OSs, such as Windows Vista, have long provided a system call to set the CPU affinity for a process. On most systems, the interface for setting the CPU affinity uses a "bitmask." Each bit indicates the binding of a given task to the corresponding processor core. In Linux, 11111111111111111111111111111111=4,294,967,295 is the default affinity mask for all processes. Because all bits are set to 1, the process can run on any processor core. Conversely, the bitmask 1 is much more restrictive. With only bit 0 being set, the process can run only on processor core 0.

In our framework, we set the CPU affinity for both of the DBPs and CDPs. For data-intensive applications, the DBPs make large data-access demands and mainly access the data in the DB. The CDPs make frequent accesses to the RI and QBM. We therefore bind the CDPs and DBPs that access different data sets to different processor cores to avoid cache-pollution problems.

3.5   Load Balancing

Our framework can not only maintain better cache locality, but also balance the loads across different processor cores. For example, there were 1,000 queries, with all of them using the same value range in a data set. Under the mechanism of static partitioning, all the queries would be dispatched to run on a single processor core, which would cause load imbalance between processor cores and result in poor performance. CARIC-DA attempts to balance the load between DBPs by modifying the value range mapped to different DBPs and to synchronize the CDPs in dispatching the queries, based on the new data-set mapping information of the RI. The load-balancing process regularly gathers the query numbers processed by each of the N DBPs as a measure of the load on each DBP (l_i) and then calculates the sum of l_i as l_{sum}. The DBP with the biggest load and the DBP with the smallest load are identified. The load difference is calculated as $l_{diff} = l_{big} - l_{small}$ and the skew is defined as $skew = \frac{l_{diff}}{l_{sum}}$. Tolerating a small load imbalance is achieved by omitting the rebalancing process whenever $skew$ is below a threshold. When the skew is bigger than a threshold, Algorithm 1 is invoked to calculate new data sets (R_i is the range of the ith data set) for the DBPs.

To describe the algorithm, let l_{ave} be the ideal load for each partition. Then the ideal load for each partition is: $l_{ave} = \frac{\sum_{i=1}^{N} l_i}{N}$. In the first part of the algorithm, we create N_{sub} subsets by dividing the ranges with $l_i > l_{ave}$ into subsets. We denote the load and range of each subset as sl_i and sR_i, respectively. We ensure that $sl_i < l_{ave}$ during the partition process and calculate the data range sR_i for the subsets. In the next part of the algorithm, we merge these subsets into N data sets that have $l_i = l_{ave}$. The new data set information will then be written back to the RI and all CDPs are synchronized to dispatch the queries according to the new RI.

Algorithm 1. Calculating new value ranges for DBPs

Create N_{sub} subsets by dividing sets with $l_i > l_{ave}$ into subsets
Let SN be the serial number of sub-sets, $SN \leftarrow 1$
for i in $[1, N]$ **do**
  if $l_i > l_{ave}$ **then**
    divide the data set into μ sub-sets where $\mu \leftarrow \lceil \frac{l_i}{l_{ave}} \rceil$;
    calculate the load and range for each sub-set SN as $sl_{SN} \leftarrow \frac{l_i}{\mu}$, $sR_{SN} \leftarrow \frac{R_i}{\mu}$,
    $SN \leftarrow SN + 1$;
  else
    $sl_{SN} \leftarrow l_i$, $sR_{SN} \leftarrow R_i$, $SN \leftarrow SN + 1$;
Merge N_{sub} subsets into N sets with $l_i = l_{ave}$, $j \leftarrow 1$
for i in $[1, N]$ **do**
  $l_i \leftarrow l_{ave}$, $R_i \leftarrow 0$
  while $l_i > sl_j$ **do**
    merge sub-set j into set i $(R_i \leftarrow R_i + sR_j)$, $l_i \leftarrow l_i - sl_j$, $j \leftarrow j + 1$
  if $l_i > 0$ **then**
    merge a part of sub-set j into set i $(R_i \leftarrow R_i + \frac{l_i}{sl_j} \times sR_j)$, $sR_j \leftarrow 1 - \frac{l_i}{sl_j} \times sR_j$

---

4   Performance Evaluation

We implemented the CARIC-DA-related functions in the C language as a middleware subsystem over the existing DBMS (PostgreSQL) and the OS (Linux). The Oprofile [13] was used to examine the hardware-level performance. We compared our CARIC-DA–PostgreSQL system against an unmodified PostgreSQL system (Baseline) to investigate the efficiency of our proposed framework.

We used a 48-core DB-server machine for answering clients' queries. The DB-server machine had four sockets, with a 12-core AMD Opteron 6174 processor per socket [5]. Each core had a clock speed of 2.2 GHz and had a 128-KB L1 cache and a 512-KB L2 cache. All 12 cores of a processor shared a 12-MB L3 cache. The server had 32 GB of off-chip memory, and two 500-GB hard-disk drives. Each of the four client machines had a Intel Xeon E5620 CPU and a 24-GB memory. To prevent the I/O subsystem from becoming a bottleneck, we set the value of *shared_buffers* to 20 GB for the PostgreSQL. This setting ensured that all DB tables in the following experiments could fit in main memory.

We first used a microbenchmark evaluation to isolate the effects and to provide in-depth analysis. We then used the TPC-C benchmark [8] to further verify the effectiveness of our proposal. Our evaluation covered seven areas.

(1) Different core-affinity strategies in our CARIC-DA system.

(2) The separate impacts of CARIC-DA on performance of select-intensive and insert-intensive workloads.

(3) The scalability of CARIC-DA.

(4) The advantages of CARIC-DA for different cache levels.

(5) The performance of CARIC-DA with data sets of different size.

(6) The efficiency of CARIC-DA when dealing with skewed data sets.

(7) The efficiency of CARIC-DA in handling the TPC-C benchmark.

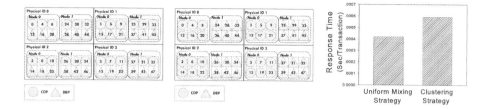

Fig. 3. Uniform mixing of CDPs and DBPs in each Node

Fig. 4. Clustering CDPs or DBPs in each Node

Fig. 5. Performance under different core-affinity settings

4.1 Core Affinity in CARIC-DA

We first decided how to set the core affinity for the various DBPs and CDPs in our CARIC-DA system. Figure 3 shows the physical location relationship for the different logical cores. There are four processors in our platform, with different physical IDs. In each processor, there are 12 cores with different logical IDs, shown as several squares located in the two Nodes [14]. We examined the performance under two core-affinity strategies for a system with 24 CDPs and 24 DBPs using the select-intensive transaction from the microbenchmark. Each transaction randomly accesses one line of a TPC-C stock table.

(1)Uniform Mixing: a mixture of DBPs and CDPs in one Node structure (Fig. 3).

(2)Clustering: all CDPs are bound to cores located in Node 0 and all DBPs are bound to cores located in Node 1 (Fig. 4).

The average response time with 24 concurrent clients is shown in Figure 5. The strategy involving the uniform mixing performs better than the strategy of clustering. Each Node is an integrated-circuit device that includes several CPU cores, up to four links for general purpose communication to other devices (such as an L3 cache, main memory interfaces). Compared with the CDP, the DBP is data intensive and makes frequent data requests to both cache and memory. In the clustering strategy, all six data-intensive DBPs are bound to the same Node, leading to intensive competition for the shared Node resources. The competition results in a longer L3-cache and memory-access latency and a comparatively worse performance. Therefore, we used the uniform-mixing strategy exclusively for the CARIC-DA system in subsequent experiments.

4.2 Microbenchmark Evaluation

In the select-intensive experiment, clients repeatedly send select transaction requests. The select transaction for the microbenchmark has only one query request for a specific line in the table. In a similar fashion, the insert transaction request comprises one record-insertion operation. We prepared a DB with only one table, namely a stock table from TPC-C comprising 100,000 lines. Both

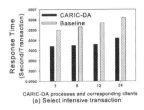 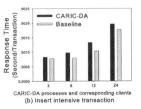 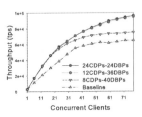

Fig. 6. Effectiveness of the CARIC-DA framework for separate select-intensive and insert-intensive transactions

Fig. 7. Throughput of CARIC-DA systems

select and insert operations randomly generated search keys within the data range of the 100,000 lines. For CARIC-DA system, we separately set up 3, 6, 12, and 24 CDPs, and 3, 6, 12, and 24 DBPs. There were also 3, 6, 12, and 24 clients, with different clients connecting to different CDPs. In Baseline system, the clients connected directly to the PostgreSQL.

Figure 6 (a) indicates that the select operation can benefit significantly from our CARIC-DA architecture. When there were 24 concurrent clients, the select transactions are executed 33% faster under CARIC-DA. These results demonstrate the effectiveness of our CARIC-DA framework in providing better performance for select-intensive workloads. However, insert-intensive operations did not benefit from our CARIC-DA framework. For each insert operation, the related DBP has to access not only the table data but also some common data, such as index data and metadata. For example, two DBPs have to update the same node of the index if the index node is already cached by different DBPs. The data in one cache will be updated, but this update operation will invalidate the copy of the data in the other cache. This kind of cache-conflict problem, caused by accessing common data, cannot be avoided in either the Baseline system or the CARIC-DA system. The extra transmission cost of queries in CARIC-DA will then lead to a worse overall performance than for the Baseline system.

4.3   Performance of CARIC-DA under Different Loads

We set up 24 CDPs and 24 DBPs in the DB-server section and repeated the select-intensive experiment. We gradually increased the number of concurrent clients to 78 clients. For 78 concurrent clients, the throughput of the CARIC-DA system is 48% higher than the Baseline system (Fig. 7). The Baseline system can linearly scale to about 36 concurrent clients, with the rising trend of overall throughput greatly reducing when there are more than 36 concurrent clients. This is because the memory bandwidth limits the performance as the number of concurrent clients increases. In contrast, the 24CDPs–24DBPs system with its well-designed cache accesses does not show such a decrease in the rising trend in overall throughput. The CDP will generate one client thread to deal with the queries in one client connection, and we use the System-V semaphores to

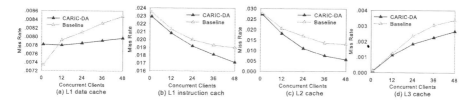

Fig. 8. Cache miss rates

synchronize the accesses to the QBM by these threads. With an increasing number of concurrent clients, any untimely scheduling of these semaphores will affect the system performance and result in a nonlinear throughput growth for the CARIC-DA system. We also set up two new CARIC-DA systems with 12CDPs–36DBPs and 8CDPs–40DBPs, respectively (Fig. 7). With reducing numbers of CDPs, there are more client threads bound to the same core. The performance of the CDPs limits the overall performance of the 8CDPs–40DBPs system. The select-intensive transaction is very short and the frequent query-and-answer transfers between clients and DBPs greatly stress the CARIC-DA middleware.

4.4 Cache Utilization

We repeated the select-intensive experiment for 24CDPs–24DBPs system, and measured the cache miss rate for the various cache levels separately to confirm that CARIC-DA is efficient because of its outstanding performance at the various cache levels. The miss rate is the percentage of misses per total number of instructions. We observed that the L1 data-cache miss rate increases from 0.73% to 0.84% as the number of concurrent client threads increases in the Baseline system (Fig. 8 (a)). However, for the CARIC-DA system, the L1 data-cache miss rate appears to be only slightly increased (from 0.78% to 0.79%). The CARIC-DA system can also improve the L1 instruction-cache performance by up to 10% (Fig.8 (b)). This is because all DBPs are restricted to running on a specific processor core, which can avoid frequent context switching in each core. The biggest cache performance improvement comes from the L2 cache, with an almost 56% reduction in cache misses compared with that for the Baseline system (Fig. 8 (c)). We also observe that the miss rate for the shared L3 cache can be reduced by 21% (Fig. 8 (d)).

4.5 Performance of CARIC-DA for Data Sets of Different Sizes

In this section, we describe the effectiveness of our proposal for a variety of data sets. We compared the average response time and cache utilization of the select-intensive application with 48 concurrent clients for a 24CDP–24DBP-based CARIC-DA system with those for the Baseline system, as shown in Figure 9. When increasing the data set to the much larger size of 1,130 MB (3,000,000 tuples), the performance gap between the CARIC-DA system and the Baseline system narrowed (Fig. 9 (a)). The diminished advantage of our proposal

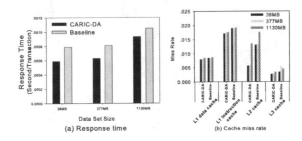

Fig. 9. Performance for data sets of different sizes

with very large data sets derives from our RI-based data-access strategy. In the CARIC-DA system, the L2 cache accesses a smaller subset, whereas the Baseline system's L2 caches each access the whole data set. However, as the size of the whole data set greatly increases to 1,130 MB, the subset size in our proposal is also greatly increased (to 47 MB). The sizes of both the whole data set and our subset are much beyond the capacity of the small L2 cache (512 KB). The size difference between the subset in our proposal and the whole data set in the Baseline system becomes less significant when compared with the size of the very small L2 cache, and the performance difference between the two systems decreases. For a real-world workload, data access is skewed and the frequently accessed data set is much smaller than the whole DB. Taking this data-access skew into account, our CARIC-DA system shows impressive advantages when dealing with GB-sized data sets, in comparison with the Baseline system.

4.6   Performance of CARIC-DA with Skewed Data Access

We use data sets of 3,000,000 tuples (table size 1,130 MB), 10,000,000 tuples (table size 3,766 MB), and 15,000,000 tuples (table size 5,649 MB). Fig. 10 plots the average response time for select-intensive transaction for the data set of 3,000,000 tuples. We calculate the average response time every 10 seconds. Initially, the distribution of the queries is uniform for the entire data set. However, at time point 50, the distribution of the load changes, with 50% of the queries being sent to 10% of the data set (Zipf (0.75) distribution). We did not use the dynamic load-balancing function of the CARIC-DA system until time point 100.

After the load change, the performance of the Baseline system improved slightly, while the performance of the non-load-balanced CARIC-DA system dropped sharply. After enabling dynamic load balancing in the CARIC-DA system, the performance improved dramatically, outperforming the Baseline system by 39%. For the other two data sets, the dynamic-load-balanced CARIC-DA system achieved improvements of 41% and 40% above the Baseline system's performance. These results confirm the efficiency of our proposal when dealing with skewed data sets and also substantiated the claim that skew favors the CARIC-DA system (the advantage is increased from 7.3% to 41%).

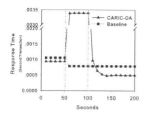

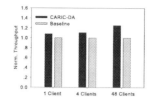

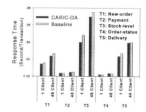

Fig. 10.   Performance
with skewed data access

Fig. 11.   Throughput
for TPC-C benchmark

Fig. 12. Performance for
different transactions

4.7   TPC-C Benchmarking

We used a 24-warehouse TPC-C data set (~4.8 GB). For the CARIC-DA system, we set up 4 CDPs and 44 DBPs and evenly partitioned the 24 warehouses at the district level. For each transaction, the CDP has only to find the appropriate buffer in the QBM for the first query of the transaction, with any follow-up queries directly accessing the same QBM. By monitoring the CPU usage of CDPs we founded out that CDPs did not need a lot of CPU and 4 CDPs were sufficient for TPC-C transactions. As shown in Figure 11, the CARIC-DA system outperforms the Baseline system by 10% to 25%.

The average response times for the five kinds of transactions are shown in Fig. 12. The CARIC-DA system can optimize the execution time for New-order transactions by 7% and can greatly optimize the Stock-level transactions by 18%. The Stock-level transactions will retrieve item information for the most recent 20 orders in a specific district. In our system, a specific DBP only processes the New-order transactions, which will also access the item information for a specific district. When a Stock-level transaction follows a New-order transaction, there is a higher possibility of the item information for recent orders existing at the private-cache level. For the Baseline system, specific DBPs have to deal with the New-order transactions from all districts. For recent-order information in a specific district, therefore, the private-cache hit possibility will be relatively low. This explains why Stock-level transactions demonstrate great performance improvements for our proposed system. In the high-concurrency experiment, the New-order transactions can be further optimized by 11%. For concurrent transactions, more New-order transactions benefit from the CARIC-DA approach, compared with the no-concurrency situation. In addition, we observed that the CARIC-DA system can achieve lower abort rates for New-order transactions. The concurrent transactions which update the same data in a specific district may cause the serialization error in the Baseline system with the isolation level of serializable. In the CARIC-DA system, these concurrent transactions will be sequentially processed and lead to less serialization error. This is the other reason for the 25% throughput improvement with the CARIC-DA system.

5   Conclusions

In this paper, we introduced CARIC-DA, which is the first work addressing the problems for DBMSs of private caches on multicore platforms. Considering the different access skews in both of the transactions and data subsets, even the whole database for the real world OLTP applications might be in different sizes, the primary working sets are small and can be captured in modern LLC. We researched on the possibility to bring the primary working set beyond the LLC and closer to processor core (in the private cache levels). We give several insights about how individual cache levels contribute towards improving performance through detailed experiments and highlight the role of the L2 cache. CARIC-DA is implemented as *pure* middleware, enabling the existing DBMS to be used without modification. Not only can it maintain better locality for each core's private cache, but it can also balance loads dynamically across different cores. Experiments show that the L2 cache miss rate can be reduced by 56% for select-intensive transactions and the throughput can be improved by 25% for TPC-C transactions by using our CARIC-DA framework. In future work, we will use other hardware platforms in additional experiments to investigate the effectiveness of our framework. Providing better cache-access patterns for decision support systems will also be a future challenge.

References

1. Adee, S.: The data: 37 years of Moore's law. IEEE Spectrum 45, 56 (2008)
2. Cieslewicz, J., Ross, K.A.: Database optimizations for modern hardware. Proceedings of the IEEE 96, 863–878 (2008)
3. Ailamaki, A., DeWitt, D.J., Hill, M.D., Wood, D.A.: DBMSs on a modern processor: Where does time go? In: VLDB, pp. 266–277 (1999)
4. Lee, R., Ding, X., Chen, F., Lu, Q., Zhang, X.: MCC-DB: Minimizing cache conflicts in multi-core processors for databases. In: VLDB, pp. 373–384 (2005)
5. AMD family 10h server and workstation processor power and thermal data sheet, http://support.amd.com/us/Processor_TechDocs/43374.pdf
6. Hardavellas, N., Pandis, I., Johnson, R., Mancheril, N.G., Ailamaki, A., Falsafi, B.: Database servers on chip multiprocessors: Limitations and opportunities. In: CIDR, pp. 79–87 (2007)
7. Salomie, T.I., Subasu, I.E., Giceva, J., Alonso, G.: Database engins on multicores, why parallelize when you can distribute? In: EuroSys, pp. 17–30 (2011)
8. Transaction processing performance council. TPC-C v5.5: On-line transaction processing (OLTP) benchmark
9. Harizopoulos, S., Ailamaki, A.: STEPS towards cache-resident transaction processing. In: VLDB, pp. 660–671 (2004)
10. Pandis, I., Johnson, R., Hardavellas, N., Ailamaki, A.: Data-oriented transaction execution. In: VLDB, pp. 928–939 (2010)
11. Foong, A., Fung, J., Newell, D.: An in-depth analysis of the impact of processor affinity on network performance. In: ICON, pp. 244–250 (2004)
12. Love, R.: Kernel korner: CPU affinity. Linux Journal (111), 8 (2003)
13. Oprofile: A system profiler for linux (2004), http://oprofile.sf.net
14. BKDG for AMD family 10h processors (2010), http://support.amd.com/en-us/search/tech-docs?k=bkdg

Approximating an Energy-Proportional DBMS by a Dynamic Cluster of Nodes

Daniel Schall and Theo Härder

DBIS Group, University of Kaiserslautern, Germany
{schall,haerder}@cs.uni-kl.de

Abstract. The most energy-efficient configuration of a single-server DBMS is the highest performing one, if we exclusively focus on specific applications where the DBMS can steadily run in the peak-performance range. However, typical DBMS activity levels—or their average system utilization—are much lower and their energy use is far from being *energy proportional*. Built of commodity hardware, *WattDB*—a distributed DBMS—runs on a cluster of computing nodes where energy proportionality is approached by dynamically adapting the cluster size. In this work, we combine our previous findings on energy-proportional storage layers and query processing into a single, transactional DBMS. We verify our vision by a series of benchmarks running OLTP and OLAP queries with varying degrees of parallelism. These experiments illustrate that WattDB dynamically adjusts to the workload present and reconfigures itself to satisfy performance demands while keeping its energy consumption at a minimum.

1 Introduction

The need for more energy efficiency in all areas of IT is not debatable. Besides reducing the energy consumption of servers, other ideas like improving the cooling infrastructure and lowering its power consumption help reducing the energy footprint of data centers. Due to their narrow power spectrum between idle and full utilization [1], the goal of satisfactory energy efficiency cannot be reached using today's (server) hardware. Reducing energy usage of servers to a sufficient level leads to a demand for energy-proportional hardware. Because such a goal seems impractical, we should at least aim at an emulation of the appropriate outcome at the system level.

Energy proportionality describes the ability of a system to reduce its power consumption to the actual workload, i.e., a system, delivering only 10% of its peak performance, must not consume more than 10% of its peak power. That goal could be approached by exploiting hardware-intrinsic properties, e.g., CPUs *automatically* entering sleep states or hard disks spinning down when idle. Unfortunately, current hardware is not energy proportional. For example, DRAM chips consume a constant amount of power—regardless of their use—and it is not possible to turn off unused memory chips in order to reduce energy consumption. Spinning down hard disks when idle conflicts with long transition times and results in slow query evaluation.

S.S. Bhowmick et al. (Eds.): DASFAA 2014, Part I, LNCS 8421, pp. 297–311, 2014.

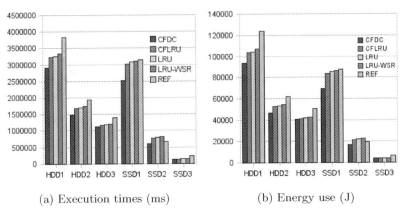

(a) Execution times (ms)   (b) Energy use (J)

Fig. 1. Performance and energy consumption running the TPC-C trace

1.1 Energy Efficiency Limits of Single-Server DBMSs

Using a key result of experiments with a single-server DBMS [10], we want to illustrate the close linkage of execution times and energy efficiency. All experiments were conducted in an identical system setting, i.e., ATX, IDE, memory size, OS, DBMS (except buffer management), and workload were left unchanged. For this reason, the details are not important in this context. Our goal was to reveal the relationship concerning performance and energy use for different external storage media[1] and buffer management algorithms[2].

To represent a realistic application for our empirical measurements, we recorded an OLTP trace (a buffer reference string using a relational DBMS) of a 20-minutes TPC-C workload with a scaling factor of 50 warehouses. The test data (as a DB file) resided on a separate magnetic disk/SSD (data disk, denoted as SATA). The data disks, connected one at a time to the system, represent *low-end* (HDD1/SSD1), *middle-class* (HDD2/SSD2), and *high-end* (HDD3/SSD3) devices. Execution times and related energy use in Figure 1 are indicative for what we can expect for single-server DBMSs by varying the storage system configurations. The storage device type dominantly determines execution time improvement and, in turn, reduction of energy use. In our experiment, the algorithmic optimizations and their relative influence to energy efficiency are noticeable, but less drastic.

The key effect identified by Figure 1 is further explained by Figure 2, where the break-down of the average working power of hardware components of interest is compared with their *idle* power values. The figures shown for HDD3 and SSD3

---

[1] We used magnetic disks (HDD1: 7.200 rpm, 70 IOPS; HDD2: 10.000 rpm, 210 IOPS; HDD3: 15.000 rpm, 500 IOPS) and flash storage (read/write IOPS) (SSD1: 2.700/50, SSD2: 12.000/130, SSD3: 35.000/3.300).

[2] CFDC [9] optimizes page caching for SSDs. Here, we cross-compared CFDC to LRU, CFLRU [11], LRU-WSR [7], and REF [15], some of which are also tailor-made for SSD use.

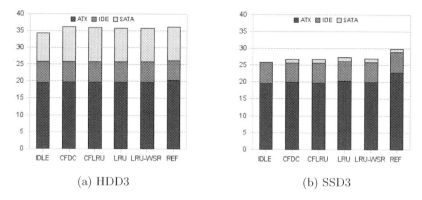

(a) HDD3   (b) SSD3

Fig. 2. Break-down of average power (W)

are indicative for all configurations; they are similar for all devices, because ATX—consuming the lion's share of the energy—and IDE remained unchanged. Ideally, utilization should determine the power usage of a component. But, no significant power variation could be observed when the system state changes from idle to working or even to full utilization. Because the time needed to run the trace is proportional to the energy consumption, the fastest algorithm is also the most energy-efficient one. This key observation was complemented by [17] with a similar conclusion that *"within a single node intended for use in scale-out (shared-nothing) architectures, the most energy-efficient configuration is typically the highest performing one"*.

In summary, energy saving is impressive, if we consider the experiment in isolation where system utilization is steadily kept very high, i.e., $> 90\%$. However, typical servers mostly reach an average CPU utilization of only $\sim30\%$ and often even less than $\sim20\%$ [2].

1.2 Varying DBMS Service Needs

We have shown in [14] that real-world workloads usually do not stress DB systems 24/7 with peak loads. Instead, the workloads alternate in patterns between high and near-idle utilization. But, DB systems have to be tailored to the peak performance to satisfy incoming queries and potential users. Therefore, DB servers usually come with big DRAM buffers and a number of disks as external storage—both components that consume a lot of energy. The majority of these resources is only needed in rare time intervals to handle peak workloads. All other times, they lie waste, thereby substantially decreasing the overall energy efficiency of the server. During times of underutilization, overprovisioned components are not needed to satisfy the workload. By adjusting the DB systems to the current workload's needs, i.e., making the system energy proportional, energy usage could be lowered while still allowing the maximum possible performance.

Both observations sketched above have strongly guided the design of WattDB: Section 2 outlines its cluster hardware, the related power consumption, and the most important aspects of its software design. In Section 3, we introduce our experimental setup, before we discuss the results of our empirical benchmark runs in Section 4. Finally, we conclude our results and give an outlook to our future work in Section 5.

2   The WattDB Approach

Based on the findings outlined above, we concluded that a single-server-based DBMS will never be able to process real-world workloads in an energy-efficient way. A cluster of lightweight (wimpy) nodes is more promising, as nodes can be dynamically switched on or off, such that the cluster can be adjusted to the current workload. Lang et al. [8] have shown that a cluster suffers from "friction losses" due to coordination and data shipping overhead and is therefore not as powerful as a comparable heavyweight server. On the other hand, for moderate workloads, i.e., the majority of real-world database applications, a scale-out cluster can exploit its ability to reduce or increase its size sufficiently fast and, in turn, gain far better energy efficiency.

In [12], we already explored the capabilities and limitations of a clustered storage architecture that dynamically adjusts the number of nodes to varying workloads consisting of simple *read-only page requests* where a large file had to be accessed via an index[3]. We concluded that it is possible to approximate energy proportionality in the storage layer with a cluster of wimpy nodes. However, attaching or detaching a storage server is rather expensive, because (parts of) datasets may have to be migrated. Therefore, such events (in appropriate workloads) should happen on a scale of minutes or hours, but not seconds.

In [13], we have focused on the query processing layer—again for varying workloads consisting of two types of *read-only SQL queries*—and drawn similar conclusions. In this contribution, we revealed that attaching or detaching a (pure) processing node is rather inexpensive. Hence, such an event can happen in the range of a few seconds—without disturbing the current workload too much.

In this paper, we substantially extended the kind of DBMS processing supported by WattDB to *complex OLAP / OLTP workloads consisting of read-write transactions*. For this purpose, we refined and combined both approaches to get one step closer to a fully-featured DBMS. Opposed to our previous work, we treat all nodes identical in this paper; hence, all nodes will act as storage and processing nodes simultaneously.

2.1   Cluster Hardware

Our cluster hardware consists of n (currently 10) identical nodes, interconnected by a Gigabit-ethernet switch. Each node is equipped with an Intel Atom D510

---

[3] Starting our WattDB development and testing with rather simple workloads facilitated the understanding of the internal system behavior, the debugging process, as well as the identification of performance bottlenecks.

CPU, 2 GB DRAM and three storage devices: one HDD and two SSDs. The configuration is considered Amdahl-balanced [16], i. e., balanced between I/O and network throughput on one hand and processing power on the other. By choosing commodity hardware with limited data bandwidth, Ethernet wiring is sufficient for interconnecting the nodes. All nodes are connected to the same Ethernet segment and can communicate with each other.

2.2 Power Consumption

As we have chosen lightweight nodes, the power consumption of each node is rather low. A single node consumes ~22 - 26 Watts when active (based on utilization) and 2.5 Watts in standby mode. The interconnecting switch needs 20 Watts and is included in all measurements.

The minimal configuration of the cluster consists of a single node, called the *master node*. All functionalities (storage, processing, and coordination) can be centralized using only a single node, thereby minimizing the static energy usage. In case, all nodes are running, the overall energy use of the cluster reaches ~270 Watts. This is another reason for choosing commodity hardware which uses much less energy compared to server-grade components. For example, main memory consumes ~2.5 Watts per DIMM module, whereas ECC memory, typically used in servers, consumes ~10 Watts per DIMM. Likewise, our HDDs need less power than high-performance drives, which makes them more energy efficient.

2.3 Software Design

A single wimpy node can quickly become a hotspot which may slow down query processing of the entire cluster. To mitigate bottlenecks, *WattDB* is designed to allow dynamic reconfiguration of the storage and query processing layers.

The *master node* accepts client connections, distributes incoming queries, and administrates metadata for all cluster nodes. This node is also responsible for controlling the power consumption of the cluster by turning nodes on and off. However, the master is not different from the rest of the nodes; it is also able to process queries and manage its own storage disks.

Storage Structures and Indexes. In WattDB, data is stored in tables, which are subdivided into partitions. Each partition is organized as a heap and consists of a set of segments, where a segment specifies a range of pages on a hard drive. Physical clustering of pages is guaranteed inside each segment. To preserve locality of data, segments are always assigned to disks on the same node managing the partition. Indexes implemented as B*-trees can be created within a partition to speed up query evaluation.

Dynamic Partitioning Scheme. From a logical point of view, database tables in WattDB are horizontally sliced into *partitions* by primary-key ranges. Each partition is assigned to a single node, possibly using several local hard disks. This node is responsible for the partition, i. e., for reading pages, performing projections and selections, and for propagating modified pages while maintaining isolation and consistency.

To support dynamic reorganization, the partitioning scheme is not static. Primary-key ranges for partitions can be changed and data can migrate among partitions on different nodes to reflect the new scheme. Hence, partitions can also be split up into smaller units, thereby distributing the data access cost among nodes, and can be consolidated to reduce its storage and energy needs.

Partitioning information is kept on the master node in an ordered, unbalanced tree to quickly identify partitions needed for specific queries. Pointers reference either inner nodes with further fragmentation of the primary-key range or point to a partition where the data is stored. Note, while moving or restructuring of a partition is in progress, its old and new state must be reachable. Therefore, each pointer field in the tree can hold two pointers. While a partition is reorganized (split or merged), the pointers may point to two different partitions: In case of a split, the first pointer refers to the new partition to which a writing transaction copies the corresponding records, whereas the second pointer references the old partition where non-moved records still remain (and vice versa in case a merge is in progress). An appropriate concurrency control scheme should enable reads and updates of the new and old partition while such a reorganization is in progress.

Figure 3 plots three stages of an exemplary partition tree while a split is processed. The first stage shows (a fraction of) the initial key distribution. In the second stage, the key range between 1,000 and 2,000 contained in partition 3.4 is split into two partitions, where a new partition 3.6 has to be created. A transaction scans the old partition 3.4 and moves records with keys between 1,000 and 1,500 to the new partition 3.6. Records with primary keys above 1,500 stay in the old partition. To allow concurrent readers to access the records, the read pointer (marked r) still points to the old partition. Thus, reading transactions will access both, the new and old partition to look for records. The write pointer already references the new partition, redirecting updates to the new location. In the last stage, the move operation is assumed to have succeeded, and read and write pointer both reference the new partition. Moving an entire partition can be considered as a special case of such a split.

By keeping records logically clustered, the query optimizer on the master node can quickly determine eligible partitions and distribute query plan operators to run in parallel. At query execution time, no additional look-ups have to go through the master node.

Concurrency Control. As every other DBMS, WattDB needs to implement mechanisms for concurrency control to ensure ACID properties [5]. Therefore, access to records needs to be coordinated to isolate read/write transactions. When changing the partitioning schema and moving records among partitions, concurrency control must also coordinate access to the records in transit. Classical pessimistic locking protocols block transactions from accessing these records until the moving transaction commits. This leads to high transaction latency, since even readers need to wait for the move to succeed. Furthermore, writers must postpone their updates to wait for the move to terminate.

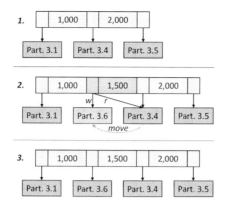

Fig. 3. Three steps of a split in the partition tree with updates of r/w pointers

Multiversion Concurrency Control (MVCC) allows multiple versions of database objects to exist. Each modification of data creates a new version of it. Hence, readers can still access old versions, even if new transactions changed the data. Each data element keeps a version counter of its creation and deletion date to enable transactions to decide which records to read by comparing the version information with the transaction's own version counter. While concurrent writers still need to synchronize access to records, readers will always have the correct version and will not get blocked by writers [3]. The obsolete versions of the records need to be removed from the database from time to time by a process we call *garbage collection* (GC).[4]

While MVCC allows multiple readers lock-free access to records, writing transactions still need to synchronize access with locks. Hence, deadlocks can arise where two or more transactions wait for each other and none can make any progress. Therefore, each node has a deadlock detection component that keeps track of locally waiting transactions in a wait-for graph (see [4]). To detect deadlocks spanning multiple nodes, a centralized detector on the master aggregates information of the individual nodes to create a global wait-for graph.

Move Semantics. Transactions need to be ACID compliant; hence, moving records among partitions must also adhere to these properties. Therefore, it is vital to move the records inside a dedicated transaction acting as follows:

1. Update the partition tree information with pending changes
2. Read records from the source partition
3. Insert the records into the target partition
4. Delete the records from the source[5]
5. Update the partition tree information with final changes
6. Commit

---

[4] PostgreSQL calls it *vacuum*.
[5] Note that the order of Insert and Delete is not important.

Because the movement is covered by a transaction, it is guaranteed that concurrent accesses to the records will not harm data consistency. Let T_{move} be the move transaction stated above, T_{old} any older, concurrently running transaction, and T_{new} any newer, concurrently running transaction. T_{old} can read all records deleted by T_{move} in the old partition until it commits and the records are finally removed by GC. T_{old} will not see the records newly created in the target partition, as the creation timestamp/version of the record is higher than its own. T_{new} will also read the records in the source partition, as the deletion version info of those records is older, but refers to a concurrently running transaction. T_{new} will not read the records in the target partition, as they were created by a concurrently running transaction. Any transactions starting after the commit of T_{move} will only see the records in the target partition. These properties follow directly from the use of MVCC.

Using traditional MVCC, writers still need to synchronize access to avoid blind overwrites. For the move transaction, we know that it will not alter the data. Therefore, blindly overwriting the newly created version in the target partition would be acceptable. We have modified the MVCC algorithm in WattDB to allow an exception from the traditional MVCC approach: Records that were moved to another partition can be immediately overwritten, i.e., they have a new version, without waiting for the move to commit.

Cost of Reorganization. In the following experiments, data is migrated in order to shutdown nodes, thus, reducing the power consumption of the cluster, or data is distributed in order to reduce query response times, which, in turn, also reduces the queries' energy consumption. Moving data is an expensive task, in terms of energy consumption and performance impact on concurrently running queries. We have observed data transfer rates of ~80 Mbit/s in parallel to the OLTP/OLAP workload, hence, it takes less than 2 minutes to move 1 GByte of data from one node to another. The overhead of the move operations should amortize by reducing the energy consumption of subsequent queries. Though it is difficult to calculate the exact energy consumption of a data move operation with respect to the impact of running queries, the energy cost can be estimated with the duration of the move operation and the (additional) power consumption. Hence, moving 1 GByte of data to a dedicated node with 25 Watts power consumption will require approximately 2.600 Joules.

Monitoring and Control. The previously described techniques allow WattDB to dynamically adjust the size of partitions on each node by moving records among them. Thus, we can control the utilization of each of the nodes.

Figure 4 sketches the feedback control loop in WattDB that monitors the cluster, processes the measurements, and takes actions to keep up performance at minimal energy cost. First, the CPU utilization, the buffer hit rate, and the disk utilization on each node is measured and sent to the master node. On the master node, the measurements are evaluated, i.e., each performance indicator is compared to a predefined high and low watermark. If the master detects a

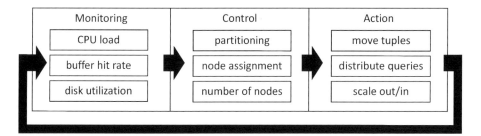

Fig. 4. Monitoring & Controlling the Cluster

workload change and a resulting imbalance in performance or energy consumption, the cluster configuration is examined and actions are evaluated to resolve the issue. Adjustment steps include the re-distribution of partitions among nodes to reduce disk utilization, re-distributing query plans to lower the CPU utilization of nodes, and powering up/down nodes in the cluster to adjust the number of available nodes. For example, the high watermark for CPU utilization is at 85%, hence, exceeding that value will induce the need for another node to help processing queries. Likewise, a disk utilization below 20 IOPS will mark this disk as underutilized and eligible for shutdown. The master node sends out change requests to the cluster nodes, which execute the desired configuration changes, e. g., re-partitioning transactions are started and nodes are powered up and down. Because the master controls transaction execution and query processing, it can rewrite query plans of incoming queries to execute operators on designated, underutilized nodes.

The *master node* is also handling incoming queries and coordinating the cluster. On the master node, a dedicated component, called *EnergyController*, monitors and controls the cluster's energy consumption. This component monitors the performance of all nodes in the cluster. Depending on the current query workload and node utilization, the *EnergyController* activates and suspends nodes to guarantee a sufficiently high node utilization depending on the workload demand. Suspended nodes do only consume a fraction of the idle power, but can be brought back online in a matter of seconds. It also modifies query plans to dynamically distribute the current workload on all running nodes thereby achieving balanced utilization of the active processing nodes.

3   Experimental Setup

The 10 nodes running WattDB and the Ethernet switch are connected to a measurement box which logs the power consumption of each device. A more detailed description of the measurement box can be found in [6]. To submit workloads to the master node, a dedicated DB-client machine, running a predefined benchmark, is used. That machine also aggregates the power measurements and the benchmark results, e. g., throughput and response time, to file. Therefore, we are able to identify the energy consumption of each benchmark run, even of each query.

In this paper, we are using an adapted TPC-H dataset[6] for the cluster. Some data types of the TPC-H specification are yet unsupported by WattDB and therefore replaced by equivalent types. For example, the *DATE* type was replaced by an *INTEGER*, storing the date as YYYYMMDD, which is functionally identical. Key constraints were not enforced because of the same reason. We pre-created the TPC-H dataset with a scale factor of 100; hence, the DB holds 100 GB of raw data initially. The data distribution strongly depends on the partitioning scheme and the number of nodes online, e. g., with 10 nodes, each node would store ~10 GB of data. Therefore, the amount of data shipped among nodes under reorganization typically ranges from 100 MB to a few GB.

The OLAP part of the benchmark is running TPC-H-like queries that access most of the records in a single query. These queries heavily rely on grouping and aggregation and are therefore memory intensive compared to OLTP queries. TPC-H is a decision support benchmark, hence, we were able to use its queries as analytical workloads. OLAP clients will select one query at a time to run from a list of queries by round-robin. For OLTP, we have taken queries from the TPC-H data generator. In addition, we created corresponding DELETE and UPDATE queries, because the generator is only using INSERT for creating the dataset. Typical OLTP queries are adding/updating/deleting customers and warehouse items; furthermore, they are submitting and updating orders.

A *workload* consists of a single DB client submitting one OLAP query per minute and a given number of DB clients sending OLTP queries. OLTP clients will wait for the query to finish, sleep for 3 seconds of "think time" and start over by submitting a new query. Every 120 seconds, a differing workload is initiated where the number of DB clients may change. Thus, WattDB will have to adjust its configuration to satisfy the changing workloads while keeping energy consumption low. Alltogether, a single *benchmark run* consists of 63 workloads, resulting in a total duration of ~2 hours.

4   Experimental Results

We have executed four different benchmarks on the cluster. First, we used the benchmark $BENCH_1$ which spawns an increasing number of DB clients sending queries to a fixed 10-node cluster without dynamic reconfiguration. The DB data was uniformly distributed to the disks of all nodes. This experiment serves as the baseline for all future measurements. Next, we ran the same benchmark against a fully dynamic cluster, where WattDB will adjust itself to fit the number of nodes and data distribution to the current workload ($BENCH_2$). Hence, initially all DB data was allocated to the disks of the master node. With growing number of DB clients, the dynamic partitioning scheme initiated a redistribution of the

---

[6] http://www.tpc.org/tpch/

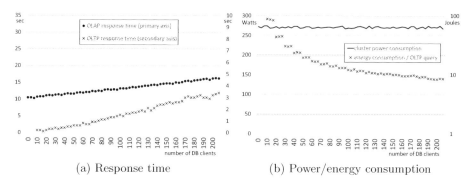

(a) Response time   (b) Power/energy consumption

Fig. 5. $BENCH_1$: Increasing load on a static cluster

DB data with each additional node activated[7], such that the data was uniformly allocated to all disks of all nodes at the end of $BENCH_2$. In the third experiment, we shuffled the workload intensities by growing and shrinking the number of (OLTP) DB clients to provide a more realistic, variable workload and, in turn, to provoke more sophisticated partitioning patterns. The cluster was able to react based on the current utilization only, as it did not have any knowledge of upcoming workloads ($BENCH_3$). Finally, we re-ran the benchmark from our third experiment but provided forecasting data to the cluster. Thus, in the last experiment, the cluster could use that information to pre-configure itself for upcoming workloads ($BENCH_4$).

Results of $BENCH_1$**:** Figure 5a plots the performance of a static database cluster with 10 nodes. All nodes were constantly online and data was uniformly distributed on them. The number of DB clients is increasing over time (x-axis from left to right). Every client is sending OLTP queries sequentially, thus, the number of DB clients defines the number of parallel queries WattDB has to handle. With no OLTP queries in parallel, the cluster takes about 10.5 seconds for an OLAP query. With rising workload, i.e., more parallel queries, response times for OLTP and OLAP queries increase. With 200 clients, the OLAP queries take 16 seconds to finish while OLTP response times increased from 0.2 to 3.3 seconds.[8] Figure 5a depicts the response times for both query types. While the performance of the static cluster is unmatched in all other experiments, its power consumption is the highest. Figure 5b shows the power consumption of the cluster (primary y-axis in Watts) and the energy consumption per query (secondary y-axis in Joules). Obviously, a static configuration will yield higher performance at the price of worse energy efficiency, especially at low utilization levels. As the plots indicate, the energy consumption per query is very high at low utilization.

---

[7] As far as repartitioning overhead is concerned, frequency and volume of data movement in $BENCH_2$ can be considered as a kind of worst case.

[8] All reported query response times are averages over 120 seconds.

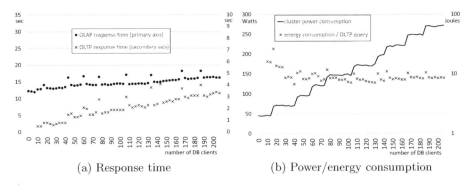

(a) Response time   (b) Power/energy consumption

Fig. 6. *BENCH₂*: Increasing load on a dynamic cluster

Results of $BENCH_2$**:** Next, in order to test dynamic reconfiguration ability of WattDB, we have re-run the same benchmark on a dynamic cluster. Figure 6a depicts the performance while increasing the number of DB clients. Starting with a low utilization, the database is running on a single node only, keeping the other 9 nodes suspended. As a consequence, (all partitions of) the entire dataset had to be allocated on the node's disks. Power consumption, as plotted in Figure 6b, is initially low (about 45 Watts for the whole cluster). By calculating the Joule integral over the differing courses of energy consumption in Figures 5b and 6b, one can an impression of the absolute energy saving possible, which is obviously substantial in this experiment.

With increasing utilization, WattDB dynamically wakes up nodes and assigns database partitions to them. Hence, re-partitioning, as previously described, needs to physically move records among nodes. Therefore, in parallel to the query workload, the movement takes up additional system resources. While the database is re-configuring, average query runtimes increase by 3 seconds for OLTP workloads and up to 6 seconds for OLAP workloads because of the extra work. Moving data among nodes takes approximately 30 to 120 seconds, depending on the amount of data to be moved. The spikes in Figure 6a visualize the degraded performance while moving data partitions. Likewise, the energy consumption per query increases for a short period of time. Still, no query is halted completely, higher runtimes are a result of increased disk/CPU utilization and wait times because of locked records.

This experiment demonstrates that WattDB is able to dynamically adjust its configuration to varying needs. While minimizing the energy consumption at low utilization, the master node re-actively switches on more nodes as the workload rises. As a result, the energy efficiency of the dynamic cluster is better than the static 10 node configuration, especially at low load levels. Query response times are not as predictable as in a static cluster, because the dynamic reconfiguration places an additional burden on the nodes. Still, the experiment shows that it is possible to trade performance for energy savings.

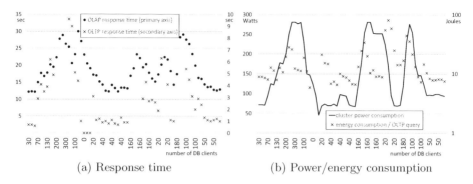

(a) Response time   (b) Power/energy consumption

Fig. 7. $BENCH_3$: Varying load on a dynamic cluster

Results of $BENCH_3$**:** The previous experiments spawned and increasing number of DB clients to submit queries, providing a steadily increasing utilization
of the cluster. Thus, the workload was rising slightly over time. Realistic benchmarks require more variance in load changes, in addition with quickly rising and
falling demands. Hence, the next experiment employs a more complex pattern
of utilization.

In Figure 7a, the x-axis exhibits the number of DB clients over time. This
experiment starts with a moderate utilization of 30 parallel clients, climbs up to
300, then quickly drops to idle. Afterwards, two more cycles are starting between
low and high utilization levels. This figure plots the performance for both OLTP
and OLAP queries, while the cluster is adjusting to the changing workloads. As
the graphs indicate, WattDB is heavily reconfiguring and, thus, query response
times vary a lot, especially when workloads are shifting.

Figure 7b illustrates the power consumption of the cluster characterized also
by high fluctuations as nodes come online and go offline again. WattDB is re-
acting to changes in the workloads based on its measurements of the nodes' utilization. Therefore, reconfiguration happens re-actively to the specific workload
changes. In this experiment with dynamic and extreme workload changes, the
reaction time of WattDB is too high to maintain proper query response times.
Consequently, energy consumption is also high, mainly due to the overhead of
cluster reconfiguration.

This experiment shows that WattDB is able to react to quickly changing
workloads, but with less satisfying results. As reconfiguration takes time and
consumes resources, a purely reactive adaptation to workloads is not sufficient.

Results of $BENCH_4$**:** To overcome the limitations of a purely reactive database cluster, WattDB should have some knowledge of the "future" in order
to appropriately pre-configure the cluster for upcoming workloads. To test this
hypothesis, we have run the same experiment as before (see $BENCH_3$), while
WattDB was continuously informed about the following three workloads, i. e.,
the number of DB clients in the next six minutes. This information can be used

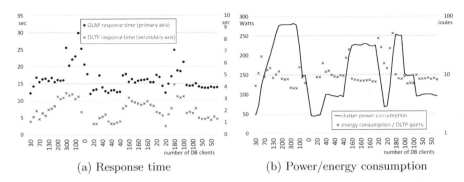

(a) Response time   (b) Power/energy consumption

Fig. 8. $BENCH_4$: Varying load on a dynamic cluster supported by forecasting

by the master node to proactively partition data to match the worst-case demands of the current and the expected future workload. Figure 8a shows the results for this experiment. The workload on the x-axis was unchanged, but the query response times are more stable compared to those shown in Figure 7a.

On one hand, power and energy consumption, depicted in Figure 8b, are higher under low utilization levels than on a purely reactive cluster (Figure 7b), because WattDB powers up nodes in advance. On the other hand, average energy consumption per query is more predictable, as the query runtimes contain less variance. Furthermore, energy consumption at high utilization levels and at workload shifts is lower, compared to the cluster operated without forecasting data.

5   Conclusion and Outlook

While we already explored opportunities for energy proportionality in the storage layer in [12] and focused on energy-proportional query execution in [13], we have substantially refined and combined both approaches in this research work. In this contribution, we have demonstrated that it is possible to deploy a database system on a dynamic cluster of wimpy nodes. We have exemplified that we can trade energy consumption for query performance and vice versa by controlling the amount of data distribution and number of nodes available to process incoming queries. At the same time, we exhibited that a dynamic cluster is more energy efficient than a statically configured one, where nodes are underutilized—particularly at low load levels.

By keeping data in logical units, partitioned by primary key, WattDB is well suited for OLTP queries, where records are typically accessed by key. The dynamic partitioning enables quick and coherent re-distribution of the data without interrupting data access or putting high overhead on look-up structures. At the same time, the massive parallelism of the cluster nodes running OLTP queries does not interfere with concurrent OLAP queries, where typically large parts of the dataset have to be scanned to calculate aggregates over groups of records.

Moving data among nodes is a time-consuming task and cannot be done very frequently, i. e., on a per-second base. As our experiments indicate, it is crucial for query response times to proactively adjust the cluster to the anticipated workload. Fortunately, workloads typically follow often an easy-to-predict pattern, e.g., workdays are similar to each other, workloads in December keep rising for e-commerce back-end DBMSs, and so on. Therefore, we will focus on forecasting workloads in future experiments.

References

1. Barroso, L.A., Hölzle, U.: The case for energy-proportional computing. IEEE Computer 40(12), 33–37 (2007)
2. Barroso, L.A., Hölzle, U.: The Datacenter as a Computer: An Introduction to the Design of Warehouse-Scale Machines. Morgan & Claypool Publishers (2009)
3. Bernstein, P.A., Goodman, N.: Multiversion concurrency control—theory and algorithms. ACM Trans. Database Syst. 8(4), 465–483 (1983), http://doi.acm.org/10.1145/319996.319998
4. Elmagarmid, A.K.: A survey of distributed deadlock detection algorithms. SIGMOD Rec. 15(3), 37–45 (1986), http://doi.acm.org/10.1145/15833.15837
5. Härder, T., Reuter, A.: Principles of transaction-oriented database recovery. ACM Computing Surveys 15(4), 287–317 (1983)
6. Hudlet, V., Schall, D.: Measuring energy consumption of a database cluster. In: BTW. LNI, vol. 180, pp. 734–737 (2011)
7. Jung, H., Shim, H., et al.: LRU-WSR: Integration of LRU and Writes Sequence Reordering for Flash Memory. Trans. on Cons. Electr. 54(3), 1215–1223 (2008)
8. Lang, W., Harizopoulos, S., Patel, J.M., Shah, M.A., Tsirogiannis, D.: Towards energy-efficient database cluster design. PVLDB 5(11), 1684–1695 (2012)
9. Ou, Y., Härder, T., Jin, P.: Cfdc: A flash-aware replacement policy for database buffer management. In: SIGMOD Workshops, DaMoN, pp. 15–20 (2009)
10. Ou, Y., Härder, T., Schall, D.: Performance and Power Evaluation of Flash-Aware Buffer Algorithms. In: Bringas, P.G., Hameurlain, A., Quirchmayr, G. (eds.) DEXA 2010, Part I. LNCS, vol. 6261, pp. 183–197. Springer, Heidelberg (2010)
11. Park, S., Jung, D., et al.: CFLRU: A Replacement Algorithm for Flash Memory. In: CASES, pp. 234–241 (2006)
12. Schall, D., Härder, T.: Towards an energy-proportional storage system using a cluster of wimpy nodes. In: BTW. LNI, vol. 214, pp. 311–325 (2013)
13. Schall, D., Härder, T.: Energy-proportional query execution using a cluster of wimpy nodes. In: SIGMOD Workshops, DaMoN, pp. 1:1–1:6 (2013), http://doi.acm.org/10.1145/2485278.2485279
14. Schall, D., Hoefner, V., Kern, M.: Towards an enhanced benchmark advocating energy-efficient systems. In: Nambiar, R., Poess, M. (eds.) TPCTC 2011. LNCS, vol. 7144, pp. 31–45. Springer, Heidelberg (2012)
15. Seo, D., Shin, D.: Recently-evicted-first Buffer Replacement Policy for Flash Storage Devices. Trans. on Cons. Electr. 54(3), 1228–1235 (2008)
16. Szalay, A.S., Bell, G.C., Huang, H.H., Terzis, A., White, A.: Low-Power Amdahl-Balanced Blades for Data-Intensive Computing. SIGOPS Oper. Syst. Rev. 44(1), 71–75 (2010)
17. Tsirogiannis, D., Harizopoulos, S., Shah, M.A.: Analyzing the energy efficiency of a database server. In: SIGMOD Conference, pp. 231–242 (2010)

APSkyline: Improved Skyline Computation
for Multicore Architectures

Stian Liknes[1], Akrivi Vlachou[1,2], Christos Doulkeridis[3], and Kjetil Nørvåg[1]

[1] Norwegian University of Science and Technology (NTNU), Trondheim, Norway
[2] Institute for the Management of Information Systems, R.C. "Athena", Athens, Greece
[3] Department of Digital Systems, University of Piraeus, Greece
stianlik@gmail.com, {noervaag,vlachou,cdoulk}@idi.ntnu.no

Abstract. The trend towards in-memory analytics and CPUs with an increasing number of cores calls for new algorithms that can efficiently utilize the available resources. This need is particularly evident in the case of CPU-intensive query operators. One example of such a query with applicability in data analytics is the skyline query. In this paper, we present APSkyline, a new approach for multicore skyline query processing, which adheres to the partition-execute-merge framework. Contrary to existing research, we focus on the partitioning phase to achieve significant performance gains, an issue largely overlooked in previous work in multicore processing. In particular, APSkyline employs an angle-based partitioning approach, which increases the degree of pruning that can be achieved in the execute phase, thus significantly reducing the number of candidate points that need to be checked in the final merging phase. APSkyline is extremely efficient for hard cases of skyline processing, as in the cases of datasets with large skyline result sets, where it is meaningful to exploit multicore processing.

1 Introduction

The trend towards in-memory analytics and CPUs with an increasing number of cores calls for new algorithms that can efficiently utilize the available resources. This need is particularly evident in the case of CPU-intensive query operators. One example of such a query with applicability in data analytics is the skyline query. Given a set P of multidimensional data points in a d-dimensional space D, the skyline query retrieves all points that are not dominated by any other point. A point $p \in P$ is said to *dominate* another point $q \in P$, denoted as $p \prec q$, if (1) on every dimension $d_i \in D$, $p_i \leq q_i$; and (2) on at least one dimension $d_j \in D$, $p_j < q_j$. The *skyline* is a set of points $SKY_P \subseteq P$ which are not dominated by any other point in P. Computing the skyline query in a multicore context is an interesting research problem that has not been sufficiently studied yet.

Previous approaches to solve this problem are variants of the *partition-execute-merge* framework where the dataset is split into N partitions (one for each core), then the local skyline set for each partition is computed, and finally the skyline set is determined by merging these local skyline sets. However, the existing works largely overlook the important phase of partitioning, and naively apply a plain random partitioning of input data to cores. As also observed in [2], a naive partitioning approach may work well

S.S. Bhowmick et al. (Eds.): DASFAA 2014, Part I, LNCS 8421, pp. 312–326, 2014.
© Springer International Publishing Switzerland 2014

in case of cheap query operators on multicore architectures, while sophisticated partitioning methods may not be beneficial. Nevertheless, in the case of skyline queries which are CPU-intensive and especially for the case of "hard" datasets having an anticorrelated distribution, which are frequent in real applications [13] and also more costly to compute, the role of partitioning is significant. Skyline queries over anti-correlated data produce a skyline set of high cardinality and a naive partitioning produces local skyline sets containing uneven portions of the final skyline set and many local skyline points not part of the skyline set. In consequence, the merging cost is significantly higher than necessary, and the final result is prohibitively long processing times.

Motivated by this observation, in this paper, we present *APSkyline*, an approach for efficient multicore computation of skyline sets. In APSkyline, we employ a more sophisticated partitioning technique that relies on angle-based partitioning [15]. Since multicore skyline processing differs from skyline processing in other parallel environments, we apply all necessary adaptations of the partitioning technique for a multicore system, which combined with additional optimizations result in significant performance gains. Even though our partitioning entails extra processing cost, the degree of pruning that can be achieved in the execute phase incurs significant savings in the subsequent phases, thus reducing the overall execution time considerably. Furthermore, we show how the partitioning task can be parallelized and efficiently utilize all cores for this problem. We provide an extensive evaluation that compares APSkyline with the state-of-the-art algorithms for multicore skyline processing, which demonstrates that our approach outperforms its competitors for hard setups of skyline processing, which are also the cases where it is most meaningful to exploit multicore processing.

Summarizing, the main contributions of this paper include:

- A new algorithm for multicore skyline computation based on angle-based partitioning that complies with the partition-execute-merge framework.
- Different partitioning methods that result in improved performance depending on the nature of the underlying datasets.
- Novel techniques for parallelization of the partitioning task, in order to reduce the cost of applying sophisticated partitioning methods.
- An extensive evaluation that compares APSkyline with the state-of-the-art algorithms for multicore skyline processing, which demonstrates that our solution outperforms its competitors for hard setups of skyline processing.

The rest of this paper is organized as follows: Sect. 2 provides an overview of related work. In Sect. 3 we describe APSkyline and the partition-execute-merge framework, while in Sect. 4 we describe in detail how partitioning is performed in APSkyline. Our experimental results are presented in Sect. 5. Finally, in Sect. 6 we conclude the paper.

2   Related Work

Since the first paper on skyline processing in databases appeared [4], there have been a large number of papers on the topic, and for a brief overview of current state-of-the-art we refer to [6]. As shown in previous papers on main-memory skyline computing with no indexes available [9, 11, 12], the best-performing algorithm is SSkyline (also known as Best) [14], and this is also used as a building block in most approaches to multicore skyline processing.

For efficient multicore computing of skylines, there are two state-of-the-art algorithms, PSkyline [9, 11] and ParallelBNL [12]. As shown in [12], PSkyline usually performs better than ParallelBNL in the case of unlimited memory available, but in a memory-constrained context, ParallelBNL performs better.

The PSkyline algorithm [9, 11], is a divide-and-conquer-based algorithm optimized for multicore processors. In contrast to most existing divide-and-conquer (D&C) algorithms for skyline computation that divide into partitions based on geometric properties, PSkyline simply divides the dataset linearly into N equal sized partitions. The local skyline is then computed for each partition in parallel using SSkyline, and the local skyline results subsequently merged. In the evaluation by Im et al. [9] of PSkyline and parallel versions of the existing algorithms BBS and SFS, PSkyline consistently had the best utilization of multiple cores.

In [12], Selke et al. suggest a parallel version of BNL using a shared linked list for the skyline window. This is a straightforward approach, where a sequential algorithm is parallelized without major modifications. However, there are some issues related to concurrent modification of the shared list. Three variants of list synchronization are suggested in the article: continuous, lazy, and lock-free. With continuous locking, each thread will acquire a lock on the next node before processing, lazy locking [7] only locks nodes that should be modified (or deleted), while lock-free is an optimistic approach that does not use any locking. Lazy and lock-free variants need to verify that iterations have been correctly performed and restart iterations that fail. In their evaluation, the lazy locking scheme is shown to be most efficient, and in the evaluation in Sect. 5, we refer to parallel BNL using the lazy locking scheme.

Relevant for our work on multicore skyline computation is also parallel and distributed skyline processing [1, 8, 15]. However, these techniques are based on the assumption that communication between processing units (servers) is very costly compared to local processing, thus making them unsuitable for direct adaption in a multicore context. There has also been other recent work on computing skylines using specialized parallel hardware, e.g., GPU [3] and FPGA [16].

3   The Partition-Execute-Merge Framework

Parallel query processing typically follows the *partition-execute-merge* framework, where three distinct phases are used to split the work to multiple workers (cores or servers, depending on the parallel setting):

1. **Partitioning phase:** The input data is usually split (partitioned) in N non-overlapping subsets S_i following the concept of horizontal partitioning, and each subset is assigned to a worker for processing.
2. **Execute phase:** Corresponds to the actual processing of each individual partition and entails a query processing algorithm that operates on the input data of the partition and produces candidate results for inclusion in the final result set.
3. **Merging phase:** Depending on the type of query, the merging phase may discard some candidate results from the final result by comparing with candidates from

---

Algorithm 1. Partition-execute-merge

Require: P is the input relation, N is number of partitions
Ensure: SKY_P is the skyline of P
  $S_1, \ldots, S_N \leftarrow Partition(P, N)$
  $L_1, \ldots, L_N \leftarrow SSkyline(S_i, \ldots, S_N)$
  $SKY_P \leftarrow SkylineMerge(L_1, \ldots, L_N)$

---

other partitions. For example, in the case of a range query, all candidate results from each partition will be part of the final result set, thus the work of the merging phase is minimal. In contrast, in the case of the demanding skyline query, points from different partitions may dominate other points, which must be removed from the skyline set. Thus, the merging phase of skyline queries entails significant processing cost.

In the following, we describe in more detail each phase of the framework in the context of skyline query processing using APSkyline, and elaborate on the individual objectives of each phase. Algorithm 1 describes our overall approach.

3.1 Partitioning Phase

The objective of the partitioning phase is to divide input data in such a way that (a) individual partitions produce local skyline sets of low cardinality, and (b) the number of points in each partition is balanced. The former goal makes intra-partition query execution particularly efficient and also reduces the cost of the merging phase, while the latter avoids having an individual worker that is assigned a larger portion of work that may lead to delayed execution and initiation of the merging phase. We will describe this in more detail in Sect. 4.

3.2 Execute Phase

In this phase, an efficient skyline processing algorithm is employed to produce the local skyline set of a given data partition. In principle, index-based skyline algorithms are not applicable due to the additional overhead that would be required by index construction. As a consequence, candidate algorithms are selected from the non-indexed family of skyline algorithms. Obviously, the objective of this phase is to rapidly produce the local skyline set of the partition.

In our approach, each thread executes an instance of SSkyline [14] using its private partition as input to compute the local skyline independently. In short, SSkyline takes an input a dataset P containing $|P|$ tuples as input, and returns the skyline set of P. The skyline is computed using two nested loops and three indices: $head$, $tail$, and i. Intuitively the inner loop searches for the next skyline tuple, while the outer loop repeats the inner loop until all skyline tuples have been found. Fig. 1 shows an example run of SSkyline. In the first iteration $head$ points to the first tuple, i to the second, and $tail$ to the last. Confirmed skyline tuples are placed left of $head$ and colored black, confirmed non-skyline tuples are placed to the right of $tail$ and marked with gray, while tuples

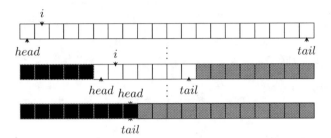

Fig. 1. SSkyline example. Black boxes are part of the skyline, white boxes are undetermined, and gray boxes are dominated (i.e., gray boxes are not part of the skyline).

in-between are still under consideration and will at some point be pointed to by *head* if they are part of the skyline. Each iteration of the outer loop confirms one skyline tuple by moving *head* to the right, and the inner loop may discard many non-skyline tuples by moving *tail* to the left. When *head* = *tail*, the algorithm terminates and returns all skyline tuples. At this point, head and tail point to the last skyline tuple, while *i* has been discarded.

3.3 Merging Phase

In the merging phase, the local skyline sets produced by the workers need to be combined in order to identify dominated points and exclude them from the final result set. In our approach, we perform parallel skyline merging using the *pmerge* algorithm from [11]. In short, the local skyline sets are merged to the final result one by one, starting with merging two of the local skyline set into a temporary skyline set, and then merging the other local skyline sets into this set one by one. The merging of the temporary skyline set and local skyline sets is performed in parallel, allowing threads to fetch tuples from the local skyline set in a round robin fashion until all local skyline sets have been merged.

4   Partitioning in APSkyline

In contrast to previous approaches that perform random partitioning naively, we intend to exploit geometric properties of the data during the partitioning. The objective of our approach is twofold: (1) reducing the time of the execute phase by assigning equally the points to threads and in a way that is beneficial for skyline processing, and (2) minimize the cost of the merge phase by reducing as much as possible the number of local skyline points. Improving the efficiency of the execute and merge phase, may entail additional processing cost for the partitioning phase, which will not dominate the overall processing cost for large and demanding datasets. In this section, we describe in more detail how to perform efficient angle-based partitioning for multicore skyline computation, using the partitioning techniques presented in [15] as a starting point.

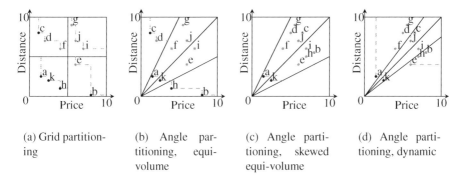

(a) Grid partition-
ing

(b) Angle par-
titioning, equi-
volume

(c) Angle parti-
tioning, skewed
equi-volume

(d) Angle parti-
tioning, dynamic

Fig. 2. Partitioning example using 4 partitions. To the left, (a) and (b) illustrate benefits of angle-based compared to grid-based partitioning (skyline points c, a, k, h, b). To the right, (c) and (d) illustrate benefits of dynamic angle-based partitioning (skyline points a and k).

In the following, in Sect. 4.1, we first describe basic partitioning schemes for sky-line processing, and then present our two variants of angle-based partitioning (*sample-dynamic* and *geometric-random*) that are appropriate for multicore systems and we employ in this work. Then, in Sect. 4.2, we show how to utilize parallel compute power not only in phases that compute the actual skyline, but also in the partitioning phase.

4.1 Partitioning Methods

Basic Techniques. Several partitioning techniques exist for parallel skyline computation. The most straightforward is simply dividing the dataset in N partitions at random without considering geometric properties of the data. In fact, this is the approach followed by the current state-of-the-art in multicore skyline processing [9, 11, 12]. However, this often results to large-sized local skyline sets with many points that do not belong to the final skyline set and leads to high processing cost for the execute and merge phase. In more details, if the points are randomly distributed, each partition follows the initial data distribution of the dataset, which means that only for some data distribution the skyline algorithm will perform efficiently [15].

A straightforward approach taking geometric properties into account is grid-based partitioning, however as illustrated in Fig. 2(a) this will result in data partitions that do not contribute to the overall skyline set, resulting in a lot of redundant processing. A more efficient approach is angle-based partitioning, first proposed by Vlachou et al. [15]. As can be seen in Fig. 2(b), (equi-volume) angle-based partitioning produces partitions whose local skyline sets are more likely to be part of the overall skyline set. For details on calculation of partition boundaries (equi-volume) based on the volume of the partitions we refer to [15].

Sample-Dynamic Partitioning. Even though equi-volume angle-based partitioning is applicable for multicore systems, as the boundaries are defined by equations and not by the data itself, the quality of the partitioning highly depends on the data distribution.

One possible problem of the equi-volume angle-based partitioning approach is the uneven sizes of produced partitions in case of skewed datasets, as illustrated in Fig. 2(c) where some partitions are empty. For this purpose, *dynamic* angle-based partitioning can be employed. Nevertheless, dynamic angle-based partitioning in multicore systems entails high processing cost and cannot be applied in a straightforward manner.

In dynamic partitioning, a maximum number of points for each partition is defined, as illustrated in Fig. 2(d). Initially, during partitioning, there is only one partition. During partitioning, when the maximum limit is reached for a partition, the actual partition is split into two. However, in a dynamic partitioning scheme, each partition split induces an expensive redistribution cost. More specifically, each time a partition is split $\frac{n_{max}}{2}$ (where n_{max} is the maximum partition size) or more tuples need to be moved from one memory location to another. This is reasonable in a parallel or distributed environment where IO-operations (including communication between nodes) are the dominating factor. However, in a shared-memory system, such a method requires a significant amount of the overall runtime. Therefore, we adopt a sample-based technique where a small sample of the data is used in order to determine the partitioning boundaries, before actual partitioning is performed.

In the *sample-dynamic* partitioning scheme, a configurable percentage s of the dataset is used to pre-compute the partitioning boundaries. To increase the likelihood of picking representative sample points, we propose to choose samples uniformly at random.

Geometric-Random Partitioning. In some cases, neither equi-volume nor sample-dynamic partitioning are able to achieve a fair workload, thus APSkyline will not be able to use the available parallel computing resources optimally. For example, input with equal rows may cause a skewed workload, and cannot be fairly distributed by geometric partitioning alone. To handle such cases, we propose a hybrid partitioning technique that utilizes geometric properties in combination with random partitioning in order to prioritize a fair workload.

The idea is to modify geometric partitioning schemes like equi-volume and sample-dynamic by having a maximum size of a partition, for example $|P|/N$. During partitioning, points mapped to a partition that is already full are allocated to another, randomly selected, partition. Note that in the case of geometric-random combined with sample-dynamic partitioning, the geometric-random modification is only applied during the final partitioning, not during processing of the sample for calculation of partition bounds.

4.2   Parallelism in the Partitioning Phase

In contrast to algorithms directed at parallel and distributed systems, a shared-memory algorithm needs to have a highly optimized partitioning phase. Thus, in the following we present techniques for parallelizing our partitioning methods. Algorithm 2 shows the parallel partitioning algorithm for the case of equi-volume partitioning. We use N threads in order to partition a relation into N partitions. Obviously, we need some way of determining which points each thread should distribute. This can be done in a round robin fashion, or using a linear partitioning strategy to define read boundaries without physically partitioning points. We use the linear strategy. To split data

---

Algorithm 2. ParallelEquiVolumePartitioning

Require: P is the input relation, N is number of partitions/threads
Ensure: $S_i, 0 \leq i < N$ contains the final partitioning
  $partitionBounds \leftarrow determinePartitionBounds()$
  parallel for $t = 0$ **to** $N - 1$ **do**   ▷ Distribute work over N threads
    for $k = t\frac{|P|}{N}$ **to** $(t + 1)\frac{|P|}{N}$ **do**
      $p \leftarrow P[k]$
      $i \leftarrow MapPointToPartition(p, partitionBounds)$
      $lock(S_i)$
      $S_i \leftarrow S_i \cup \{p\}$
      $unlock(S_i)$

Algorithm 3. ParallelDynamicPartitioning

Require: P is the input relation, N is number of partitions/threads s is sample fraction
Ensure: $S_i, 0 \leq i < N$ contains the final partitioning
  $nextPartID \leftarrow 0$
  $partitionBounds \leftarrow determineInititialBounds()$
  for $t = 0$ **to** $s|P|$ **do**   ▷ Sample size
    $p \leftarrow P[rand(0, |P| - 1|)]$ ▷ Random point
    $i \leftarrow MapPointToPartition(p, partitionBounds)$
    $S_i \leftarrow S_i \cup \{p\}$
    if S_i is full **then**
      $nextPartID \leftarrow nextPartID + 1$
      split S_i into S_i, S_{next}
      $partitionBounds \leftarrow determineNewBounds()$
  parallel for $t = 0$ **to** $N - 1$ **do**   ▷ Distribute work over N threads
    for $k = t\frac{|P|}{N}$ **to** $(t + 1)\frac{|P|}{N}$ **do**
      $p \leftarrow P[k]$
      $i \leftarrow MapPointToPartition(p, partitionBounds)$
      $lock(S_i)$
      $S_i \leftarrow S_i \cup \{p\}$
      $unlock(S_i)$

---

into 2 partitions, thread 1 processes tuples $[1, \ldots, |P|/2]$ and thread 2 processes tuples $[|P|/2 + 1, \ldots |P|]$. This ensures no need for locking during read. All threads place tuples into multiple shared collections (partitions) so here locks are needed. However, we emphasize that in contrast to locks used for concurrency control in, e.g., database systems, this is low-overhead locks, and typically implemented as spinlocks. In practice, some collisions will occur and threads will sometimes have to wait for locks. However, because the time needed to write result tuples is much smaller than the time required to perform the calculations for mapping a point to partition, lock waiting time is not significant.

In contrast to the equi-volume scheme where the boundaries are pre-determined, for the sample-dynamic partitioning scheme the boundaries must be computed first. In this case we suggest a two-step process, where partitioning boundaries are calculated sequentially using a certain percentage of the input relation (i.e., a sample) before parallel partitioning is performed, as described in Algorithm 3. In the first step, the bounds are computed and then in the second step the data points are partitioned. Finally, it is straightforward to modify Algorithm 3 for supporting the geometric-random partitioning scheme.

5   Experimental Evaluation

In this section, we present the results of the experimental evaluation. All our experiments are carried out on a machine with two Intel Xeon X5650 2.67GHz six-core processors, thus providing a total of 12 physical cores at each node. Each processor can run up to 12 hardware threads using hyper-threading technology. Each core is equipped with private L1 and L2 caches, and all cores on one die share the bigger L3 cache. All algorithms are implemented in Java.

5.1   Experimental Setup

Datasets. For the dataset P we employ both real and synthetic data collections. The synthetic sets are: (1) uniform (UN), (2) correlated (CO), and (3) anti-correlated (AC). For UN dataset, the values for all d dimensions are generated independently using a uniform distribution. The CO and AC datasets are generated as described in [4]. The real datasets are: (1) NBA which is a 5-dimensional dataset containing approximately 17K entries, where each entry records performance statistics for a NBA player, and (2) ZILLOW5D, a 5-dimensional dataset containing more than 2M entries about real estate in the United States based on crawled data (from www.zillow.com), and where each entry includes number of bedrooms and bathrooms, living area, lot area, and year built.

Algorithms. We compare our new algorithms against the current state-of-the-art multicore algorithms PSkyline [9, 11] and ParallelBNL [12] (implementations are based on source code made available by Selke et al. [12]). We implemented three variants of our algorithm (APSkyline) that differ based on the variant of angle-based partitioning employed:

- APSEquiVolume: Equi-volume angle-based partitioning.
- APSSampleDynamic: Sample-dynamic partitioning with sample size $s = 1\%$.
- APSSampleDynamic+: Sample-dynamic partitioning in combination with the geometric-random modification, with limit for each partition $\frac{|P|}{N}$, sample size $s = 1\%$.

Measurements. Our main metric is the runtime of each algorithm. Each test is executed 10 times and median values are used when reporting results. We perform one dry-run before taking any measurements in every experiment. Additionally, we measure variance, minimum, and maximum values in order to ensure that tests are sufficiently accurate. For synthetic datasets, new input is generated for each of the 10 executions, in order to factor out the effect of randomization.

Parameters. We vary the following parameters: dimensionality (d=2-5), cardinality ($|P|$=50K-15M), number of threads (partitions) (N=1-1024), and data distribution

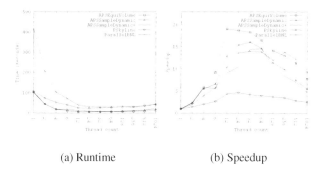

(a) Runtime   (b) Speedup

Fig. 3. Comparison of algorithms running with a different number of threads

(AC,CO,UN,NBA,ZILLOW5D). Default parameters are size of 5M 5-dimensional tuples of AC distribution. As also observed in [15], the anti-correlated dataset is most interesting, since the skyline operator aims to balance contradicting criteria [10, 15]. Moreover, anti-correlated distributions are closer to many real-life datasets according to [13]. Combined with the fact that multicore processing is typically employed for expensive setups of query processing, AC distribution is the setup that makes sense in the multicore context.

5.2 Experimental Results

Effect of Thread Count. Fig. 3(a) shows the results in the case of an AC dataset as we increase the number of threads from 1 to 1024. We expect to reach peak performance at 24 threads, which is the maximum number of hardware threads available (12 physical cores + hyper-threading). As number of threads increase beyond 24, we expect that performance will gradually decrease due to increased synchronization costs without additional parallel compute power. A main observation is that the APSkyline variants consistently outperform the competitor algorithms, which is a strong witness for the merits of our approach. For a low thread count ParallelBNL is inefficient compared to all the other D&C based algorithms. This is most likely due to the fact that D&C based algorithms use SSkyline for local skyline computation, which uses an array for storing results. The linked list used in ParallelBNL is not as memory-efficient as an array structure for sequential computations. Unsurprisingly, there is no significant performance difference between variations of APSkyline (Fig. 3(a)).

In Fig. 3(b), we measure the speedup of each algorithm. The speedup for each algorithm is relative to the same algorithm run with one thread, not to a common reference point, in order for the results to be easily compared to related work [11, 12]. We observe that APSkyline achieves super-linear speedup for up to 12 threads. Obviously, super-linear speedup cannot be explained by parallelism alone, we therefore attribute positive results to a combination of an increased parallel compute power, an increase in high-level cache (each core contributes with its private cache), and smaller input cardinalities for SSkyline. ParallelBNL has a good speedup as thread count is increased.

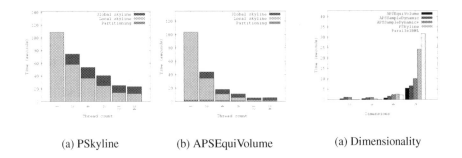

(a) PSkyline   (b) APSEquiVolume   (a) Dimensionality

Fig. 4. Segmented runtime for PSkyline and APSEquiVol-   **Fig. 5.**   Comparison   with
ume   varying dimensionality

This is in line with results presented in [12], and shows that a basic algorithm can be quite effective when parallel compute power and low-cost synchronization constructs are available. PSkyline shows a modest speedup compared to the other algorithms.

Fig. 4 shows the segmented runtime for PSkyline and APSEquiVolume (the two best-performing algorithms). Local skyline computation shows diminishing performance gains as the available parallel compute power increases. Partitioning refers to the time needed for the partitioning phase (Sec. 3.1), local skyline for the time of the execute phase (Sec. 3.2) and global skyline to the time of the merging phase (Sec. 3.3). Local skyline processing gets more efficient as the number of threads increase for both approaches, but in the case of APSEquiVolume the processing cost reduces more rapidly. This is due to the employed geometric partitioning, that alters the data distribution of the points assigned to each thread in a way that the required domination tests are fewer. In contrast, a random partitioning technique is expected to assign points that follow the anti-correlated data distribution to each thread, which leads to a more demanding local processing. Due to the geometric partitioning, APSEquiVolume results also to fewer local skyline tuples, which in turn leads to a smaller cost of the merging phase compared to PSkyline.

In summary, APSkyline clearly outperforms all other algorithms. When all cores are in use (24 threads), APSkyline is *4.2 times faster than PSkyline* and *5.2 times faster than ParallelBNL*. Fig. 4 also shows that the partitioning technique used in APSkyline is more expensive than the one used in PSkyline. Nevertheless, the time spent for partitioning is negligible compared to benefits attained in the subsequent phases. APSkyline is able to utilize parallel compute power in every phase as shown by Fig. 4.

Effect of Data Dimensionality. Fig. 5 shows the obtained results when increasing the number of dimensions from 2 to 5. For a dimensionality of 3 or less, ParallelBNL is the most efficient algorithm, while APSSampleDynamic+ is the least efficient. For a dimensionality of 4, APSEquiVolume and APSSampleDynamic outperform other algorithms by a small margin. Finally, all variations of APSkyline significantly outperform ParallelBNL and PSkyline for 5-dimensional datasets. In contrast to earlier algorithms, APSkyline scales well with dimensionality. It should be emphasized that size of the

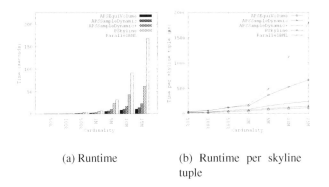

(a) Runtime

(b) Runtime per skyline tuple

Fig. 6. Comparison of algorithms running for varying data cardinality

skyline set increases rapidly with the dimensionality of the dataset, making skyline processing for higher dimensional data more demanding. This experiment verifies that for hard setups, as in the case of high dimensionality, our algorithms outperform the competitors and (more importantly) the benefit increases for higher values of dimensionality.

Effect of Data Cardinality. In Fig. 6(a), we examine how the algorithms scale for increased size of dataset. We observe that APSkyline achieves the best runtime for all input sizes. In particular, we notice that APSEquiVolume is *15.8 times faster than ParallelBNL* and *5.9 times faster than PSkyline* with 15M input tuples. Fig. 6(a) clearly depicts that APSkyline variants are robust when increasing the data cardinality. Moreover, this experiment shows that even for small-sized AC datasets (which contain a large percentage of skyline tuples), the setup is challenging thus both ParallelBNL and PSkyline demonstrate sub-optimal performance.

Fig. 6(b) shows the time used per skyline tuple by each algorithm. It is evident that ParallelBNL and PSkyline do not handle high dataset cardinalities well. Time used per skyline tuple should ideally be unchanged as cardinality increase. However, the merging phase requires pairwise comparisons between local skyline tuples, which in turn lead to a quadratic and not linear behavior of the skyline algorithms. As the number of local skyline tuples increase with the dataset cardinality, it is expected that the time per skyline tuple increases due to the pairwise comparisons. In this regard, APSkyline is quite successful. Processing time per skyline tuple increases very slowly compared to ParallelBNL and PSkyline, and this is an excellent example of the ability of an angle-based partitioning scheme to eliminate non-skyline tuples early.

Effect of Data Distribution. Fig. 7 shows the results for different synthetic data distributions and depicts the segmented runtime for each algorithm. First, we observe that skyline processing over the AC dataset is much more demanding for any algorithm (5000-30000 msec) than skyline processing over the UN or CO dataset (100-1300 msec). This verifies our claim that the AC dataset is a more typical use-case of multicore processing, which primarily targets the case of expensive query operators. AP-

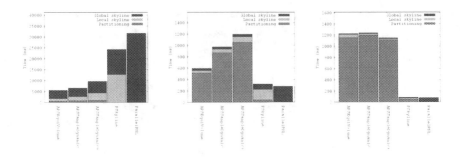

Fig. 7. Segmented runtime for synthetic datasets (anti-correlated to the left, independent in middle, and correlated to the right)

Skyline is significantly faster than its competitors for the challenging AC dataset. The most efficient variant (APSEquiVolume) is almost 5 times better than PSkyline and 6 times better than ParallelBNL. Despite the fact that APSkyline is significantly faster for hard setups (e.g., AC) where the number of skyline points is high, in the case of easy setups (e.g., UN or CO) the cost of partitioning of APSkyline dominates its runtime thus rendering the competitor algorithms more efficient.

When comparing the variants of APSkyline, we observe that equi-volume partitioning is most efficient. Due to the synthetic data generation, the dataset is fairly distributed by the equi-volume scheme for the AC and UN datasets. In addition, the partitioning is clearly more efficient for APSEquiVolume, since no sampling is used, and the partition boundaries are simply determined by equations independently of the underlying data. Thus, the small gain in the performance of computing the local and (global) overall skyline sets is dominated by the additional cost of the partitioning, rendering APSEquiVolume the best variant for synthetic datasets.

Real Datasets. Fig. 8 shows the obtained results for the real-life datasets. First, Fig. 8(a) depicts the statistics for the NBA dataset. Recall that this dataset is a small dataset containing approximately only 17K tuples. Combined with the fact that NBA is fairly correlated [5] means that NBA is not a very challenging case for skyline computation. Thus, Fig. 8(a) clearly depicts that the overhead of partitioning is too high compared to the total processing cost. PSkyline achieves the best performance for NBA, even though D&C based algorithms show similar performance for the local skyline computation and the merging phase. However, APSkyline spends much time partitioning and is therefore less efficient than PSkyline. Moreover, the equi-volume partitioning scheme is outperformed by all other partitioning schemes. This is attributed to the fact that the dynamic strategy is better tailored for real-life datasets which are not symmetric as the synthetic datasets. In case of symmetries, an equi-volume partitioning can distribute the work fairly. In contrast, in a real-life dataset, a (sample-)dynamic partitioning scheme is more robust than a fixed scheme that does not adapt to its input.

For the ZILLOW5D dataset, the best-performing algorithm is APSSampleDynamic+, thus demonstrating the usefulness of the geometric-random modification. ParallelBNL is outperformed in an order of magnitude by all other algorithms. In fact, APSSample-

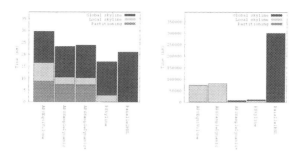

Fig. 8. Segmented runtime for real-life datasets (NBA to the left, ZILLOW5D to the right)

Dynamic+ is 36 times faster than ParallelBNL in this case. Additionally, we observe that PSkyline and APSSampleDynamic+ spend significantly less time in the local skyline computation phase than APSEquiVolume and APSSampleDynamic. The reason for APSEquiVolume and APSSampleDynamic spending so much time in the local skyline computation phase when processing the ZILLOW5D dataset is lack of a fair work distribution. We observed that most points were placed in only a few partitions, causing the majority of threads being idle. APSSampleDynamic and PSkyline do not have this problem, as they always divide data fairly. Nevertheless, APSSampleDynamic+ is able to outperform PSkyline with a factor of 1.4 using an angle-based partition technique in combination with random partitioning.

Discussion. In summary, our novel algorithm APSkyline based on angle-partitioning outperforms existing approaches for the most time-consuming datasets, while ParallelBNL and PSkyline excelled for simpler cases. By taking into account that the anti-correlated dataset is most interesting for skyline queries, since the skyline operator aims to balance contradicting criteria combined with the fact that multicore processing is typically employed for expensive setups of query processing, it highlights the value of our approach. Moreover, PSkyline performed slightly better than APSkyline variants for a small real-life dataset and was quite efficient for a large real-life dataset. Nevertheless APSSampleDynamic+ was able to reduce runtime by approximately 30% compared to PSkyline for the large real-life dataset. Thus, the APSkyline variants outperformed the existing approaches for all demanding datasets.

6   Conclusions

In this paper, we have presented APSkyline, a new approach for multicore skyline computing. The use of angle-based partitioning increases the degree of pruning that can be achieved in the execute phase, thus significantly reducing the number of candidate points that need to be checked in the final merging phase. As shown by our experimental evaluation, APSkyline is extremely efficient for hard cases of skyline processing, where we significantly outperform the previous state-of-the-art approaches.

References

1. Afrati, F.N., Koutris, P., Suciu, D., Ullman, J.D.: Parallel skyline queries. In: Proc. of ICDT (2012)
2. Blanas, S., Li, Y., Patel, J.M.: Design and evaluation of main memory hash join algorithms for multi-core CPUs. In: Proc. of SIGMOD (2011)
3. Bøgh, K.S., Assent, I., Magnani, M.: Efficient GPU-based skyline computation. In: Proc. of DaMoN (2013)
4. Börzsönyi, S., Kossmann, D., Stocker, K.: The skyline operator. In: Proc. of ICDE (2001)
5. Chan, C.-Y., Jagadish, H.V., Tan, K.-L., Tung, A.K.H., Zhang, Z.: On high dimensional skylines. In: Ioannidis, Y., Scholl, M.H., Schmidt, J.W., Matthes, F., Hatzopoulos, M., Böhm, K., Kemper, A., Grust, T., Böhm, C. (eds.) EDBT 2006. LNCS, vol. 3896, pp. 478–495. Springer, Heidelberg (2006)
6. Chomicki, J., Ciaccia, P., Meneghetti, N.: Skyline queries, front and back. SIGMOD Record 42(3), 6–18 (2013)
7. Heller, S., Herlihy, M.P., Luchangco, V., Moir, M., Scherer III, W.N., Shavit, N.N.: A Lazy Concurrent List-Based Set Algorithm. In: Anderson, J.H., Prencipe, G., Wattenhofer, R. (eds.) OPODIS 2005. LNCS, vol. 3974, pp. 3–16. Springer, Heidelberg (2006)
8. Hose, K., Vlachou, A.: A survey of skyline processing in highly distributed environments. VLDB J. 21(3), 359–384 (2012)
9. Im, H., Park, J., Park, S.: Parallel skyline computation on multicore architectures. Inf. Syst. 36(4), 808–823 (2011)
10. Morse, M., Patel, J.M., Jagadish, H.: Efficient skyline computation over low-cardinality domains. In: Proc. of VLDB (2007)
11. Park, S., Kim, T., Park, J., Kim, J., Im, H.: Parallel skyline computation on multicore architectures. In: Proc. of ICDE (2009)
12. Selke, J., Lofi, C., Balke, W.T.: Highly scalable multiprocessing algorithms for preference-based database retrieval. In: Kitagawa, H., Ishikawa, Y., Li, Q., Watanabe, C. (eds.) DASFAA 2010. LNCS, vol. 5982, pp. 246–260. Springer, Heidelberg (2010)
13. Shang, H., Kitsuregawa, M.: Skyline operator on anti-correlated distributions. PVLDB 6(9), 649–660 (2013)
14. Torlone, R., Ciaccia, P.: Finding the best when it's a matter of preference. In: Proc. of SEBD (2002)
15. Vlachou, A., Doulkeridis, C., Kotidis, Y.: Angle-based space partitioning for efficient parallel skyline computation. In: Proc. of SIGMOD (2008)
16. Woods, L., Alonso, G., Teubner, J.: Parallel computation of skyline queries. In: Proc. of FCCM (2013)

Greedy Filtering: A Scalable Algorithm
for K-Nearest Neighbor Graph Construction

Youngki Park[1], Sungchan Park[1], Sang-goo Lee[1], and Woosung Jung[2]

[1] School of Computer Science and Engineering, Seoul National University
{ypark,baksalchan,sglee}@europa.snu.ac.kr
[2] Department of Computer Engineering, Chungbuk National University
wsjung@cbnu.ac.kr

Abstract. Finding the k-nearest neighbors for every node is one of the most important data mining tasks as a primitive operation in the field of Information Retrieval and Recommender Systems. However, existing approaches to this problem do not perform as well when the number of nodes or dimensions is scaled up. In this paper, we present *greedy filtering*, an efficient and scalable algorithm for finding an approximate k-nearest neighbor graph by filtering node pairs whose large value dimensions do not match at all. In order to avoid skewness in the results and guarantee a time complexity of $O(n)$, our algorithm chooses essentially a fixed number of node pairs as candidates for every node. We also present a faster version of greedy filtering based on the use of inverted indices for the node prefixes. We conduct extensive experiments in which we (i) compare our approaches to the state-of-the-art algorithms in seven different types of datasets, and (ii) adopt other algorithms in related fields (similarity join, top-k similarity join and similarity search fields) to solve this problem and evaluate them. The experimental results show that greedy filtering guarantees a high level of accuracy while also being much faster than other algorithms for large amounts of high-dimensional data.

Keywords: k-nearest neighbor graph, similarity join, similarity search.

1 Introduction

Constructing a k-Nearest Neighbor (k-NN) graph is an important data mining task which returns a list of the most similar k nodes for every node [1]. For example, assuming that we constructed a k-NN graph whose nodes represent users, we can quickly recommend items to user u by examining the purchase lists of u's nearest neighbors. Furthermore, if we implement an enterprise search system, we can easily provide an additional feature that finds k documents most similar to recently viewed documents.

We can calculate the similarities of all possible pairs of k-NN graph nodes by a brute-force search, for a total of $n(n-1)/2$. However, because there are many nodes and dimensions (features) in the general datasets, not only does calculating the similarity between a node pair require a relatively long execution

S.S. Bhowmick et al. (Eds.): DASFAA 2014, Part I, LNCS 8421, pp. 327–341, 2014.

Y. Park et al.

time, but the total execution time will be very large. The inverted index join algorithm [2] is much faster than a brute-force search in sparse datasets. It is one of the fastest algorithms among those producing exact k-NN graphs, but it also requires $O(n^2)$ asymptotic time complexity and its actual execution time grows exponentially.

Another way to construct a k-NN graph is to execute a k-nearest neighbor algorithm such as locality sensitive hashing (LSH) iteratively. LSH algorithms [3, 4, 5, 11] first generate a certain number of signatures for every node. When a query node is given, the LSH compares its signatures to those of the other nodes. Because we have to execute the algorithm for every node, the graph construction time will be long unless one query can be executed in a short time.

As far as we know, NN-Descent [6] is the most efficient approach for constructing k-NN graphs. It randomly selects k-NN lists first before exploiting the heuristic in which a neighbor of a neighbor of a node is also be a neighbor of the node. This dramatically reduces the number of comparisons while retaining a reasonably high level of accuracy. Although the performance is adequate as the number of nodes grows, it does not perform well when the number of dimensions is scaled up.

In this paper, we present greedy filtering, an efficient, scalable algorithm for k-NN graph construction. This finds an approximate k-NN graph by filtering node pairs whose large value dimensions do not match at all. In order to avoid skewness in the results and guarantee a time complexity of $O(n)$, our algorithm selects essentially a fixed number of node pairs as candidates for every node. We also present a faster version of greedy filtering based on the use of inverted indices for the prefixes of nodes. We demonstrate the effectiveness of these algorithms through extensive experiments where we compare various types of algorithms and datasets. More specifically, our contributions are as follows:

- We propose a novel algorithm to construct a k-NN graph. Unlike existing algorithms, the proposed algorithm performs well as the number of nodes or dimensions is scaled up. We also present a faster version of the algorithm based on inverted indices (Section 3).
- We present several ways to construct a k-NN graph based on the top-k similarity join, similarity join, and similarity search algorithms (Section 4.1). Additionally, we show their weaknesses by analyzing their experimental results (Section 4.2).
- We conduct extensive experiments in which we compare our approaches to existing algorithms in seven different types of datasets. The experimental results show that greedy filtering guarantees a high level of accuracy while also being much faster than the other algorithms for large amounts of high dimensional data. We also analyze the properties of the algorithms with the TF-IDF weighting scheme (Section 4.2).

v_1	d_1 0.5	d_3 0.37	d_8 0.33	d_4 0.31	d_9 0.23	...
v_2	d_1 0.73	d_2 0.55	d_5 0.37	d_3 0.1	d_8 0.05	...
v_3	d_2 0.4	d_7 0.29	d_6 0.27	d_1 0.25	d_{10} 0.1	...
v_4	d_5 0.8	d_4 0.35	d_3 0.3	d_2 0.27	d_1 0.25	...
v_5	d_3 0.48	d_5 0.37	d_7 0.34	d_4 0.32	d_{10} 0.2	...

Fig. 1. Example of Greedy Filtering [1]: The prefixes of vectors are colored. We assume that the hidden elements (as described by the ellipse) have a value of 0 and $k = 2$.

2 Preliminaries

2.1 Problem Formulation

Let G be a graph with n nodes and no edges, V be the set of nodes of the graph, and D be the set of dimensions of the nodes. Each node $v \in V$ is represented by a vector, which is an ordered set of elements $e_1, e_2, ..., e_{|v|-1}, e_{|v|}$ such that each has a pair consisting of a dimension and a value, $\langle d_i, r_j \rangle$, where $d \in D$ and $0 \leq r_j \in \mathbb{R} \leq 1$. The values are normalized by L_2-norm such that the following equation holds:

$$\sum_{\langle d_i \in D, r_j \in \mathbb{R} \rangle \in V} r_j^2 = 1. \tag{1}$$

Definition 1 (k-NN Graph Construction): Given a set of vectors V, the k-NN graph construction returns for each vector $x \in V$, $argmax_{y \in V \wedge x \neq y}^k (sim(x, y))$, where $argmax^k$ returns the k arguments that give the highest values.

We use the cosine similarity as the similarity measure for k-NN graph construction. The cosine similarity is defined as follows:

$$sim(v_i \in V, v_j \in V) = \frac{v_i \cdot v_j}{\|v_i\| \|v_j\|} = v_i \cdot v_j. \tag{2}$$

Example 1: In Figure 1, if we assume that the hidden elements (as described by the ellipse) have a value of 0 and $k = 2$, the k-nearest neighbors of v_1 are v_2 and v_4, because $sim(v_1, v_2)$, $sim(v_1, v_3)$, $sim(v_1, v_4)$, and $sim(v_1, v_5)$ are 0.42, 0.13, 0.34, 0.28, respectively. The k-NN graph is obtained by finding k-nearest neighbors for every vector: $\{v_2, v_4\}, \{v_4, v_1\}, \{v_2, v_4\}, \{v_2, v_1\}, \{v_1, v_4\}$.

2.2 Related Work

The k-NN graph construction is closely related to other fields, such as the *similarity join*, *top-k similarity join*, and *similarity search* fields. First, we introduce the similarity join problem as follows:

Definition 2 (Similarity Join): Given a set of vectors V and a similarity threshold ϵ, a similarity join algorithm returns all possible pairs $\langle v_i \in V, v_j \in V \rangle$ such that $sim(v_i, v_j) \geq \epsilon$.

The inverted index join algorithm [2] for similarity join builds inverted indices for all dimensions and then exploits them to calculate the similarities. While it performs much faster than the brute-force search algorithm for sparse datasets, it still has to calculate all of the similarities between the vectors. On the other hand, prefix filtering techniques [2, 7, 8] effectively reduce the search space. They sort the elements of all vectors by their dimensions and set the prefixes such that the similarity between two vectors is below a threshold when their prefixes do not have a common dimension. As a result, we can easily prune many vector pairs by only looking at their prefixes.

The top-k similarity join is identical to the similarity join with regards to finding the most similar pairs. The difference is that it is based on a parameter k instead of ϵ. The top-k similarity join is defined as follows:

Definition 3 (Top-k Similarity Join): Given a set of vectors V and a parameter k, a top-k similarity join algorithm returns $argmax^k_{(x \in V, y \in V) \wedge x \neq y} (sim(x, y))$, where $argmax^k$ returns the k arguments that give the highest values.

The most common strategy is to calculate the similarities of the most *probable* vector pairs first and then to iterate this step until a stop condition occurs. For example, Kim *et al.* [9] estimates a similarity value ϵ corresponding to the parameter k, selects the most probable candidates, and continues to select candidates until it can be guaranteed that all vector pairs excluding those that were already selected as the candidates have similarity values of less than ϵ. Similarly, Xiao *et al.* [10] stops its iteration when it can be guaranteed that the similarity value of the next probable vector pair is not greater than that of any candidate that has been selected.

Lastly, we define the similarity search problem as follows. This is usually referred to as the approximate k-Nearest Neighbor Search problem (k-NNS) [3].

Definition 4 (Similarity Search): Given a set of vectors V, a parameter k and a query vector $x \in V$, the similarity search returns $argmax^k_{y \in V \wedge x \neq y} (sim(x, y))$, where $argmax^k$ returns the k arguments that give the highest values.

The Locality-Sensitive Hashing (LSH) scheme is one of the most common approaches for similarity search. It initially generates for every node a certain number m of signatures, $s_1, s_2, ..., s_{m-1}, s_m$. When a query vector is given, the LSH compares its signatures to those of the other vectors. Because the degree of signature match between two vectors indicates the similarity between them, we can find the k-nearest neighbors based on the results of the matches. The method used for generating signatures mainly depends on the target similarity measure. For example, Broder *et al.* [11] represents a *shingle* vector-based approach for the jaccard similarity measure; while Charikar *et al.* [5] presents a random hyperplane-based approach for cosine similarity.

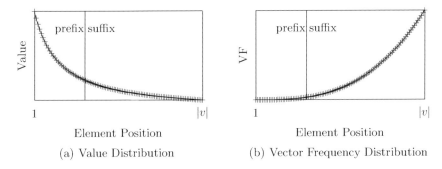

(a) Value Distribution

(b) Vector Frequency Distribution

Fig. 2. Typical Distributions After Performing Our Pre-Processing Steps

Note that the problem definitions in related work are analogous to that of k-NN graph construction such that the abovementioned solutions can also be applied to constructing k-NN graphs. For example, if we know all of the similarities between vectors by the inverted index join algorithm, we can obtain the k-NN graph by taking the most similar k vectors for each vector and throwing the rest away. However, these types of approaches do not perform well as the number of nodes or dimensions is scaled up. In Section 4, we discuss these issues in detail, present several ways to construct a k-NN graph based on the algorithms of these fields, and analyze their performance results.

A few approaches have been presented for the purpose of the k-NN graph construction. Examples include kNN-overlap [12], kNN-glue [12], and NN-Descent [6]. The process of the first two algorithms is as follows: (i) all of the vectors are divided into subsets, (ii) k-NN graphs for the subsets are recursively computed, and (iii) the results are formed into a final k-NN graph. Then a heuristic in which a neighbor of a neighbor of a vector is also the neighbor of the vector is additionally applied to improve the accuracy further. On the other hand, the NN-Descent algorithm exploits only the above heuristic, but in a more sophisticated way by adopting four additional optimization techniques: the local join, incremental search, sampling, and early termination techniques. The experimental results show that although NN-Descent outperforms the other algorithms in terms of accuracy and execution time, it does not perform well as the number of dimensions scales up [6]. In the rest of this paper, we present novel approaches to cope with this problem and compare the approaches to existing algorithms.

3 Constructing a k-Nearest Neighbor Graph

3.1 Greedy Filtering

Before presenting our algorithms, we introduce several distributions of datasets, as follows: Figure 2(a) shows the value of each element of a vector $v \in V$, where the value of the ith element is larger than that of $(i+1)^{th}$ element. Figure 2(b) shows the *vector frequency* of the ith element of a vector $v \in V$, where the vector frequency of the ith element is smaller than that of $(i+1)^{th}$ element. Let $dim(e)$ be the dimension of element e. Then the vector frequency of the element e is defined as the number of vectors that have the element of dimension $dim(e)$.

An interesting finding is that regardless of the dataset used, the dataset often follows distributions similar to those shown in Figure 2(a) and Figure 2(b) by performing some of the most common pre-processing steps. If we sort the elements of each vector in descending order according to their values, their value distributions will be similar to the distribution shown in Figure 2(a). Note that this pre-processing step does not change the similarity values between vectors. Furthermore, if we weigh the value of each element according to a scheme that adds weights to the values corresponding to sparse dimensions, such as IDF, TF-IDF, or BM25, then the vector frequency distributions will be similar to the distribution shown in Figure 2(b). These weighting schemes are widely used in Information Retrieval and Recommender Systems along with popular similarity measures [13].

Let v' and v'' be the prefix and suffix of vector $v \in V$, respectively. The prefix v' consists of the first n elements and the suffix v'' consists of the last m elements such that $|v| = n + m$. Then $sim(v_i \in V, v_j \in V) = sim(v_i', v_j') + sim(v_i', v_j'') + sim(v_i'', v_j') + sim(v_i'', v_j'')$. Our intuition is as follows: assuming that the elements of each vector follows the distributions shown in Figure 2 and that the prefix and suffix of each vector is determined beforehand, $sim(v_i, v_j)$ would not have a high value when $sim(v_i', v_j') = 0$ because, first, $sim(v_i', v_j'')$ would not have a high value; the vector frequencies of the elements in v_i' are so small that there is a low probability that v_i' and v_j'' have a common dimension. Although there are some common dimensions in their elements, the values of the elements in v_j'' are so small that they do not increase the similarity value significantly. For a similar reason, $sim(v_i'', v_j')$ and $sim(v_i'', v_j'')$ would not have a high value: The elements of high values have low vector frequencies and the elements of high vector frequencies have low values. When $sim(v_i', v_j') \neq 0$, on the other hand, $sim(v_i, v_j)$ would have a relatively high value because v_i' and v_j' have the highest values.

If we generalize this intuition, we can assert that two vectors are not one of the k-nearest neighbors of each other if their prefixes do not have a single common dimension. That is to say, we would obtain an approximate k-NN graph by calculating the similarities between vectors that have at least one common dimension in their prefixes. Note that the vector frequencies of the prefixes are so small that they usually do not have a common dimension. Thus we can prune many vector pairs without computing the actual similarities. Because this approach initially checks whether the dimensions of large values match, we call it *greedy filtering*.

Definition 5 (Match): Let v_i and v_j be the vectors in V, and let e_i and e_j be any of the elements of v_i and v_j, respectively. We hold that e_i and e_j match if $dim(e_i) = dim(e_j)$. We also say that v_i and v_j match if any $e_i \in v_i$ and $e_j \in v_j$ match.

Definition 6 (Greedy Filtering): Greedy filtering returns for each vector $x \in V$, $argmax^k_{y \in V \wedge x \neq y \wedge match(x,y)}(sim(x,y))$, where $argmax^k$ returns the k arguments that give the highest values, and $match(x,y)$ is true if and only if x and y match.

Example 2: Figure 1 shows an example of greedy filtering, where the prefixes are colored. If we assume that the hidden elements (described by ellipse) have a value of 0 and $k = 2$, greedy filtering calculates the similarities of $\langle v_1, v_2 \rangle$, $\langle v_1, v_3 \rangle$, $\langle v_1, v_4 \rangle$, $\langle v_1, v_5 \rangle$, $\langle v_2, v_3 \rangle$, $\langle v_4, v_5 \rangle$, and $\langle v_1, v_4 \rangle$, filters out $\langle v_2, v_4 \rangle$, $\langle v_2, v_5 \rangle$, $\langle v_3, v_4 \rangle$ and $\langle v_3, v_5 \rangle$, and returns k-nearest neighbors for every vector: $\{v_2, v_4\}$, $\{v_3, v_1\}$, $\{v_2, v_1\}$, $\{v_5, v_1\}$, $\{v_4, v_1\}$.

Note that the result of Example 2 is slightly different from that of Example 1, because greedy filtering is an approximate algorithm. If the dataset follows the distributions similar to those of Figure 2, the algorithm will be more accurate. In Section 4, we will justify our intuition in more detail.

3.2   Prefix Selection Scheme

If we set the prefix such that $|v'_i| = |v_i|$, $\forall v_i \in V$, then greedy filtering generates the exact k-NN graph though its execution time will be very long. On the other hand, if we set the prefix such that $|v'_i| = 0$, $\forall v_i \in V$, then greedy filtering returns a graph with no edges while the algorithm will terminate immediately. Note that the elapsed time of greedy filtering and the quality of the constructed graph depend on the prefix selection scheme. In general, there is a tradeoff between time and quality.

Assume that greedy filtering can find the approximate k-nearest neighbors for $v_i \in V$ if the number of matched vectors of v_i is equal to or greater than a small value μ. Then if for each vector v_i we find v'_i such that $|v'_i|$ is minimized and the number of matched vectors of v_i is at least μ, then we can expect a rapid execution of the algorithm and a graph of good quality.

Algorithm 1 describes our prefix selection scheme, where $e^j_{v_i}$ denotes the jth element of vector v_i and $dim(e)$ denotes the dimension of element e. In line 2, we initially prepare an empty list for each dimension. Because one list $L[d_i]$ contains vectors that have the dimension of d_i in their prefixes, if any list has the two different vectors v_i and v_j, then greedy filtering will calculate the similarity between them. In lines 7-8, we insert the vectors in R into the lists, meaning that we increase the prefixes of the vectors in R by 1. In lines 10-13, we estimate the number of matched vectors, denoted by M, for each $v_i \in R$. In lines 14-16, we check the stop conditions for each vector and determine which vectors will increase their prefixes again.

Note that Algorithm 1 sacrifices two factors for the performance and ease of implementation. First, it allows the duplicate execution of the brute-force search (lines 19-20 of Algorithm 1 and lines 3-5 of Algorithm 2). If the two vectors v'_i and v'_j have the d number of dimensions that match, we will calculate the d number of calculations of $sim(v_i, v_j)$. Although we can avoid these redundant computations by exploiting a hash table, this is not good for scalability in general. Second, we overestimate the value μ for a similar reason: if the two vectors v'_i and v'_j have

---

Algorithm 1. Greedy-Filtering (V, μ)

Input: a set of vectors V, a parameter μ
Output: k-NN queues Q

```
1  begin
2  │   L[d_i] ⟵ φ, ∀d_i ∈ D /* candidates */
3  │   C ⟵ 1 /* an iteration counter */
4  │   R ⟵ V /* vectors to be processed */
5  │   repeat
6  │   │   /* find candidates */
7  │   │   for v_i ∈ R do
8  │   │   │   add v_i to L[dim(e_{v_i}^C)]
9  │   │   /* check stop conditions */
10 │   │   for v_i ∈ R do
11 │   │   │   M ⟵ 0
12 │   │   │   for j ← 1 to C do
13 │   │   │   │   M ⟵ M + |L[dim(e_{v_i}^j)]|
14 │   │   │   if M ≥ μ or C ≥ |v_i| then
15 │   │   │   │   P[v_i] ⟵ C
16 │   │   │   │   remove v_i from R
17 │   │   C ⟵ C + 1
18 │   until |R| > 0
19 │   if default algorithm then
20 │   │   return Brute-force-search(L)
21 │   else
22 │   │   return Inverted-index-join(V, P)
```

---

d number of dimensions that match, then M increases by d instead of 1. Also, we calculate the value of M only once per iteration; this makes M slightly larger.

Example 3: Figure 1 shows the result of our prefix selection scheme when $\mu = 2$. Let $M(v)$ be the value M of the vector v. Initially, the prefix size of each vector is 1: $M(v_1) = M(v_2) = 1$ and $M(v_3) = M(v_4) = M(v_5) = 0$, because only $\langle v_1, v_2 \rangle$ match. As the next step, we increase the prefix sizes of all vectors by 1, as $M(v_i) < \mu, \forall v_i \in V$. Then $\langle v_1, v_5 \rangle$, $\langle v_2, v_3 \rangle$ and $\langle v_4, v_5 \rangle$ match. At this point, $M(v_1) = M(v_2) = M(v_5) = 2$ and $M(v_3) = M(v_4) = 1$. Thus we increase the prefix sizes of v_3 and v_4. As we continue until the stop condition is satisfied, our prefix selection scheme selects the colored elements shown in Figure 1.

Our prefix selection scheme has $O(|V||D|^2)$ time complexity, and the brute-force search has to compare each vector v to M number of other vectors. However, our preliminary results show that the prefix sizes are so small that we can regard D as a constant. Furthermore, we set the variable M close to μ; empirically, M is not twice as large as μ. Assuming that D is a constant and $M = 2\mu$, the total complexity of greedy filtering is $O(|V| + 2\mu|V|) = O(|V|)$.

Algorithm 2. Brute-force-search (L)

    Input: lists for dimensions L
    Output: k-NN queues Q
1 **begin**
2     $Q[v_i] \longleftarrow \phi, \forall v_i \in V$ /* empty queues */
3   **for** $d_i \in D$ **do**
4   compare all vector pairs $\langle v_x, v_y \rangle$ in $L[d_i]$
5   update the priority queues, $Q[v_x]$ and $Q[v_y]$
6   **return** Q

3.3 Optimization

Our algorithm uses a brute-force search a constant number of times for each vector. Because the execution times of the brute-force search highly dependent on the sizes of the vectors, it will take a relatively long time when a dataset contains very large vectors. For instance, experimental results show that the execution time of datasets whose vector sizes are relatively large, such as TREC 4-gram, is longer than that of other datasets.

We present one variation of greedy filtering, called *fast greedy filtering*. The main idea of this approach is that if $sim(v_i', v_j')$ is relatively high, then $sim(v_i, v_j)$ will also be relatively high. Then we can formulate an approximate k-NN graph by calculating the similarities between prefixes. Algorithm 1 and Algorithm 3 describe the process of this algorithm in detail: $e_{v_i}^j$ denotes the jth element of vector v_i, and $dim(e)$ and $value(e)$ denote the dimension and value of element e, respectively. In Algorithm 1, we set the prefix of each vector according to the abovementioned prefix selection scheme and invoke Algorithm 3. Then in lines 6-12 of Algorithm 3, we calculate the similarities between the current vector $v_i \in V$ and the other vectors already indexed and update the k-nearest neighbors of v_i and the indexed vectors. In lines 14-15, we put the current vector v_i into the inverted indices. Unlike greedy filtering, the execution time of fast greedy filtering is highly dependent on the number of dimensions and the vector frequencies of the datasets rather than the vector sizes.

Definition 7 (Fast Greedy Filtering): For each vector $x \in V$, fast greedy filtering returns $argmax_{y \in V \wedge x \neq y \wedge match(x,y)}^k \left(sim(x', y') \right)$, where $argmax^k$ returns the k arguments that give the highest values, and where $match(x, y)$ is true if and only if x and y match.

Example 4: If we apply fast greedy filtering to the example in Figure 1, the algorithm returns slightly different results: $\{v_2, v_5\}$, $\{v_3, v_2\}$, $\{v_2, v_1\}$, $\{v_5, v_1\}$, $\{v_4, v_5\}$ when $k = 2$.

Algorithm 3. Inverted-index-join (L, P)

Input: a set of vectors V, prefix sizes P
Output: k-NN queues Q

```
1 begin
2  │  Q[v_i] ⟵ φ, ∀v_i ∈ V /* empty queues */
3  │  I[d_i] ⟵ φ, ∀d_i ∈ D /* empty indices */
4  │  for v_i ∈ V do
5  │  │     /* verification phase */
6  │  │     C[v_j] ⟵ 0, ∀v_j ∈ V /* sim(v_i, v_j) = 0 */
7  │  │     for l ⟵ 1 to P[v_i] do
8  │  │     │    for ⟨v_j, r_j⟩ ∈ I[dim(e^l_{v_i})] do
9  │  │     │    └    C[v_j] ⟵ C[v_j] + r_j * value(e^l_{v_i})
10 │  │     for v_j ∈ V do
11 │  │     │    if C[v_j] > 0 then
12 │  │     │    └    update the queues, Q[v_x] and Q[v_y]
13 │  │     /* indexing phase */
14 │  │     for l ⟵ 1 to P[v_i] do
15 │  │     └    add ⟨v_i, value(e^l_{v_i})⟩ to I[dim(e^l_{v_i})]
16 │  return Q
```

---

4   Experiments

4.1   Experimental Setup

Algorithms. We considered eight types of algorithms for a comparison. Three algorithms among them adopt the similarity join (abbreviated by SIM) [2], the top-k similarity join (TOP) [10], and similarity search (LSH) [5] approaches. Two algorithms among them are NN-Descent (DE1) and Fast NN-Descent (DE2) [6], originally developed for the purpose of constructing k-NN graphs. The other two algorithms are greedy filtering (GF1) and fast greedy filtering (GF2) algorithms as proposed in this paper. Finally, we use the inverted index join (IDX) [2], which calculates all similarities with inverted indices, as a baseline algorithm. In all experiments, we set the number of neighbors to 10 (k=10).

We adopted the similarity join algorithm for k-NN graph construction. First, we implement the vector similarity join algorithm, *MM-join* [2], which outperforms the All-pairs algorithm [7] in various datasets. Then, we iterate the execution of the algorithm while decreasing the threshold ϵ by δ until either at least $s\%$ of vectors find k-nearest neighbors or until the elapsed time is higher than that of inverted index join. We used the following values in the experiments: $\epsilon = 1.00$ (the initial value), $\delta = 0.05$ and $s = 30$.

Adapting the top-k similarity join algorithm [10] for the k-NN graph construction process is along the same lines as that of the similarity join algorithm, except (1) we increase the parameter k at each iteration instead of decreasing δ,

Table 1. Datasets and Statistics

Dataset Statistics	$|V|$	$|D|$	Avg. Size	Avg. VF
DBLP	**250,000**	163,841	5.14	7.85
TREC	125,000	484,733	79.83	20.59
Last.fm	125,000	56,362	4.78	10.60
DBLP 4-gram	150,000	279,380	27.97	15.02
TREC 4-gram	50,000	**731,199**	**509.20**	34.82
Last.fm 4-gram	100,000	194,519	20.77	10.68
MovieLens	60,000	10,653	141.23	**795.44**

and (2) because the top-k similarity join algorithm uses *sets* as data structures, we need to transform the data structures into vectors and set new upper bounds for the suffixes of vectors using the prefix filtering and length filtering conditions. We set $s = 70$ for the top-k similarity join algorithm.

We also adopted the similarity search algorithm for k-NN graph construction by executing the algorithm N times. We used random hyperplane-based locality sensitive hashing for cosine similarity [5]. We cannot adopt other LSH algorithms, such as those in Broder et al. [11] or Gionis et al. [3], as they were originally developed for other similarity measures. We set the number of signatures for each vector to 100.

Datasets. We considered seven types of datasets for a comparison. There are two document datasets (DBLP[1] and TREC[2]), one text dataset that consists of music metadata (Last.fm[3]), three artificial text datasets (DBLP 4-gram, TREC 4-gram and Last.fm 4-gram), and one log dataset that consists of the movie ratings of users (MovieLens[4]). Note DBLP 4-gram, TREC 4-gram, and Last.fm 4-gram are derived from DBLP, TREC and Last.fm, respectively. We remove whitespace characters in the original vectors and extracted the 4-gram sequences from them. Table 1 shows their major statistics, where $|V|$ denotes the number of vectors and $|D|$ is the number of dimensions, *Avg. Size* denotes the average size of all vectors, and *Avg. VF* is defined as the average vector frequencies of all dimensions.

Evaluation Measures. We use the execution time and the scan rate as the measures of performance. The execution time is measured in seconds; it does not include the data preprocessing time, which accounts for only a minor portion. The scan rate is defined as follows:

$$Scan\ Rate = \frac{\#\ similarity\ calculations}{|V|\,(|V|-1)/2} \tag{3}$$

The *similarity calculation* expresses the exact calculation of the similarity between a pair. Thus, the brute-force search and the inverted index join always

---

[1] http://dblp.uni-trier.de/xml/

[2] http://trec.nist.gov/data/t9_filtering.html/

[3] http://www.last.fm/

[4] http://grouplens.org/datasets/movielens/

have a scan rate of 1, as they calculate all of the similarities between vectors. On
the other hand, fast greedy filtering has a scan rate of 0 because this algorithm
only estimates the degrees of similarity.

We use the level of accuracy as the measure of quality. Assuming that an
algorithm returns k neighbors for each vector, the accuracy of the algorithm is
defined as follows:

$$Accuracy = \frac{\#\ correct\ k\text{-}nearest\ neighbors}{k\,|V|} \tag{4}$$

Weighting Schemes. The value of each element can be weighted by the popular
weighting scheme, such as *TF-IDF*. Let v be a vector in V and e_i be an element
in v. Then, we define the TF-IDF as follows:

$$tf\text{-}idf(e_i, v) = \left(0.5 + \frac{0.5 * value(e_i)}{max\,\{value(e_j) : e_j \in v\}}\right) * \left(log\frac{|V|}{VF(e_i)}\right), \tag{5}$$

Here, $value(e)$ is the initial value of e. In the text datasets, the initial values are
the term frequencies; in the MovieLens dataset, the values are the ratings.

4.2   Performance Comparison

Comparison of All Algorithms. Figure 3 and Table 2 show the execution
time, accuracy, and scan rate of all algorithms with a small number of TREC
nodes. We do not specify the accuracy and scan rate of inverted index join in
Table 2, as its accuracy is always 1 and its scan rate is always 0. By the same
token, the scan rates of LSH and GF2 are left blank. We set $\mu = 300$ for our
greedy filtering algorithms.

The experimental results show that the greedy filtering approaches (GF1 and
GF2) outperform all other approximate algorithms in terms of the execution
time, accuracy and scan rate. The second best algorithms behind GF1 and GF2
are the NN-Descent algorithms (DE1 and DE2). However, as already descrbed
in work by Dong et al., the accuracy of the algorithms significantly decreases as
the number of dimensions scales up. The other algorithms require either a long
execution time or return results that are not highly accurate. The top-k similarity join and similarity join algorithms require a considerable amount of time to
construct k-NN graphs, and locality sensitive hashing based on random hyper-
planes requires many signatures (more than 1,000 signatures in our experimental
settings) to ensure a high level of accuracy.

Comparison of All Datasets. Table 3 shows the comparison results of the two
outperformers, greedy filtering and NN-Descent, over the seven types of datasets
with the TF-IDF weighting scheme. The results of their optimized versions are
specified within the parentheses. In this table, we define a new measure, *time*, as
the execution time divided by the execution time of inverted index join. We set
the parameters μ such that the accuracy of GF1 is at least 90%. The experimental
results show that GF1 outperforms the NN-Descent algorithms in all of the
datasets except for DBLP and MovieLens. Although the execution time of GF1

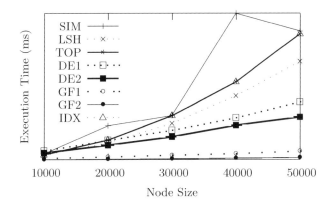

Fig. 3. Execution Time of All Algorithms (TREC)

Table 2. Accuracy and Scan Rate of All Algorithms

Node	Accuracy (TREC)							Scan Rate (TREC)						
	SIM	LSH	TOP	DE1	DE2	GF1	GF2	SIM	LSH	TOP	DE1	DE2	GF1	GF2
10K	0.00	0.01	0.68	0.54	0.43	**0.96**	0.65	0.05	-	0.06	0.27	0.19	**0.05**	-
20K	0.00	0.01	0.76	0.50	0.38	**0.95**	0.63	0.07	-	0.08	0.15	0.11	**0.03**	-
30K	0.00	0.01	0.76	0.48	0.36	**0.94**	0.62	0.05	-	0.08	0.10	0.08	**0.03**	-
40K	0.00	0.01	0.78	0.47	0.34	**0.93**	0.61	0.08	-	0.08	0.08	0.06	**0.02**	-
50K	0.00	0.01	0.79	0.46	0.33	**0.93**	0.59	0.05	-	0.08	0.07	0.05	**0.02**	-

is slower than the times required by the the NN-Descent algorithms for the two datasets, its accuracy is much higher.

Note that while fast greedy filtering exploits inverted index join instead of brute-force searches, it is not always faster than greedy filtering. Fast greedy filtering can be more effective in a dataset for which the vector sizes are relatively large and the number of dimensions and the vector frequencies are relatively small. For example, fast greedy filtering outperforms the other algorithms in the TREC 4-gram datasets, which have the largest average size.

Performance Analysis. Recall that before executing the greedy filtering algorithm, we utilize some of the most common pre-processing steps, as described in Section 3. First, we weigh the value of each element according to a weighting scheme, and then we sort the elements of each vector in descending order according to their values. Figure 4 shows the distributions of all of our datasets after performing these pre-processing steps. Note that the distributions after pre-processing are similar to those in Figure 2. Note also that Figure 4 and the experimental results are in accord with our intuition as presented in Section 3: For example, the distributions of DBLP 4-gram, Last.fm, and TREC in Figure 4 are very similar to those shown in Figure 2; moreover, the experimental results in Table 3 show that their execution time is better than those of the other

Table 3. Comparison of All Datasets

Datasets (TF-IDF)	DE1 (DE2)			GF1 (GF2)		
	Time	Accu-racy	Scan Rate	Time	Accu-racy	Scan Rate
DBLP	0.015 **(0.013)**	0.14 (0.11)	0.005 **(0.004)**	0.242 (0.076)	**0.98** (0.90)	0.102 (-)
TREC	0.190 (0.140)	0.43 (0.27)	0.030 (0.021)	0.030 **(0.009)**	**0.90** (0.56)	0.007 (-)
Last.fm	0.322 (0.189)	0.69 (0.68)	0.014 (0.008)	**0.063** (0.149)	**0.98** (0.80)	0.003 (-)
DBLP 4-gram	0.066 (0.046)	0.52 (0.34)	0.019 (0.011)	**0.004** (0.006)	**0.93** (0.59)	0.001 (-)
TREC 4-gram	0.228 (0.163)	0.60 (0.42)	0.066 (0.047)	0.106 **(0.003)**	**0.90** (0.48)	0.035 (-)
Last.fm 4-gram	1.207 (0.800)	0.65 (0.65)	0.013 (0.008)	**0.139** (0.204)	**0.90** (0.59)	0.001 (-)
MovieLens	0.244 (0.161)	0.55 (0.38)	0.046 (0.028)	0.302 **(0.013)**	**0.90** (0.19)	0.073 (-)

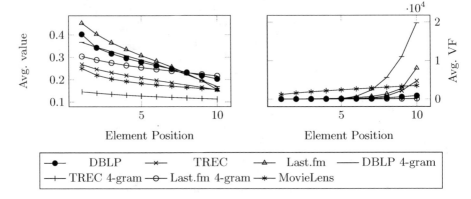

Fig. 4. Distributions of All Datasets

datasets. As another example, the distributions of MovieLens with TF-IDF are relatively less similar to those shown in Figure 2 in that the element position does not greatly affect the vector frequency. Thus, their performance is slightly worse than the performance levels of the other datasets.

5   Conclusion

In this paper, we present *greedy filtering*, an efficient and scalable algorithm for finding an approximate k-nearest neighbor graph by filtering node pairs whose large value dimensions do not match at all. In order to avoid skewness in the results and guarantee a linear time complexity, our algorithm chooses essentially a fixed number of node pairs as candidates for every node. We also present *fast greedy filtering* based on the use of inverted indices for the node prefixes. We demonstrate the effectiveness of these algorithms through extensive experiments in which we compare various types of algorithms and datasets.

The limitation of our approaches is that they are specialized for high dimensional sparse datasets, weighting schemes that add weight to the values corresponding to sparse dimensions, and cosine similarity measure. In future work, we would like to extend our approaches to more generalized algorithms.

Acknowledgment. This work was supported by the National Research Foundation of Korea(NRF) grant funded by the Korea government(MSIP) (No. 20110017480). This research was supported by Basic Science Research Program through the National Research Foundation of Korea(NRF) funded by the Ministry of Science, ICT & Future Planning(NRF-2012R1A1A1043769).

References

[1] Park, Y., Park, S., Lee, S., Jung, W.: Scalable k-nearest neighbor graph construction based on Greedy Filtering. In: WWW 2013, pp. 227–228 (2013)

[2] Lee, D., Park, J., Shim, J., Lee, S.-g.: An efficient similarity join algorithm with cosine similarity predicate. In: Bringas, P.G., Hameurlain, A., Quirchmayr, G. (eds.) DEXA 2010, Part II. LNCS, vol. 6262, pp. 422–436. Springer, Heidelberg (2010)

[3] Gionis, A., Indyk, P., Motwani, R.: Similarity search in high dimensions via hashing. In: VLDB 1999, pp. 518–529 (1999)

[4] Durme, B., Lall, A.: Online generation of locality sensitive hash signatures. In: ACL 2010, pp. 231–235 (2010)

[5] Charikar, M.: Similarity estimation techniques from rounding algorithms. In: STOC 2002, pp. 380–388 (2002)

[6] Dong, W., Moses, C., Li, K.: Efficient k-nearest neighbor graph construction for generic similarity measures. In: WWW 2011, pp. 577–586 (2011)

[7] Bayardo, R., Ma, Y., Srikant, R.: Scaling up all pairs similarity search. In: WWW 2007, pp. 131–140 (2007)

[8] Xiao, C., Wang, W., Lin, X., Yu, J., Wang, G.: Efficient similarity joins for near-duplicate detection. ACM Trans. on Database Systems 36(3), 15–41 (2011)

[9] Kim, Y., Shim, K.: Parallel top-k similarity join algorithms using MapReduce. In: ICDE 2012, pp. 510–521 (2012)

[10] Xiao, C., Wang, W.: X Lin, and H. Shang. Top-k set similarity joins. In: ICDE 2009, pp. 916–927 (2009)

[11] Broder, A., Glassman, S., Manasse, M., Zweig, G.: Syntactic clustering of the web. Computer Networks and ISDN Systems 29(8), 1157–1166 (1997)

[12] Chen, J., Fang, H., Saad, Y.: Fast approximate kNN graph construction for high dimensional data via recursive lanczos bisection. The Journal of Machine Learning Research 10, 1989–2012 (2009)

[13] Said, A., Jain, B., Albayrak, S.: Analyzing weighting schemes in collaborative filtering: Cold start, post cold start and power users. In: SAC 2012, pp. 2035–2040 (2012)

On Mining Proportional Fault-Tolerant Frequent Itemsets*

Shengxin Liu[1] and Chung Keung Poon[2]

[1] Department of Computer Science, City University of Hong Kong, Hong Kong
sx.liu@my.cityu.edu.hk
[2] Department of Computer Science and Center for Excellence,
Caritas Institute of Higher Education, Hong Kong
ckpoon@cihe.edu.hk

Abstract. Mining robust frequent itemsets has attracted much attention due to its wide applications in noisy data. In this paper, we study the problem of mining proportional fault-tolerant frequent itemsets in a large transactional database. A fault-tolerant frequent itemset allows a small amount of errors in each item and each supporting transaction. This problem is challenging since the anti-monotone property does not hold for candidate generation and the problem of fault-tolerant support counting is known to be NP-hard. We propose techniques that substantially speed up the state-of-the-art algorithm for the problem. We also develop an efficient heuristic method to solve an approximation version of the problem. Our experimental results show that the proposed speedup techniques are effective. In addition, our heuristic algorithm is much faster than the exact algorithms while the error is acceptable.

1 Introduction

Large transactional databases are ubiquitous nowadays, thanks to the advancement in computing technologies and the ever-expanding applications of computers in different areas. A transactional database, whether storing market-basket data [2], stock market data [19], web graph [7] or gene expression data [6], can be abstracted as a collection of transactions over a set of items. The *support* of an itemset (set of items) X, denoted as $sup(X)$, is the number of transactions that contain all the items in X. A *frequent itemset* is one with support larger than or equal to a given minimum threshold (minsup). The well-known Frequent Itemset Mining (FIM) problem [2] is to elicit all the frequent itemsets in a transactional database. It is a fundamental problem in data mining and is essential to many important tasks such as association rule discovery [3,10,23,9], subspace clustering [1,12], etc. It is also well-known that $sup(X)$ is anti-monotonic, i.e., $sup(X) \geq sup(X')$ whenever $X \subseteq X'$. Thus, a popular and successful approach [3] to the Frequent Itemset Mining Problem is to enumerate the candidate itemsets in a systematic manner (for example, a candidate X is generated before

---

* This work was fully supported by a grant from the Research Grants Council of the Hong Kong Special Administrative Region, China [Project No. CityU 122512].

S.S. Bhowmick et al. (Eds.): DASFAA 2014, Part I, LNCS 8421, pp. 342–356, 2014.

any of its supersets) and prune the generation of any superset of X whenever $sup(X)$ is found to be less than minsup.

Real-life data, however, are often noisy due to measurement errors, missing values or simply vagaries of human behavior. As observed in [22,15,14], large patterns are easily fragmented by noise so that traditional frequent itemset mining can only report the fragments, which are much less interesting and informative than the original "true" pattern. Thus there have been different proposals of fault-tolerant frequent itemset (FTFI) with the aim of enabling the recovery of the original patterns/itemsets.

1.1 Previous and Related Work

Yang et al. [22] introduced two versions of fault-tolerant frequent itemsets, called *strong* and *weak ETI* (*error-tolerant itemsets*). An itemset X is a strong ETI if there is a large enough set of transactions, *each* of which missing no more than a specified fraction of items in X. A weak ETI, on the other hand, just requires the existence of a set of transactions in which an *average* transaction misses no more than a specified fraction of items. Note that an ETI, strong or weak, may contain a free-rider (an item that is absent in many, or even all, of the supporting transactions) so long as other items in the ETI appear frequent enough in these transactions. The Approximate Frequent Itemset (AFI) of Liu et al. [14] avoids the problem by also requiring each item to be absent from at most a fraction of supporting transactions. Later, Poernomo and Gopalkrishnan [17] gave a more general definition of proportional FTFI's, allowing proportional error on the whole matrix (transaction-set × itemset) as well, thus putting both the AFI of [14] and the weak/strong ETI of [22] as special cases.

Another line of investigation pioneered by Pei et el. [15] studied fault-tolerant patterns in which each transaction is allowed to miss a constant number of items in the itemset while each item is required to appear in at least a constant number of its supporting transactions (i.e., each item is allowed to be absent in all but a constant number of its supporting transactions). Wang and Lee [21], and Koh and Yo [11] proposed more efficient algorithms. Later Zeng et al. [24] and Lee et al. [13] extended the constant transaction relaxation to the proportional transaction relaxation.

In general, by allowing some errors, an itemset can have a larger *fault-tolerant support* (FT-support, to be defined formally in Section 2), which is the number of supporting transactions under the relaxed criteria, and a longer itemset may then become qualified as a fault-tolerant frequent itemset. However, the mining of proportional FTFI poses big challenges. First, the FT-support, $sup_{FT}(X)$, of an itemset X is not anti-monotone (i.e., $X' \subseteq X$ does not imply $sup_{FT}(X') \geq sup_{FT}(X)$). Thus, the aforementioned approach of generating and pruning candidate itemsets [3] cannot be used here directly. Second, the computation of $sup_{FT}(X)$ given a candidate itemset X turns out to be NP-hard even for constant error tolerance for each item [16].

To avoid the first problem, there are definitions of FTFI's that have more stringent requirements to ensure that the anti-monotone property is valid. For example, Poernomo and Gopalkrishnan [16] studied constant FTFI's (i.e., only constant amount of error is allowed per transaction and per item). Cheng et al. [5] introduced more constraints on AFI while Seppänen and Mannila [18] introduced the concept of *dense itemset* which can be viewed as a recursive version of Yang et al's weak ETI. The *fuzzy frequent pattern* proposed by Wang et al. [20] provides fault-tolerance by giving each transaction with an initial budget to pay for subsequent insertions of missing items. When the remaining budget falls below a user-specified threshold, the transaction no longer contributes to the support. Besson et al. [4] proposed the notion of consistent and relevant patterns to reduce the number of patterns. For a discussion and comparison of various definitions of FTFI's, readers are referred to [8]. These definitions are either somewhat less intuitive or inflexible.

Liu et al. [14] overcame the first problem (without strengthening the definition) by deriving an anti-monotonic upper-bounding function for $sup_{FT}(X)$. Later, Poernomo and Gopalkrishnan [17] proposed better upper-bounding functions. They showed that one of their functions, $f_o(X)$, computes the tightest upper-bound on $sup_{FT}(X)$ when no other information about items outside X is known. They also gave a more efficient upper-bounding function $f_a(X)$.

For the second challenge, Poernomo and Gopalkrishnan [16] formulated the problem of $sup_{FT}(X)$ computation (for constant error tolerance) as an integer linear program (ILP) and [17] extended it to an iterative method that solves for proportional error tolerance. As solving an ILP is time-consuming, it is desirable to design more efficient algorithms. In view of the NP-hardness of the problem, a natural question that suggests itself is the approximation of the FT-support of an itemset. Liu et al. [14] proposed a heuristic method to approximate the FT-support of candidate AFI's.

1.2   Our Contribution

In this paper, we study the problem of mining proportional FTFI. We devise a number of speedup techniques to make the mining more efficient compared with the algorithm of Poernomo and Gopalkrishnan [17] (which will be referred to as the PG Algorithm in the sequel). First, we exploit the *periodic anti-monotone* property of $sup_{FT}(X)$ to obtain a more effective upper bounding function than that in [17]. For the computation of the FT-support of a given itemset X, we propose a technique of *super-frequent items* in order to reduce the number of calls to the ILP solver. Experimental results show that our speedup techniques are effective, especially when the data density is high.

We also consider an approximation version of the proportional FTFI mining problem and propose a greedy algorithm based on a natural idea combined with previous techniques for exact proportional FTFI mining. Our algorithm may slightly under-estimate the FT-support count and hence identify slightly fewer FTFI's. Experimental results show that our greedy algorithm is much faster than the exact algorithms while the error is acceptable. Moreover, they are much more accurate compared with the heuristics proposed in Liu et al. [14].

1.3 Paper Organization

In the next section, we state some useful definitions and explain the framework of the state-of-the-art algorithm ([17]). This is followed by a description of our speed up techniques and our approximation algorithm in sections 3 and 4 respectively. We present our experiments in section 5 and conclude the paper in section 6.

2 Preliminaries

A pattern P is a pair (X, Y) where X is an itemset and Y is a set of transactions.

Definition 1. *A pattern (X, Y) is said to satisfy a* transaction relaxation *with parameter α if for all transaction t in Y, t contains at least $\lceil (1-\alpha)|X| \rceil$ items of X, i.e., $|t \cap X| \geq \lceil (1-\alpha)|X| \rceil$. Similarly, it is said to satisfy an* item relaxation *with parameter β if for all item x in X, x is contained in at least $\lceil (1-\beta)|Y| \rceil$ transactions in Y. Finally, a pattern is said to satisfy a* pattern relaxation *with parameter γ if $\sum_{t \in Y} |t \cap X| \geq \lceil (1-\gamma)|X| \cdot |Y| \rceil$.*

Definition 2. *Given relaxation parameters α, β and γ, the (α, β, γ)-fault-tolerant support of X, denoted by $sup_{FT}(X, \alpha, \beta, \gamma)$, is the maximum size of Y in a pattern (X, Y) that satisfies all the relaxation requirements. Given a minimum support threshold $\sigma > 0$, X is said to be an (α, β, γ)-fault-tolerant frequent itemset (FTFI) if $sup_{FT}(X, \alpha, \beta, \gamma) \geq \sigma$.*

Definition 3. *Given a database $\mathcal{D}$, a minimum support threshold σ, as well as relaxation parameters α, β and γ, the* **Fault-Tolerant Frequent Itemset Mining Problem** *is to identify the set of all (α, β, γ)-fault-tolerant frequent itemsets.*

In this paper, we focus on the case of $\gamma = 1$, i.e., when there is no restriction on the pattern relaxation. For convenience and when the context is clear, we write $sup_{FT}(X, \alpha, \beta, 1)$ as $sup_{FT}(X)$.

In [14] and [17], upper-bounding functions $f_u(X)$ are derived such that (1) $sup_{FT}(X) \leq f_u(X)$ for all itemset X, and (2) $f_u(X)$ is anti-monotone. Then if we have determined that $f_u(X) < \sigma$ for some X, we can deduce that X and any of its superset cannot be FTFI's. Hence we need not compute $sup_{FT}(X')$ for any superset X' of X. With such bounding functions, a framework (shown in Algorithm 1) similar to that of [3] can still be applied.

The framework first initializes the set C of candidate itemsets in line 1. As long as C is not empty, the framework executes the while-loop (line 2-8). The framework enumerates an itemset X in C (line 3). The itemset enumeration of both PG and our algorithms is done by a depth-first search (DFS) on a standard enumeration tree (like the Eclat Algorithm [23]) so that an itemset is generated before any of its supersets. After processing the itemset X (line 4-7), the framework updates the set of candidate itemsets C in line 8.

In general, beside the itemset enumeration strategy, there are two other critical modules for bound checking and support counting. In line 4, the framework

---

Algorithm 1. The framework

1 *initialize(C)* ;   // C is a set of candidate itemsets
2 **while** *C is not empty* **do**
3      $X = nextItemset(C)$;
4   **if** $bound(X) \geq \sigma$ **then**
5          $sup = computeFTsupport(X)$; // compute $sup_{FT}(X)$
6   **if** $sup \geq \sigma$ **then**
7   **output** (X, sup) ; // X is an FTFI
8      $update(C, X)$;

---

computes an upper bound, $bound(X)$, on $sup_{FT}(X)$. Based on this $bound(X)$, the framework can decide whether X and its supersets can be safely pruned. Details of the bound functions will be explained in Section 2.1. The calculation of $sup_{FT}(X)$ (line 5) can be formulated as an ILP problem, details in Section 2.2. Finally, the framework outputs X if $sup_{FT}(X) \geq \sigma$ in line 7.

2.1   Bound Checking Module

For the bound checking in line 4 of Algorithm 1, Poernomo and Gopalkrishnan [17] considered two anti-monotonic functions, $f_o(X) = sup_{FT}(X, 1, \beta, 1)$ and $f_a(X) = sup_{FT}(X, 1, 1, \beta)$, both of which being upper bounds on $sup_{FT}(X)$. It is proved in [17] that $f_o(X)$ is the tightest possible upper bound on $sup_{FT}(X) = sup_{FT}(X, \alpha, \beta, 1)$ when no other information about the items not in X is known. However, computing $f_o(X)$ is quite time-consuming. For efficiency purpose, Poernomo and Gopalkrishnan [17] proposed using $f_a(X)$ instead.

2.2   Support Counting Module

Fix an itemset $X = \{a_1, \ldots, a_q\}$ and consider the problem of computing $sup_{FT}(X)$. Let $N(X) = \{t_1, \ldots, t_p\}$ be the set of transactions satisfying the transaction relaxation constraint. Clearly, in any fault-tolerant frequent patten (X, Y), the set of transactions Y must be a subset of $N(X)$ and we seek for the largest such Y that satisfies the item relaxation constraint. In [17], $N(X)$ is partitioned into equivalence classes, $C_1, C_2, \ldots, C_k$, based on their intersections with X. A class C is said to be in *level* ℓ if every transaction t in C misses ℓ items of X.

For each item $a_i \in X$ and equivalence class C_j, define $b_{i,j}$ to be the *cost* on item a_i when a transaction in C_j is chosen. That is, $b_{i,j} = 1$ if transactions in C_j does not contain item a_i; and $b_{i,j} = 0$ otherwise. With respect to itemset X, any transaction from the same equivalence class incurs the same cost. Define v_j to be the number of transactions chosen from C_j. Then any feasible solution Y must satisfy $\sum_j b_{i,j} \cdot v_j \leq \beta \sum_j v_j$ for all i. We can find a largest Y by solving an ILP that maximizes $\sum_j v_j$ subject to the above constraints.

Algorithm 2. $ComputeFTsupport(X)$

---

1 set s to be an initial upper bound on $sup_{FT}(X)$;
2 **do**
3   apply preprocessing;
4   solve the ILP specified by (1, 2, 3); denote the optimal solution by V^*;
5   **if** $|V^*| = s$ **then**
6 | V^* is optimal ;
7   **else**
8   // $|V^*| < s$
9   reduce s to $s = |V^*|$;
10 **while** V^* *not yet optimal*;
11 **return** $|V^*|$;

In [17], an iterative framework (Algorithm 2) is proposed in which the following ILP is solved:

$$\max \sum_j v_j \tag{1}$$

$$\text{s.t.} \sum_j b_{i,j} \cdot v_j \le \beta s \qquad \forall i \tag{2}$$

$$v_j \in \{0, \ldots, |C_j|\}. \tag{3}$$

Here s is a guess on the maximum $|Y|$. It is shown in [17] that if the initial s in line 1 is an upper bound on maximum $|Y|$, Algorithm 2 can compute $sup_{FT}(X)$ exactly. In [17], they set $s = sup_{FT}(X, \alpha, 1, \beta)$ initially.

Since solving ILP is time-consuming, the PG Algorithm employs a number of preprocessing techniques (line 3) to reduce the number of ILP calls and the number of variables in an ILP call. For example, they observed that transactions missing at most one item can always be included without losing the optimality of the final solution:

Lemma 1 (Property 6 in [17]). *Let C_1 and C_2 be any two equivalence classes corresponding to transactions in $N(X)$ that miss itemsets X_1' and X_2' of X respectively. Further assume that $X_1' \subset X_2'$. Then, there is an optimal solution to the ILP (1,2,3) in which $v_1 < |C_1|$ implies $v_2 = 0$.*

This lemma implies that one should consider transactions in increasing order of the number of missed items in X. In particular, in [17] all transactions in the (unique) class in level 0 are first chosen, followed by as many transactions from classes in level 1 as possible. If a constraint is already saturated by such a choice, we can set all other variables in that constraint to zero. In other words, we can remove these variables from our ILP, thus reducing the number of variables. We also consider the quantity $\lfloor \alpha|X| \rfloor$, which represents the maximum number of errors each transaction can have. If it is zero, this becomes the standard support counting problem, i.e., $sup_{FT}(X) = sup(X)$. If it is one, it suffices to consider only equivalence classes in level 0 and 1 due to Lemma 1. The optimal choice can be determined easily without solving an ILP.

3   Our Algorithm

We now describe our speedup techniques. We first observe and exploit the *periodic anti-monotone* property to derive an effective upper bounding function for $sup_{FT}(X)$ in Section 3.1. Then in Section 3.2, we propose a technique of *super-frequent items* in order to reduce the number of calls to the ILP solver. Our whole algorithm follows the framework as described in Algorithm 1, except that in line 5 our algorithm calls a modified version of support counting algorithm (Algorithm 4 in Section 3.2) instead of Algorithm 2.

3.1   Periodic Anti-monotone Property

Our bounding function $bound(X)$ uses a different approach based on the following simple lemma (proof omitted):

Lemma 2. *Let X' and X be itemsets where $X' \subseteq X$. If $\lfloor \alpha |X| \rfloor = \lfloor \alpha |X'| \rfloor$, then $sup_{FT}(X) \leq sup_{FT}(X')$.*

Now, consider the quantity $\lfloor \alpha |X| \rfloor$ where α is the transaction relaxation parameter. As we increase $|X|$ in unit steps, the bound will increase by 1 once every $1/\alpha$ steps approximately. Within a sequence of steps in which $\lfloor \alpha |X| \rfloor$ does not increase, Lemma 2 implies that the function $sup_{FT}(X)$ is still anti-monotone. We say that $sup_{FT}(X)$ possesses the *periodic anti-monotone* property. Thus, for those steps where $\lfloor \alpha |X| \rfloor = \lfloor \alpha |X'| \rfloor$, we use $sup_{FT}(X')$ as the upper bound for $sup_{FT}(X)$. For the other steps, we just use $f_a(X)$.

	a	b	c	d
1	1	1	1	1
2	1	0	1	1
3	1	0	1	0
4	0	1	1	1
5	0	1	0	1
6	0	0	0	1

Fig. 1. Example for Periodic Anti-Monotone Property

Consider an example in Figure 1. Each row (column) represents a(n) transaction (item). The entry is '1' if the corresponding transaction contains the corresponding item; '0', otherwise. Suppose $\alpha = \beta = 0.4$ and $\sigma = 4$. We want to derive a tight upper bound on $sup_{FT}(X)$, where itemset $X = abcd$ is extended from itemset $X' = abc$. According to the definition, $f_o(X) = sup_{FT}(X, 1, \beta, 1) = 5$ and $f_a(X) = sup_{FT}(X, 1, 1, \beta) = 6$. Because $\lfloor \alpha |X| \rfloor = \lfloor \alpha |X'| \rfloor$, X is *periodic anti-monotone*. Assume that we have already computed $sup_{FT}(X') = 3$. Then $bound(X) = 3 < \sigma$. We need not calculate $sup_{FT}(X)$ for X by using our derived upper bounding function $bound(X)$, while other bound checking functions proposed by Poernomo and Gopalkrishnan [17] cannot have such pruning.

In section 5, we will show that the above example is not a rare situation and the bound $sup_{FT}(X')$, whenever applicable, is tighter than $f_a(X)$ most of the

---

Algorithm 3. $bound(X)$

---

1 let X' be the parent of X;
2 **if** $\lfloor \alpha(|X|) \rfloor = \lfloor \alpha(|X'|) \rfloor$ **then**
3   // applying the periodic anti-monotone property
4   **return** $sup_{FT}(X')$;
5 **else**
6   **return** $f_a(X)$;

---

time. Thus our bounding function $bound(X)$ is more effective in reducing the number of calls to $sup_{FT}(X)$ computation. Moreover, it also provides a tighter initial guess for the computation of $sup_{FT}(X)$. The pseudocode of the bound checking module is shown in Algorithm 3. Note that the value of $sup_{FT}(X')$ required in line 4 can be obtained almost for no time since it has been computed before.

Now, we explain the update of C in line 8 of Algorithm 1. First, X is removed from C. Since $bound(X)$ is not anti-monotone, we cannot prune the children of X even if $bound(X) < \sigma$. Instead, we check if $f_a(X) < \sigma$. Recall that $f_a(X)$ is an anti-monotonic upper bound on $sup_{FT}(X)$. Therefore, we add the children of X into C only if $f_a(X) \geq \sigma$.

3.2   Super-Frequent Items Technique

We introduce a technique that makes use of *super-frequent* items (which are items present in a large fraction of transactions) to reduce ILP calls. This technique is especially effective when the dataset is dense.

Definition 4. *An item a is* super-frequent *w.r.t. X if item a is not contained in at most $\sigma\beta$ transactions of $N(X)$.*

Intuitively, if we extend an itemset X' into X by adding an item $a_{i'}$, there will be a new constraint for item $a_{i'}$ in constraints (2) (see Section 2.2): $\sum_j b_{i',j} \cdot v_j \leq \beta s$. If item $a_{i'}$ is super-frequent w.r.t. X, we can always treat the new constraint as satisfied. This is because any subset Y of $N(X)$ contains at most $\beta\sigma$ transactions that miss item $a_{i'}$. If X is an FTFI, we can find a Y with size at least σ. Hence the relative error for item $a_{i'}$ is at most β, meeting the item relaxation criteria. If in addition, $N(X) = N(X')$, we can conclude that $sup_{FT}(X) = sup_{FT}(X')$ without solving an ILP:

Theorem 1. *If an item a is super-frequent and $N(X) = N(X')$, then $sup_{FT}(X, \alpha, \beta, 1) = sup_{FT}(X', \alpha, \beta, 1)$.*

Proof. Suppose item a is super-frequent w.r.t. X and $N(X) = N(X')$. For any set $Y \subseteq N(X')$ such that (X', Y) satisfies all the relaxation requirements, $Y \subseteq N(X)$ also holds. Thus, the pattern (X, Y) satisfies the transaction relaxation by definition of $N(X)$. Since (X', Y) satisfies the item relaxation requirement, the item relaxation requirement is satisfied for each item in X'. The item relaxation for item a is also satisfied because a is super-frequent w.r.t. X.   □

Algorithm 4. $Modified\_ComputeFTsupport(X)$

---

1 **if** $isSuperfrequent(X)$ **then**
2 | **return** $sup_{FT}(X')$;
3 **else**
4 |__ **return** $ComputeFTsupport(X)$;

Instead of checking if $N(X') = N(X)$, we just need to see if they have the same size due to the lemma 3 (proof omitted).

Lemma 3. *Let X' and X be itemsets as defined above. Then either $N(X) \subseteq N(X')$ or $N(X') \subseteq N(X)$.*

Using our modified Algorithm 4, the exact $sup_{FT}(X)$ can be computed. Suppose the current candidate itemset X is generated from parent X' by adding item a. That is, $X = X' \cup \{a\}$. In line 1 we check if the new item a is super-frequent w.r.t. X and $|N(X)| = |N(X')|$. If this is the case, we directly obtain that $sup_{FT}(X) = sup_{FT}(X')$. Otherwise, if the above technique fails to produce the FT-support, we invoke the iterated ILP solver in line 4.

The aim of the *super-frequent* technique is to avoid calling the expensive ILP solver while we can still obtain $sup_{FT}(X)$. Moreover, there are more *super-frequent* items in dense dataset than that in sparse one. In section 5, our experimental results confirm that the *super-frequent* technique is effective, especially when the data density is high.

4   Estimating Fault-Tolerant Support

Since the running time of Algorithm 4 is dominated by ILP in line 4, both AFI and our greedy approximation algorithms try to estimate the FT-support without ILP. The following heuristic algorithms have the same structure as shown in Algorithms 1, 3 and 4 but we will modify line 4 of Algorithm 4 as described below.

4.1   AFI Heuristic Methods

The heuristics of Liu et al. [14] starts with set Y including all the transactions and examines the transactions one by one. Transactions with more misses are removed, beginning from those whose zeroes aligned with low density items. We call this heuristics **AFI-trans**. In order to speed up, we also modify the above procedure by removing unpromising transactions based on the equivalence classes. This modified variant is called **AFI-class**. These heuristics do not make use of the iterative framework to gradually zero-in to the correct FT-support (to be explained in next section).

4.2   Our Greedy Heuristic Methods

Inspired by Lemma 1 ([17]), a natural greedy heuristics goes as follows: Starting with an empty set Y, repeatedly add transactions into Y one by one in increasing order of the number of missed items of X until no more transactions can be added without violating some constraints.

This could be slow when there are many transactions, as in typical transactional databases. We speed up the process by considering the transactions class by class according to the equivalence classes $C_1, C_2, \ldots, C_k$. More precisely, we initially set all v_j's to 0. Then we consider the equivalence classes C_j in increasing order of their level with ties broken arbitrarily. Suppose the current class under consideration is C_j. We will include as many transactions from the class as possible without violating any constraint. We stop the process when either (1) no more classes can be added, or (2) all classes have been considered. Note also that this modification has implicitly incorporated the preprocessing techniques described in Section 2.2.

Besides Lemma 1, our greedy heuristic algorithm also makes use of the iterative framework used in Algorithm 2 for computing exact FT-support. Processing the transactions in the order mentioned above, a one-pass greedy algorithm has to decide whether the current transaction should be added into set Y or not immediately. The tricky part is, however, we have no idea about the final size of Y, exact or approximate. Suppose current transaction t violates some fault-tolerant criteria and we have to discard it. However, as we include more and more transactions into Y, we can tolerate more errors, reflected in the fault-tolerant criteria becoming less stringent. It is then possible for the transaction t to satisfy the new criteria at some point. However, as we just examine the transactions once, transaction t is excluded forever. Our iterative framework, on the other hand, scan the transactions multiple times. Starting with an initial upper bound on the size of Y, a greedy iteration obtains a tighter upper bound so that the next iteration has more accurate fault-tolerant criteria. Eventually, the iterative framework returns an underestimated (approximate) solution.

Our experimental results show that our greedy heuristics is much faster than our exact algorithm while its error is still acceptable and is much better than that of AFI heuristics [14].

5   Empirical Evaluation

In this section, we perform empirical evaluation of FTFI mining algorithms. We separate the experiments into two groups, one for the exact algorithms and one for the approximate algorithms. All the experiments are performed on a 64-bit machine equipped with Intel Core i7-3770 3.4GHz CPU and 16GB main memory. Moreover, all algorithms are implemented using C++ and compiled using g++.

The executable of the PG algorithm can be downloaded from the authors' website[1]. However we do not know the exact itemset enumeration strategy and

---

[1] http://www.cais.ntu.edu.sg/~vivek/pubs/ftfim09/

Table 1. Characteristics and Default Parameters of the Datasets

Dataset	# of Trans.	# of Items	Ave. Len.	Density	Min. Sup.	Trans. Relax.	Item Relax.
Kosarak	990,002	41,270	8.1	0.02%	0.3%	0.3	0.2
Pumsb	49,046	2,113	74	3.50%	90%	0.3	0.02
Mushroom	8,124	119	23	19.33%	20%	0.3	0.04
Chess	3,196	75	37	49.33%	70%	0.3	0.02

many other optimization details in their implementation. Therefore we implemented the PG Algorithm according to the description in [17] and aimed at demonstrating the effectiveness of our new techniques. We used the same linear programming solver *LPSolve*[2] package as in the PG implementations.

The datasets we use are classical benchmarks available from the FIMI repository[3]. We choose four real datasets based on the experimental study of previous work. There are one sparse dataset, Kosarak, one dense dataset, Chess, and two medium density datasets, Mushroom and Pumsb. We summarize some characteristics and the baseline settings of parameters for each dataset in Table 1. We perform 5 repeated runs per experiment and show the averages for all the tests.

5.1   Exact Algorithms

First, we study the efficiency of our exact algorithms under different parameter settings, namely, minimum support threshold σ, transaction relaxation α and item relaxation β. Then we present the effectiveness of our proposed techniques: the *periodic anti-monotone* technique and the *super-frequent items* technique.

Efficiency Comparisons. We compared four algorithms for the exact proportional FTFI mining problem. They are **PG**, which is our implementation of the PG Algorithm, **Periodic** and **Super**, which are **PG** augmented with the *periodic anti-monotone* bounding function and the *super-frequent items* technique respectively, and **All**, which incorporate both the *periodic anti-monotone* and *super-frequent items* techniques.

Figure 2 shows the running time of the algorithms with respect to σ, α and β for the datasets, Kosarak, Mushroom, Pumsb and Chess. In the first column of Figure 2, we observe that the running time of all the algorithms increases as the minimum support σ decreases. The second and third columns of Figure 2 show that a higher relaxation in α or β increases the running time of all the algorithms. Moreover, **PG** and **All** are respectively the slowest and fastest in most cases. It is reasonable because **All** utilizes both speed-up techniques while **PG** uses none of them.

When data density is not high, **Super** has similar behavior as **PG**. When density is high, however, **Super** performs well. **Periodic** potentially considers fewer candidate itemsets and gives a tighter upper-bound for Algorithm 2. Hence it improves the time performance in all datasets. We list the speedup ratio with respect to our baseline implementation **PG** in the Table 2.

---

[2] http://lpsolve.sourceforge.net/5.5/
[3] http://fimi.cs.helsinki.fi/data/

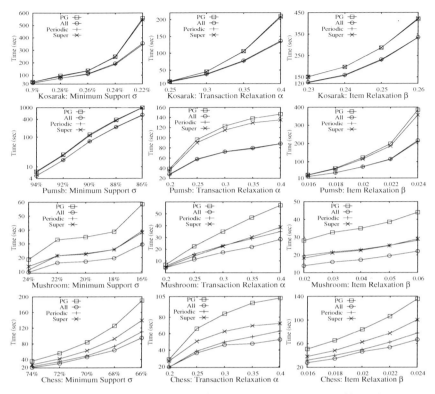

Fig. 2. Efficiency Performance of the Exact FTFI Mining Algorithms

Effectiveness of Proposed Techniques. Table 3 shows the percentage of FTFI's that can be identified using the *super-frequent items* technique (line 1-2 of Algorithm 4). Due to space limitation, we only show the average percentage on each dataset. We can observe that the *super-frequent items* technique performs well when the dataset density is high.

As mentioned in Section 3.1, there are two advantages for using *periodic anti-monotone* technique. We compare **PG** with **Periodic**. First, our bounding function $bound(X)$ is more effective in reducing the number of calls to $sup_{FT}(X)$. Second, a tighter initial upper bound for Algorithm 2 results in fewer iterations. From our experiments, we found that in more than 99% of the candidate itemsets X that are tested, our $bound(X)$ is tighter than $f_a(X)$. Table 4 shows the number of calls to $sup_{FT}(X)$ in the column "# Calls" and the average iterations per call in the column "# Iters.". Again, we take the average number on each dataset due to space limitation.

We can see that the number of calls and the number of iterations per call are reduced by using our proposed *periodic anti-monotone* technique most of the time. In the Kosarak dataset, however, the number of iterations is higher in **Periodic** than that of **PG**. One possible reason is that the number of calls in **Periodic** is 20% fewer than that in **PG** and those calls with fewer iterations are already pruned.

Table 2. Relative Time Performance Summary for Exact Algorithms

Dataset	Periodic	Super	All
Kosarak	1.20	1.06	1.16
Pumsb	1.65	1.06	1.62
Mushroom	1.52	1.59	1.97
Chess	1.66	1.35	1.90

Table 3. The Effect of Super-frequent Technique

Dataset	% Super-frequent
Kosarak	0.0%
Pumsb	0.6%
Mushroom	26.32%
Chess	16.7%

Table 4. The Effect of Periodic Anti-monotone Technique

Dataset	PG		Periodic	
	# Calls	# Iters.	# Calls	# Iters.
Kosarak	26,989.5	4.62	21,522.5	4.74
Pumsb	142,854	3.36	142,239.6	2.08
Mushroom	34,610.3	2.85	34,232.6	1.95
Chess	129,055.6	3.10	127,892.3	1.85

Table 5. Relative Time Performance for Approximate Algorithms

Dataset	AFI-trans	AFI-class	Greedy
Kosarak	1.76	4.53	4.09
Pumsb	4.34	20.92	9.20
Mushroom	7.28	17.10	14.47
Chess	9.91	16.12	8.92

5.2 Approximate Algorithms

As stated before, the greedy heuristics is always under-estimating the true FT-support of each itemset. Algorithm **AFI-trans** removes the unpromising transactions one by one following the procedure described by Liu et al. [14] while **AFI-class** improves the time performance by deleting unpromising transactions class by class.

We can see from Figure 3 that all greedy approaches are much faster than the **All**. Moreover, by removing unpromising transactions based on classes, our improved AFI greedy method **AFI-class** is much faster than the original one proposed by Liu et al. [14]. We also observe that **AFI-class** is slightly faster than **Greedy** since our method needs to perform a small number of iterations to get the final solution. In the Table 5, we present the ratio of the running time of our fastest exact algorithm **All** to that of the other greedy heuristic methods, i.e. **AFI-trans**, **AFI-class** and **Greedy**.

To compare the accuracy of different greedy heuristics, we present the percentage of FTFI's successfully identified with correct FT-support over the total number of FTFI's but excluding those that can be computed via preprocessing directly as described in Section 3.1.

Tables 6 and 7 compare the accuracy of different greedy heuristics under various parameter settings in the Kosarak and Mushroom datasets. We find that **AFI-trans** is more accurate than **AFI-class** since **AFI-class** more aggressively removes transactions if there exists a violation. Based on these results, we conclude that our algorithm **Greedy** is substantially more accurate than **AFI-class** and **AFI-trans**. An explanation is that the AFI-based heuristics decide if a transaction can be included without an accurate guess on the size of final Y while our greedy heuristics **Greedy** obtains better and better guesses iteratively and select the transactions more wisely.

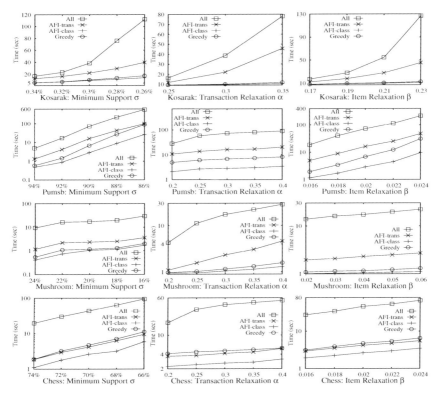

Fig. 3. Efficiency Performance of the Approximate FTFI Mining Algorithms

Table 6. Accuracy in Kosarak

σ	α	β	AFI-class	AFI-trans	Greedy
0.3%	0.3	0.2	8.29%	26.8%	77.2%
0.26%	0.3	0.2	22.1%	66.5%	91.8%
0.3%	**0.35**	0.2	4.7%	22.1 %	80.7%
0.3%	0.3	**0.23**	9.8%	35.5%	71.6%

Table 7. Accuracy in Mushroom

σ	α	β	AFI-class	AFI-trans	Greedy
20%	0.3	0.04	11.9%	28.1%	86.2%
16%	0.3	0.04	7.9%	35.4%	89.9%
20%	**0.4**	0.04	2.9%	19.9%	79.6%
20%	0.3	**0.06**	15.6%	32.5%	83.1%

6   Conclusion

In this paper, we propose techniques that substantially speed up the state-of-the-art algorithm for mining proportional FTFI. We also develop an efficient heuristic method to solve an approximation version of the problem. The efficiency and effectiveness of both algorithms are verified in our experimental evaluation. In the future, we plan to derive approximation algorithms with provable bounds on the error and consider the case of sparse datasets.

References

1. Agrawal, R., Gehrke, J., Gunopulos, D., Raghavan, P.: Automatic subspace clustering of high dimensional data for data mining applications. In: SIGMOD 1998, pp. 94–105 (1998)

2. Agrawal, R., Imieliński, T., Swami, A.: Mining association rules between sets of items in large databases. In: SIGMOD 1993, pp. 207–216 (1993)
3. Agrawal, R., Srikant, R.: Fast algorithms for mining association rules in large databases. In: VLDB 1994, pp. 487–499 (1994)
4. Besson, J., Pensa, R.G., Robardet, C., Boulicaut, J.-F.: Constraint-based mining of fault-tolerant patterns from boolean data. In: Bonchi, F., Boulicaut, J.-F. (eds.) KDID 2005. LNCS, vol. 3933, pp. 55–71. Springer, Heidelberg (2006)
5. Cheng, H., Yu, P.S., Han, J.: Approximate frequent itemset mining in the presence of random noise. In: Soft Computing for Knowledge Discovery and Data Mining, pp. 363–389 (2008)
6. Cong, G., Tung, K., Anthony, Xu, X., Pan, F., Yang, J.: FARMER: finding interesting rule groups in microarray datasets. In: SIGMOD 2004, pp. 143–154 (2004)
7. Dourisboure, Y., Geraci, F., Pellegrini, M.: Extraction and classification of dense implicit communities in the web graph. ACM Trans. Web 3(2), 7:1–7:36 (2009)
8. Gupta, R., Fang, G., Field, B., Steinbach, M., Kumar, V.: Quantitative evaluation of approximate frequent pattern mining algorithms. In: KDD 2008, pp. 301–309 (2008)
9. Han, J., Cheng, H., Xin, D., Yan, X.: Frequent pattern mining: current status and future directions. Data Mining and Knowledge Discovery 15(1), 55–86 (2007)
10. Han, J., Pei, J., Yin, Y.: Mining frequent patterns without candidate generation. In: SIGMOD 2000, pp. 1–12 (2000)
11. Koh, J.-L., Yo, P.-W.: An efficient approach for mining fault-tolerant frequent patterns based on bit vector representations. In: Zhou, L.-z., Ooi, B.-C., Meng, X. (eds.) DASFAA 2005. LNCS, vol. 3453, pp. 568–575. Springer, Heidelberg (2005)
12. Kriegel, H.-P., Kröger, P., Zimek, A.: Clustering high-dimensional data: A survey on subspace clustering, pattern-based clustering, and correlation clustering. TKDD 3(1), 1:1–1:58 (2009)
13. Lee, G., Peng, S.-L., Lin, Y.-T.: Proportional fault-tolerant data mining with applications to bioinformatics. Information Systems Frontiers 11(4), 461–469 (2009)
14. Liu, J., Paulsen, S., Sun, X., Wang, W., Nobel, A., Prins, J.: Mining approximate frequent itemsets in the presence of noise: algorithm and analysis. In: SDM 2006, pp. 405–416 (2006)
15. Pei, J., Tung, A.K.H., Han, J.: Fault-tolerant frequent pattern mining: Problems and challenges. In: DMKD 2001, pp. 7–12 (2001)
16. Poernomo, A.K., Gopalkrishnan, V.: Mining statistical information of frequent fault-tolerant patterns in transactional databases. In: ICDM 2007, pp. 272–281 (2007)
17. Poernomo, A.K., Gopalkrishnan, V.: Towards efficient mining of proportional fault-tolerant frequent itemsets. In: KDD 2009, pp. 697–706 (2009)
18. Seppänen, J.K., Mannila, H.: Dense itemsets. In: KDD 2004, pp. 683–688 (2004)
19. Sim, K., Li, J., Gopalkrishnan, V., Liu, G.: Mining maximal quasi-bicliques to co-cluster stocks and financial ratios for value investment. In: ICDM 2006, pp. 1059–1063 (2006)
20. Wang, X., Borgelt, C., Kruse, R.: Fuzzy frequent pattern discovering based on recursive elimination. In: ICMLA 2005, pp. 391–396 (2005)
21. Wang, S.-S., Lee, S.-Y.: Mining fault-tolerant frequent patterns in large databases. In: International Computer Symposium 2002 (2002)
22. Yang, C., Fayyad, U., Bradley, P.S.: Efficient discovery of error-tolerant frequent itemsets in high dimensions. In: KDD 2001, pp. 194–203 (2001)
23. Zaki, M.J., Parthasarathy, S., Ogihara, M., Li, W.: New algorithms for fast discovery of association rules. In: KDD 1997, pp. 283–286 (1997)
24. Zeng, J.-J., Lee, G., Lee, C.-C.: Mining fault-tolerant frequent patterns efficiently with powerful pruning. In: SAC 2008, pp. 927–931 (2008)

An Efficient K-means Clustering Algorithm
on MapReduce

Qiuhong Li[1,2], Peng Wang[1,2], Wei Wang[1,2], Hao Hu[1,2], Zhongsheng Li[3], and Junxian Li[1,2]

[1] School of Computer Science, Fudan University, Shanghai, China
[2] Shanghai Key Laboratory of Data Science, Fudan University
[3] Jiangnan Institute of Computing Technology
{09110240012,pengwang5,weiwang1,huhao,09110240011}@fudan.edu.cn,
lizhsh@yeah.net

Abstract. As an important approach to analyze the massive data set, an efficient k-means implementation on MapReduce is crucial in many applications. In this paper we propose a series of strategies to improve the efficiency of k-means for massive high-dimensional data points on MapReduce. First, we use locality sensitive hashing (LSH) to map data points into buckets, based on which, the original data points is converted into the weighted representative points as well as the outlier points. Then an effective center initialization algorithm is proposed, which can achieve higher quality of the initial centers. Finally, a pruning strategy is proposed to speed up the iteration process by pruning the unnecessary distance computation between centers and data points. An extensive empirical study shows that the proposed techniques can improve both efficiency and accuracy of k-means on MapReduce greatly.

1   Introduction

Clustering large-scale datasets becomes more and more popular with the rapid development of massive data processing needs in different areas [16]. With the rapid development of Internet, huge amount of web documents appear. As these documents contain rich semantics, such as text and medias, they can be represented as multi-dimensional vectors. To serve the searching and classification of these documents, we need efficient high-dimensional clustering technology. The clustering algorithms for massive data should have the following features:

- It should have good performance when the number of clusters is large. The number maybe more than several thousands, because many applications need to depict the feature of the data in a fine granularity. However, it will increase the cost significantly.
- It can deal with the high-dimensional data efficiently. Computing the distance between high-dimensional centers and data points is time consuming, especially when the large number of clusters is large.

S.S. Bhowmick et al. (Eds.): DASFAA 2014, Part I, LNCS 8421, pp. 357–371, 2014.

In this paper, we select k-means algorithm to resolve the clustering problem for massive high-dimensional datasets. The k-means algorithm has maintained its popularity for large-scale datasets clustering, it is among the top 10 algorithms in data mining [20]. The k-means algorithm is used in many applications such as multimedia data management, recommendation system, social network analysis and so on.

However, the execution time of k-means is proportional to the product of the number of clusters and the number of data points per iteration. Clustering the massive high-dimensional data set into a large number of clusters is time consuming, even executing on MapReduce. To solve this and other related performance problems, Alsabti et al. [3] proposed an algorithm based on the data structure of the k-d tree and used a pruning function on the candidate centroid of a cluster. However tree structure is inefficient for high-dimension space. For high-dimensional similarity search, the best-known indexing method is locality sensitive hashing (LSH) [15]. The basic method uses a family of locality-sensitive hash functions to hash nearby objects in the high-dimensional space into the same bucket.

In this paper, we use LSH from a different angle, instead of using it as an index. We use it to partition data points into buckets, based on which, the original data points is converted into the weighted representative points as well as the outlier points, named data skeleton. The benefit is two-folds. First, the number of data points used to initialize the centers is reduced dramatically, which can improve its efficiency. Second, during the iteration phase of k-means, we use it to prune off the unnecessary distance computation. In addition, the output of LSH is integer numbers, which makes it a natural choice to generate the "Key" for MapReduce.

A high-quality initialized centers are important for both accuracy and efficiency of k-means. Some initialization algorithms [5] exploit the fact that a good clustering is relatively spread out. Based on it, these approaches prefer the points far away from the already selected centers. However, they need to make k passes over the whole data set sequentially to find k centers, which limits their applicability to massive data: The scalable k-means++ [2] overcomes the sequential nature by selecting more than one centers at an iteration. However, it still faces huge computation when updating the weights for all points. In this paper, we propose an efficient implement of scalable k-means++ on MapReduce, which improve both efficiency and accuracy of center initialization.

Finally, a pruning strategy is proposed to speed up the iteration process of $k-means$. The basic idea is as follows. First, we use the representative points, p, to find the nearest center, say c. If their distance is apparently smaller than that between the representative points and the second nearest center, c is also the nearest center for all data points represented by p. Second, the locality property of LSH is used to make pruning. In other words, for a data point p, we only compute the distance between it and the small number of centers which stays in the buckets near to that of p.

Our contributions are as follows:

- The LSH-based data skeleton is proposed to find representative points for similar points, which can speed up the center initialization phase and iteration phase.
- We propose an efficient implement of scalable k-means++ on MapReduce, which improve both efficiency and accuracy of center initialization.
- A pruning strategy is proposed to speed up the iteration process of k-means.
- We implement our method on MapReduce and evaluate its performance against the implementation of scalable k-means++ in [2]. The experiments show we get better performance than scalable k-means++, both in initialization phase and iteration phase.

The rest of this paper is organized as follows. We discuss related work in Section 2. Section 3 gives the preliminary knowledge. We propose our LSH-based k-means method in Section 4. A performance analysis of our methods is presented in Section 5. We conclude the study in section 6.

2   Related Work

Clustering problems have attracted interests of study for the past many years by data management and data mining researchers. The k-means algorithm keeps popular for its simplicity. Despite its popularity, k-means suffers several major shortcomings such as the need of specified k value and proneness of the local minima. There are many variants of naive k-means algorithm. Ordonez and Omiecinski [18] studied disk-based implementation of k-means, taking into account the requirements of a relational DBMS. The X-means [19] extends k-means with efficient estimation of the number of clusters. Joshua Zhexue Huang [14] proposes a k-means type clustering algorithm that can automatically calculate variable weights. Alsabti et al. [3] proposed an algorithm based on the data structure of the k-d tree and used a pruning function on the candidate centroid of a cluster. The k-d tree fulfils the space partitioning and a partition is treated a unit for processing. The processing in batch can reduce the computation substantially.

The k-means algorithm has also been considered in a parallel environment. Dhillon and Modha [12] considered k-means in the message-passing model, focusing on the speed up and scalability issues in this model. MapReduce [10] as a popular massive-scale parallel data analysis model gains more and more attention and a lot of enthusiasm in parallel computing communities. Hadoop [1] is a famous open-source implementation of MapReduce model. There are many applications on top of Hadoop. Mahout [17] is a famous Apache project which serves as a scalable machine learning libraries, including the k-means implementation on Hadoop. Robson L.F.Cordeiro [8] proposed a method to cluster multi-dimensional datasets with MapReduce. Yingyi Bu proposes HaLoop [7], which is a modified version of Hadoop, and gives the implementation of k-means algorithm on it. Ene et al.[13] considered the k-median problem in MapReduce and gave a constant-round algorithm that achieves a constant approximation.

D. Arthur and S. Vassilvitskii propose k-means++ [5], which can improve the initialization procedure. Scalable k-means++ [6] is proposed by Bahman Bahmani and Benjamin Moseley, which can cluster massive data efficiently.

Yi-Hsuan Yang[21] presented an empirical evaluation of clustering for music search result. The dataset is sampled first to decide the partitions, then m mappers read the data, ignore the elements from the clusters found in the sample and send the rest to r reducers. r reducers use the plug-in to find clusters in the received elements and send the clusters descriptions to one machine which merges the clusters received and get the final clusters.

3   Preliminary Knowledge and Background

3.1   K-means Algorithm and Its Variants

In data mining, k-means is a method of cluster analysis which aims to partition n observations into k clusters in which each observation belongs to the cluster with the nearest distance. $X = \{x_1, x_2, ..., x_n\}$ be a set of observations in the d-dimensional Euclidean space. $\|x_i - x_j\|$ denote the Euclidean distance between x_i and x_j. $C = \{c_1, c_2, ..., c_k\}$ be k centers. We denote *the cost of X with respect to C* as

$$\phi_X(C) = \sum_{x \subset X} d^2(x, C) = \sum_{x \subset X} \min_{1 \le i \le k} \|x - c_i\|^2 \tag{1}$$

where $d^2(x, C)$ is the smallest distance between x and all points in C.

The goal of k-means is to find C such that the cost $\phi_X(C)$ is minimized. Clustering is achieved by an iterative process that assigns each observation to its closest center, constantly improving the centers according to the points assigned to each cluster. The process stops when a maximum number of iterations is achieved or when a quality criterion is satisfied. The quality criterion is generally set that $\phi_X(C)$ is less than a predefined threshold.

3.2   MapReduce and Hadoop

MapReduce [11] was introduced by Dean et. al. in 2004. It is a software architecture proposed by Google. The kernel idea of MapReduce is *map* and *reduce*. The *map* phase calls *map* functions iteratively, each time processing an key/value input record to generate a set of intermediate key/value pairs, and a *reduce* function merges all intermediate values associated with the same intermediate key. It is a simplified parallel programming model and it is associated with processing and generating large data sets. Hadoop implements a computational paradigm named MapReduce, where the application is divided into many small fragments of works, every fragment is processed by a *map* function first on any node in the cluster, then the results are grouped by the key and processed by a *reduce* function. Hadoop provides a distributed file system (HDFS) that stores data on the compute nodes.

3.3 Locality Sensitive Hashing

The indexing technique called locality sensitive hashing(LSH) [4] emerged as a promising approach for high-dimensional data similarity search. The basic idea of locality sensitive hashing is to use hash functions that map similar objects into the same hash buckets with high probability. LSH function families have the property that objects that are close to each other have a higher probability of colliding than objects that are far apart. More formally, assume S be the domain of objects, and D be the distance measure between objects.

Definition 1. *A function family $\mathcal{H}=\{h : S \to U\}$ is called $(r; cr; p_1; p_2)$- sensitive for D if for any $v; q \in S$*

- *if $v \in B(q, r)$ then $P_{rH}[h(q) = h(v)] \geq p_1$,*
- *if $v \notin B(q, cr)$ then $P_{rH}[h(q) = h(v)] \leq p_2$.*

In this paper, we adopt hash function in Eqution 1, considering its simplicity.

Different LSH families can be used for different distance functions. D. Datar et al [9] have proposed LSH families for l_p norms, based on p-stable distributions. When p is 2, it is Eclidean space. Here, each hash function is defined as:

$$h_{a,b}(v) = \left\lfloor \frac{a.v + b}{r} \right\rfloor \tag{2}$$

4 LSH-kmeans for Massive Datasets

4.1 Overview

In this session we propose LSH-kmeans which includes several optimization strategies for large-scale high-dimensional data clustering. Our goal is to optimize both the initialization phase and the iteration phase for k-means on MapReduce. First, we use LSH to get the data skeleton by which the similar points are reduced to a weighted point (Session 4.2). Secondly, we propose an efficient implement of scalable k-means++ on MapReduce, which improve both efficiency and accuracy of center initialization. Furthermore, we reduce the intermediate messages for the MapReduce implementation by adopting coarse granularity of input (Session 4.3). Thirdly, we make use of the low bound property of LSH to prune off the unnecessary comparisons which can guarantee the correctness of the clustering results (Session 4.4).

4.2 Data Skeleton

For massive datasets, it is quite time-consuming for k-means to compute the distance between any point and center pair, especially when k is large. One intuitive idea is that we can group the similar points together first, and so all points in the same group may belong to the same cluster with high probability. We illustrate it in Fig. 1. Let c_1 and c_2 are two center points, and p_1 and p_2 are two points. r_1, $r_{1'}$, r_2 and $r_{2'}$ are the distances between p_1, p_2 and c_1, c_2 respectively. d is the distane between p_1 and p_2. If we know that $r_1 < r_2$ and d is small, it is very likely that $r_{1'} < r_{2'}$. We can generalize it to Theorem 1.

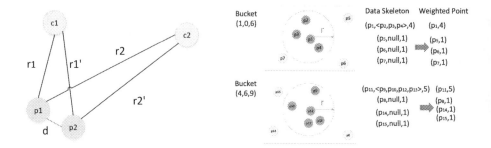

Fig. 1. Points and Centers   **Fig. 2.** Data Skeleton

Theorem 1. *Given c_1 and c_2 as two centers, p_1 and p_2 as two points with the distance d, r_1, $r_{1'}$, r_2 and $r_{2'}$ are the distances between p_1, p_2 and c_1, c_2 respectively. If $r_1 < r_2$ and $r_2 - r_1 > 2 * d$, then it holds that $r_{1'} < r_{2'}$.*

Proof. According to triangle inequality, we have $r_1 - d < r_{1'} < r_1 + d$ and $r_2 - d < r_{2'} < r_2 + d$. Therefore, we have

$$
\begin{aligned}
r_{1'} &< r_1 + d \\
&< r_2 - 2d + d & (r_2 - r_1 > 2 * d) \\
&= r_2 - d \\
&< r_{2'} & (r_2 - d < r_{2'})
\end{aligned}
$$

∎

According to Theorem 1, if we group similar points together, we have large probability to save some unnecessary distance computation. In this paper, we propose data skeleton to achieve this goal.

Each element of data skeleton is a triple $< p_r, L_p, weight >$, where L is the list of points which are represented by p. Formally, for point p, if the distance between p and p_r is smaller than a user-specified threshold, denoted as ε, we add it to L_p. $weight$ is set to $|L_p| + 1$. Data skeleton is computed as follows. First, we use m number of LSH functions to divide the original data set into buckets. All points in each point share the same hash values. From each bucket, we select the center of all points in it as a representative data point, p_r. The distances between all points in this bucket and p_r are computed, and all of them with distance smaller than ε are added into L_p. Other points are named as outlier points, each of which is also an element in data skeleton. L_p of an outlier point is an empty set, and the weight is set to 1. Fig. 2 illustrate it.

Considering that LSH may miss some similar points due to its probabilistic property, we use an iterative process to compute data skeleton with MapReduce. In each iteration, we use a different set of hash functions to find the representative points and outlier points. For the first iteration, the input is the whole dataset. For the rest iterations, the input is the outlier points in the last round. m hash functions in each iteration are selected independently.

Note that, different from the traditional k-means, in which all points have equal weights, in data skeleton, different points have different weights. For representative point, the more points it represents, the higher the weight. For outlier points, the weight is 1. We will use the weighted points to initialize the centers in next section.

4.3 Improving the Seeding Quality

Improving the initialization procedure of k-means is quite important in terms of the clustering quality. The state-of-the-art k-means variants are k-means++ and k-means‖. We introduce them respectively.

K-means++ and K-means‖. The k-means++ is proposed by Arthur and Vassilvitskii[7], which focuses on improving the quality of the initial centers. The main idea is to choose the centers one by one in a controlled fashion, where the current set of chosen centers will stochastically bias the choice of the next center. The sampling probability for a point is decided by the distance between the point and the center set(Line 3). The distances are considered as weights for sampling. After an iteration, the weights should be changed because of the new centers added into the center set. The details of k-means++ are presented in Algorithm 1. The advantage of k-means++ is that the initialization step itself obtains an $(8 \log k)$-approximation to the optimization solution in expectation. However, its inherent sequential nature makes it unsuitable for massive data set.

Algorithm 1. k-means++ initialization

Require: X : the set of points, k: number of centers;
 1: $C \leftarrow$ sample a point uniformly at random from X
 2: **while** $|C| \leq k$ **do**
 3:     Sample $x \in X$ with probability $\frac{d^2(x,c)}{\phi_X(C)}$
 4:     $C \leftarrow C \bigcup x$
 5: **end while**

The second variant of k-means is k-means‖, which is the underlying algorithm for the scalable k-means++ in [6]. The k-means‖ improves the parallelism of k-means++ by selecting l centers at one iteration. The k-means‖ picks an initial center and computes the initial cost of the clustering. It then proceeds in $\log \psi$ iterations. In each iteration, given the current set of centers C, it samples each x with probability $\frac{l \cdot d^2(x,c)}{\phi_X(C)}$ and obtain l new centers. The sampled centers are then added to C, the quantity $\phi_X(C)$ updated, and the iteration continued. The details are presented in Algorithm 2.

The MapReduce implementation of k-means‖ is given in [6], which is called the scalable k-means++. In the rest of this paper, for simplicity, we denote the MapReduce implementation of k-means‖ as k-means++ directly.

Algorithm 2. k-means$\|$ initialization

Require: X : the set of points,l: the number of centers sampled for a time, k: number of centers;

1: $C \leftarrow$ sample a point uniformly at random from X
2: $\psi \leftarrow \phi_X C$
3: **for** $o(\log \psi)$ **do**
4:     Sample C' each point $x \in X$ with probability $\frac{l \cdot d^2(x,C)}{\phi_X(C)}$
5:     $C \leftarrow C \bigcup C'$
6: **end for**
7: For $x \in C$, set w_x to be the number of points in X closer to x than any other point in C
8: Recluster the weighted points in C into k

Our Approach. We improve k-means++ with the help of data skeleton. There are two major improvements. First, we use the weighted points in data skeleton to sample the centers. The advantage of sampling centers on the weighted points produced from the data skeleton is that we can acquire a more spread-out center set because the close points are treated as a weighted point. Second, we propose a new sampling implementation on MapReduce.

Our approach includes three steps:

- **Step** 1: Data partitioning and weight initialization
- **Step** 2: Sampling l centers
- **Step** 3: Updating the weights

In first step, we divide the points in data skeleton into $|B|$ disjoint blocks, $(B_1, B_2, \cdots, B_{|B|})$ randomly. Then a random data point is selected as the first center in C. Based on it, we compute the initial sampling weights both each block as well the all points in it. Each block has a unique ID, which is a number within the interval $[1, |B|]$. For each block, the initial sampling weight is the sum of the weights for points in this block. The sampling weight for a weighted point $< x, w_x >$ is $w_x * d^2(x, C)$, denoted as wp_x. It is the smallest distance between x and all centers in C, multiplied by w_x. The sampling weight for block B_i is defined as $\sum_{x \in B_i} wp_x$, denoted as wb_i. Moreover, we compute $\phi_x(C)$, which is the sum of all wb_i's.

A MapReduce job is utilized to partition the data. In Map phase, each point in data skeleton is assigned to a block randomly and the distances for this point to the current centers are calculated. In Reduce phase, we compute the sum of the weights for the points in this block.

Step 2 and 3 are an iterative process. In each round, step 2 generate l centers, and step 3 update the weights of block and points. The iteration continues until we obtain k centers.

In second step, we generate l random numbers randomly within the range $(0, \phi_X(C)]$. For the i-th random number num_i, we can get the corresponding block and points. The corresponding block is B_{n_b} such that $\sum_{j<n_b} wb_j < num_i \leq \sum_{j \leq n_b} w_j$. Within block B_{n_b}, the point p_{n_p} which satisfies

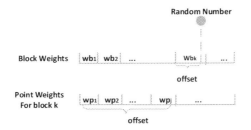

Fig. 3. Sample blocks and points

$$\sum_{j<n_b} wb_j + \sum_{j<n_p} wp_j < num_i \leq \sum_{j<n_b} wb_j + \sum_{j\leq n_p} wp_j$$

is selected as the i-th center in this iteration. Fig. 3 illustrates it. Also, a MapReduce job is executed in step 2.

Since l new centers are added, the weights for both the block and the points are changed. In third step, we update the weights for blocks and points. This can be implemented by a MapReduce job only with Map phase to update the weights for blocks and for points.

4.4 Pruning Unnecessary Comparisons Using LSH

After k centers are initialized, the iterative phase is conducted to adjust the centers. When k is large, and the data points are high-dimensional, this phase is very time consuming. In this section, we propose a pruning strategy to prune off the unnecessary distance comparisons, which contains two types of pruning. Next, we introduce them respectively.

The first pruning strategy is based on Theorem 1 to reduce the number of data points which needs to find the nearest centers. For each representative point in data skeleton p_r, we first compute its distance to all centers. Let c_1 is the nearest center and c_2 is the second nearest one. If it holds that $d(p_r, c_2) - d(p_r, c_1) > 2\varepsilon$, we can conclude that c_1 is exactly the nearest center for all points represented by p_r. In this case, we need not to compute the distance for all these points.

The second strategy is utilized the local property of LSH to reduce the number of centers to be compared for each data points. Specifically, for a data point, we only compute the distance between it and a few centers which stays in the buckets near to that of this point. It is based on Theorem 2.

Theorem 2. *Given a LSH function:* $h_{a,b}(v) = \lfloor \frac{a \cdot v + b}{r} \rfloor$. *If* $| h_{a,b}(v_1) - h_{a,b}(v_2) | \geq \delta_h$, *then we have* $d(v_1 - v_2) \geq \frac{(\delta_h - 1) \cdot r}{|a|}$.

Proof. According to definition of LSH, we have $| h_{a,b}(v_1) - h_{a,b}(v_2) | = | \lfloor \frac{a \cdot v_1 + b}{r} \rfloor - \lfloor \frac{a \cdot v_2 + b}{r} \rfloor | \geq \delta_h$. *We can conclude that* $| \frac{a \cdot v_1 + b}{r} - \frac{a \cdot v_2 + b}{r} + 1 | \geq \delta_h$. *Therefore, we have* $| \frac{a \cdot (v_1 - v_2)}{r} | \geq \delta_h - 1$. *We have* $| v_1 - v_2 | \geq \frac{r(\delta_h - 1)}{|a \cos \theta|} \geq \frac{r \cdot (\delta_h - 1)}{|a|}$. *Here* θ *is the angle between point* a *and vector* $v_1 - v_2$. $\blacksquare$

We only use one hash function to map all data points and centers into buckets. We use $h(p)$ to denote the bucket ID for point p. For each point p, we first find the bucket, say b_i, which satisfies

- b_i is closest to $h(p)$
- b_i contains at least one center.

In b_i, We randomly select one center, denoted as c. Note that although c and p belong to the nearest buckets, c may not be the closest center for p, due to the probability property of LSH. However, we can use c to prune centers based on Theorem 2.

We calculate the distance from p to c, denoted as $d(p, c)$. From Theorem2, we can get the threshold δ_h to guarantee if the difference of LSH function value is greater than δ_h, the real distance is greater than $d(p, c)$. So we can use only need to compute the distance between p and centers in set $\{c || h(c) - h(p)| < \delta_h\}$.

In fact, these two pruning strategies can be combined. Algorithm 3 shows the pseudo-code of MapReduce job to combine these two pruning strategies. For Map phase, the $< key, value >$ pairs of input represent $< p_r, L_p, weight >$.

We can get the representative point p_r and the points in L_p (Line 3). We record the centers in ordered buckets according to the hash values by a given LSH function when mapper initializes. The results can be shared by all *map* functions. We use binary search to find the closest bucket from the given point p_r and compute the distance from p_r to a point in the bucket. From Theorem 2, we can get the safe pruning threshold δ_h (Line 5 $\sim$ Line 7). We can prune the centers by the threshold δ_h (Line 8 $\sim$ Line 14). From left centers, we get the closet center c' for p_r (Line 15). From Theorem 1, we can conclude that the points in L shares the same center if the distance from p_r to c' is less than $min + 2\varepsilon$. We denote the set of candidate centers as *closeSet*. Because if the distance is greater than $min + 2\varepsilon$, they share the same closest center with p_r. Here min is the distance from p_r to its closest center c' (Line 17 $\sim$ Line 24). We find the closest centers for the points represented by p_r in *closeSet* (Line 25 $\sim$ Line 35). In Reduce phase, new centers are calculated (Line 38 $\sim$ Line 41);

5   Experiments

In this section we present the experimental results of LSH-kmeans. The experiment environment includes a cluster of 14 computers, each of which has two Pentium(R) Dual-Core (2.70GHz) CPU E5400 and 4GB of memory, using Linux. Hadoop version 0.20.3 and Java 1.6 are used as the MapReduce system.

5.1   Dataset and Baseline

We use two datasets. The first is KDDCUP1999 dataset which is publicly available from the UC Irvine Machine Learning repository. The original dataset is comprised of 41 attributes and one class label.

Algorithm 3. LSH-based pruning

Require: Set[1:k] C, parameter a, b, r, ε
1: Pruning-Map(Key k, Value v)
2: **begin**
3:     Set $< p_r, L_p, weight > \leftarrow v$
4:     $hash1 = h(p_r)$
5:     get $so - far - closest$ from the closest bucket from $hash1$ using binary search;

6:     $dis = d(p_r, so - far - closest)$
7:     $\delta_h = \frac{|a| \cdot dis}{r} + 1$
8:     Set $C' = null$
9:     **for** c in C **do**
10:        $hash2 = \lfloor \frac{a \cdot c + b}{r} \rfloor$
11:        Set $diff \leftarrow abs(hash1 - hash2)$
12:        **if** $diff \leq \delta_h$ **then**
13:            $C' = C' + \{c\}$
14:        **end if**
15:     **end for**
16:     get closest center c' for p_r in C'
17:     $min = distance(p_r, c')$
18:     $closeSet = null$
19:     **for** c in C' **do**
20:        $dis2 = d(p_r, c)$
21:        **if** $| dis2 - min | \leq 2\varepsilon$ **then**
22:            $closeSet = closeSet + \{c\}$
23:        **end if**
24:     **end for**
25:     **if** $closeSet = null$ **then**
26:        **for** p in L_p **do**
27:            $Output(c', p)$
28:        **end for**
29:     **else**
30:        **for** p in L_p **do**
31:            get closest center cen' from $closeSet$
32:            $Output(cen', p)$
33:        **end for**
34:     **end if**
35:     $Output(c', p_r)$
36: **end**
37: Pruning-Reduce(Key k, Set $values$)
38: **begin**
39:     $mean = (\sum_{v \in values} v)/sizeof(values)$
40:     $center = $ nearest point from $mean$
41:     $Output(center, null)$
42: **end**

The second dataset is from our music database, which consists of about 1000 MP3 songs downloaded from the Internet, in which most of them are pop songs and the rest are classical and folk music. We extract the key features from the audible data and get the 26-dimension set of points. One point represents a frame of the song. Totally the dataset includes 919711 26-dimensional vectors.

5.2   Experiment Results

Data Reduction of Data Skeleton. In this experiment, we compare the number of data points in original dataset and data skeleton. We execute this process for three iterations. For KDDCUP1999, the time for an iteration about 60s, much less than an k-means iteration (above 600s for k=1500). For Music Frames, the time for an iteration about 130s, much less than an k-means iteration (above 1567s for k=1500). The results are shown in Fig. 4. We see that after data skeleton, the number of data points is reduced dramatically.

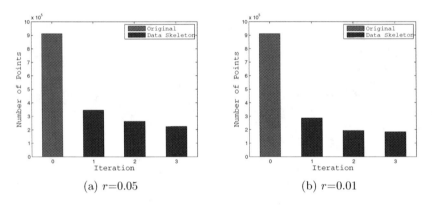

(a) r=0.05   (b) r=0.01

Fig. 4. Data Skeleton For KDDCUP1999

The Center Initialization. We evaluate the quality by two factors with the cost $\phi_X(C)$ in Equation 1. One is the clustering cost after seeding phase, and the other is the convergence property. We use KDDCUP1999 dataset for this evaluation. First we analysis the time performance, which is shown in Fig 7. The time for seeding using LSH-kmeans is about 1/3 the seeding phase of k-means++. The cost comparison is shown in Table 1. It can be seen that our approach generates better cost than k-means++.

The LSH Pruning Performance. To evaluate the performance of the LSH pruning in the iteration phase, we compare the time performance with and without LSH pruning on KDDCUP1999 dataset. The results are shown in Fig. 6. In each iteration, the time using LSH pruning is only 1/3 of the time without pruning. Theorem 1 provides the guarantee for the correctness of the pruning. It can be seen that when k is 3000, the time cost is reduced by about 68%.

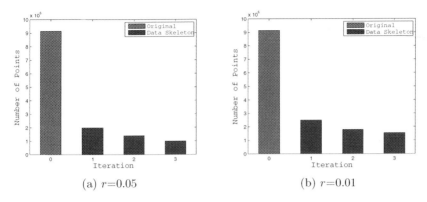

(a) $r=0.05$ (b) $r=0.01$

Fig. 5. Data Skeleton For Music Frames

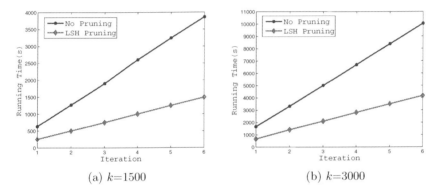

(a) $k=1500$ (b) $k=3000$

Fig. 6. LSH Pruning Performance

Table 1. Comparison of Clustering Cost (k=3000)

Iteration	Cost of Original Dataset	Cost of Data Sleleton
1	47824.77	47664.18
2	40292.91	40200.01
3	38318.60	38222.02
4	37474.73	37355.58
5	37019.76	36950.85
6	36714.02	36672.52

The Overall Performance Comparisons. We analyze the overall performance of LSH-kmeans compared with k-means++. LSH-kmeans includes three phases, skeleton, sampling centers and iteration. k-means++ includes two phases, sampling centers and iterations. The running time for different phases are shown in Fig. 7. Fig. 7(a) shows four groups of data, which are the running time for different phases in terms of LSH-kmeans and k-means++, with k as 1500 and

3000 respectively. The dataset is KDDCUP1999. As we can see, LSH-kmeans outperforms k-means++ greatly, especially for a larger k ($k = 3000$). The time cost is reduced by 67% when k is 1500, and 76% when k is 3000. We can get the same conclusion for Music Frames from Fig. 7(b). The time cost is reduced by 57% when k is 1500, and 64% when k is 3000.

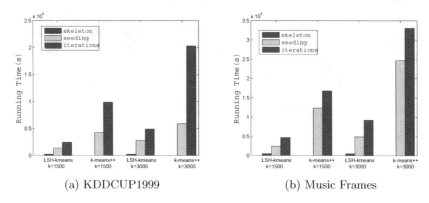

(a) KDDCUP1999   (b) Music Frames

Fig. 7. Overall Performance Comparisons

6   Conclusion and Future Work

In this paper we propose an improved k-means algorithm to cluster high-dimensional data on MapReduce with the LSH technology. With the increase of k, the number of clusters, the computation cost increases rapidly for high-dimension data. Our method improves the performance of k-means both in the initialization phase and the iteration phase. We implement LSH-kmeans on MapReduce and evaluate its performance on several datasets. The experiment results show that our method can improve the performance dramatically without decreasing the quality.

Acknowledgments. This work was supported by NSFC under grants 60673133, 61103009, 61033010, and the Key Project of Shanghai Municipal Science and Technology Commission (Scientific Innovation Act PlanGrant No.13511504804).

References

1. http://hadoop.apache.org/
2. http://mahout.apache.org/
3. AlSabti, K., Ranka, S.: An efficient space-partitioning based algorithm for the K-means clustering. In: Zhong, N., Zhou, L. (eds.) PAKDD 1999. LNCS (LNAI), vol. 1574, pp. 355–360. Springer, Heidelberg (1999)
4. Andoni, A., Indyk, P.: Near-optimal hashing algorithms for approximate nearest neighbor in high dimensions. In: FOCS, pp. 459–468 (2006)

5. Arthur, D., Vassilvitskii, S.: k-means++: the advantages of careful seeding. In: SODA, pp. 1027–1035 (2007)
6. Bahmani, B., Moseley, B., Vattani, A., Kumar, R., Vassilvitskii, S.: Scalable k-means++. CoRR, abs/1203.6402 (2012)
7. Bu, Y., Howe, B., Balazinska, M., Ernst, M.: Haloop: Efficient iterative data processing on large clusters. PVLDB 3(1), 285–296 (2010)
8. Cordeiro, R.L.F., Traina Jr., C., Traina, A.J.M., López, J., Kang, U., Faloutsos, C.: Clustering very large multi-dimensional datasets with mapreduce. In: KDD, pp. 690–698 (2011)
9. Datar, M., Immorlica, N., Indyk, P., Mirrokni, V.S.: Locality-sensitive hashing scheme based on p-stable distributions. In: Symposium on Computational Geometry, pp. 253–262 (2004)
10. Dean, J., Ghemawat, S.: Mapreduce: simplified data processing on large clusters. In: OSDI (2004)
11. Dean, J., Ghemawat, S.: Mapreduce: simplified data processing on large clusters. In: OSDI (2004)
12. Dhillon, I.S., Modha, D.S.: A data-clustering algorithm on distributed memory multiprocessors. In: Zaki, M.J., Ho, C.-T. (eds.) KDD 1999. LNCS (LNAI), vol. 1759, pp. 245–260. Springer, Heidelberg (2000)
13. Ene, A., Im, S., Moseley, B.: Fast clustering using mapreduce. In: KDD, pp. 681–689 (2011)
14. Huang, J.Z., Ng, M.K., Rong, H., Li, Z.: Automated variable weighting in k-means type clustering. IEEE Trans. Pattern Anal. Mach. Intell. 27(5), 657–668 (2005)
15. Indyk, P., Motwani, R.: Approximate nearest neighbors: Towards removing the curse of dimensionality. In: STOC, pp. 604–613 (1998)
16. Kriegel, H.-P., Kröger, P., Zimek, A.: Clustering high-dimensional data: A survey on subspace clustering, pattern-based clustering, and correlation clustering. TKDD 3(1) (2009)
17. Mondal, A., Lifu, Y., Kitsuregawa, M.: P2PR-tree: An R-tree-based spatial index for peer-to-peer environments. In: Lindner, W., Fischer, F., Türker, C., Tzitzikas, Y., Vakali, A.I. (eds.) EDBT 2004. LNCS, vol. 3268, pp. 516–525. Springer, Heidelberg (2004)
18. Ordonez, C., Omiecinski, E.: Efficient disk-based k-means clustering for relational databases. IEEE Trans. Knowl. Data Eng. 16(8), 909–921 (2004)
19. Pelleg, D., Moore, A.W.: X-means: Extending k-means with efficient estimation of the number of clusters. In: ICML, pp. 727–734 (2000)
20. Wu, X., Kumar, V., Quinlan, J.R., Ghosh, J., Yang, Q., Motoda, H., McLachlan, G.J., Ng, A.F.M., Liu, B., Yu, P.S., Zhou, Z.-H., Steinbach, M., Hand, D.J., Steinberg, D.: Top 10 algorithms in data mining. Knowl. Inf. Syst. 14(1), 1–37 (2008)
21. Yang, Y.-H., Lin, Y.-C., Chen, H.H.: Clustering for music search results. In: ICME, pp. 874–877 (2009)

Efficient Mining of Density-Aware Distinguishing Sequential Patterns with Gap Constraints[*]

Xianming Wang[1], Lei Duan[1,3,**], Guozhu Dong[2],
Zhonghua Yu[1], and Changjie Tang[1]

[1] School of Computer Science, Sichuan University, China
[2] Department of Computer Science & Engineering, Wright State University, USA
[3] State Key Laboratory of Software Engineering, Wuhan University, China
xmwang.scu@gmail.com, {leiduan,yuzhonghua,cjtang}@scu.edu.cn,
guozhu.dong@wright.edu

Abstract. Distinguishing sequential patterns are useful in characterizing a given sequence class and contrasting that class against other sequence classes. This paper introduces the density concept into distinguishing sequential pattern mining, extending previous studies which considered gap and support constraints. Density is concerned with the number of times of given patterns occur in individual sequences; it is an important factor in many applications including biology, healthcare and financial analysis. We present gd-DSPMiner, a mining method with various pruning techniques, for mining density-aware distinguishing sequential patterns that satisfy density and gap, as well as support, constraints. With respect to computational speed, when the procedures related to density are masked gd-DSPMiner is substantially faster than previous distinguishing sequential pattern mining methods. Experiments on real data sets confirmed the effectiveness and efficiency of gd-DSPMiner in the general setting and the ability of gd-DSPMiner to discover density-aware distinguishing sequential patterns.

Keywords: contrast mining, sequential pattern, density-aware.

1 Introduction

Sequence data mining has attracted considerable attention for its usefulness in a wide range of applications. For example, the mining of protein and nucleotide sequences can help scientists identify significant biological features including those that distinguish protein families with different functions and those that characterize important biological sites (such as transcription start sites) in DNA sequences, and the mining of sequences that describe the spread of infectious diseases can help in developing early warning and prevention strategies.

---

[*] This work was supported in part by NSFC 61103042, SRFDP 20100181120029, and SKLSE2012-09-32.

[**] Corresponding author.

S.S. Bhowmick et al. (Eds.): DASFAA 2014, Part I, LNCS 8421, pp. 372–387, 2014.

One important task of sequence data mining is discovering interesting sequential patterns from sequence data. Several types of sequential patterns have been extensively studied, including frequent sequential pattern [1], closed sequential pattern [2], distinguishing sequential pattern [3], periodic pattern [4], and partially ordered pattern [5]. Among these types, distinguishing sequential patterns with gap constraints are very useful for identifying important features for discriminating one class of sequences from sequences of other classes. In this study, we incorporate *density* into the concept of distinguishing sequential patterns with gap constraints and study the efficient mining of density-aware distinguishing sequential patterns with both density and gap constraints.

Our study is motivated by the observation that density of patterns is an important factor in the field of biology and genomics [6]; indeed, the density concept has often been indicated by statements such as "the sequences are rich in C & G" and "an XYZ pattern is over-represented." The well known "CpG islands," which are intervals of sequences that are unusually dense in C, G and CG [7–9], are a concrete example. Density of patterns is also important in healthcare and financial applications, among others.

While the traditional concept of "frequency" of a sequential pattern is concerned with the number of sequences that match the sequential pattern, the concept of "density" of a sequential pattern is concerned with the number of occurrences of the pattern in individual sequences. Technically, the density of a pattern P in a sequence S is the ratio of the count of P's occurrences to its maximum number of possible occurrences in a sequence of length-$|S|$. (Details are given in Section 2.) Only sequences S where P is dense will contribute to the frequency of P in given dataset of sequences. For example, in sequence $S = abcab$, subsequence 'ca' occurs once, and its maximum number of occurrences in any length-5 sequence is 4; so the density of 'ca' in S is $1/4$. On the other hand, the density of subsequence 'ab' in S is $2/4$. (In the above example, we assume that gaps are not allowed in subsequences matching given patterns for the sake of simplicity.) To the best of our knowledge, density has not been considered as a factor in previous studies on distinguishing/contrast pattern mining.

Challenges: Considering the density of a pattern in a sequence essentially relies on the count of the pattern's occurrences in the sequence. This leads to new challenges that are not present in traditional distinguishing pattern mining, caused by two issues. Firstly, the number of occurrences of a pattern may be large. For example, the number of occurrences of pattern $P = ba$ in sequence $S = bbbaabbabaaa$ (without gap constraint) equals the number of combinations of b's and subsequent a's, which is $3 \times 6 + 2 \times 4 + 1 \times 3 = 29$. Although it may be smaller when considering the gap constraint, the number of the occurrences will still be large in that situation. Secondly and a tougher challenge: the Apriori property, which states that "the support of a pattern cannot exceed the support of any of its sub-patterns," does not hold. For example, in sequence $S = abbccc$, there are 6 instances of pattern $P = abc$, while there are 2 instances of P's subpattern ab. As a result, traditional Apriori-based pruning strategies cannot be adapted to solve our problem.

Contributions: This paper makes the following main contributions on mining minimal density-aware distinguishing sequential patterns with both density and gap constraints: (1) introducing the concept of density in minimal distinguishing sequential patterns mining; (2) presenting an efficient algorithm for discovering minimal density-aware distinguishing sequential patterns. When the density factor is masked, the algorithm is a faster solution than the algorithm of [3] for mining minimal distinguishing sequential patterns with gap constraints; (3) reporting extensive experiments on sequences concerning human genes and on sequences on protein families, to evaluate our minimal density-aware distinguishing sequential pattern mining algorithm, and to demonstrate that the proposed algorithm can find interesting patterns, and it is effective and efficient.

Paper Outline: The rest of the paper is organized as follows. Section 2 provides the detailed definitions, as well as the necessary terminologies and notations. Section 3 summarizes the related work in the area. Section 4 presents our gd-DSPMiner algorithm. Section 5 reports an experimental study on the data sets concerning human genes and protein families. Finally, Section 6 discusses the future work and concluding remarks.

2   Problem Definition

In this section we give a formal definition for density-aware distinguishing sequential patterns. To that end, we will also need to define two concepts concerning sequence matching with respect to "gap" and "density" constraints.

We use Σ to denote the alphabet, which is a finite set of distinct symbols, of the sequences of interest. For example, Σ is $\{A, C, G, T\}$ for DNA sequences, and Σ is the set of symbols for the 20 amino acids for protein sequences.

A sequence S over Σ is a list of the form $S = e_1 e_2 ... e_n$, where $e_i \in \Sigma$ for $1 \leq i \leq n$. Each integer $i \in \{1, ..., n\}$ will be called a position of S. We will use $S_{[i]}$ to denote the symbol of S at position i, and use $|S|$ to denote the length of S. For example, given $S = abacd$, $S_{[3]} = a$ and $|S| = 5$.

We will use λ to denote the empty sequence. As usual, the concatenation of λ with any sequence S is equal to S, i.e. $\lambda S = S\lambda = S$.

Given a sequence S with $|S| = n$, a position subsequence is a list of the form $< i_1, ..., i_m >$ such that $1 \leq i_1 < ... < i_m \leq n$. For example, given $S = abacdc$, $< 1, 2, 5 >$ is a position subsequence.

We say a sequence S' is contained in a sequence S (also S' is a subsequence of S and S is a supersequence of S'), denoted by $S' \sqsubseteq S$, if there is a position subsequence $< k_1, ..., k_m >$ of S such that $S' = S_{[k_1]}...S_{[k_m]}$.

A gap is an interval of skipped positions between two consecutive positions in a position subsequence, and the size of a gap is the number of distinctive positions in the interval. For example, the interval between positions 2 and 5 in the position subsequence $< 1, 2, 5 >$ is a gap and its size is 2.

A **gap constraint** γ is specified by two natural numbers M and N satisfying $M \leq N$. A position subsequence $< k_1, ..., k_p >$ of a sequence is said to satisfy γ if $M \leq k_{m+1} - k_m - 1 \leq N$, for all m satisfying $1 \leq m < p$. Given γ, a pattern P is said to *match* a sequence $S = e_1 e_2 ... e_n$, if there exists a position subsequence $< k_1, ..., k_{|P|} >$ satisfying γ and satisfying $P = e_{k_1} e_{k_2} ... e_{k_{|P|}}$. Each position subsequence $< k_1, ..., k_{|P|} >$ satisfying the conditions given above is called an *instance* of P in S.

For example, for the gap constraint $[2, 3]$, pattern $P = db$ matches sequence $S = ddcba$ and $< 1, 4 >$ is an instance of P in S; on the other hand, P does not match sequence $edcba$ (since the gap between d and b is 1).

We write $N_{\ell, \gamma}^{S}$ to denote the number of all possible length-ℓ position subsequences satisfying a gap constraint γ in a given sequence S. To illustrate, consider $S = edcabc$ and $\gamma = [1, 3]$. The set of length-2 position subsequences is $\{< 1, 3 >, < 1, 4 >, < 1, 5 >, < 2, 4 >, < 2, 5 >, < 2, 6 >, < 3, 5 >, < 3, 6 >, < 4, 6 >\}$, and the set of length-3 ones is $\{< 1, 3, 5 >, < 1, 3, 6 >, < 1, 4, 6 >, < 2, 4, 6 >\}$. Thus, $N_{2, [1,3]}^{S} = 9$, and $N_{3, [1,3]}^{S} = 4$. Observe that distinct length-ℓ position subsequences can share common positions, as illustrated above.

The count of instances of a pattern P in a sequence S with respect to a gap constraint γ, denoted as $count(S, P, \gamma)$, is the number of distinct instances of P in S satisfying γ. The **density** of P in S with respect to γ is defined as

$$density(S, P, \gamma) = \frac{count(S, P, \gamma)}{N_{|P|, \gamma}^{S}} \tag{1}$$

Given a density threshold δ, a gap constraint γ, and a set of sequences D, the *gd-support*[1] of a pattern P in D, denoted by $sup_{\gamma}^{D}(P, \delta)$, is the fraction of sequences S whose density of P is $> \delta$, namely

$$sup_{\gamma}^{D}(P, \delta) = \frac{|\{S \in D \mid density(S, P, \gamma) > \delta\}|}{|D|} \tag{2}$$

Given a gd-support threshold α, if $sup_{\gamma}^{D}(P, \delta) \geq \alpha$, we say P is *gd-frequent* in D with gap constraint γ and density constraint δ.

As shown below, the traditional support of sequential pattern with gap constraint [10] is a special case of the gd-support when the density threshold is zero, and that traditional support is always greater than or equal to the gd-support.

Lemma 1. *Given a density threshold δ, a gap constraint γ, a set of sequences D, and a pattern P, we have (a) the traditional support of P is equal to $sup_{\gamma}^{D}(P, 0)$ and (b) $sup_{\gamma}^{D}(P, 0) \geq sup_{\gamma}^{D}(P, \delta)$.*

Since patterns are also sequences, the concept of sequence containment can be extended to patterns. Hence we can say a pattern P' is a subpattern of another pattern P, and write $P' \sqsubseteq P$.

Definition 1. (gd-DSP) *Given two classes of sequences pos (the positive) and neg (the negative), a gap constraint γ, a density threshold δ, and two gd-support thresholds α and β, a pattern P is called a Minimal gd-Constrained Distinguishing Sequential Pattern (gd-DSP), if conditions (1–3) are all true:*

---

[1] The g in *gd* stands for gap and the d in *gd* stands for density.

Table 1. A sequential database example

Id	Sequence	Class
1	aabb	pos
2	aaabbb	pos
3	abbababc	pos
4	edcba	neg
5	ddcba	neg
6	ccba	neg

Table 2. Summary of notations

Notation	Description
Σ	Sequence alphabet
λ	Empty sequence
$\gamma : [M, N]$	Gap constraint specified by two natural numbers M and N ($M \leq N$)
$N_{\ell,\gamma}^S$	Number of all possible length-ℓ subsequences satisfying gap constraint γ in sequence S
δ	Density threshold
$sup_\gamma^D(P,\delta)$	gd-support of pattern P in sequence set D with respect to γ under density threshold δ
pos, neg	The positive, negative sequence classes resp.

1. *Frequent in pos:* $sup_\gamma^{pos}(P,\delta) \geq \alpha$;
2. *Infrequent in neg:* $sup_\gamma^{neg}(P,\delta) \leq \beta$;
3. *Minimality condition: There is no subpattern of P satisfying 1 and 2.*

Given δ, α, β and γ, the minimal density-aware distinguishing sequential pattern mining problem is to find all the gd-DSPs from pos and neg.

Example 1. In the data shown in Table 1, suppose $\delta = 0.55$, $\alpha = 0.333$, $\beta = 0$, and $\gamma = [0,5]$. Then $sup_\gamma^{pos}('b',\delta) = sup_\gamma^{pos}('a',\delta) = 0 < \alpha$. However, $sup_\gamma^{pos}('ab',\delta) = 0.667$ and $sup_\gamma^{neg}('ab',\delta) = 0$. Thus, pattern ab is a gd-DSP.

Table 2 lists the frequently used notations of this paper.

3   Related Work

Sequential pattern mining [1] has attracted a lot of attention in research and development, since sequence data is ubiquitous and plays an important role in science, medicine, and other fields. A comprehensive review of the abundant literature in sequential pattern mining is out of the capacity of this paper. Several recent surveys [11, 12] and monographs [10, 13] on the topic provide thorough treatments. It is worth mentioning that there exist a large number of articles concerning applications of sequential pattern mining on biological sequences, such as [14–16].

As gap constraints lend flexibility in sequence and pattern matching, one well studied problem of sequential pattern mining is mining sequential patterns with gap constraints [14, 16–18]. Gap constraint is important in biology analysis, and has been used in alignment of DNA or protein sequences [19].

Zhang *et al.* [4] considered mining repetitive pattern with gap constraints in a single long sequence. The concept of *repetitive* defined in [4] is similar to the concept of *density*, although it was limited to one sequence. Furthermore, He *et al.* [20] studied the situation of inexact matching of a pattern in a sequence, with a focus on how to count the inexact matchings (instances) of a pattern in a given sequence efficiently. There are several essential differences between [4, 20] and this study. First, [4] and [20] focus on mining patterns from a single sequence, while we consider sets of multiple sequences. Second, we also consider classes in this paper.

Distinguishing/contrast patterns have been used in many interesting applications in bioinformatics and computational biology, since such patterns characterize a set of sequences and capture contrast information between different classes of sequences. For example, She et al. [15] studied the problem of using subsequences that discriminate outer membrane proteins (OMPs) from non-OMPs for outer membrane prediction. Shah et al. [21] built a classifier, which takes the contrast patterns as features for peptide folding prediction. Their experimental results indicate that the contrast pattern based classification outperforms simple secondary structure prediction based approaches.

Ji et al. [3] introduced ConSGapMiner to find the minimal distinguishing subsequences, which are minimal subsequences that occur frequently in one class of sequences but infrequently in sequences of another class. To the best of our knowledge, [3] is the most related published paper to our paper. However, ConSGapMiner does not consider the density of the minimal distinguishing subsequences. In contrast, our method is more general, as it employs gd-support to measure both frequency and density of each candidate pattern. In fact, [3] solves a problem that is a special case of our problem, namely the case where the density threshold (δ) is set to zero, and experimental study (Section 5) indicates that our method is more efficient in the setting of $\delta = 0$.

4   gd-DSPMiner Algorithm

In this section we present our gd-DSPMiner algorithm, for discovering gd-DSPs from a given sequential database. The gd-DSPMiner algorithm uses the following three main steps in an iterative manner: (i) candidate generation, (ii) minimality test, and (iii) gd-support checking. In each iteration, gd-DSPMiner finds all gd-DSPs of some length-ℓ ($\ell \geq 1$), as well as the patterns that will be used for generating candidates of length-($\ell + 1$). Below we discuss each of those three main steps, with a focus on the new ideas that make the algorithm efficient.

4.1   Candidate Generation

Our gd-DSPMiner uses a candidate-generation-and-test approach instead of the popular depth-first approach (which was used in distinguishing subsequence pattern mining [3]). We made this choice because (a) there is no anti-monotone property of gd-support (so the traditional support-based pruning method cannot be applied in gd-DSP mining), (b) some minimal patterns cannot be discovered in time to help prune wasted computation in depth-first search, and (c) a candidate-generation-and-test approach allows us to more effectively prune the search space and reduce the computation cost.

To illustrate (a), for the data given in Example 1, we have $sup^{pos}_{[0,5]}(`a`, 0.55) = 0$, $sup^{pos}_{[0,5]}(`b`, 0.55) = 0$, and $sup^{pos}_{[0,5]}(`ab`, 0.55) = 0.667 > 0$. To illustrate (b),

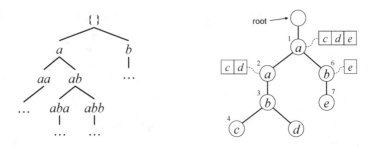

Fig. 1. Example of a lexicographic tree   **Fig. 2.** Example of a prefix-leaf tree

consider the lexicographic sequence tree given in Figure 1. Suppose pattern b is a gd-DSP. Then all patterns containing b are not minimal, so the subtree rooted at ab can be pruned. In a depth-first search pattern b is examined later than the patterns that start with a and hence ab cannot be pruned.

We now describe the candidate-generation step of our algorithm. We use $\mathbb{C}^\ell$ to denote the set of *candidate patterns* with length-ℓ ($\ell \geq 1$). Observe that $\mathbb{C}^1$ consists of all symbols in Σ.

To minimize computation cost, we partition $\mathbb{C}^\ell$ into four disjoint subsets:

- $\mathbb{U}^\ell$: a set of patterns P where our algorithm can determine that P cannot be contained by any gd-DSP;
- $\mathbb{M}^\ell$: the set of patterns P where P is a super-pattern of a discovered gd-DSP;
- $\mathbb{P}^\ell$: the set of minimal gd-DSPs with length-ℓ;
- $\mathbb{R}^\ell$: the set of patterns P such that P is not gd-DSP but P might be a subpattern of a longer gd-DSP (complementing $\mathbb{U}^\ell \cup \mathbb{M}^\ell \cup \mathbb{P}^\ell$).

The contents of $\mathbb{U}^\ell$ and $\mathbb{R}^\ell$ will be determined by our algorithm. We want to efficiently determine these sets without losing any valid gd-DSP and to maximize the $\mathbb{U}^\ell$ set. To determine $\mathbb{U}^\ell$, we need a definition, and a lemma.

Definition 2. (Max-Prefix & Max-Suffix) *Given a pattern* $P = P_{[1]}P_{[2]}...P_{[n]}$ *with* $n \geq 2$, *the max-prefix of* P, *denoted by* $pre^m(P)$, *is* $P_{[1]}P_{[2]}...P_{[n-1]}$, *and the max-suffix of* P, *denoted by* $suf^m(P)$, *is* $P_{[2]}...P_{[n-1]}P_{[n]}$. *For the case* $|P| = 1$, $pre^m(P) = suf^m(P) = \lambda$ *(the empty sequence).*

For example, given $P = acefr$, $pre^m(P) = acef$ and $suf^m(P) = cefr$.

Lemma 2. *Given* δ, γ, α *and* β, *a pattern* P *and its superpatterns cannot be gd-DSPs if* $sup_\gamma^{pos}(pre^m(P), 0) < \alpha$ *or* $sup_\gamma^{pos}(suf^m(P), 0) < \alpha$.

Based on Lemma 2, we define $\mathbb{U}^\ell = \{P \in \mathbb{C}^\ell \mid sup_\gamma^{pos}(pre^m(P), 0) < \alpha$ or $sup_\gamma^{pos}(suf^m(P), 0) < \alpha\}$.

After removing $\mathbb{U}^\ell$ from $\mathbb{C}^\ell$, gd-DSPMiner discovers $\mathbb{M}^\ell$ from $\mathbb{C}^\ell - \mathbb{U}^\ell$ by performing minimality test (see Section 4.2), and $\mathbb{P}^\ell$ from $\mathbb{C}^\ell - \mathbb{U}^\ell - \mathbb{M}^\ell$ by performing gd-support check (see Section 4.3). Then, $\mathbb{R}^\ell$ is defined by

$$\mathbb{R}^\ell = \mathbb{C}^\ell - \mathbb{U}^\ell - \mathbb{M}^\ell - \mathbb{P}^\ell. \tag{3}$$

gd-DSPMiner uses $\mathbb{R}^\ell$ to generate $\mathbb{C}^{\ell+1}$. Since any pattern containing a pattern in $\mathbb{U}^\ell \cup \mathbb{M}^\ell \cup \mathbb{P}^\ell$ cannot be a gd-DSP, this will not lead to the loss of any gd-DSP. gd-DSPMiner will generate candidate patterns with length-$(\ell+1)$ by "joining" patterns in $\mathbb{R}^\ell$.

Definition 3. *Let P and Q be two patterns in $\mathbb{R}^\ell$. If $pre^m(P) = suf^m(Q)$, a length-$(\ell+1)$ pattern in $\mathbb{C}^{\ell+1}$ is composed by concatenating $Q_{[1]}$ with P (or Q with $P_{[\ell]}$), denoted by $P \oplus Q$:*

$$P \oplus Q = Q_{[1]} \underbrace{P_{[1]} P_{[2]} ... P_{[\ell]}}_{P} = \underbrace{Q_{[1]} Q_{[2]} ... Q_{[\ell]}}_{Q} P_{[\ell]} \tag{4}$$

Suppose $P = cefr$, $Q = acef$, and $P, Q \in \mathbb{R}^4$. As $pre^m(P) = suf^m(Q) = cef$, $P \oplus Q = acefr \in \mathbb{C}^5$.

By using Equation 4 for every pair of patterns in $\mathbb{R}^\ell$ such that the max-prefix of one pattern is the same as the max-suffix of the other, all patterns in $\mathbb{C}^{\ell+1}$ are generated.

For efficiency, all discovered gd-DSPs $(\cup_{l=1}^\ell \mathbb{P}^l)$ are organized in a modified prefix tree. The modifications will be illustrated in the next subsection. The main idea is to store symbols of leaf-node descendants at certain internal nodes, to improve the efficiency of the minimality test (which is a well recognized and repeatedly used expensive operation for pattern mining).

4.2   Minimality Test

The goal of minimality test is to guarantee that all superpatterns of discovered gd-DSPs ($\mathbb{M}^\ell$) will be excluded in the set of gd-DSPs.

If we compare patterns in $\mathbb{C}^\ell - \mathbb{U}^\ell$ with patterns in $\cup_{i=1}^{\ell-1} \mathbb{P}^i$ directly, the complexity is $O(|\cup_{i=1}^{\ell-1} \mathbb{P}^i| \times |\mathbb{C}^\ell - \mathbb{U}^\ell| \times \ell)$. To improve the efficiency, we develop Lemma 3.

Lemma 3. *Let $P \in \mathbb{C}^\ell - \mathbb{U}^\ell$ and $Q \in \cup_{l=1}^{\ell-1} \mathbb{P}^l$. If $Q \sqsubseteq P$, then $Q_{[1]} = P_{[1]}$ and $Q_{[|Q|]} = P_{[|P|]}$.*

Lemma 3 is important as it allows gd-DSPMiner to improve significantly its efficiency in the minimality test. For a pattern $P \in \mathbb{C}^\ell - \mathbb{U}^\ell$, it is unnecessary to perform minimality test between P and every pattern in $\cup_{l=1}^{\ell-1} \mathbb{P}^l$. Instead, we only need to consider the patterns in $\cup_{l=1}^{\ell-1} \mathbb{P}^l$ that share the identical first and last symbols with P. For example, to perform the minimality test on pattern $P = acefr$, gd-DSPMiner only considers the patterns in $\cup_{l=1}^4 \mathbb{P}^l$ whose first symbol is 'a' and the last symbol is 'r'.

To make use of Lemma 3, gd-DSPMiner needs to efficiently access each pattern's first and last symbols. To this end, we design a modified prefix tree, called *prefix-leaf tree*.

A **prefix-leaf tree** is a special prefix tree, where each internal node whose parent is the root node or whose parent has more than one child, also stores symbols of its leaf-node descendants. Figure 2 gives an example of a prefix-leaf tree, which contains 3 patterns $(aabc, aabd, abe)$. Among all internal nodes, node 1 (the child of the root node) records the symbols of its leaf-node descendants, namely (c, d, e). As node 1 has two children, its children, namely nodes 2 and 6, record the symbols of their leaf-node descendants respectively. Note that, our implementation uses a bit vector to record the leaf-descendant symbols, so it is efficient to check whether or not a given symbol is among the symbols of leaf-node descendants.

gd-DSPMiner builds a prefix-leaf tree to record all discovered gd-DSPs (namely $\cup_{l=1}^{\ell-1} \mathbb{P}^l$) so far. For a given pattern $P \in \mathbb{C}^\ell - \mathbb{U}^\ell$, the efficiency of gd-DSPMiner for checking whether there exists a pattern $Q \in \cup_{l=1}^{\ell-1} \mathbb{P}^l$ satisfying $Q \sqsubseteq P$ can benefit from the prefix-leaf tree in two aspects. On one hand, by the leaf-descendants of the root node's children it is efficient to check whether $P_{[|P|]} \neq Q_{[|Q|]}$. On the other hand, in case $P_{[1]} = Q_{[1]}$ and $P_{[|P|]} = Q_{[|Q|]}$, gd-DSPMiner performs depth-first transversal on the prefix-leaf tree to check whether P contains Q. It is efficient to prune the branches which are not contained by P by using the leaf-descendants of the internal nodes.

For example, consider the patterns represented by the prefix-leaf tree in Figure 2. Suppose pattern $P = affce$. As symbol e ($P_{[5]}$) is a leaf-descendant of node 1 ($P_{[1]} = a$), there must exist at least one pattern whose first symbol is a and last symbol is e. A depth-first transversal is performed on the prefix-leaf tree to check whether P is minimal. As symbol e is not a leaf-descendant of node 2, there is no need to check any children of node 2.

The complexity to build a prefix-leaf tree representing all patterns in $\cup_{l=1}^{\ell-1} \mathbb{P}^l$ is $O(|\cup_{l=1}^{\ell-1} \mathbb{P}^l| \times \ell)$.

4.3   gd-Support Checking

After removing $\mathbb{U}^\ell$ and $\mathbb{M}^\ell$ from $\mathbb{C}^\ell$, we need to check the gd-support of each candidate pattern (Equation 2). To that end, we need to calculate the density of each candidate pattern in each sequence , and therefore we need to compute $count(S, P, \gamma)$ and $N_{|P|, \gamma}^S$, for all patterns P and sequences S, and for given gap constraint γ $(= [M, N])$.

One naïve way to compute $count(S, P, \gamma)$ is recording all position subsequences of P in S. However, this will induce a huge memory consumption. Luckily, since $count(S, P, \gamma)$ is the number of instances of P in S, it is unnecessary to record the position subsequences of each pattern. Below we present a method, which stores a reduced amount of computed results and which scans each sequence only once for each given pattern. We begin the discussion of our method with Definition 4.

Definition 4. (last-match-position) *Let* $S_{[j_1]} S_{[j_2]} ... S_{[j_n]}$ *(*$1 \leq j_1 < j_2 < ... < j_n \leq |S|$*) be an instance of pattern* P *in sequence* S. *Then,* j_n *is a last-match-position of* P *in* S.

Note that, when considering gap constraint, there may be several instances at an identical last-match-position. For example, there are 4 instances of S' in S at last-match-position 5 for $S' = ab$, $S = aaaabbbb$, and gap range $\gamma = [0, 4]$.

We use a triple, denoted as $< id(S), lmp, count >$, to record the number of instances of P in S at each last-match-position, where $id(S)$ is the sequence identifier of S, lmp is a last-match-position of P in S, and $count$ is the number of instances of P in S at lmp. We call the triple a *match-triple*. For sequence set D, let *match-set*(P) denote the set of match-triples $(\{< id(S), lmp, count >|$ $S \in D\})$; this set gives the number of instances of P at each last-match-position in every sequence of D.

Next, we present our method for constructing the match-sets of candidate patterns:

Step 1: Construct the match-set of each symbol in Σ.

Step 2: For pattern P with length-ℓ ($\ell > 1$), let $P_{[1,\ell]} = P_{[1]}P_{[2]}...P_{[\ell]}$ denote the first l ($1 \leq l < \ell$) symbols in P. We now discuss how to compute the match-set of $P_{[1,l+1]}$ from those of $P_{[1,l]}$ and $P_{[l+1]}$. Let t' be a match-triple of $P_{[1,l]}$ and t'' be a match-triple of $P_{[l+1]}$ such that $t'.id = t''.id$. If $M \leq (t''.lmp - t'.lmp - 1) \leq N$, a match-triple t of $P_{[1,l+1]}$ is created/updated as follows:

- $t.id = t'.id$
- $t.lmp = t''.lmp$
- $t.count = t.count + t'.count$.

Here, we set $t.count = 0$ if t did not exist (i.e., t is created due to t' and t'').

The match-set of each symbol x (viewed as a one-symbol pattern) in a sequence S is simply $\{< id(S), i, 1 >| x = S_{[i]}\}$.

Consider sequences in Table 1 and $\gamma = [1, 3]$. The match-triple of pattern bb in sequence 3 at last-match-position 5 is $< 3, 5, 2 >$. As the last-match-position of pattern c in sequence 3 is 8 (hence $< 3, 8, 1 >$ is the match-triple of c in sequence 3 at last-match-position 8), using the above procedure we see that the match-triple of pattern bbc in sequence 3 is $< 3, 8, 2 >$.

For efficiency purposes, we build a prefix tree to organize all candidate patterns. By traversing the tree in a depth-first manner, the match-set of each candidate pattern (a path from the root to the leaf) is constructed. Correspondingly, we can get the number of instances of the pattern in each sequence.

As candidate patterns with length-$(\ell + 1)$ are expanded by length-ℓ patterns, it appears that it makes sense to use incremental computation for length-$(\ell + 1)$ patterns by storing the match-sets of length-ℓ patterns. However, our experiments show that this may induce a considerable amount of memory consumption, which makes our method impractical when the length of candidate patterns grows larger than 10 in the computer used for our experiments. (There are possibly millions of candidate patterns and each has a large match-set.)

Instead, in our implementation, as soon as we know that the match-triples of a node will not be needed for computing the match-triples of other patterns, we discard those match-triples to free memory. In this way, gd-DSPMiner makes a trade-off between computation time and memory consumption.

For computing $N_{|P|,\gamma}^{S}$ we use formulae given in [4]. Details of those formulae can be found in [4].

Algorithm 1. $gd\text{-}DSPMiner(pos, neg, \delta, \alpha, \beta, \gamma)$

Input: pos: a class of sequences, neg: another class of sequences, δ: density threshold,
      α: minimal gd-support for pos, β: maximal gd-support for neg, γ: gap constraint
Output: dsp: the set of gd-DSPs of pos against neg

```
 1: set ℂ ← Σ; // length-1 candidate patterns
 2: while ℂ ≠ ∅ do
 3:    U ← ∅, M ← ∅, P ← ∅;
       // remove unpromising patterns by Lemma 2
 4:    for each pattern P ∈ ℂ do
 5:       if supᵧᵖᵒˢ(preᵐ(P),0) < α or supᵧᵖᵒˢ(sufᵐ(P),0) < α then
 6:          U ← U ∪ {P};
 7:       end if
 8:    end for
 9:    CU ← ℂ − U;
       // perform minimality test
10:    for each pattern P ∈ CU do
11:       if there exists pattern Q ∈ dsp satisfying Q ⊑ P then
12:          M ← M ∪ {P};
13:       end if
14:    end for
15:    CUM ← CU − M;
       // check gd-support
16:    for each pattern P ∈ CUM do
17:       if supᵧᵖᵒˢ(P,δ) ≥ α and supᵧⁿᵉᵍ(P,δ) ≤ β then
18:          P ← P ∪ {P};
19:       end if
20:    end for
21:    dsp ← dsp ∪ P; R ← CUM − P; ℂ ← ∅;
       // generate candidates with longer length
22:    for each pair of patterns P, Q ∈ R do
23:       if preᵐ(P) = sufᵐ(Q) then
24:          ℂ ← ℂ ∪ {P ⊕ Q};
25:       end if
26:    end for
27: end while
28: return dsp;
```

4.4   gd-DSPMiner

This subsection gives the pseudo-code of gd-DSPMiner (Algorithm 1) and some technical ideas used in our implementation.

Recall that each pattern in $\mathbb{C}^{\ell+1}$ is composed from a pair of patterns $P, Q \in \mathbb{R}^\ell$ satisfying $pre^m(P) = suf^m(Q)$ (Definition 3). To speed up the computation process, we employ a hash-join like method. The hash function hashes a pattern P to an integer by: $H(P) = \left(\sum_{i=1}^{|P|} h(P_{[i]}) \times 31^{|P|-i+1} \right) \bmod L_{table}$ where $h(P_{[i]})$ is the index of $P_{[i]}$ in Σ $(0 \le h(P_{[i]}) < |\Sigma|)$, and L_{table} (a prime number) is the size of the hash table.

We use two hash tables, a "prefix table" and a "suffix table." Given a pattern X, we store it in prefix table at position $H(pre^m(X))$ and in suffix table at position $H(suf^m(X))$, respectively. (At each hash position in a table a linked list is used.) We only attempt to join two patterns P and Q to generate a new candidate pattern if P and Q are hashed to the same position in the prefix table and the suffix table, respectively.

5   Empirical Evaluation

In this section, we report a systematic empirical study on real data sets to verify the effectiveness and efficiency of gd-DSPMiner and the ability of gd-DSPMiner to find density-aware distinguishing patterns. All experiments were conducted on a computer with an Intel Xeon Core X5660 2.80 GHz CPU, and 4 GB main memory, running the CentOS 5.9 operating system. gd-DSPMiner was implemented in C++ and compiled by GCC 4.1.2.

5.1   gd-DSP Discovery from Human Genes

One important task in bioinformatics is identifying human genetic variations that can be used to identify biological sites in sequences. We extracted 100 human genes from http://dbtss.hgc.jp/, and applied gd-DSPMiner to find the contrast patterns between the upstream regions of TSSs (Transcription Start Sites) and downstream regions after the end of genes. Specifically, for each gene, we took the sequence containing the last 100 base pairs in the upstream of the TSS as an instance in pos, and took the first 100 base pairs in the downstream of the gene as an instance in neg.

Table 3 lists the discovered gd-DSPs, when δ (min density threshold) = 0.01, α (min gd-support for pos) = 0.3, β (max gd-support for neg) = 0.1, and γ (gap constraint) = $[1, 2]$. We can see that most gd-DSPs in the result are composed by C and G, which is consistent with the famous "CpG islands" observation in genetics. The importance of "CpG islands" in genetics indicates that gd-DSPMiner can discover useful patterns. We also observe that our discovered patterns are more specific than what is stated by the concept of "CpG islands," and they are associated with density measures, which offer more specific and potentially useful information. Moreover, we notice that some discovered gd-DSPs, such as $CCTC$, are not implied by the concept of "CpG island," which may also provide new insight for biologists.

We now discuss the effect of the density threshold (δ) on gd-DSPs mining. For a pattern P, the gd-$support$ of P increases by 1 for each sequence S such that the density of P in S (defined by Equation 1) is larger than δ. Take $CCTC$ in Table 3 as an instance; since $N^S_{4,[1,2]} = 740$, there are respectively 30 positive sequences and 5 negative sequences containing at least 8 ($> 740 \times 0.01$) distinct instances of $CCTC$. Moreover, we observed that the larger the value of δ, the smaller the average length of the discovered gd-DSPs.

Table 3. gd-DSPs ($\delta = 0.01$, $\alpha = 0.3$, $\beta = 0.1$, $\gamma = [1, 2]$)

gd-DSP	sup_γ^{pos}	sup_γ^{neg}	gd-DSP	sup_γ^{pos}	sup_γ^{neg}	gd-DSP	sup_γ^{pos}	sup_γ^{neg}
CCCC	0.47	0.06	CCCG	0.53	0.02	CCGC	0.53	0.03
CCGG	0.48	0.01	CCTC	0.30	0.05	CGCC	0.57	0.02
CGCG	0.53	0.02	CGGC	0.47	0.01	CGGG	0.49	0.04
GCCC	0.49	0.03	GCCG	0.51	0.03	GCGC	0.49	0.01
GCGG	0.50	0.02	GGCC	0.51	0.02	GGCG	0.49	0.01
GGGC	0.50	0.03	GGGG	0.46	0.08			

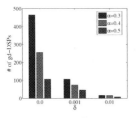

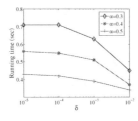

Table 4. Density comparison

		gd-DSP	non gd-DSP
pos	Avg	1.3×10^{-2}	3.2×10^{-3}
	Max	2.1×10^{-2}	7.6×10^{-3}
	Min	7.9×10^{-3}	1.1×10^{-3}
neg	Avg	1.6×10^{-3}	4.1×10^{-3}
	Max	2.7×10^{-3}	1.9×10^{-2}
	Min	1.1×10^{-3}	1.1×10^{-3}

Fig. 3. # of gd-DSPs w.r.t. δ **Fig. 4.** Runtime w.r.t. δ

Table 4 lists the average, maximum and minimum densities of gd-DSPs as well as non gd-DSPs with length-4 in *pos*, *neg*, respectively for the setting used for Table 3. We observe the following: (1) the average, maximum and minimum density values of gd-DSPs in *pos* are respectively are 8.13, 7.78 and 7.18 times larger than their counterparts in *neg*, but (2) for non gd-DSPs the former are all ≤ 1 times the latter; (3) the densities of gd-DSPs are much larger than their counterparts of non gd-DSPs in the average, maximum and minimum cases. Thus the gd-DSPs are indeed over represented in the upstream of the TSSs than elsewhere. This table also indicates that $\delta = 0.01$ is still a relatively large density value (approximately 3.1 times larger than the average of non gd-DSPs in *pos*).

Figure 3 illustrates the numbers of gd-DSPs with different density thresholds (δ), when $\beta = 0.1$ and $\gamma = [1, 2]$. Not surprisingly, the number of gd-DSPs decreases when the density threshold increases. We note that the number of gd-DSPs without density constraint ($\delta = 0$) is much larger than that of gd-DSPs with density constraints ($\delta > 0$). Thus, considering density as a factor in distinguishing sequential pattern mining is helpful for finding more distinctive patterns and removing distinguishing patterns that do not have density preference to the classes. Figure 4 shows the runtime of gd-DSPMiner with respect to density threshold. As density threshold increases, the runtime decreases. Clearly, our heuristic pruning techniques speed up the search in practice.

5.2 Efficiency Test on Protein Families

To the best of our knowledge, there were no existing methods tackling exactly the same problem as gd-DSPMiner. The most related previous work is Con-SGapMiner [3], whose mining results are the same as gd-DSPMiner when the density threshold $\delta = 0$.

Table 5. Protein family pairs

pos	$|pos|$	neg	$|neg|$	avg. len. (pos)	avg. len. (neg)
CbiA	80	CbiX	76	205	106
DUF1694	16	DUF1695	5	123	186
SrfB	5	Spheroidin	4	954	1021
TatC	75	TatD_DNase	119	205	262

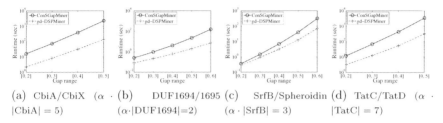

(a) CbiA/CbiX ($\alpha \cdot$ (b) DUF1694/1695 (c) SrfB/Spheroidin (d) TatC/TatD ($\alpha \cdot$
$|$CbiA$|$ = 5) ($\alpha \cdot |$DUF1694$|$=2) ($\alpha \cdot |$SrfB$|$ = 3) $|$TatC$|$ = 7)

Fig. 5. Efficiency test w.r.t. gap range on protein family pairs

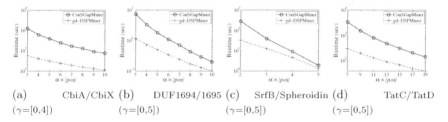

(a) CbiA/CbiX (b) DUF1694/1695 (c) SrfB/Spheroidin (d) TatC/TatD
(γ=[0,4]) (γ=[0,5]) (γ=[0,5]) (γ=[0,5])

Fig. 6. Efficiency test w.r.t. α on protein family pairs

Like the authors did in [3], we applied our methods as well as ConSGap-Miner on the protein families to discover the minimal distinguishing sequences which have zero support in the *neg* data set ($\beta = 0$), setting $\delta = 0$ in gd-DSPMiner. Statistics about the protein families, selected from PFam: Protein Family Database (http://pfam.sanger.ac.uk/), are list in Table 5. As the value of M in the gap constraint γ is fixed as 0 in ConSGapMiner, we set $M = 0$ in this empirical study.

Figure 5 shows the efficiency test on the four protein family pairs with varying γ for a fixed α. The runtime of both ConSGapMiner and gd-DSPMiner increases with the gap range. gd-DSPMiner runs (approximately 10 times) faster than ConSGapMiner, since gd-DSPMiner employs a more efficient pruning method to speed up the search.

Figure 6 shows the runtime of both ConSGapMiner and gd-DSPMiner with varying α for a fixed γ. We can see that the runtime of both ConSGapMiner and gd-DSPMiner decreases with α, since the number of candidate patterns is smaller when α is larger. Again, gd-DSPMiner is faster than ConSGapMiner.

6   Conclusions

In this paper, we studied a novel and interesting problem of mining density-aware distinguishing sequential patterns with both density and gap constraints. We presented a mining method with various pruning techniques. Our experiments on real data sets show that mining density-aware distinguishing sequential patterns is interesting and has potential, and our method is effective and efficient to discover density-aware distinguishing sequential patterns with gap constraints.

As future work, we intend to explore the use of our method in various biological, genetic, and financial applications.

References

1. Agrawal, R., Srikant, R.: Mining sequential patterns. In: ICDE, pp. 3–14 (1995)
2. Yan, X., Han, J., Afshar, R.: Clospan: Mining closed sequential patterns in large databases. In: SDM 2003 (2003)
3. Ji, X., Bailey, J., Dong, G.: Mining minimal distinguishing subsequence patterns with gap constraints. Knowl. Inf. Syst. 11(3), 259–286 (2007)
4. Zhang, M., Kao, B., Cheung, D.W., Yip, K.Y.: Mining periodic patterns with gap requirement from sequences. ACM Trans. Knowl. Discov. Data 1(2) (2007)
5. Pei, J., Wang, H., Liu, J., Wang, K., Wang, J., Yu, P.S.: Discovering frequent closed partial orders from strings. IEEE Trans. on Knowl. and Data Eng. 18(11), 1467–1481 (2006)
6. Durbin, R., Eddy, S., Krogh, A., Mitchison, G.: Biological Sequence Analysis: Probabilistic Models of Proteins and Nucleic Acids. Cambridge University Press, Cambridge (1998)
7. Gardiner-Garden, M., Frommer, M.: CpG islands in vertebrate genomes. Journal of Molecular Biology 196(2), 261–282 (1987)
8. Bock, C., Paulsen, M., Tierling, S., Mikeska, T., Lengauer, T., Walter, J.: CpG island methylation in human lymphocytes is highly correlated with DNA sequence, repeats, and predicted DNA structure. PLoS Genetics 2(3), e26 (2006)
9. Jabbaria, K., Bernardi, G.: Cytosine methylation and CpG, TpG (CpA) and TpA frequencies. Gene 333 (2004)
10. Dong, G., Pei, J.: Sequence Data Mining. Springer, Heidelberg (2007)
11. Mabroukeh, N.R., Ezeife, C.I.: A taxonomy of sequential pattern mining algorithms. ACM Comput 43(1), 3:1–3:41 (2010)
12. Mooney, C.H., Roddick, J.F.: Sequential pattern mining – approaches and algorithms. ACM Comput. Surv. 45(2), 19:1–19:39 (2013)
13. Kumar, P., Krishna, P.R., Raju, S.B.: Pattern Discovery Using Sequence Data Mining: Applications and Studies, 1st edn. IGI Publishing, Hershey (2011)
14. Ferreira, P.G., Azevedo, P.J.: Protein sequence pattern mining with constraints. In: Jorge, A.M., Torgo, L., Brazdil, P.B., Camacho, R., Gama, J. (eds.) PKDD 2005. LNCS (LNAI), vol. 3721, pp. 96–107. Springer, Heidelberg (2005)
15. She, R., Chen, F., Wang, K., Ester, M., Gardy, J.L., Brinkman, F.S.L.: Frequent-subsequence-based prediction of outer membrane proteins. In: KDD 2003, pp. 436–445 (2003)
16. Wu, X., Zhu, X., He, Y., Arslan, A.N.: PMBC: Pattern mining from biological sequences with wildcard constraints. Comput. Biol. Med. 43(5), 481–492 (2013)

17. Li, C., Yang, Q., Wang, J., Li, M.: Efficient mining of gap-constrained subsequences and its various applications. ACM Trans. Knowl. Discov. Data 6(1), 2:1–2:39 (2012)
18. Xie, F., Wu, X., Hu, X., Gao, J., Guo, D., Fei, Y., Hua, E.: MAIL: Mining sequential patterns with wildcards. Int. J. Data Min. Bioinformatics 8(1), 1–23 (2013)
19. Gusfield, D.: Algorithms on Strings, Trees, and Sequences: Computer Science and Computational Biology. Cambridge University Press, New York (1997)
20. He, D., Zhu, X., Wu, X.: Approximate repeating pattern mining with gap requirements. In: ICTAI 2009, pp. 17–24 (2009)
21. Shah, C.C., Zhu, X., Khoshgoftaar, T.M., Beyer, J.: Contrast pattern mining with gap constraints for peptide folding prediction. In: FLAIRS 2008, pp. 95–100 (2008)

Identifying Top k Dominating Objects over Uncertain Data

Liming Zhan, Ying Zhang, Wenjie Zhang, and Xuemin Lin

University of New South Wales, Sydney NSW, Australia
{zhanl,yingz,zhangw,lxue}@cse.unsw.edu.au

Abstract. Uncertainty is inherent in many important applications, such as data integration, environmental surveillance, location-based services (LBS), sensor monitoring and radio-frequency identification (RFID). In recent years, we have witnessed significant research efforts devoted to producing probabilistic database management systems, and many important queries are re-investigated in the context of uncertain data models. In the paper, we study the problem of top k dominating query on multi-dimensional uncertain objects, which is an essential method in the multi-criteria decision analysis when an explicit scoring function is not available. Particularly, we formally introduce the top k dominating model based on the state-of-the-art top k semantic over uncertain data. We also propose effective and efficient algorithms to identify the top k dominating objects. Novel pruning techniques are proposed by utilizing the spatial indexing and statistic information, which significantly improve the performance of the algorithms in terms of CPU and I/O costs. Comprehensive experiments on real and synthetic datasets demonstrate the effectiveness and efficiency of our techniques.

1 Introduction

Ranking query is an essential analytic method which focuses on retrieving the top k most important answers from massive quantity of data according to an user's preference. In many applications, users need to make decision against multiple features of the objects. For instance, cost, comfort, safety, and fuel economy may be some of the main criteria we consider when purchasing a car. Therefore, each object can be described by a point in a multi-dimensional space where each dimension corresponds to a particular selection feature. If there is a scoring (utility) function (e.g., additive linear function) which can quantify the preference of a user, objects can be immediately ranked based on their corresponding scores. However, in many scenarios users cannot find a proper scoring function due to various reasons such as the lack of domain knowledge. By utilizing the dominance relationship, the **top k dominating query** provides a simple and intuitive way to rank objects when an explicit scoring function is not available. The dominance relationship has been widely used in the multi-criteria decision analysis where an object A *dominates* another object B if A is not worse than B on every dimension and A is strictly better than B on at least one dimension. The goodness of an object A, namely *dominance score*, can be naturally

S.S. Bhowmick et al. (Eds.): DASFAA 2014, Part I, LNCS 8421, pp. 388–405, 2014.

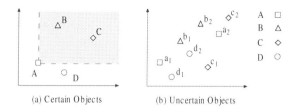

(a) Certain Objects   (b) Uncertain Objects

Fig. 1. Certain Objects and Uncertain Objects

measured by the number of other objects dominated by A in a set of objects, and the top k objects with highest *dominance scores* are returned as the top k dominating objects.

Example 1. As shown in Figure 1(a), there is a set $\mathcal{O}$ of 4 objects. The *dominance score* of the object A is 2 which is the number of other objects within the shaded region. Similarly, the *dominance scores* of B, C and D are 0, 0 and 1 respectively. Therefore, the results of the top 2 dominating query on $\mathcal{O}$ are A and D.

Motivation. In many applications such as data integration, environmental surveillance, location based service, the uncertainty is inherent due to many factors including limitation of measuring equipments, probabilistic model based data integration, noise, delay or loss of data updates, etc. Consequently, the data in the above applications are usually described by probabilistic model (i.e., uncertain objects) instead of deterministic points in a multi-dimensional space where an uncertain object may be described by probabilistic density function or a set of instances (points). For example, in the meteorology system, sensors collect the temperature and relative humidity at a large number of sites. The readings may be uncertain due to the noise or the limit of the sensor reader. Another example is the performance evaluation of NBA players. A player attends multiple games and statistics (e.g. score and #rebounds) of each game are recorded. Consequently, the performance of a player may be naturally modeled as an uncertain object where each record corresponds to an instance. In these applications, the top k dominating query plays an important role in multi-criteria decision analysis when the scoring function is not available. For instance, users may identify the top k risky observation sites in the meteorology system, or find the top k valuable players based on their game-by-game performance. Due to the inherent differences between uncertain data and traditional data, many important queries are re-investigated in the context of uncertain data models. In the paper, we study the problem of the top k dominating query on uncertain data.

Challenges. The main challenges of the paper are two-fold. Firstly, we need to develop a model to properly identify the top k dominating objects. Secondly, efficient computation algorithm is required to support the new top k dominating query on uncertain data.

     As shown in Figure 1(b), each uncertain object may consist of multiple instances (points) , and each instance will appear with a particular probability.

For example, the uncertain object A consists of two instances a_1 and a_2. Due to the existence of multiple instances of the uncertain objects, it is non-trivial to measure the number of other uncertain objects dominated by an uncertain object (i.e., *dominance score*) since the traditional dominance relationship is defined against two points. Intuitively, we can derive the *dominance score* of an instance u based on the probability mass of the instances which are dominated by u. Then the problem of the top k dominating query can be mapped to the problem of the top k query on uncertain data since each multi-dimensional uncertain object corresponds to a score distribution based on the *dominance score* and appearance probabilities of their instances. The problem of the top k query on uncertain data has been extensively studied in recent years, and many models are proposed. Particularly, as shown in [9] the *parameterized ranking* function can unify most popular ranking semantics, and hence it is widely used as the de facto top k semantics for uncertain data. Although there are some existing work [11,10,20] investigating the top k dominating query on uncertain data, none of them supports the *parameterized ranking* semantics. Moreover, as shown in [3] the top k semantics adopted in [11,10,20] cannot properly capture the ranking of both probabilities and values.

In the paper, we adopt the *parameterized ranking* semantics to formally define the top k dominating query on multi-dimensional uncertain objects and propose an effective and efficient algorithm to support the top k dominating query by developing novel pruning techniques based on popular R-tree based indexing structure and some simple statistics information.

Contributions. Our principal contributions in this paper can be summarized as follows.

- We formally introduce a top k dominating query for multi-dimensional uncertain objects based on the state-of-the-art top k semantics on uncertain data.
- We propose effective and efficient top k dominating computation algorithms on multi-dimensional uncertain objects based on novel pruning techniques.
- We further improve the performance of the algorithm by utilizing statistics information of the uncertain objects.
- Comprehensive experiments on real and synthetic datasets demonstrate the efficiency and scalability of our techniques.

Roadmap. The rest of the paper is organized as follows. Section 2 introduces the problem and some preliminary knowledge. Section 3 develops efficient algorithms to support the top k dominating query by utilizing the spatial indexing and statistics information. The experimental results are reported in Section 4. This is followed by the related work presented in Section 5. We conclude our paper in Section 6.

2   Preliminary

We present problem definition and necessary preliminaries in this section. Table 1 summarizes notations frequently used throughout the paper.

Table 1. The summary of notations

Notation	Meaning
U, V	uncertain objects
u, v	instances of the uncertain objects
$u \prec v$	u *dominates* v
p_u	occurrence probability of the instance u
$s(u)$	*dominance score* of the instance u
$\varUpsilon(U)/\varUpsilon(u)$	the *rank score* of an object U/instance u
U_{mbr}	minimal bounding rectangle of U
U_{mbr}^{-} (U_{mbr}^{+})	lower (upper) corner point of U_{mbr}
U_s	*dominance score* distribution of U
U_s^{-} (U_s^{+})	lower (upper) bound of the *dominance scores* for instances in U

2.1 Problem Definition

A point (instance) p is in a d-dimensional space and the i-th dimensional co-ordinate value of p is denoted by $p.D_i$. Without loss of generality, we assume smaller coordinate values are preferred. For two points p and q, p *dominates* q, denoted by $p \prec q$, if $p.D_i \leq q.D_i$ for all dimension $i \in [1, d]$ and there is at least one dimension $j \in [1, d]$ with $p.D_j < q.D_j$. Meanwhile, we use $p \preceq q$ to denote that p *dominates or equals* q.

Uncertain Object Model. An uncertain object can be described either *continuously* or *discretely*. In the paper, we focus on the *discrete* case. Note that we can discretize a continuous probability density function (PDF) of an uncertain object by sampling methods. In the *discrete* case, an uncertain object U consists of a set $\{u_1, u_2, \ldots, u_m\}$ of instances (points). For $1 \leq i \leq m$, an instance u_i occurs with probability p_{u_i} $(p_{u_i} > 0)$, and $\sum_{i=1}^{m} p_{u_i} = 1$. We assume that the uncertain objects are *independent* to each other. In the following paper, we use *object* to denote *multi-dimensional uncertain object* whenever there is no ambiguity. Given an object U, U_{mbr} denotes the minimal bounding rectangle which contains all of the instances of U. Let U_{mbr}^{-} (U_{mbr}^{+}) denote the lower (upper) corner of U_{mbr}, we have $U_{mbr}^{-} \preceq u$ and $u \preceq U_{mbr}^{+}$ for any instance $u \in U_{mbr}$.

Dominance Score Distribution. Based on the dominance relationship against two points (instances), we can easily measure the goodness of an instance as follows.

Definition 1 (Dominance Score). *Given a set $\mathcal{O}$ of objects, the dominance score of an instance u of the object U, denoted by $s(u)$, is*

$$s(u) = \sum_{V \in \mathcal{O} \setminus U} \sum_{v \in V \ \wedge \ u \prec v} p_v \tag{1}$$

where $\sum_{v \in V \wedge u \prec v} p_v$ represents the probability that the object V is dominated by the instance u.

Given a set $\mathcal{O}$ of objects, we can derive the *dominance score distribution* for each object $U \in \mathcal{O}$, denoted by U_s, where $U_s = \{(s(u_1), p_{u_1}), \ldots, (s(u_i), p_{u_i}), \ldots,$

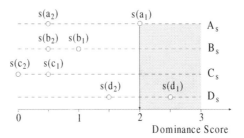

Fig. 2. Dominance Score Distributions

$(s(u_m), p_{u_m})\}$. We use $Pr(U_s > c)$ to denote the probability that U_s is larger than the value c, where $Pr(U_s > c) = \sum_{u_i \in U \wedge s(u_i) > c} p_{u_i}$.

Example 2. Regarding the example in Figure 1(b) and Definition 1, *dominance score distributions* of four objects A, B, C and D are depicted in Figure 2 where we assume each instance has appearance probability 0.5. Particularly, as a_1 dominates four instances (b_1, b_2, d_2, and c_2), and a_2 only dominates c_2, we have $s(a_1) = 2$, $s(a_2) = 0.5$. Since $p_{a_1} = p_{a_2} = 0.5$, we get $A_s = \{(2, 0.5), (0.5, 0.5)\}$. Similarly, we have $B_s = \{(1, 0.5), (0.5, 0.5)\}$, $C_s = \{(0.5, 0.5), (0, 0.5)\}$, $D_s = \{(2.5, 0.5), (1.5, 0.5)\}$, and $Pr(D_s > 2) = p_{d_1} = 0.5$.

Parameterized Ranking. The *dominance score distributions* of a set of objects can be ranked by the top k semantics studied for uncertain data in the literature. In the paper, we focus on the *parameterized ranking* function (PRF^ω) proposed in [9] since it can unify other popular ranking functions.

For a set $\mathcal{O}$ of objects, a possible world W proposed in [4] is a set of instances with one instance from each uncertain object. Given a set of uncertain objects $\{U_1, U_2, \ldots, U_n\}$, a possible world $W = \{u_1, u_2, \ldots, u_n\}$ is a set of instances sequentially sampled from each object. Assume the uncertain objects are independent to each other, and the probability of W to appear is $Pr(W) = \prod_{i=1}^{n} p_{u_i}$. In each world W, an object is ranked based on the score (value) of its corresponding instance in W. In the paper, we use $r_W(U)$ to denote the rank of an object in the possible world W which is abbreviated to $r(U)$ whenever there is no ambiguity. Let $\mathcal{W}$ denote the set of all possible worlds, and we have $\sum_{W \in \mathcal{W}} Pr(W) = 1$ where $Pr(W)$ is the occurring probability of the possible world W. Then we have the formal definition of *parameterized ranking* function.

Definition 2 (PRF^ω). *Let ω be a weighted function which maps an object-rank pair to a complex number, the **rank score** of an object U, denoted by $\Upsilon(U)$, is defined as follows.*

$$\Upsilon(U) = \sum_{i \in [1, n]} \omega(i) \times Pr(r(U) = i) \tag{2}$$

where $\omega(i)$ denotes the weight of the i-th position, and $Pr(r(U) = i)$ denotes the probability of U ranked at the i-th position, i.e., $Pr(r(U) = i) = \sum_{W \in \mathcal{W} \wedge r(U) = i} Pr(W)$. Recall that $r(U)$ denotes the rank of U in the possible world W.

In the paper, we assume $w(i) \geq w(j)$ for any two ranking positions i and j where $i < j$. This is very intuitive as a higher position is usually at least as desirable as those behind it and thus should be given a higher weight.

Problem Statement. Given a set $\mathcal{O}$ of objects, we aim to return the top k dominating objects based on their *dominance score distributions* and *parameterized ranking* semantics; that is, we retrieve the top k objects with the highest *rank scores* regarding their *dominance score distributions*.

2.2   Computing Rank Score $\Upsilon(U)$

As shown in [9], the *rank score* of an object U can be calculated by the summation of the *rank scores* of its instances. For an instance $u \in U$, we can use the following *generating function* $\mathcal{F}(\mathbf{x}, u)$ to calculate the *rank score* of u, where $P_{V,u}$ is the probability that V_s is larger than $s(u)$ (i.e., $P_{V,u} = Pr(V_s > s(u)) = \sum_{v \in V \wedge s(v) > s(u)} p_v$). Recall that V_s is the *dominance score distribution* of the object V. Intuitively, a small $P_{V,u}$ is in favor of the *rank score* of the instance u.

$$\mathcal{F}(\mathbf{x}, u) = \left(\prod_{V \in \mathcal{O} \setminus U} (1 - P_{V,u} + P_{V,u} \cdot \mathbf{x}) \right) (p_u \cdot \mathbf{x}) \tag{3}$$

Then we have $Pr(r(u) = i) = c_i$ where $r(u)$ is the rank position of instance u and c_i is the coefficient of $\mathbf{x}^i$ in $\mathcal{F}(\mathbf{x}, u)$. Therefore, after applying Equation 2, we have

$$\Upsilon(u) = \sum_{1 \leq i \leq n} w(i) \times c_i \tag{4}$$

$$\Upsilon(U) = \sum_{u \in U} \Upsilon(u) \tag{5}$$

Example 3. In Figure 2, we have $s(a_1) = 2$ and hence $Pr(B_s > s(a_1)) = 0.0$, $Pr(C_s > s(a_1)) = 0.0$, and $Pr(D_s > s(a_1)) = 0.5$ (i.e., the probability mass of the instances of D in the shaded area). According to Equation 3, $\mathcal{F}(\mathbf{x}, a_1) = (1) \times (1) \times (0.5 + 0.5 \mathbf{x}) \times 0.5 \mathbf{x} = 0.25 \mathbf{x} + 0.25 \mathbf{x}^2$. Therefore, we have $Pr(s(a_1) = 1) = 0.25$ and $Pr(s(a_1) = 2) = 0.25$. Similarly, $\mathcal{F}(\mathbf{x}, a_2) = (0.5 + 0.5 \mathbf{x}) \times (1) \times (\mathbf{x}) \times 0.5 \mathbf{x} = 0.25 \mathbf{x}^2 + 0.25 \mathbf{x}^3$, and hence $Pr(s(a_2) = 1) = 0$, $Pr(s(a_2) = 2) = 0.25$,and $Pr(s(a_2) = 3) = 0.25$. Suppose $w(1) = 4$, $w(2) = 3$, $w(3) = 2$ and $w(4) = 1$ in PRF^ω, then $\Upsilon(A) = \Upsilon(a_1) + \Upsilon(a_2) = (0.25 \times 4 + 0.25 \times 3) + (0.25 \times 3 + 0.25 \times 2) = 3$ according to Equation 4 and 5. Similarly, we have $\Upsilon(B) = 2.5$, $\Upsilon(C) = 1.5$ and $\Upsilon(D) = 3.75$. Therefore, $\Upsilon(D) > \Upsilon(A) > \Upsilon(B) > \Upsilon(C)$.

3   Approach

A straightforward solution for the problem of top k dominating query is to first calculate the *dominance score* for each instance of the objects by conducting dominance checks against the instances of other objects, then compute the *rank scores* of the objects. The k objects with the highest *rank scores* are the top k dominating objects. However, it is cost-inhibitive because the cost of *dominance score* computation is not cheap and the total number of instances may be huge. In this section, we propose efficient algorithms to identify the top k dominating uncertain objects. Specifically, Section 3.1 presents the framework of our algorithms following the *filtering and verification* paradigm. Section 3.2 proposes efficient algorithms for the computation of the *dominance score*. Section 3.3 introduces the spatial pruning technique based on the MBRs of the objects. Rank score based pruning technique is proposed in Section 3.4. In Section 3.5, statistics based approach further improves the performance of the algorithms by utilizing the probabilistic inequalities.

To facilitate the top k dominating query, we assume uncertain objects are organized by R-tree indexing techniques. Given a set of uncertain objects, the MBRs of the objects are indexed by R-tree [7], which is called the *global R-tree*.As to each uncertain object U, an *aggregate R-tree* [14] is employed to organize the instances where the aggregate value of each intermediate entry is the probability mass of the instances in the entry, which is called a *local R-tree* for an uncertain object U.

3.1   Compute Top k Dominating Objects

In this subsection, we introduce the framework of the top k dominating algorithm based on the *filtering and verification* paradigm in Algorithm 1. Pruning techniques are developed to reduce computational cost by eliminating non-promising objects. Assume MBRs of the objects are organized by the *global R-tree* $\mathcal{R}$, Line 1 derives the lower and upper bounds of the *dominance scores* for each object U, denoted by U_s^- and U_s^+ respectively (See Section 3.2). Then Line 2 applies the spatial based pruning technique to identify the candidate objects which are kept in a set $\mathcal{C}$. In addition to the candidate objects, $\mathcal{L}$ keeps the objects whose instances may contribute to the *rank scores* computation of candidate objects. As shown in Section 3.3, we can safely remove remaining objects (i.e., $\mathcal{O} \setminus \mathcal{L}$) from *rank scores* computation [1]. A max-heap $\mathcal{H}$ is employed to guarantee that instances are accessed by decreasing order, which is initialized by objects in $\mathcal{L}$ where the upper bounds of their *dominance scores* are key values in the heap(Line 4). The *generating function* $\mathcal{F}(x, u)$ is maintained to compute the *rank scores* for instances accessed (Line 8 and 10). Algorithm 1 maintains a *rank score* threshold λ to prune non-promising objects in Line 11 and 17 by utilizing *rank score* based pruning techniques (See details in Section 3.4). Line 18 loads the instances of an object and calculate their *dominance scores*, where an efficient *dominance scores* computation algorithm is presented in Section 3.2. Then Line 19 pushes these instances into the heap $\mathcal{H}$ for further processing. Line 5 terminates the algorithm when there is no candidate object for further exploration or the heap is empty. Finally the k objects with the highest *rank scores* are returned as the top k dominating objects.

---

[1] These objects may be involved in the *dominance score* computation of other objects.

Algorithm 1. Top k Dominating Objects $(\mathcal{R},\ k)$

    Input   : $\mathcal{R}$: the *global* R-tree of $\mathcal{O}$, k: objects retrieved

    Output : Top k dominating objects

1 $\lambda := 0$; Compute U_s^+ and U_s^- for all objects $U \in \mathcal{O}$;

2 $\mathcal{L}, \mathcal{C} \leftarrow$ spatial based pruning against $\mathcal{O}$;

3 **for** each $U \in \mathcal{L}$ **do**

4   Push U into $\mathcal{H}$ with key value U_s^+ ;

5 **while** $\mathcal{H} \neq \emptyset$ **or** $\mathcal{C} \neq \emptyset$ **do**

6      $E \leftarrow deheap(H)$;

7   **if** E is an instance u from object U **then**

8   Update the *generating function* $\mathcal{F}(x, u)$;

9   **if** $U \in \mathcal{C}$ **then**

10   Compute $\Upsilon(u)$ based on $\mathcal{F}(x, u)$;

11               $\mathcal{C} := \mathcal{C} \setminus U$ if U can be pruned by λ ;

12   **if** u is the last instance of U **then**

13   Compute $\Upsilon(U)$; Update λ; $\mathcal{C} := \mathcal{C} \setminus U$;

14   **else**

15   E corresponds to the object U ;

16   **if** U can be pruned based on λ **then**

17               $\mathcal{C} := \mathcal{C} \setminus U$;

18   Compute *dominance scores* for instances of U ;

19   Push every $u \in U$ into $\mathcal{H}$ with key value $s(u)$;

20 **return** Top k objects with highest *rank scores*

3.2   Compute Dominance Scores

In this subsection, we introduce efficient *dominance scores* computation algorithms based on the R-tree structure.

Dominance Relationships for Rectangles. A pair of uncertain objects may have three relationships as follows. Let R^+ and R^- denote the upper and lower corners of a rectangle R, we say the rectangle R_1 *fully dominates* another rectangle R_2 if $R_1^+ \prec R_2^-$. Similarly, we have R_1 *partially dominates* R_2 if $R_1^- \prec R_2^+$ and $R_1^+ \not\prec R_2^-$. Otherwise, we say R_1 does not dominate R_2. It is immediate that we have $x \prec y$ for any points $x \in R_1$ and $y \in R_2$ if R_1 *fully dominates* R_2. Similarly, $x \not\prec y$ for any points $x \in R_1$ and $y \in R_2$ if R_1 does not *dominate* R_2.

Compute the Lower and Upper Bounds $(U_s^-,\ U_s^+)$
Based on the definition of the *dominance score* (Definition 1) and the *dominance relationship* between two rectangles, we can derive the lower and upper bounds of the *dominance scores* of the instances from an object U, denoted by U_s^+ and U_s^- respectively, based on the MBRs of the objects. In this subsection, U_s^- equals the number of other objects whose MBRs are *fully dominated* by U_{mbr}. Similarly, U_s^+ is the number of other objects whose MBRs are *fully dominated* or *partially dominated* by U_{mbr}.

Motivation. We may issue two dominating range queries based on the lower and upper corners of the MBR of each object against the *global R-tree* $\mathcal{R}$ in Algorithm 1 to compute their lower and upper bounds of the *dominance scores*. Nevertheless, we can improve the computational cost by conducting a spatial join based computation. We conduct the dominance checks in a level-by-level fashion such that the dominance count can be calculated at higher level, and hence significantly reduce the number of dominance checks.

Algorithm. Algorithm 2 illustrates the details of the *dominance score* bounds computation based on MBRs of the objects, which follows the *synchronized R-tree traversal paradigm* used in spatial join. For each entry E (data entry or intermediate entry) of the *global R-tree* $\mathcal{R}$, we use a tuple T to record its lower and upper bounds of its *dominance score*, denoted by T_s^- and T_s^+ respectively, while $T.owner$ refers to the entry E. Clearly, we do not need to further explore another entry E_1 regarding E if E *fully dominates* E_1 or E does not *dominate* E_1. We use $T.set$ to keep a set of entries which are *partially dominated* by E. A FIFO queue $\mathcal{Q}$ is employed to maintain the tuples, and $\mathcal{Q}$ is initialized by a tuple T where $T.owner$ and $T.set$ are set to the root of the *global R-tree* $\mathcal{R}$ (Line 1-2). For each tuple T popped from $\mathcal{Q}$, if the entries from $T.owner$ and $T.set$ are data entries, then we have $U_s^+ = T_s^+$ and $U_s^- = T_s^-$ where $T.owner$ refers to the MBR of the object U. Otherwise, Line 7-17 expand $T.owner$ and $T.set$ for *dominance score* computation of the lower level entries. For presentation simplicity, the child entries of a data entry are referred to the data entry itself (Line 7 and 11). Specifically, for each pair of entries $t.owner$ and e at Line 12 and 13, the lower bound t_s^- and upper bound t_s^+ are increased by $e.cn$ if t_{mbr} *fully dominates* e_{mbr}, where $e.cn$ is the aggregate number of objects in entry e (Line 15). Otherwise, we only increase t_s^+ and keep e in $t.set$ for further computation if t_{mbr} *partially dominates* e_{mbr} (Line 17). Algorithm 2 terminates when $\mathcal{Q}$ is empty, and the lower and upper *dominance score* bounds of the objects are ready for the spatial pruning in Section 3.3.

Compute the Dominance Score

In the top k dominating algorithm (Line 18 of Algorithm 1) we need to compute the *dominance scores* of the instances for the object U. We can come up with an efficient *dominance scores* computation algorithm for an object U by slightly modifying Algorithm 2. Suppose U_s^- is calculated, we only need to consider a set $\mathcal{S}$ of objects which are *partially dominated* by U. For each object $V \in \mathcal{S}$, we set $T.owner$ and $T.set$ to the roots of R_U and R_V respectively at Line 1 where R_U and R_V are the local R-tree of the objects U and V respectively. When algorithm terminates, for each instance $u \in U$ we have the probability mass of instances from V which are dominated by u, denoted by $P(u, V)$. According to Definition 1, the *dominance score* of the instance u can be calculated as follows.

$$s(u) = U_s^- + \sum_{V \in \mathcal{S}} P(u, V). \tag{6}$$

3.3 Spatial Pruning Technique

Considering that the cost is expensive to compute *dominance scores*, we propose effective spatial pruning techniques to reduce the number of candidate objects based on the *global R-tree*; that is, we aim to prune a set of objects from *dominance score* computation without accessing the instances of the objects such that the CPU and I/O costs can be significantly reduced.

Theorem 1. *Given two objects U and V, we have $\Upsilon(U) > \Upsilon(V)$ if $U_s^- > V_s^+$.*

Proof. Since $U_s^- > V_s^+$, we have $s(u) > s(v)$ for any $u \in U$ and $v \in V$. Therefore, we have $Pr(r(U) \leq i) > Pr(r(V) \leq i)$ where $Pr(r(U) \leq i)$ denotes the probability that U is ranked not lower than the i-th position. According to Equation 2 and the monotonic property of the weight function (i.e., $\omega(i) \geq \omega(j)$ for any two positions i and j with $i < j$), we have $\Upsilon(U) > \Upsilon(V)$.

Spatial Based Pruning. Let f_c denote the k-th largest lower bound of the *dominance scores* regarding objects in $\mathcal{O}$, according to Theorem 1, we have $\mathcal{C} = \{U | U \in \mathcal{O} \text{ and } U_s^+ \geq f_c\}$ in Algorithm 1; that is, only the object U with $U_s^+ \geq f_c$ can become the top k candidate objects. Let f_s denote the smallest lower bounds of *dominance scores* regarding objects in $\mathcal{C}$, we have $\mathcal{L} = \{U | U \in \mathcal{O} \text{ and } U_s^+ \geq f_s\}$ in Algorithm 1. Note that, as shown in Equation 3 the *rank score* of an instance can only be affected by other instances with larger *dominance scores*.

Algorithm 2. Compute Dominance Score Bounds ($\mathcal{R}$)

    Input   : $\mathcal{R}$: the *global R-tree* of $\mathcal{O}$
    Output : Objects with lower and upper *dominance scores* bounds
1 $T.owner :=$ root of $\mathcal{R}$; $T.set :=$ root of $\mathcal{R}$;
2 Push T into FIFO queue $\mathcal{Q}$;
3 **while** $\mathcal{Q} \neq \emptyset$ **do**
4 $T \leftarrow dequeue(\mathcal{Q})$;
5   **if** $T.owner$ is not a data entry **or** entries in $T.set$ are not data entries **then**
6 $\mathcal{L} := \emptyset; \mathcal{W} := \emptyset$;
7   **for** each child entry e of $T.owner$ **do**
8 $t.owner = e; t_s^- := T_s^-; t_s^+ := t_s^-$;
9 $\mathcal{L} := \mathcal{L} \cup t$;
10   **for** each entry e in $T.set$ **do**
11 $\mathcal{W} := \mathcal{W} \cup$ child entries of e ;
12   **for** each tuple t in $\mathcal{L}$ **do**
13   **for** each entry e in $\mathcal{W}$ **do**
14   **if** t_{mbr} *fully dominates* e_{mbr} **then**
15 $t_s^- := t_s^- + e.cn; t_s^+ := t_s^+ + e.cn$;
16   **else if** t_{mbr} *partially dominates* e_{mbr} **then**
17 $t_s^+ := t_s^+ + e.cn$; $t.set := t.set \cup e$;
18   Push t into $\mathcal{Q}$;

3.4   Rank Score Based Pruning Technique

In this subsection, we propose effective pruning techniques based on the *rank scores* of the objects. Let λ in Algorithm 1 denote the k-th largest *rank scores* for objects accessed, clearly we can safely remove an object U from the top k candidates if we can claim the upper bound of $\Upsilon(U)$ is smaller than λ.

Rank Score Based Pruning. In Algorithm 1, we invoke the *rank score* based pruning technique at Line 11 and 17. Let u_i be the i-th visited instance of U, we can calculate $\Upsilon(U)^+$ by setting $p_{u_i} = \sum_{i \leq j \leq m} u_j$ (i.e., move probability mass of the unvisited instances of U to u_i), and prune the object U from candidate set $\mathcal{C}$ at Line 11 if $\Upsilon(U)^+ \leq \lambda$. Let u_0 denote an instance where $s(u_0) = U_s^+$ and $p_{u_0} = 1.0$,i.e., an object constructed by pushing all instances of U to the lower corner of U_{mbr}. Consequently, at Line 17 of Algorithm 1, we can calculate $\Upsilon(U)^+$ based on U_s^+ without loading instances of U and remove U from candidate set $\mathcal{C}$ if $\Upsilon(U)^+ \leq \lambda$. It is immediate that we can also remove any unvisited object $V \in \mathcal{C}$ (i.e., the object V with $V_s^+ < U_s^+$) in $\mathcal{C}$ from the top k candidates since we have $\Upsilon(V) \leq \Upsilon(V)^+$ and $\Upsilon(V)^+ \leq \Upsilon(U)^+$.

3.5   Enhance the Performance with Statistics

Assume some statistics information (*mean* and *variance*) of the objects are available, we can further enhance the performance of the top k dominating query by utilizing probabilistic inequalities. Specifically, given two objects U and V, we use $\Delta^-(V, U)$ ($\Delta^+(V, U)$) to denote the contribution of U towards the lower (upper) bound of the *dominance score* of V. In Section 3.2 we have $\Delta^-(V, U) = 0$ and $\Delta^+(V, U) = 1$ if V *partially dominates* U. This subsection shows that we can derive tighter lower and upper bounds for the *dominance scores* of the objects based on their MBRs and statistics information.

Motivation. As shown in Figure 3(a), given the MBR of an object U and a point p, we use rectangle A to denote the area of U_{mbr} on the left side of p (shaded area). Similarly, in Figure 3(b), we use rectangle B to denote the area of U_{mbr} on the bottom of p (shaded area). Moreover, the rectangle R_p (the rectangle with thick line) represents the area of U_{mbr} which is dominated by the point p. Let $P(R)$ denote the probability mass of U within the rectangle R, then we should have $0 \leq \Delta^-(V, U) \leq P(R_p)$ if p corresponds to the upper corner of V_{mbr} (i.e., V_{mbr}^+). Since $P(R_p) \geq 1 - (P(A) + P(B))$ in Figure 3(b), we may have $\Delta^-(V, U) = 1 - (P(A) + P(B))$ .As we cannot have exact $P(A)$ ($P(B)$) value without accessing instances of U, this subsection shows that we can derive the upper bound of $P(A)$ ($P(B)$), denoted by $P^+(A)$ ($P^+(B)$) based on statistics information. Then we can come up with tight $\Delta^-(V, U)$ and $\Delta^+(V, U)$ values without loading instances of U. Specifically, we may have $\Delta^-(V, U) = 1 - (P^+(A) + P^+(B))$ (p is the upper corner of V_{mbr}) and $\Delta^+(V, U) = min(P^+(A), P^+(B))$ (p is the lower corner of V_{mbr}). With similar rationale, we may have $\Delta^+(V, U) = min(P^+(A), P^+(B))$ when p corresponds to the lower corner of V_{mbr} in Figure 3(d).

Example 4. Suppose we have $P^+(A) = 0.2$ and $P^+(B) = 0.3$ in Figure 3. We have $P(R_p) \geq 1 - (P^+(A) + P^+(B)) = 0.5$. Therefore, we can set $\Delta^-(V, U) = 0.5$ in Figure 3(b). Similarly, in Figure 3(d) we have $P(R_p) \leq min(P^+(A), P^+(B))$ and hence $\Delta^+(V, U)$ can be set to 0.2.

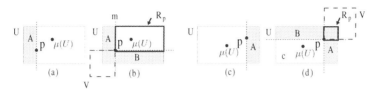

Fig. 3. Statistic based Pruning

Definitions and Lemmas. We formally define two statistics information of an object U as follows.

Definition 3 (*mean* $\mu(U)$). *We use $\mu(U)$ to denote the mean of an object U, where $\mu(U).D_i = \sum_{u \in U}(u.D_i \times p_u)$.*

Definition 4 (*variance* $\sigma^2(U)$). *$\sigma^2(U)$ denotes the variance of an object U on each dimension; that is, $\sigma_i^2(U) = \sum_{u \in U}((u.D_i - \mu(U).D_i)^2 \times p_u)$.*

Let $\delta(x, y)$ be $\frac{1}{1 + \frac{x^2}{y^2}}$ if $y \neq 0$, 1 if $x = 0$ and $y = 0$, and 0 if $x \neq 0$ and $y = 0$.

Lemma 1 (Cantelli's Inequality [13]). *Suppose that t is a random variable in 1-dimensional space with mean $\mu(t)$ and variance $\sigma^2(t)$, $Prob(t - \mu(t) \geq a) \leq \delta(a, \sigma(t))$ for any $a \geq 0$, where $Prob(t - \mu(t) \geq a)$ denotes the probability of $t - \mu(t) \geq a$.*

Note that Lemma 1 extends the original Cantellis Inequality [13] to cover the case when $\sigma = 0$ and/or $a = 0$. Then we can come up with another version of Cantelli's Inequality [12], which provides an upper-bound for $Prob(t \preceq b)$ when $b < \mu$.

Lemma 2. *Assume that $0 < b < \mu(t)$. Then, $Prob(t \leq b) \leq \delta(\mu(t) - b, \sigma(t))$.*

Proof. Let $t' = 2\mu(t) - t$. It can be immediately verified that $\sigma^2(t') = \sigma^2(t)$ and $\mu(t) = \mu(t')$. The theorem holds by applying Cantelli's Inequality on t'.

Compute $\Delta^-(V, U)$. Given two objects U and V, the following theorem indicates that we can derive $\Delta^-(V, U)$ based on the statistics information.

Theorem 2. *Given two objects U and V, let p denote the upper corner of V_{mbr}, we have $\Delta^-(V, U) = 1 - \sum_{1 \leq i \leq d}(\delta(\mu_i(U) - p.D_i, \sigma_i(U)))$ if $p \prec \mu(U)$. Note that we set $\delta(\mu_i(U) - p.D_i, \sigma_i(U))$ to 0 if $p.D_i < U_{mbr}^-.D_i$.*

Proof. Let R_i denote the left side of the rectangle U_{mbr} divided by the point p on the i-th dimension (e.g., $R_1 = A$ and $R_2 = B$ in Figure 3(b)). We use $P(R_i)$ to record the probabilistic mass of the instances in U contained by R_i. Clearly, we have $P(R_i) = 0$ if $p.D_i < U_{mbr}^-.D_i$ since R_i corresponds to an empty rectangle.

Otherwise, we have $P(R_i) \leq \delta(\mu_i(U) - p_i, \sigma_i(U))$ according to Theorem 2. Let R_p denote the rectangle whose lower (upper) corner is p (U_{mbr}^+), and $P(R_p)$ records the probabilistic mass of the instances in U contained by R_p. Then we have $P(R_p) \geq 1 - \sum_{1 \leq i < d}(\delta(\mu_i(U) - p.D_i, \sigma_i(U)))$, which implies that we can set $\Delta^-(V,U)$ to $1 - \sum_{1 \leq i < d}(\delta(\mu_i(U) - p.D_i, \sigma_i(U)))$ since all instances within R_p are dominated by p which is the upper corner of V_{mbr}. Therefore, the theorem holds.

Compute $\Delta^+(V,U)$. The following theorem indicates that we can derive $\Delta^+(V,U)$ based on the statistics information.

Theorem 3. *Given two objects U and V, let p denote the lower corner of V_{mbr}, we have $\Delta^+(V,U) = min(\{\delta(p.D_i - \mu_i(U), \sigma_i(U))\})$ with $1 \leq i \leq d$ if $\mu(U) \prec p$ and $p \prec U_{mbr}^+$.*

We omit the details of the proof since it is similar to Theorem 2.

Achieve Better Dominance Score bounds. Let $\hat{V}_s^-$ ($\hat{V}_s^+$) denote the new lower (upper) bound for the *dominance score* of the object V, and $\mathcal{S}$ is the set of objects which are *partially dominated* by V, we have $\hat{V}_s^- = V_s^- + \sum_{U \in \mathcal{S}} \Delta^-(V,U)$ and $\hat{V}_s^+ = V_s^- + \sum_{U \in \mathcal{S}} \Delta^+(V,U)$.

Suppose the statistics information of the objects are kept with the data entries of the objects in the *global* R-tree, we can use the new lower and upper bounds of *dominance scores* in Algorithm 1. Our empirical study shows that although the statistics information slightly increase the index size, the gain is significant since the tighter *dominance scores* bounds lead to smaller candidate size, and hence reduce the CPU and I/O costs.

4   Experiment

In this section, we present results of a comprehensive performance study to evaluate the efficiency and scalability of the proposed techniques in the paper. As there is no existing work on top k dominating query on uncertain data following the *parameterized ranking* semantics, we only evaluate the techniques proposed in the paper. Following algorithms are implemented for performance evaluation.

- **NAIVE:** The *straightforward solution* is introduced in the Section 3.
- **NAIVES:** Algorithm 1 proposed in Section 3 where only the spatial based pruning technique (Section 3.3) is employed.
- **BAS:** Algorithm 1 proposed in Section 3 where spatial based pruning technique and spatial join based *dominance score* computation algorithms (Section 3.2) are employed. It is employed as the baseline algorithm in our empirical study.
- **TKDOM: BAS** Algorithm which also applies the *rank score* based pruning technique (Section 3.4).
- **TKDOM*: TKDOM** Algorithm which also applies statistics based techniques (Section 3.5).

We employ a specific *parameterized ranking linear* function $PFR^e(\alpha)$ to rank the objects. Like the setting in [9], we use $PFR^e(\alpha = 0.95)$ in the experiments.

Datasets. We evaluate our techniques on both synthetic and real datasets. Synthetic datasets are generated by using the methodologies in [1] regarding the following parameters. Dimensionality d varies from 2 to 5 with default value 3. Data domain in each dimension is [0, 10000]. The number n of objects in each dataset varies from 10K to 50K with default value 10K. The number m of instances per object varies from 50 to 300 with the default value 100. The value k varies from 10 to 50 with default value 20. The edge length h of object MBRs varies from 100 to 600 with default value 200. Centers of objects follow either Equally(E), Correlated(C) or Anti-correlated(A) distribution where default is A distribution.The instances of an uncertain object follow popular distributions Normal(N) and Uniform(U) where N distribution is default. And two real datasets, Forest CoverType dataset (COV) (http://archive.ics.uci.edu/ml) and Household (HOU) (http://www.ipums.org), are employed to represent the centers of the uncertain objects. In COV, we select the horizontal and vertical distances of each observation point to the Hydrology as well as the elevation of the point. In HOU, each record represents the percentage of an American family's annual income spent on 3 types of expenditures (e.g., gas, etc.). We choose 20,000 objects in COV and HOU respectively. For each object, we generate the instances according to the default setting above. Then with the default setting, the total number of instances in synthetic and real datasets are 1 millions and 2 millions respectively.

All algorithms are implemented in standard C++ and compiled with GNU GCC. Experiments are run on a PC with Intel Xeon 2.40GHz dual CPU and 4G memory under Debian Linux. In the paper, we evaluate the I/O performance of the algorithms by measuring the number of uncertain objects explored.

Performance Evaluation

In the first experiment, we vary the value of k and evaluate the performance of the five algorithms against the default synthetic dataset where k varies from 10 to 50 in Figure 4. As expected, all techniques proposed are effective since the performance of them degrades slowly against the growth of k. The NAVIE and NAIVES algorithms are much slower than the other algorithms, and the I/O costs of them are also much higher than the others. Then for better report on the performance of the algorithms, we exclude the NAIVE and NAIVES algorithms in the following experiments since they have been significantly outperformed by BAS, TKDOM and TKDOM* algorithms.

Impact of Data Distribution. We evaluate the performance of BAS, TKDOM and TKDOM* against 8 datasets in Figure 5, where C_N denotes the 3-dimensional synthetic data whose centers and instances follow the Correlated and Normal distributions respectively, and similar definitions go to C_U, E_N, E_U, A_N and A_U. It is observed that the distribution of the instances (N and U) does not noticeably affect performance of the algorithms, so we only perform tests on normal distribution in the following experiments. On the other

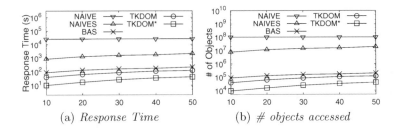

(a) *Response Time*   (b) *# objects accessed*

Fig. 4. Diff. top k

side, all algorithms are very sensitive to the distribution of the object centers. This is because of the nature of dominating query. It is easy to distinguish and sort the *dominance scores* of objects to get a small size of candidate set under the correlated distribution.Whereas anti-correlated distribution leads to more computation time because each object only fully dominate a limited number of other objects, so we use anti-correlated distribution as a default setting for locations. As expected, TKDOM* significantly outperforms other algorithms under all data distributions.

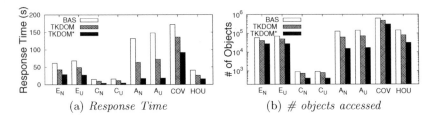

(a) *Response Time*   (b) *# objects accessed*

Fig. 5. Impact of Data Distributions

Evaluating Impacts by Different Setting. We study the scalability of our algorithms regarding different number of objects (n) in one dataset, number of instances (m), length of MBR edges (h), and the dimensionality (d) in Figure 6. The response time and the number of I/O increase with the increase of number of objects, instance number, MBR edge length. Clearly, the dataset size increases with objects and instances number thus the filtering processing becomes more expensive. Larger MBR edge length makes it difficult to prune object pairs as there is larger overlap in their MBRs. While the results decrease when d arises. This is because the average number of objects *dominated* or *partially dominated* by each object decreases against the growth of the dimensionality. That means with the increase of dimensionality d, the average area of MBRs gets smaller compared to the whole data space; consequently, the power of the pruning rules becomes more significant. Note that the edge lengths of the objects remains unchanged when the dimensionality grows in the experiments. The results also demonstrate that each pruning rule is very effective and significantly reduces the processing time and I/O cost.

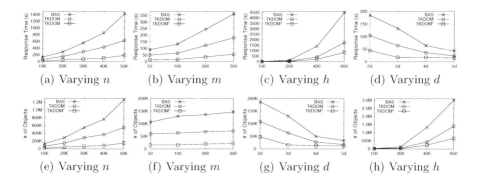

(a) Varying n (b) Varying m (c) Varying h (d) Varying d

(e) Varying n (f) Varying m (g) Varying d (h) Varying h

Fig. 6. Varying Parameters

5   Related Work

With the emergence of many recent important and novel applications involving uncertain data, there has been a great deal of research attention dedicated to this field. Particularly, top k queries are important in analyzing uncertain data. Unlike a top k query over certain data which returns the k best alternatives according to a ranking function, a top k query against uncertain data has inherently more sophisticated semantics. A large amount of work has been dedicated to top k queries with different semantics such as U-Top$k$ and U-kRank [16], PT-k [8], expected rank top k [3], c-Typical-Top$k$ [6] and unified top k [9] semantics. The top k semantics can be widely utilized in various queries, including range query [21], nearest neighbor query [22], reverse nearest neighbor query [2], etc.

Top k dominating queries are introduced by Papadias *et al.* [15] to retrieve the k points that dominate the largest number of other points. Unlike skyline query, the result does not necessarily contain skyline points. Top k dominating queries are widely studied on certain data [18,19,17].

Recently, uncertain query processing has received an increasing attention in many applications. Top k dominating query on uncertain data has been studied in [10,11]. They propose probabilistic top k dominating (PTD) query to retrieve k uncertain objects. Zhang *et al.* [20] formalize threshold-based probabilistic top k dominating queries to overcome some inherent computational deficiency in an exact computation. Nevertheless, none of them supports the *parameterized ranking* semantics. Recently, Feng *et al.* investigate the problem of probabilistic top k dominating query over sliding windows [5]. However, they only consider the x-tuple uncertain model. Techniques proposed in [10,11,20,5] cannot be applied to the top k dominating model proposed in our paper due to the inherent difference of the models.

6   Conclusion

We investigate the problem of identifying top k dominating objects over uncertain data. We formally define a new model for the top k dominating query on multi-dimensional uncertain data. By utilizing the popular R-tree indexing techniques as well as spatial based and rank score based pruning techniques, we develop an effective and efficient algorithm following the *filtering and verification* paradigm. We further improve the performance of the algorithm based on some simple statistics information of the objects. Our experiments convincingly demonstrate the effectiveness and efficiency of our techniques. In the future, we can further investigate the top k dominating query over large dataset.

Acknowledgement. Ying Zhang is supported by ARC DE140100679 and DP130103245. Wenjie Zhang is supported by ARC DE120102144 and DP120104168. Xuemin Lin is supported by NSFC61232006, NSFC61021004, ARC DP120104168 and DP110102937.

References

1. Börzsönyi, S., Kossmann, D., Stocker, K.: The skyline operator. In: ICDE 2001 (2001)
2. Cheema, M.A., Lin, X., Zhang, W., Zhang, Y.: Influence zone: Efficiently processing reverse k nearest neighbors queries. In: ICDE, pp. 577–588 (2011)
3. Cormode, G., Li, F., Yi, K.: Semantics of ranking queries for probabilistic data and expected ranks. In: ICDE (2009)
4. Dalvi, N.N., Suciu, D.: Efficient query evaluation on probabilistic databases. VLDB J. 16(4), 523–544 (2007)
5. Feng, X., Zhao, X., Gao, Y., Zhang, Y.: Probabilistic top-k dominating query over sliding windows. In: Ishikawa, Y., Li, J., Wang, W., Zhang, R., Zhang, W. (eds.) APWeb 2013. LNCS, vol. 7808, pp. 782–793. Springer, Heidelberg (2013)
6. Ge, T., Zdonik, S., Madden, S.: Top-k queries on uncertain data: On score distribution and typical answers. In: SIGMOD (2009)
7. Guttman, A.: R-trees: A dynamic index structure for spatial searching. In: SIGMOD Conference, pp. 47–57 (1984)
8. Hua, M., Pei, J., Zhang, W., Lin, X.: Ranking queries on uncertain data: A probabilistic threshold approach. In: SIGMOD (2008)
9. Li, J., Saha, B., Deshpande, A.: A unified approach to ranking in probabilistic databases. VLDB J. 20(2) (2011)
10. Lian, X., Chen, L.: Top-k dominating queries in uncertain databases. In: EDBT, pp. 660–671 (2009)
11. Lian, X., Chen, L.: Probabilistic top-k dominating queries in uncertain databases. Inf. Sci. 226, 23–46 (2013)
12. Lin, X., Zhang, Y., Zhang, W., Cheema, M.A.: Stochastic skyline operator. In: ICDE, pp. 721–732 (2011)
13. Meester, R.: A Natural Introduction to Probability Theory (2004)
14. Papadias, D., Kalnis, P., Zhang, J., Tao, Y.: Efficient olap operations in spatial data warehouses. In: Jensen, C.S., Schneider, M., Seeger, B., Tsotras, V.J. (eds.) SSTD 2001. LNCS, vol. 2121, pp. 443–459. Springer, Heidelberg (2001)

15. Papadias, D., Tao, Y., Fu, G., Seeger, B.: Progressive skyline computation in database systems. ACM Trans. Database Syst. 30(1) (2005)
16. Soliman, M.A., Ilyas, I.F., Chang, K.C.: Top-k query processing in uncertain databases. In: ICDE 2007 (2007)
17. Tiakas, E., Papadopoulos, A.N., Manolopoulos, Y.: Progressive processing of subspace dominating queries. VLDB J. 20(6), 921–948 (2011)
18. Yiu, M.L., Mamoulis, N.: Efficient processing of top-k dominating queries on multi-dimensional data. In: VLDB, pp. 483–494 (2007)
19. Yiu, M.L., Mamoulis, N.: Multi-dimensional top-k dominating queries. VLDB J. 18(3), 695–718 (2009)
20. Zhang, W., Lin, X., Zhang, Y., Pei, J., Wang, W.: Threshold-based probabilistic top-k dominating queries. VLDB J. 19(2), 283–305 (2010)
21. Zhang, Y., Lin, X., Tao, Y., Zhang, W.: Uncertain location based range aggregates in a multi-dimensional space. In: ICDE, pp. 1247–1250 (2009)
22. Zhang, Y., Lin, X., Zhu, G., Zhang, W., Lin, Q.: Efficient rank based knn query processing over uncertain data. In: ICDE (2010)

Probabilistic Reverse Top-k Queries

Cheqing Jin, Rong Zhang*, Qiangqiang Kang, Zhao Zhang, and Aoying Zhou

Software Engineering Institute, East China Normal University, China
{cqjin,rzhang,zhzhang,ayzhou}@sei.ecnu.edu.cn, kangqiang1107@126.com

Abstract. Ranking-aware query is one of the most fundamental queries in the database management field. The ranking query that returns top-k elements with maximal ranking scores according to a ranking function has been widely studied for decades. Recently, some researchers also focus on finding all customers who treat the given query object one of their top-k favorite elements, namely reverse top-k query. In such applications, each customer is described as a vector. However, none of the existing work has considered the uncertain data case for reverse top-k query, which is our focus. In this paper, we propose two methods to handle probabilistic reverse top-k query, namely BLS and ALS. As a basic solution, BLS approach checks each pair of user and product to find the query result. While as an advanced solution, ALS approach uses two pruning rules and historical information to significantly improve the efficiency. Both detailed analysis and experiments upon real and synthetic data sets illustrate the efficiency of our proposed methods.

1 Introduction

Ranking-aware query has been studied extensively for several decades. Given a ranking function, the ranking query will return k elements with maximal ranking scores by using that ranking function. Representative work includes TA approach [1]. To some extent, the ranking score computed by a ranking function describes the preference of a customer to an element, and the top-k query returns k most favorite elements to that customer.

Recently, some researchers also study reverse ranking query [2–5]. Given a customer set, an element set and a query object, reverse ranking query tends to return a set of customers who like the query object very much. In some work, the preference of each customer is represented as his (/her) most favorite object (which may be virtual). In this way, reverse ranking query turns to become the reverse Nearest Neighbor query (NN query). In other words, it tries to find all customers whose *favorite object* is the nearest neighbor of the query object.

A popular way to describe the preference of a customer is to use a weight vector w. The value at each entry is non-negative, and the sum of all entries is 1. Given a query product q, reverse top-k query will return all customers who treat q as one of its k most favorite products. Here, we use the inner product of q and w to describe the degree that a user likes the product.

---

* Corresponding author.

S.S. Bhowmick et al. (Eds.): DASFAA 2014, Part I, LNCS 8421, pp. 406–419, 2014.

Vlachou et al. have proposed two methods to handle reverse top-k query in [2], including RTA and GRBA. The RTA approach needs to check all weights to verify whether the query object is one of its top-k favorite objects or not. The advantage is that we can use the information gathered in the previous weight to gain better performance. The GRBA approach handles this issue in the other aspect. It maintains some information offline, based on which the query processing can be hastened. In [5], Vlachou et al. continue to study how to identify the most influential objects with reverse top-k queries. Note that the most influential objects are at the top-k list for most users. In [4], Vlachou et al. study how to monitor reverse top-k queries over mobile devices.

However, above work only considers deterministic data application. Data uncertainty widely exists in many applications since data may be gathered from many different sources. In general, an uncertain object is described as a probabilistic distribution function (pdf), not single value. Although uncertain data management has been studied widely in recent years, to the best of our knowledge, there is no work on reverse top-k query upon uncertain data.

Reverse top-k query processing upon uncertain data is not easy to handle. The main challenge is that the possible world space may expand exponentially. In this paper, we made the following contributions.

– We define probabilistic reverse top-k query on uncertain data set. In the new semantics, each object is described as a pdf function. It will try to find all customers who treat the given query object as one of the most favorite objects with probability at least τ, where τ is pre-defined.
– We propose two methods to handle this issue. The first method, Basic Linear Scan (BLS), as a basic solution, checks all weights one by one independently. The second method, Advanced Linear Scan (ALS), uses two pruning rules along with historical information to improve the overall performance.
– We also implement a series of experiments upon real-life and synthetic data sets to evaluate the efficiency and effectiveness of the proposed methods.

The rest of the paper is organized as follows. In Section 2, we define the problem formally. In Section 3, we review some related work. In Section 4 and 5, we describe our two methods in detail. In Section 6, a series of experiments are conducted to evaluate the proposed methods. We conclude the paper in brief in the last section.

2   Problem Definition

Let W denote a set containing n users. The preference of each user (e.g. the ith user) is described as a weight vector w_i. For all j, $w_i^{(j)} \geq 0$, and $\sum_{j=1}^{d} w_i^{(j)} = 1$. Here, $w_i^{(j)}$ refers to the j-th entry of w_i. The scoring function on a user w_i and a deterministic object o, $f(w_i, o)$, is computed as the inner product of w_i and o, i.e., $f(w_i, o) = \langle w_i, o \rangle$. User preferences are mutually independent.

We extend the concepts to uncertain data environment. Let D denote a set containing m uncertain objects in d-dimensional space. Each uncertain object

o is described as a probabilistic density function (pdf) over an uncertain region
$(\text{region}(o))$ [1]. Formally, $\int_{x \in \text{region}(o)} pdf(x)\mathrm{d}x = 1$. But for any $x \notin \text{region}(o)$,
$pdf(x) = 0$. Uncertain objects in D are also mutually independent.

Definition 1 (Score Confidence). *Given an uncertain object o, a user preference w, and a score threshold ξ, the Score Confidence, $SC_{w,o,\xi}$, describes the
probability that w prefers o with a score at most ξ, as shown in Equation (1).*

$$SC_{w,o,\xi} = \Pr[f(w,x) \leq \xi \wedge x \in \text{region}(o)] \tag{1}$$

Definition 2 (Probabilistically Domination). *Given a weight w, two uncertain objects o_i and o_j. We say o_i probabilistically dominates o_j under w if the
following equation holds.*

$$\begin{cases} SC_{w,o_i,\xi} \geq SC_{w,o_j,\xi}, & \forall \xi \geq 0; \\ SC_{w,o_i,\xi} > SC_{w,o_j,\xi}, & \exists \xi \geq 0. \end{cases} \tag{2}$$

We next consider a simple situation that one object must probabilistically dominate another one for any preference vector. Assume each uncertain object o is
enclosed by a hyper-rectangle with the bottom-left corner $o.L$ and the top-right
corner $o.U$. Since each entry in the preference vector is positive, the ranking score
of any point in o is between that of $o.L$ and $o.U$. Formally, we use *domination*
to describe this kind of relationship.

Definition 3 (Domination). *Given two points p_1 and p_2, p_1 dominates p_2 if:
(i) $\forall i, p_1^{(i)} \leq p_2^{(i)}$; and (ii) $\exists j, p_1^{(j)} < p_2^j$. We denote it as $p_1 \prec p_2$.*

Lemma 1. *Given two uncertain objects o_i and o_j. If $o_i.U \prec o_j.L$, then o_i probabilistically dominates o_j for any preference vector.*

The correctness of this lemma comes from Definition 2 directly. We then define
probabilistic threshold reverse top-k query (PTR-k query) as returning a set
of users who treat the given object q as one of its top k favorite objects with
probability at least τ, where τ is a pre-defined threshold parameter.

Definition 4 (Probabilistic Threshold Reverse top-k, PTR-k). *Given a
product set D, a user set W, a query object q, a number k, and a threshold
parameter τ, PTR-k query returns all weights w from W with $PT_{w,k}(q) \geq \tau$,
where $PT_{w,k}(q)$ is defined below.*

$$PT_{w,k}(q) = \int_{\forall x \in \text{region}(q)} pdf\{x\} \cdot \left(\sum_{\forall T = \{p_1,\cdots,p_s\} \subseteq D,\ s \leq k} \left(\prod_{\forall p_i \in T} SC_{w,p_i,f(w,x)} \right. \right.$$
$$\left. \left. \cdot \prod_{\forall p_j \in D \setminus T} (1 - SC_{w,p_j,f(w,x)}) \right) \right) \mathrm{d}x \tag{3}$$

---

[1] The symbol o represents an uncertain object from now on.

3   Related Work

Data uncertainty widely exists in many applications, such as sensor networks, financial applications, and Web [6]. Probabilistic ranking-aware query processing has been studied extensively in the past decade. In order to meet different application requirements, some uncertain ranking queries are defined, including U-top$k$ [7], U-$k$ranks [7], PT-k [8], global top-k [9], ER-top$k$ [10], ES-top$k$ [10], UTop-Set$k$ [11], c-Typical-Top$k$ [12], unified top$k$ [13] and U-Pop$k$ [14]. Some researchers also extend the work to probabilistic data stream applications [15].

Reverse ranking query is an important kind of ranking-aware queries. Given a query object q and a preference function, it returns the rank of q in a data object set [16]. Most recently, Vlachou et al. study a variant of reverse ranking query, namely reverse top-k query, and work out a series of papers [2–5]. Given a customer set (each customer has a preference weight vector), a product set and a query object q, the goal is to return all customers who treat q as one of the top-k products. In [2], they propose novel solutions to handle monochromatic and bichromatic reverse top-k queries. Subsequently, they extend their work by adding the support of executing monochromatic reverse top-k query in d-dimensional space $(d > 2)$ [3]. They study how to find the most influential object that is preferred by most customers [5]. They also consider how to monitor reverse top-k queries over mobile devices [4].

Reverse ranking query on uncertain data has also been studied in recent years [17, 18]. In [17], Lee et al. first propose a notion of expected reverse ranks for imprecise data, and then an incremental method for efficient processing. In [18], Lian et al. use another way to define inverse ranking query. Given an object q and a threshold value τ, it will return all ranks of q with probability at least τ. However, such work cannot be used to solve our issue directly. In [17], the user preference is not described as a weight vector. In [18], the focus is to return a few ranks, while this paper aims at returning a series of weights.

4   Basic Linear Scan

In this section and the next section, we study how to handle PTR-k query efficiently. First, we will introduce a basic linear scan approach (BLS) that needs to

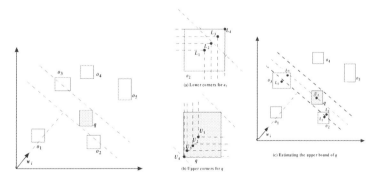

Fig. 1. An example of the BLS approach

Fig. 2. Upper bound estimation

compute the top-k result for each customer one by one. The advantage is that no information needs to be prepared in advance. We then provide another advanced method, which is capable of pruning parts of customers based on the computation results of the former customer in the next section. With the help of a little information computed in advance, this method outperforms BLS significantly.

Algorithm 1. Basic Linear Scan (abbr. BLS) (W, k, τ, D, q)

1: $W' \leftarrow \emptyset$;
2: **for each** $w \in W$ **do**
3:    **if** $compTopkProb(w, q, k, D) \geq \tau$ **then** $W' \leftarrow W' \cup \{w\}$;
4: **return** W';

Algorithm 1 illustrates the detailed steps of Basic Linear Scan approach (BLS). Let W' denote a result set. For each weight w in W, we test whether it needs to be output to W' or not. If w treats q as one of the top-k objects, then we add w to W'. The subroutine compTopkProb (Algorithm 2) aims at computing the probability that the user w treats q as one of his (/her) top-k favorite objects.

To obtain better performance, Algorithm 2 utilizes *probabilistic domination* to hasten the processing. According to Lemma 1, if $f(w, q.L) > f(w, o.U)$, then o probabilistically dominates q. We use a counter c to record such objects. Hence, if $c \geq k$, q cannot be one of the top-k favorite objects to w (at Lines 3-6). Otherwise if $c < k$, we must check all candidate weights in S to get the final result, where Monte-Carlo technique, a widely adopted approximate solution, will be employed (at Line 7).

Example 1. Figure 1 illustrates how to execute the BLS approach. Let's consider a small uncertain data set D of 5 objects, denoted as $o_1, \cdots, o_5$. The rectangle denotes the range of each object. The goal is to test whether q is one of the top-k ($k = 3$) elements for w_i. We have drawn the vector for w_i in the figure. After visiting all objects, we find o_1 probabilistically dominates q, and q probabilistically dominates o_4 and o_5. Hence, $c = 1$ and $S = \{o_2, o_3\}$. Since $c < k$, we still need to explore S subsequently. At this stage, we invoke Monte-carlo method to compute the approximate probability of being a member of top-2 products for q. If the probability is greater than the predefined τ, then w_i is treated as one of its top-k elements.

The performance of BLS depends on the sizes of W and D. In general, when the size of W or D increases, the executing cost also increases. In addition, the executing cost is dependent on the position of the query object q. Although invoking compTopkProb is quite expensive, this part of work can be avoided if $c \geq k$. Even when $c \geq k$, the execution work can also be lowered since the size of S is generally far smaller than that of D, i.e., $|S| \ll |D|$.

5   Advanced Linear Scan

Except for simplicity, BLS also has two weaknesses. First, it needs to invoke expensive Monte-carlo method frequently to compute the exact answer when

---

Algorithm 2. compTopkProb (w, q, k, D)

1: $c \leftarrow 0;$ $S \leftarrow \emptyset$; $res \leftarrow 0;$
2: **for each** object o in D
3:    **if** $f_w(q.L) > f_w(o.U)$ **then**
4:        $c \leftarrow c + 1;$
5:        **if** $(c \geq k)$ **then break**;
6:    **else if** $f_w(q.U) > f_w(o.L)$ **then** $S \leftarrow S \cup \{o\};$
7: **if** $(c < k)$ **then** $res \leftarrow$ compute the prob. that is top-$(k - c)$ of S;
8: **return** res;

$c < k$. Second, it does not consider the order of weights carefully. In other words, the performance remains the same when the order changes.

In this section, we propose a novel approach, namely Advanced Linear Scan (ALS), to overcome the above two weaknesses. At first, we devise two efficient pruning rules to decide whether a weight belongs to the query result or not when $c < k$, so that the frequency of invoking compTopkProb is reduced. Second, after using a buffer to store a small number of elements, the execution efficiency can be improved if all weights have been ordered by Cosine similarity.

5.1   Two Pruning Rules

Given an object o, let $o.U$ and $o.L$ denote the top-right and the bottom-left points of its hyper-rectangle. We define two special kinds of points for an uncertain object, namely *Lower Corners* and *Upper Corners*.

Definition 5 (Lower corners). *Given an uncertain object o and a count b, the lower corners of o are b points, denoted as $o.L_1, \cdots, L_b$. $\forall l, 1 \leq l \leq b$, we have: (i) $o.L_l$ is at the line from $o.L$ to $o.U$, and (ii) $\int_{x \in region(o) \wedge x \prec o.L_l} dx = l/b$.*

Definition 6 (Upper corners). *Given an uncertain object o and a count b, the upper corners of o are b points, denoted as $o.U_1, \cdots, U_b$. $\forall l, 1 \leq l \leq b$, we have: (i) $o.U_l$ is at the line from $o.L$ to $o.U$, and $\int_{x \in region(o) \wedge o.U_l \prec x} dx = l/b$.*

The ALS algorithm contains two phases: *off-line* phase and *online* phase. The value of b is assigned in advance. During the *off-line* phase, it generates the lower corners for all uncertain objects in D. When a query object q arrives, it then generates the upper corners for q. Figure 2(a)-(b) illustrates four lower corners for o_2 and upper corners for q respectively when $b = 4$. Note that $o_2.L_4$ is just $o_2.U$, and $q.U_4$ is $q.L$.

Given a weight w and an upper corner of q ($q.U_l$), there exists a unique hyperplane (or a line in 2-D space) across $q.U_l$ and perpendicular to w. Since there are b upper corners, we can draw b hyperplanes for all upper corners of q. Let S denote a set containing all objects whose hyper-rectangles are "across" the hyperplane. The size of S is computed as: $s = |S|$. Let $p_{l,1}, \cdots, p_{l,s}$ denote the accumulative probability of each object in S below the l-th hyperplane, and P_l the sum of probabilities, i.e., $P_l = \sum_{i=1}^{s} p_{l,i}$. Let c represent the number of objects with a ranking score smaller than $q.L$.

Pruning Rule 1. *For $l = 1, \cdots, b$, if $P_l \geq k - c$, the upper bound of the probability that q is one of the top-k objects for w is* $\min_{l=1}^{b} \left((1 - \frac{l}{b}) + \frac{l}{b} \cdot \exp \left(-\frac{(P_l - k + c + 1)^2}{2P_l} \right) \right)$.

Proof. The correctness comes from the following observations. First, according to Definition 6, the accumulative probability of q below its l-th hyperplane is no greater than $1 - \frac{l}{b}$. Under the ultimate situation, this part of q is always at the top-k list. Hence, we add it to the upper bound. Second, since c objects in D are definitely below q, at most $k - c - 1$ objects in S can co-exist at the same time to make q a top-k object. According to Chernoff bound, this probability is no greater than $\exp \left(-\frac{(P_l - k + c + 1)^2}{2P_l} \right)$ (see Lemma 2 below). Finally, since we can compute an upper bound for each hyperplane in this way, the minimal one is treated as the final upper bound.

Lemma 2. *Consider a series of Poisson trails, each with a success confidence of $p_1, p_2, \cdots$. Let P denote the sum of the success probability, $P = \sum_i p_i$. If $P > k - 1$, the probability that at most $k - 1$ events happen simultaneously is no greater than $\exp \left(-\frac{(P - k + 1)^2}{2P} \right)$.*

Proof. Let $X_1, X_2, \cdots$ denote a series of random variables. For any X_i, we have: $\Pr[X_i = 1] = p_i$, and $\Pr[X_i = 0] = 1 - p_i$. Let X denote the sum of all above random variables, i.e., $X = \sum_i X_i$. $E[X] = P$. Hence, $E[X] = E[\sum_i X_i] = \sum_i E[X_i] = P$. Let $\epsilon = 1 - \frac{k-1}{P}$. According to Chernoff bound, we have:

$$\Pr\{X \leq k - 1\} = \Pr\{X \leq (1 - \epsilon)E[X]\} \leq \exp \left(\frac{-\epsilon^2 \cdot E[X]}{2} \right) = \exp \left(-\frac{(P - k + 1)^2}{2P} \right)$$

The condition $P > k - 1$ must hold, since $\epsilon \in (0, 1)$.

Pruning rule 1 is time-efficient since it only considers the sum of probabilities of all objects in a small set S. When P is significantly greater than $k - c$, the effectiveness of this pruning rule is quite excellent. However, the pruning capability is reduced when the value of P is close to $k - c$. To handle such situation, we can use another pruning rule based on dynamic programming (DP) technique.

The definitions of $p_{l,1}, p_{l,2}, \cdots, p_{l,s}$ are as same as before. Let $r[i]^{(j)}$ denote the probability that i of the first j elements co-exist at the same time. Initially, $r[0]^{(1)} = 1 - p_{l,1}$, $r[1]^{(1)} = p_{l,1}$. When the j-th object comes, the array $r[\cdot]$ is maintained as follows.

$$r[i]^{(j)} = r[i - 1]^{(j-1)} \cdot p_{l,j} + r[i]^{(j-1)} \cdot (1 - p_{l,j}) \tag{4}$$

The probability that at most $k - c - 1$ objects in S co-exist is computed as $P'_l = \sum_{i=1}^{k-c-1} r[i]^{(s)}$. Now, we describe the second pruning rule below.

Pruning Rule 2. *The upper bound is no greater than $\min_{l=1}^{b} ((1 - \frac{l}{b}) + \frac{l}{b} \cdot P'_l)$.*

Proof. The proof of this pruning rule is similar to that of the first pruning rule. The main difference is that we use P_l' to represent the probability that at most $k - c - 1$ objects in S are below the hyperplane at the same time.

The space complexity pruning rule 2 is $O(k)$, and the time complexity is $O(k \cdot s)$. Generally, this upper bound is tighter than the first pruning rule when P_l is not large.

The premier of these pruning rules is the computation of the probability array under the hyperplane, i.e., $p_{l,1}, \cdots, p_{l,s}$. However, computing such information precisely is expensive since we don't know the hyperplane till the query object q is given. Fortunately, we can use the lower corners of objects collected in the *off-line* phase to estimate such values. For example, $\forall o_i \in S$, if $o_i.L_j$ and $o_i.L_{j+1}$ are below and above the hyperplane respectively, we can use $\frac{j}{b}$ to estimate $p_{l,i}$. In other words, $\frac{j}{b} \leq p_{l,i}$.

Example 2. Figure 2(c) illustrates an example of using these two pruning rules. We set $b = 4$, $k = 2$, and $\tau = 0.8$. Three dash lines represent the line perpendicular to w and across $q.U_4$, $q.U_2$, and $q.U$ respectively. Let's first consider the line across $q.U_4$. In this case, $c = 1$ and $S = \{o_2, o_3\}$. The probabilities of o_2 and o_3 below q are estimated as 0.5 and 0.25 respectively. Since $k - c > 0.5 + 0.25$, we cannot obtain an upper bound by using Pruning rule 1. But according to Pruning rule 2, the upper bound is: $0.375 (= (1 - 0.5) \times (1 - 0.25))$. We next consider the line across $q.U_2$. Since $c = 2$, according to Pruning rule 2, the upper bound is $0.5 (= 1 - 24)$. Finally, we consider the line across $q.U$, the upper bound is 1. In conclusion, the upper bound is 0.375 based on Pruning rule 2, while Pruning rule 1 does not return a meaningful upper bound.

5.2   The Novel Approach Using These Two Pruning Rules

We then describe our novel approach, named ALS (Algorithm 3), to handle PTR-k query by integrating two pruning rules. Initially, the result set W' and the buffer S are emptied. Subsequently, we process all weights in W one by one. For each weight w, we will estimate the upper bound of the probability to be in the result set by invoking the subroutine estimateUB (Algorithm 4, to be introduced later). If the upper bound is smaller than τ, it is no need to process w any more; otherwise, we need to evaluate the actual probability by invoking the subroutine compTopkProb (Algorithm 2). It is worth noting that before invoking

---

Algorithm 3. $ALS(W, D, k, q, b, \tau)$

1: $W' \leftarrow \emptyset$; $D' \leftarrow \emptyset$;
2: **for each** $w \in W$ **do**
3:    **if** $(estimateUB(w, D', q, b, \tau) \geq \tau)$ **then**
4:        $D' \leftarrow genBuffer(w, D, k, \tau, q)$;
5:        **if** $(compTopkProb(w, q, k, D') \geq \tau)$ **then** $W' \leftarrow W' \cup \{w\}$;
6: **return** W';

Algorithm 4. $estimateUB(w, S, q, b, \tau)$

1: Compute $\{f_w(q.U_l)\}$ for all upper corners of q;
2: **for each** object o in S
3:     Compute $\{f_w(o.L_l)\}$ for all lower corners of o;
4: $res \leftarrow$ upper bound by pruning rule 1;
5: **if** $res > \tau$ **then** $res \leftarrow$ upper bound by pruning rule 2;
6: **return** res;

Algorithm 5. $genBuffer(w, D, k, \tau, q)$

1: $D' \leftarrow \emptyset$;
2: Compute $f_w(o.U)$ for each object o in D;
3: $f^k \leftarrow$ the k-th smallest $f_w(o.U)$ value;
4: **for each** object o in O **do**
5:     **if** $(f_w(o.L) \leq f^k)$ **then** $D' \leftarrow D' \cup \{o\}$;
6: **return** D';

---

compTopkProb, we need to generate a buffer of a small number of objects by using genBuffer (Algorithm 5, to be introduced later). Finally, it returns W' as the result.

Algorithm estimateUB (Algorithm 4) illustrates how to estimate the upper bound by using two pruning rules. At first, we need to compute the ranking scores for the lower corners of objects in S and the upper corners of q. Subsequently, we begin to utilize these two pruning rules one by one to obtain the result.

Recall that the second optimization of ALS is to use historical information. We use a buffer to achieve this goal. When processing a weight w_i, we will maintain a buffer D' that contains some selected objects in D. Then, when "seeing" the next weight w_{i+1}, we will implement validation by checking D' at first. Only when this step fails do we conduct the validation by checking the whole D. Note that the validation on D' is much cheaper than that of D since $|D'| \ll |D|$.

An important observation is that if w does not treat q as one of its top-k objects in D', neither will it treat q as a top-k membership in D. Moreover, even if q is treated as a top-k element for D', q may still not be a top-k element of D for that weight.

If the subsequent weights are more similar, the buffer can be much more useful. Hence, we order all weights by the Cosine similarity in the off-line phase. First, we randomly select one object in D. Subsequently, we repeat to select one object with the maximal Cosine similarity to the previous weight. Finally, all weights are stored in the system for on-line processing.

Algorithm genBuffer (Algorithm 5) illustrates how to generate a buffer D'. For an object o, the ranking score of $o.U$ ($f_w(o.U)$) is greater than any other point in region(o). We scan all objects in D to obtain the k-th smallest $f_w(o_i.U)$ value, denoted as f^k. All objects (say, o') with $f_w(o'.L) < f^k$ are reserved in D'.

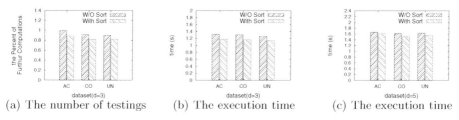

(a) The number of testings   (b) The execution time   (c) The execution time

Fig. 3. The impact of preference sorting

6   Experiments

We report experimental results in this section. All codes are written in Java and run in a computer with Intel 2.4GHz Core CPU and 4GB memory.

6.1   Data Sources

There are two kinds of data sources in our paper, including product data sets and user preference data sets.

We use synthetic and real product datasets, as described below.

- **Synthetic datasets:** We generate three synthetic datasets, named independent(*indep*), correlated(*corr*) and anti-correlated(*anti*). In *indep* dataset, all attribute values are generated independently using a uniform distribution. We create the *corr* dataset by repeating the following two steps for many times. First, select a plane perpendicular to the line from $(0, \cdots, 0)$ to $(1, \cdots, 1)$ using a normal distribution. Second, select a point in the plane by using a normal distribution again. The *anti* dataset can be generated in a similar way, except that the line in the first step may be one of the other diagonal lines. Such data sets are also used in [19] [2].
- **Real-life datasets:** We also use two real-life datasets, including a bank marketing (BM) dataset [2] and a taobao (TB) dataset [3]. The BM data set (45,211 tuples) is related with direct marketing campaigns (phone calls) of a Portuguese banking institution, including the customer's age, balance, etc. The TB dataset (51,026 tuples) is collected from taobao, the most popular E-commerce site in China. It includes product's properties, such as product's current price, sold amount and the lowest price. We choose 3 and 5 numeric dimensions from BM and TB datasets to construct our experimental datasets respectively.

We generate an uncertain data set by adding uncertainties to an existing deterministic dataset. Let S denote a deterministic data set with L attributes, $\xi_{\max}^{(l)}$ and $\xi_{\min}^{(l)}$ the maximal and minimal values of the l-th attribute of S, and λ a parameter to control the uncertainty added in. For each point p in S, we generate an

---

[2] http://archive.ics.uci.edu/ml/datasets/Bank+Marketing
[3] http://www.taobao.com

uncertain tuple u (hyper-rectangle) with the range of the l-the attribute denoted as $[p^{(l)} - r_l, p^{(l)} + r_l]$, where r_l is randomly selected from $\left[\frac{\xi^{(l)}_{\max} - \xi^{(l)}_{\min}}{4\lambda}, \frac{\xi^{(l)}_{\max} - \xi^{(l)}_{\min}}{\lambda} \right]$. We name the new dataset as S-λ.

Next, we also generate two synthetic user preference data sets, including a uniform set and a clustered set.

- **Uniform distribution**: We repeatedly select one vector from d-dimensional space, and then normalized it to a standard form. We short the name as UNI.
- **Clustered distribution**: The weight set satisfies clustered distribution. For clustered weight, similar to literature[24], we first randomly select n cluster centroids which belong to $(d-1)$-dimensional hyperplane defined by $\sum \vec{w}_i = 1$. Then, we generate each coordinate on the $(d-1)$-dimensional hyperplane by following a nor-mal distribution for each axis with variance σ^2, and the mean is equal to the corresponding coordinate of the centroid. We short the name as CLU.

6.2   Experimental Reports

Impact of Preference Sorting. We test the impact of preference sorting in ALS approach. As mentioned above, we maintain a small buffer D' in memory so as to avoid executing the evaluation task for the whole data set D. To utilize this function well, all preference vectors are sorted by their Cosine similarity in the off-line phase.In this part, we set $|W| = 1,000, |D| = 1,000$, and $\tau = 0.7$. Note that we set $\tau = 0.7$ in the following context.

Figure 3 illustrates the effects of preference sorting under different product data sets in 3-D space. Figure 3(a) shows how many times the compTopkProb subroutine (as well as the genBuffer subroutine) has been invoked when using the preference sorting or not. Since invoking such a subroutine is quite expensive, we expect that this kind of cost can be reduced significantly with the help of sorting. The x-axis represents three different synthetic data sets. The y-axis is a factor computed by $\mu/|W|$, where μ represents the number of compTopkProb subroutine invoked, and $|W|$ the size of the weight data set W. It is observed that at least 10% performance improvement has been gained.

Moreover, we also test the overall execution time with or without sorting operator. The data sets in Figure 3(b) and (c) are in 3-D and 5-D space respectively. In all situations, using sorting operator can make the method run more efficiently.

Efficiency Testing. In Figure 4, we study the performance of BLS and ALS algorithms by varying the dimensions d upon three product data sets (UN, AC, and CO) and a uniform weight data set W. We set $|D| = 1,000, |W| = 1,000$ and $k = 10$. Notice that the x-axis represents the number of dimensions, and y-axis represents the execution time. In all situations, ALS outperforms BLS significantly by one or two orders of magnitude. For ALS approach, the execution

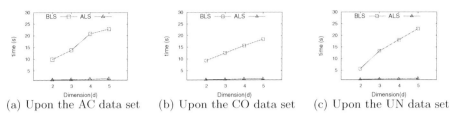

Fig. 4. Executing cost upon different product data set

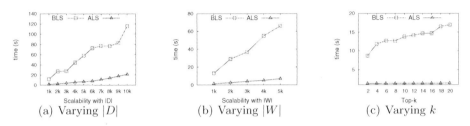

Fig. 5. Scalability test

time is still below 1.0 for all data sets. However, For BLS approach, the execution time increases to around 20 seconds in 5-D space.

The reason is that ALS employs two novel technologies for better performance. First, it uses two pruning rules based on Chernoff bound and dynamic programming. Second, it uses historical information: each pair of consecutive weights are very similar due to the off-line computation. As a result, the frequency of executing compTopkProb subroutine has been greatly reduced.

Scalability Testing. We test the scalability of two approaches through three factors, including the size of product data set $|D|$, the size of weight data set $|W|$, and parameter k. Figure 5(a) shows the execution time under different $|D|$. we can observe that the execution time increases linearly to the size of D. Figure 5(b) shows the execution time under different size of weight data set W. Similarly, we can see that the processing time almost increases linearly to the size of preference data set. Finally, Figure 5(c) shows the execution time when varying the parameter k. We can see that this parameter does not change very much for the two approaches. For BLS method, the execution time is nearly doubled when k increases 10 times. For ALS method, the execution time is almost unchanged.

The main execution cost of two approaches are closely related to the computation of the inner product between a product and a user's preference. Such kind of computation is distributed in Algorithms 2, 4, and 5. Figure 6 illustrates the amount of such kind of computation under different conditions. We can observe that the gap becomes larger when either the size of product data set increases, or the size of weight data set increases. When k increases, the computation also increases, but the amount is not as significant as the rest two.

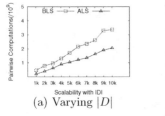
(a) Varying $|D|$

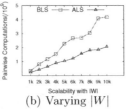
(b) Varying $|W|$

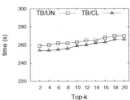
(c) Varying k

Fig. 6. Scalability test

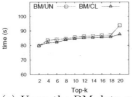
(a) Upon the BM data set

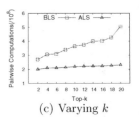
(b) Upon the MGT data set

Fig. 7. Performance evaluation upon the real data set

The Performance on Real Data Set. Finally, we study the performance of ALS for two real data sets: BM and TB. For each real data set, we compare the execution time upon the uniform and clustered weight data set. We fix —W— to 1,000 and change k from 2 to 20. Figure 7(a) shows the experimental results for the BM data set. we can see that clustered data set needs less time than uniform data set, because the clustered data is more easily correlated after sorting. In other words, the buffer used by the previous weight may still be very useful for the next weight, so as to reduce the frequency of executing the compTopkProb subroutine. Since the size and dimension of the TB data set is larger than BM, so that TB needs more execution time than BM.

7   Conclusion

Recommending correct products to a given customer is critical in many applications. Recently, reverse top-k query is proposed to handle this issue after using a preference weight vector to describe each customer's preference. Since data uncertainty widely exists in many real applications, we extend the semantics to the uncertain data environments, and propose two novel solutions to process this issue, including BLS and ALS. Experimental results show the efficiency of the proposed methods.

Acknowledgement. Our research is supported by the 973 program of China (No. 2012CB316203), NSFC (61370101, 61232002, 61103039, and 61021004), Shanghai Knowledge Service Platform Project (No. ZF1213), and Innovation Program of Shanghai Municipal Education Commission(14ZZ045).

References

1. Nepal, S., Ramakrishna, M.V.: Query processing issues in image (multimedia) databases. In: Proc. of ICDE (1999)
2. Vlachou, A., Doulkeridis, C., Kotidis, Y., Nørvåg, K.: Reverse top-k queries. In: ICDE, pp. 365–376 (2010)
3. Vlachou, A., Doulkeridis, C., Kotidis, Y., Nørvåg, K.: Monochromatic and bichromatic reverse top-k queries. IEEE Trans. Knowl. Data Eng. 23(8), 1215–1229 (2011)
4. Vlachou, A., Doulkeridis, C., Nørvåg, K.: Monitoring reverse top-k queries over mobile devices. In: Proc. of MobiDE (2011)
5. Vlachou, A., Doulkeridis, C., Nørvåg, K., Kotidis, Y.: Identifying the most influential data objects with reverse top-k queries. PVLDB 3(1), 364–372 (2010)
6. Aggarwal, C.C.: Managing and mining uncertain data. Springer (2009)
7. Soliman, M.A., Ilyas, I.F., Chang, K.C.-C.: Top-k query processing in uncertain databases. In: Proc. of ICDE (2007)
8. Hua, M., Pei, J., Zhang, W., Lin, X.: Ranking queries on uncertain data: A probabilistic threshold approach. In: Proc. of ACM SIGMOD (2008)
9. Zhang, X., Chomicki, J.: On the semantics and evaluation of top-k queries in probabilistic databases. In: Proc. of DBRank (2008)
10. Cormode, G., Li, F., Yi, K.: Semantics of ranking queries for probabilistic data and expected ranks. In: Proc. of ICDE (2009)
11. Soliman, M.A., Ilyas, I.F.: Ranking with uncertain scores. In: Proc. of ICDE (2009)
12. Ge, T., Zdonik, S., Madden, S.: Top-k queries on uncertain data: on score distribution and typical answers. In: Proc. of SIGMOD (2009)
13. Li, J., Saha, B., Deshpande, A.: A unified approach to ranking in probabilistic databases. In: Proc. of VLDB (2009)
14. Yan, D., Ng, W.: Robust ranking of uncertain data. In: Proc. of DASFAA (2011)
15. Jin, C., Yi, K., Chen, L., Yu, J.X., Lin, X.: Sliding-window top-k queries on uncertain streams. Proc. of the VLDB Endowment 1(1), 301–312 (2008)
16. Li, C.: Enabling data retrieval: by ranking and beyond. PhD thesis, University of Illinois at Urbana-Champaign (2007)
17. Lee, K.C.K., Ye, M., Lee, W.-C.: Reverse ranking query over imprecise spatial data. In: COM.Geo (2010)
18. Lian, X., Chen, L.: Probabilistic inverse ranking queries in uncertain databases. VLDB J. 20(1), 107–127 (2011)
19. Börzsöddotonyi, S., Kossmann, D., Stocker, K.: The skyline operator. In: Proc. of ICDE (2001)

Monitoring Probabilistic Threshold SUM Query Processing in Uncertain Streams

Nina Hubig[1], Andreas Züfle[2], Tobias Emrich[2], Matthias Renz[2],
Mario A. Nascimento[3], and Hans-Peter Kriegel[2]

[1] Helmholtz Center Munich, TU Munich
nina.hubig@helmholtz-muenchen.de
[2] Ludwig Maximilians University, Munich
{zuefle,emrich,renz,kriegel}@dbs.ifi.lmu.de
[3] University of Alberta, Edmonton, Canada
mario.nascimento@ualberta.ca

Abstract. Many sources of data streams, e.g. geo-spatial streams derived from GPS-tracking systems or sensor streams provided by sensor networks are inherently uncertain due to impreciseness of sensing devices, due to outdated information, and due to human errors. In order to support data analysis on such data, aggregation queries are an important class of queries. This paper introduces a scalable approach for continuous probabilistic SUM query processing in uncertain stream environments. Here we consider an uncertain stream as a stream of uncertain values, each given by a probability distribution among the domain of the sensor values. Continuous probabilistic sum queries maintain the probability distribution of the sum of possible sensor values actually derived from the streaming environment. Our approach is able to efficiently compute the probabilistic SUM according to the possible world semantics, i.e., without any loss of information. Furthermore, we show the query's answer can be efficiently updated in dynamic environments where attribute values change frequently. Our experimental results show that our approach computes probabilistic sum queries efficiently, and that processing queries incrementally instead of performing computation from scratch further boosts the performance of our algorithm significantly.

1 Introduction

In many applications it is becoming increasingly necessary to consider uncertain data, e.g., information that is incomplete, unreliable, noisy, or deliberately obfuscated for privacy reasons. Such uncertain data is prevalent in sensor data such as geo-spatial data derived from GPS tracking systems or other sensor measurements like temperature, evaporation or surface melting of ice data, due to impreciseness of sensing devices and due to obsoleteness of information. Furthermore, since such data is often crowd-sourced, uncertainty is also caused by human-errors. For this reason, the management of uncertain data has moved into the focus of database research in the last years. However, most proposed algorithms and index structures are tailored towards static data, where objects are assumed to not, or only rarely, change their (uncertain) attribute values. In sensor stream applications however, such assumptions do not hold, as there

S.S. Bhowmick et al. (Eds.): DASFAA 2014, Part I, LNCS 8421, pp. 420–435, 2014.

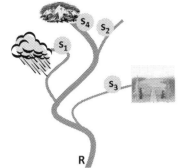

ID	Parameter	Time	Distribution
s_1	Precipitation	11:40	0.2[25-50], 0.3[10-24], 0.5[0-10]
s_2	River inflow	11:40	0.1[25-35], 0.3[10-42], 0.6[0-45]
s_3	Lake inflow	11:40	0.1[0], 0.5[10-55], 0.4[0-32]
s_4	Snow melt	11:40	0.4[0], 0.3[0-3], 0.2[3-5], 0.1[5-10]
s_1	Precipitation	11:42	0.0[25-50], 1.0[10-24], 0.0[0-10]
s_2	River inflow	11:45	0.1[25-35], 0.7[10-42], 0.2[0-45]
...	...	...	...

Fig. 1. Example Application

exists an essential temporal component. For instance, the position of moving objects may change continuously, and the uncertain parameters measured by sensor devices may also change frequently, due to imprecision of hardware and due to fluctuation of the environment. In consideration of such a dynamic setting, we do not only assume that attribute values of objects are uncertain, but also assume that the attribute values may change frequently over time. In this paper, we focus on solutions for efficiently answering one particular type of aggregate query, namely the SUM query. Since uncertain data is involved, the sum of numeric attribute values becomes a random variable. We will show how to efficiently compute the distribution of this random variable according to possible world semantics, i.e., in an exact way considering all possible instantiations of a stream database and their respective probabilities. Since objects are assumed to change frequently, we propose a technique to update this distribution quickly when objects change their values - thus avoiding re-computation from scratch. As an application, consider Figure 1 depicting an application where a set of sensors $\{s_1, ..., s_4\}$ measures sources of water flowing into a single river R. Such sources may include precipitation, glacial meltdown and bodies of water (i.e., other rivers, lakes) feeding the river. Depending on the source, each sensor may have a different degree of uncertainty, as for example the impact of precipitation on the river may be based on a highly uncertain geological model, whereas the inflow of other rivers may be fairly accurate. In such an example, it may be of interest to measure the total inflow to R, and trigger a warning signal if this inflow exceeds some critical threshold τ as exceeding such threshold may indicate the danger of a potential flooding of a settlement located at R.

The table shown in Figure 1 corresponds to exemplary sensor reading in this application, where the uncertain attribute value of an object is defined on a *continuous* domain, i.e., each attribute is given by an infinite number of possible values. In contrast, objects defined on a *discrete* domain have a finite number of possible values. Such domains are relevant for example in applications counting the number of vehicles located on a road-network. This work focuses on objects defined on a continuous domain, whereas the discrete case can be seen as a special case of the continuous case. Since continuity is not possible to measure directly, a common approach is to discretize the continuum in intervals, associated with the probability of the attribute value to be in this interval. In the example in the table of Figure 1, the environment is monitored by a set $S = \{s_1, s_2, s_3, s_4\}$ of four sensors at different points in time. Each sensor reading contains a set of value ranges, each associated with respective probabilities. In this example, sensor s_1, which measures the current amount of precipitation, reports that with 30% probability the true amount of water inflow caused by precipitation is in the range [10-24]. The uncertainty in the measured values may be quite large, because sensor s_1 may be located at some distance from river R, and thus, the actual effect on R may be a parameter for which a geological estimation model is used. Such models often return probability distributions, since many of the involved parameters are only estimated with some degree of confidence. Clearly, the value of these sensors may change over time, as the weather and other environmental parameters change. In this example, the uncertain attribute values of sensors change in irregular intervals, depending on the sensor type. In such a stream scenario, a solution where the probabilistic SUM has to be recomputed from scratch each time a sensor changes its value is not viable: An incremental approach that is able to update the probability distribution of the sum efficiently is required.

While existing techniques borrowed from the field of (static) uncertain database management allow to compute the probability that the sum of probabilistic streams exceeds a given threshold at a single point of time, the main contribution of this work is to present an efficient approach to *monitor* probabilistic threshold sum queries over time. The main challenge here is to update a probabilistic sum, such that outdated values are removed from the result, without having to recompute the whole sum from scratch. The particular difficulty that arises here, is that existing approaches discard any probability density information for the event of having a sum of exactly n for the case of $n > \tau$. Our proposed approach shows how to efficiently reconstruct previously discarded information. For this purpose, we adapt existing approaches to represent probability mass functions using generating polynomials, and adapt the common concept of polynomial division to allow to restore information that has previously been pruned. Furthermore, we show how our approach can be implemented efficiently in a wireless sensor network setting.

2   Related Work

Uncertain databases have received plenty attention in recent years. The main challenges here are data representation [1] and efficient query processing [2, 3]. A general classification of probabilistic queries and evaluation algorithms over uncertain datasets was provided by Cheng [4]. One of the first works addressing aggregation queries in uncertain databases was [5]. However the work does not consider SUM queries, but focuses only on COUNT queries. The *Trio* system [6] at exponential run-time. These works interpret an uncertain database using *possible world semantics*, thus interpreting an uncertain database as a random variable whose distribution is defined by the distributions of all uncertain objects in the database. Use possible worlds semantics, in order to avoid exponentially-sized query results caused by the exponentially large number of possible alternatives of an uncertain database, efficient solutions have been developed for a large number of queries, including range queries [7], nearest-neighbor queries [8, 9] and ranking queries [2, 10]. Aggregate queries using possible world semantics have been studied in [11–13]. However, these works either yield run-times exponential in the number of uncertain streams [11], only yield approximate results [12] and/or offer no update mechanics to update a sum query in the case where an uncertain object changes its distribution [11–13]. An extension to this case is non-trivial due to the complexity of uncertain data, where an aggregate is not given by a single scalar value, but rather by a probability density function that maps possible aggregate values to probability values.

Over the past few years stream processing has emerged as an important area for data management. Today many applications exist where data is modelled by transient data streams. Examples of such applications include network monitoring, security, web applications, sensor networks etc. State-of-the-art systems are becoming mature at managing such streams and monitoring these aggregates on high speed streams such as in IP traffic analysis [14], financial and scientific data streams [15], and others [16]. See [17] for surveys of systems and algorithms for data stream management. Together with the component of uncertainty added by data types like sensor data or geo-spatial data we have to deal with massive data sets that arrive online and have to be monitored, managed and mined in real time. Therefore a probabilistic approach of data stream models proposed by Jayram et. al.[18] was developed. In this work, we will borrow the uncertain stream model of [18] to describe uncertain objects that change their distribution over time.

3   Problem Definition

3.1   Probabilistic Stream Model

In this section we recapitulate the *probabilistic stream model* which has been used in recent publications [18, 19] and will also be utilized by this work.

Definition 1. *(Probabilistic Stream). A probabilistic stream is a data stream $A = \langle A(0), A(1), ..., A(T) \rangle$ in which each data item $A(t) \in A$ encodes a random variable that takes a value in $[n] := \{0, ..., n\}$ at time t. In particular each $A(t)$ consists of a set of at most l tuples of the form (j, p_j^t) for some $j \in [n]$ and $p_j^t \in [0, 1]$. These tuples define the random variable $X(t)$ where $X(t) = j$ with probability p_j^t.*

This model describes each stream as an uncertain object with *attribute uncertainty*, i.e., the attribute of a stream is a random variable.

Definition 2. *A probabilistic stream database $\mathcal{D} = \{A_1, ..., A_N\}$ stores a total of N probabilistic streams. Each stream A_i has an initial distribution $A_i(0)$ at time $t = 0$. In each round $t \in [T]$, the distribution of A_i is set to $A_i(t)$.*

A probabilistic stream database naturally gives rise to a distribution over deterministic streams. Specifically we consider stream $A_i(t)$ at time t to be determined according to the random variable $X_i(t)$ and each element of the stream to be determined independently. Each instantiation of random variables $x_1, ..., x_N$ such that $X_1(t) = x_1, ..., X_N(t) = x_N$ is called a world at time t. Assuming stochastic independence between streams, the joint probability of such a world is given by

$$P(X_1(t) = x_1, X_2(t) = x_2, ..., X_N(t) = x_N) = \prod_{i \in [N]} P[X_i(t) = x_i] \qquad (1)$$

Any world such that $P(X_1(t) = x_1, ..., X_N(t) = x_N) > 0$ is called a possible world at time t. In the following, we let $\mathcal{W}(t)$ denote the set of all possible worlds at time t. For each possible world $w \in \mathcal{W}(t), sum(w)$ denotes the SUM of stream values in world w. The assumption of stochastically independent probabilistic streams may seem to be a violation of Tobler's First Law of Geography [20], which states that all things are correlated - but closer things are more correlated. For example, two sensors measuring precipitation in close vicinity are likely to measure similar values. Using our model, this may very well be the case: Our independence assumption only requires the uncertainty, i.e., the measurement error, to be independent between sensing sites. Thus, two co-located sensors may measure the same expected value, and may have the same sensor specific measurement variance, but these measurement errors are assumed to be independent. A main drawback of this probabilistic stream model is the restriction of the co-domain of each probability density function to natural numbers. In practise, random variables generally have a continuous distribution, e.g., a normal distribution defined on real numbers. For example, in the mentioned application, the measured water flow in a river may be modelled using the measured value as expectation, associated with a sensor specific gaussian error. Furthermore, the distribution may be non-parametric, such as a distribution estimated by kernel estimation, as may be the result of complex geological precipitation models. However, applying possible world semantics for continuous data yields an infinite number of possible worlds even in the case where the database contains only one stream object. Thus, in the following, we propose a different probabilistic stream model that discretizes an uncertain stream by a set of value intervals each associated with the probability that the stream takes a value in this interval. Each value interval is a conservative bound of the true value. An example of this model is shown in the example corresponding to the table of Figure 1. This interval approximation can be arbitrarily tight depending on the number of intervals used for approximation, at the cost of processing a large number of intervals.

Definition 3. *(Probabilistic Range Stream). A probabilistic range stream is a data stream $A = \langle A(0), A(1), ..., A(T) \rangle$ in which each data item $A(j) \in A$ encodes a random variable that takes an interval $[min \in [n], max \in [n]], max \geq min$. In particular each $A(j)$ consists of a set of at most l tuples of the form (j, p_j^i) for some interval j and $p_j^i \in [0, 1]$. These tuples define the random variable $X_i(t)$ where $X_i(t) = j$ with probability $P(X_i(t) = j)$.*

We note that a probabilistic stream as defined in Definition 1 is a special case of a probabilistic range stream where it always holds that for each possible interval $[min, max]$ we have $min = max$.

A probabilistic range stream database and the joint probability distribution over such streams is defined analogously to the discrete case (c.f. Definition 2 and Equation 1), except that each instantiation of a random variable X_i corresponds to an interval of values rather than a single discrete value. The main challenge of query processing on this model, is that in each possible world, streams only return an interval bound for their true value. This inherent uncertainty has to be considered. We will show in Section 3.3 how to efficiently answer probabilistic sum queries in consideration of this additional uncertainty. In particular, the aim is to compute the exact distribution of the probabilistic SUM of a set of streams, given the above data representation.

3.2   Probabilistic SUM Queries

The result of a probabilistic query q is a random variable. The distribution of this random variable is determined by running q on all possible worlds $\mathcal{W}$. Since possible worlds are deterministic, each query will return a deterministic result value v. All result values are grouped (by value) and the probabilities of all worlds resulting in this value are aggregated. However, such a naive approach is not viable due to the exponential number of possible worlds. Furthermore, the number of possible results may grow exponentially large in the number of probabilistic streams, thus making the problem of probabilistic sum query processing #P-hard.[1]

Therefore, this work considers a special type of probabilistic sum query, the probability threshold sum query, which returns the probability that the sum of stream values at any time t exceeds a given threshold τ.

Definition 4 (Probabilistic threshold SUM Query). *Let $\mathcal{D}$ be a probabilistic stream database consisting of a set of probabilistic streams S, let $\mathcal{W}(t)$ denote the set of all possible worlds defined by streams S at a time t and let τ be a real value. A* probabilistic threshold SUM query *is defined as:*

$$PSQ(\mathcal{D}, \tau) = \sum_{w \in \mathcal{W}(t), SUM(w) \geq \tau} P(w), \qquad (2)$$

Clearly, the sum of random variables is also a random variable. Thus, the sum of stream values at a time t is a random variable, given by a distribution of potential answers. In the case of probabilistic range streams (c.f. Definition 3), there may be worlds $w \in W(t)$ where τ is in between the sum of minimum interval and the sum of maximum interval values. Thus, it is not possible to decide whether world w satisfies the query predicate, and the resulting probability $PSQ(\mathcal{D}, \tau)$ will be uncertain, i.e. given by a probability interval. Thus, a probabilistic sum query on probabilistic range streams may yield an answer such as "The sum of stream values exceeds threshold τ with a probability of 40% to 50%". In a concrete application, it may be possible to make a decision (e.g. initiate a flood warning) based on this probability interval, or it may be possible to signal the streams to perform a more detailed reading to yield a more refined query result.

The problem of computing the probabilistic sum, in both the case discrete and the continuous case has been solved in [13] in the context of wireless sensor networks. However, [13] proposes a solution for the static case only - where objects are not assumed to change their values frequently. In this work, we will extend this work to an incremental approach, that allows to efficiently update the probabilistic sum given a new stream update, without having to recompute the result from scratch. For self-containment of this work, a brief summary of the techniques of [13] is given in Section 3.3. Then, our main contribution of proposing an incremental approach for this problem is then given in Section 4.

---

[1] This theoretical result can be shown easily by constructing a worst-case example, where $A_i \in \{0, 2^i\}$. In this case, each $k \in \{0, 2^N\}$ has a non zero probability to be the probabilistic sum, yielding a $O(2^N)$ space complexity to store the result, and thus yielding a time complexity of at least $O(2^N)$.

3.3   Probabilistic Threshold Sum Query Processing

In order to allow an approach more efficient than the naive method, in both the discrete and continuous case, [13] proposes to process probabilistic threshold SUM queries using *uncertain generating functions* as follows.

Consider the following generating function

$$UGF(\mathcal{D}) = \prod_i \sum_j p_j^i x^{lb_j^i} y^{ub_j^i - lb_j^i}. \tag{3}$$

In a nutshell, Equation 3 is an alternative representation of a probabilistic stream using value ranges: Each possible interval $([lb_j^i, ub_j^i], p_j^i)$ of a stream S_i is represented as a monomial $p_j^i x^{lb_j^i} \cdot y^{ub_j^i}$. A sum is used to connect all alternative of a single stream S_i, and multiplication is used as connection between streams.

Expansion yields a sum of monomials of the form

$$UGF(\mathcal{D}) = \sum_{i,j} c_{i,j} x^i y^j \tag{4}$$

where each monomial $c_{i,j} x^i y^j$ corresponds to an equivalent set W_i of possible worlds such that for each possible world $w \in W_i$ it holds that the sum equals at least i, and may possibly have an additional value of j, and thus, equals at most $i + j$. The coefficient $c_{i,j}$ corresponds to the total probability of all worlds in W_i. This property holds by exploiting that polynomial multiplications leads to a multiplication of coefficients (probabilities), and a summation of exponents (possible values).

The expansion in Equation 4 allows to unify worlds having exactly the same upper and lower bound of the sum, i.e., worlds having the same approximated probabilistic sum. This unification can be performed iteratively, at each partial expansion of Equation 4, allowing high performance gain if unification is applicable frequently. But clearly, the size of the expansion of Equation 3 may be exponential in the number of sensors - as in the worst case, no monomials in the expansion (see Equation 4) can be unified by exploiting identical exponents. Thus, the choice of intervals is of high importance to allow effective unification, and thus higher efficiency to reduce the size of the resulting polynomial. If the bounds are chosen with high precision, the number of bounds having non-zero probability becomes large, and the probability of allowing unification of monomials, i.e., the probability that two worlds have the same bounds, approaches zero. In contrast, if a very coarse discretization is used, unification becomes highly applicable, but the accuracy of the chosen bounds suffers.

The parameter τ is used to significantly reduce the number of monomials without loss of information with respect to a probabilistic threshold sum query. Any monomial $c_{i,j} x^i y^j$ can be pruned such that $i \geq \tau$, since in any of these worlds, the sum cannot possibly be less than τ. Furthermore, all monomials such that $x < \tau$ and $x + y \geq \tau$ can be combined into a single monomial $p_{i\infty} x^i y^\infty$, where $p_{i\infty}$ is the sum of all combined monomials. The rationale here is that by performing polynomial multiplications to incorporate further streams, the exponents of x and y must be non-decreasing. Thus, since the upper bound of the sum in this world is already greater than τ, it cannot possible drop below τ and can no longer be used to make any decision.

Example 1. Assume two streams S_1 and S_2, where S_1 has a value in the interval $[0, 1]$ with a probability of 0.6 and a value in $[1, 2]$ with a probability of 0.4, while stream S_2 has value of 0 with a probability of 0.3 and a value of 1 with a probability of 0.7. The joint distribution of the sum of the values of these streams is computed according to Section 3.3:

$$GF(S_1) \cdot GF(S_2) = (0.4x^1y^1 + 0.6x^0y^1) \cdot (0.7x^1y^0 + 0.3x^0y^0) =$$

$$0.28x^2y^1 + 0.54x^1y^1 + 0.18x^0y^1.$$

Semantically, a monomial such as $0.28x^2y^1$ implies that with a probability of 0.28, the sum of streams S_1 and S_2 is at least two, and at most three. Now assume that $\tau = 1.5$, such that the monomial $0.28x^2y^1$ is pruned. The reasoning for this pruning is that the corresponding set of possible worlds of this monomial must satisfy the query predicate of having a sum greater or equal to τ. Since we assume streams to have non-negative values, these worlds are guaranteed to remain true hits, as their sum may not drop below τ, and thus these worlds can be excluded from computation. The main research challenge of this paper, is to adapt the above polynomial representation of the sum of probabilistic streams, in the case where some sensor S change their probabilistic distributions. This task requires to remove the effect of old distributions of all sensors in S. Since the sum was derived by polynomial multiplication, the effect can be removed by its inverse, polynomial division. However, the fact that monomials have been pruned, and thus, are no longer stored by the database makes straightforward polynomial multiplication inapplicable. The research result presented in the next section is an adapted polynomial division approach, that allows to restore pruned information, by exploiting special properties of polynomials in this context.

4   Incremental Sum Queries

The solutions proposed in Section 3.3 work well for the snapshot case, that is, for the case in which the state of the stream database only matters at a single point of time t. In most practical applications however, we are interested in monitoring a SUM value of the database. Naively, we can apply the solutions of Section 3.3 to each individual point of time in $[T]$. In practice, only a small set of streams will update their *pdf* in any given round, as for example, many sources may be sleeping to preserve energy. For the first phase, recall that the probabilistic SUM of $\mathcal{D} = \{S_1, ..., S_N\}$ at time t can be computed using the product of generating functions of each stream:

$$GF(\mathcal{D}(t)) = \prod_{1 \leq i \leq N} GF(\{S_i(t)\})$$

Due to the associativity of polynomial multiplication, this can be rewritten as

$$GF(\mathcal{D}(t)) = GF(\{S_k(t)\}) \cdot \prod_{1 \leq i \neq k \leq n} GF(\{S_i(t)\}),$$

where $S_k(t) \in \mathcal{D}$ is the stream whose effect we want to remove from $GF(\mathcal{D}(t))$. To remove the stream $S_k(t)$, an option is to perform a simple polynomial division:

$$GF(\mathcal{D}(t) \setminus \{S_k(t)\}) = GF(\mathcal{D}(t)) \div GF(\{S_k(t)\})$$

Unfortunately, this approach of polynomial division is **not** viable, since in the polynomial $GF(\mathcal{D}(t))$, many monomials $c_{i,j}x^iy^j$ have already been pruned (cf. Section 3.3) and are no longer available, in particular polynomials having $i > \tau$. Using simple polynomial division, these monomials of $GF(\mathcal{D}(t))$, may effect the result of the division. Semantically, this means that we no longer have any access to any possible worlds of $\mathcal{D}(t)$, where the SUM certainly exceeds τ. However, in some of these worlds, the SUM of streams $S \setminus \{s_k\}$ excluding sensor $S_k(t)$ may still be less than τ. These worlds (and their corresponding probabilities) would be missing in the interpretation of $GF(S \setminus \{s_k\})$, thus the result may be incorrect, as shown by continuation of Example 1:

Example 2. Assuming the setting of Example 1 where the distribution of the probabilistic sum of streams S_1 and S_2 is given by:

$$0.28x^2y^1 + 0.54x^1y^1 + 0.18x^0y^1.$$

Assume that $\tau = 1.5$, such that the monomial $0.28x^2y^1$ is pruned. Now, stream S_1 changes its value to a distribution having value 2 with a probability of 0.5 and value 3 with a probability of 0.5. To update the polynomial $0.54x^1y^1 + 0.18x^0y^1$ according to this update, we need to remove the effect of the old value of S_1 by using the following division:

$$(0.54x^1y^1 + 0.18x^0y^1) \div (0.4x^1y^1 + 0.6x^0y^1)$$

Straightforward polynomial division yields a result having a remainder, i.e. a result that is not equal to $GF(S_2) = (0.7x^1y^0 + 0.3x^0y^0)$.

Fortunately, we can compute $GF(S \setminus \{s_k\})$ correctly, in a different way, as described next.

4.1   Reverse Polynomial Division

As shown in the previous section, straightforward polynomial division is not possible in the case where monomials having a degree less than τ are pruned. This pruning however is essential for the efficiency of the previously proposed algorithms. In the following, we will propose an adaption of the polynomial division algorithm called *reverse polynomial division*.

For this approach, we need a factorized representation of polynomials:

$$GF(\mathcal{D}) = \sum_{i,j \in R} c_{i,j}x^iy^j = x^{X_{min}}y^{Y_{min}} \cdot \sum_{i,j \in R} c_{i,j}x^{i-X_{min}}y^{j-Y_{min}} \qquad (5)$$

where $X_{min} = argmin_i(GF(\mathcal{D}))$ and $Y_{min} = argmin_j(GF(\mathcal{D}))$ denote the lowest x and y exponents of $GF(\mathcal{D})$, respectively. To compute probabilistic SUM of the database

$\mathcal{D} \setminus S_k(t)$ without consideration of stream S_k at time t, for the case where $GF(\mathcal{D})$ and $GF(\{S_k(t)\} \in \mathcal{D})$ are truncated according to τ, we can divide the polynomial representation of $\mathcal{D}$ (c.f. Equation 5) by the polynomial representation $GF(\{S_k(t)\} = x^{X'_{min}} y^{Y'_{min}} \cdot \sum_{i,j \in R} c'_{i,j} x^{i-X'_{min}} y^{j-Y'_{min}}$ of S_k. We obtain

$$G(\mathcal{D} \setminus S_k(t)) = \frac{GF(\mathcal{D})}{GF(\{S_k(t)\} \in \mathcal{D})} = \frac{x^{X_{min}} y^{Y_{min}} \cdot \sum_{i,j \in R} c_{i,j} x^{i-X_{min}} y^{j-Y_{min}}}{x^{X'_{min}} y^{Y'_{min}} \cdot \sum_{i,j \in R} c'_{i,j} x^{i-X'_{min}} y^{j-Y'_{min}}}.$$

$$= x^{X_{min}-X'_{min}} y^{Y_{min}-X'_{min}} \cdot \frac{\sum_{i,j \in R} c_{i,j} x^{i-X_{min}} y^{j-Y_{min}}}{\sum_{i,j \in R} c'_{i,j} x^{i-X'_{min}} y^{j-Y'_{min}}}.$$

The first factor $x^{X_{min}-X'_{min}} y^{Y_{min}-X'_{min}}$ of above term is trivial to compute. The challenge is to compute the second factor, comprising of a polynomial division correctly, despite the information that has been lost due to pruned monomials.

For this polynomial division, we can make the following assumptions:

Lemma 1. *Let A and B be polynomials and let $C = A \cdot B$ be the product of two polynomials, then the quotient $C \div B$ is also a polynomial (without remainder).*

Proof. The above lemma is evident considering that polynomial multiplication and polynomial division are the inverse of each other. Thus, the quotient $C \div B$ equals A, and A is a polynomial by definition. Since, in our setting, we only remove the effect of streams which have previously been incorporated into the full polynomial, the second assumption holds as well.

Lemma 2. *Both the numerator and the denominator polynomials must have at least one monomial having an x-exponent of zero.*

Proof. The above lemma holds due to the factorized prolynomial representation: For each polynomial P consider the monomial $M := argmin_{M \in P} i$ having the smallest exponent i of x in P. After factorization (c.f. Equation 5), this monomial has the form $c_{i,j} x^{i-X_{min}} y^{j-Y_{min}}$. By definition of M, is holds that $i = X_{min}$.

To solve this problem, we slightly adapt the straightforward polynomial division algorithm: The straightforward algorithm divides, in each iteration, the monomial of the dividend having the highest degree, by the monomial of the divisor having the highest degree. Instead, we divide, in each iteration, the monomial of the dividend having the "lowest degree", by the monomial of the divisor having the "lowest degree", where we define the monomial having the lowest degree as the monomial having the lowest x exponent, and in case of ties, having the lowest y exponent. Due to Lemma 1, this operation will always yield a result having a non-negative exponent. Similar to traditional polynomial division, in each iteration, only the numerator polynomial changes, such that Lemma 2 is guaranteed to hold for the denominator in each iteration.

To illustrate this change of straightforward polynomial division, we continue the running example.

Example 3. Again, given the probabilistic sum of streams S_1 and S_2 that is given by the polynomial $GF(\mathcal{D}_{old}) = (0.54x^1 y^1 + 0.18x^0 y^1)$, we wish to remove the effect of sensor S_1 given by its generating polynomial $GF(S_1) = (0.4x^1y^1 + 0.6x^0y^1)$. Factorization yields $GF(\mathcal{D}_{old}) = y \cdot (0.54x^1 + 0.18x^0)$ and $GF(S_1) = y \cdot (0.4x^1 + 0.6x^0)$. For the polynomial division

$$\frac{y}{y} \cdot \frac{0.54x^1 + 0.18x^0}{0.4x^1 + 0.6x^0}$$

we now use the reverse approach, by dividing monomials $0.18x^0$ and $0.6x^0$, yielding a first result of 0.3. Analogous to straightforward polynomial division, the polynomial $0.3 \cdot (0.4x^1 + 0.6x^0)$ is substracted from the numerator polynomial, yielding a new numerator polynomial of $(0.54x^1 + 0.18x^0) - (0.12x^1 + 0.18x^0) = 0.42x^1$, completing the first iteration. In the next iteration, the monomial $0.42x^1$ is the numerator monomial having the lowest degree. Thus, dividing the monomials $0.42x^1$ and $0.6x^0$ yield $0.7x^1$ as the second part of the result. Then, the monomial $0.7x^1 \cdot (0.4x^1 + 0.6x^0)$ is substracted from the numerator polynomial, yielding a new numerator polynomial of $0.42x^1 - 0.7x^1(0.4x^1 + 0.6x^0) = -0.28x^2$. All monomials of the numerator now have an exponent greater than τ and can be ignored, yielding an empty numerator and thus terminating the polynomial division. To summarize polynomial division, the two result monomials 0.3 and $0.7x^1$ have been obtained, yielding the result polynomial $(0.7x^1 + 0.3)$ which equals the generating polynomial of S_2. Thus, the effect of S_1 has been successfully removed.

5   Experiments

We used the following setup: Each sensor is modelled by a probabilistic stream as described in Section 3. At initial time t_0, the value of each stream is zero with a probability of one, semantically meaning the stream does not yet exist.

When a probabilistic stream object changes its value, according to Definition 1, the new value is given by a number of l (per default $l = 3$) possible intervals. The range of each interval is generated randomly by uniformly sampling two values in the interval $[0, v]$ (by default $v = 10$), using the smaller sampled value as lower bound, and using the larger value as upper bound of the corresponding interval. The probabilities of each interval are generated by uniformly drawing values in $[0, 1]$ and normalizing these values. Each stream object may change its value in each of a series of 1000 updates rounds. In each update round, each probabilistic stream has a probability of δ (per default $\delta = 0.5$) to change its value to a new random value. The result of a probabilistic SUM query with threshold parameter τ is updated in each round using reverse polynomial division as described in Section 4. The performance of the implemented algorithms is evaluated in measuring the total runtime.

In the first set of experiments, we compare two approaches: The **naive** approach computes, in each update round, the probabilistic sum from scratch using polynomial multiplication as proposed in [13]. In contrast, the **incremental** approach uses reverse polynomial multiplication to avoid heavy re-computation by performing computation only for stream objects that have changed their distribution in a given update round.

5.1   Runtime Evaluation

All runtime experiments were done on the same 64bit system using an Intel core quad 2 CPU Q9450 (clock speed 2,66GHz) and 4GB RAM.

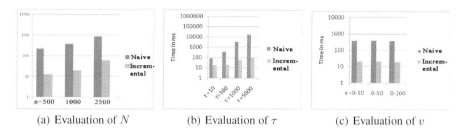

(a) Evaluation of N (b) Evaluation of τ (c) Evaluation of v

Fig. 2. Experimental Runtime Results

Evaluation of the Number of Steam Objects N. The runtime evaluation with respect to the number of stream objects is illustrated in Figure 2(a). For both algorithms, it can be observed that the total runtime increases seemingly linear with respect to the number of streams N. Clearly, since the fraction of stream objects that change their value in each iteration is a constant fraction of N, the average number of stream objects that has to be processed increases linear in N. Furthermore, the generating polynomial representing the probabilistic sum increases its size linear in N, leading for each object that has to be processed, to a computational cost increasing linear in N. In total, both aspects lead to a run-time quadratic in N. However, the size of the generating polynomial is capped by the threshold parameter τ, since any monomial becoming too large can be pruned as explained in Section 4. Thus, the size of the generating polynomial of the probabilistic sum quickly approaches τ, thus becoming constant in N, and thus explaining the total run-time linear in N.

Increasing the Threshold τ. The threshold τ limits the size of the result, as well as the size of intermediate results that are computed. However, if τ is chosen too small, queries become meaningless, since the probability that the sum is greater than τ will approach one. For this experiment, we scale τ up to 5000. Note that a value of $\tau = 5000$ is extremely large, as the value of each of the 1000 streams is between zero and ten, such that the maximal possible sum is $10,000$, but only a fraction of all possible worlds exceeds $\tau = 5000$.

Figures 2(b) shows the results of our experiment varying τ. Their foremost trend is that the runtime of both algorithms increase linearly. Clearly, the incremental algorithms have a much better performance. With higher τ the runtime of the incremental-centralized algorithm approaches the runtime of the incremental-in-network algorithm.

Increasing the Value Range v. Increasing the value range v corresponds to increasing the values of the exponents of the polynomials, and thus the degree of the resulting polynomials. The higher the number of exponents of individual streams, the smaller the likelihood that monomials can be unified after polynomial expansions, thus leading to larger polynomials. Figure 2(c) shows two interesting behaviors. First the runtime of the naive algorithms appears to be affected by varying the value range v. The reason is that the central algorithm always has to perform one thousand (one for each stream) polynomial multiplications between the current total polynomial (which is growing in size after each multiplication, up to a maximum size restricted by τ) and a polynomial of size l. The actual exponents of these polynomials have little impact on the run time. Also, the fact that in early iterations, less monomials can be unified is negligible, since the maximum polynomial size is quickly reached.

Increasing the Probability of Change. We further evaluated the fraction δ of streams changing their distribution in each update round. We observed a linear increase in run-time, which is not surprising considering that our approach considers each changed stream individually. Furthermore, our experiments have shown that the run-time of our polynomial division is comparable to the run-time of a polynomial multiplication. Thus the run-time of an update of a single stream (which requires a single division and a single multiplication) is equivalent to the run-time of performing two polynomial mul-tiplications. Thus, our approach requires $2 \cdot \delta$ times the runtime of a from-scratch re-computation, thus beating this approach if less than half of the streams change their distribution in each update round.

5.2 Application: Wireless Sensor Networks

In addition to run-time experiments, we simulated a wireless sensor network (WSN), which is a common application of probabilistic streams. In WSNs, data is collected at sensors and transmitted, at the cost of limited energy supply, to a sink node. Thus it is important to reduce as much as possible the amount of data transmitted by nodes. Using a WSN as a source of streaming data we show in the following that our proposed ideas also serve to minimize the number of messages transmitted, hence maximizing the lifetime of the WSN. The locations of the sensors were randomly chosen within a $100m \times 100m$ area and each sensor node was assumed to have a fixed wireless radio range of 30m. All generated sensor instances of the WSNs used a hop-wise shortest-path tree as routing topology. We assume in all experiments that messages are delivered using a multi-hop setup. Since the query is only sent once from the root to all child nodes and will be amortized over time, we only measure nodes-to-root messages.

In the following experiments, we measure the total number of messages that have to be sent throughout the stream network in order to answer a probabilistic threshold sum

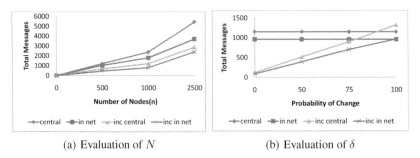

(a) Evaluation of N (b) Evaluation of δ

Fig. 3. Network Traffic Experiments

query. Therefore, we scale the following parameters: The number of streams within the network (N) and the percentage of changing measurements of a sensor, i.e., where the new value (value distribution) differs from the previous one (δ). The size of a messages is fixed at 4kb in this experiment. If the size of the information that has to be sent from one node to another node exceeds this value, then the information has to be distributed to multiple messages. We evaluate the following approaches:

The **central** algorithm performs all computation at a single central node without any incremental approach. Thus, at each point of time, all node of the network have to propagate their new value to the root node, where all computations are performed using the techniques described in Section 3.3. The **incremental** algorithm performs incremental query processing at a single central node. At each subsequent round, streams only send information if they have changed their value. For this approach, probabilistic stream objects must propagate both their old distribution as well as their new distribution to the central node as the old distribution is no longer stored explicitly in the generating polynomial sum. The **in-network algorithm** aims at distributing computational overhead within the stream network. In each update round, each node of the logical connection tree maintains the probabilistic sum of its sub-tree using polynomial multiplication. Upon applying the pruning rules of Section 4, the resulting polynomial is propagated to its logical parent node. This way, the root node requires to compute the product of a small set of polynomials, rather than for N polynomials. Finally, the **incremental in-network algorithm** runs similar to the in-network algorithm, except that at each update round, each network node only propagates values that have changed. Therefore, each node propagates two polynomials: The polynomial of all old values, i.e. the joint polynomial of all streams that have changed in the corresponding subtree of the network, and the polynomial of all new values in the corresponding subtree. At the root node, the polynomials of old values are removed using reverse polynomial division as shown in Section 4, and new values are incorporated using polynomial multiplication.

Increasing the Number of Nodes N. In the first experiment, we evaluate the performance of the four algorithms by varying the size of the network N. The results are shown in Figure 3(a). As expected, the number of messages increases with the number N of streams in the network. Similar to cpu-cost, the communication cost grows super linear to the number of nodes for all four approaches. Again, the reason is that the number of nodes not only influences the number of nodes that have to send messages, but also the number of messages that have to be transmitted through the network per node. Both approaches, the incremental approach and the in-network approach, each improves significantly in performance compared to the simple centralized approach. However, the incremental in-network query processing approach is the overall best solution saving up to 50% of the communication cost compared to the centralized approach.

Varying the Probability of Change. We evaluate the fraction of sensors that change their values per measurement round. The result illustrated in Figure 3(b). As we can see, the two non-incremental algorithms are not affected by the number of changing sensor values. This is obvious, since the non-incremental approaches require all streams to send their values, regardless of whether the sensors have changed their value in the last iteration or not. The two incremental algorithms show exceptional good performance in the case where the fraction of streams changing their value in each update round is small. Increasing δ, the performance of these approaches degrades. In the worst case, where all streams update their distribution in each update round, the overall communication cost exceeds the cost of the non-incremental solutions. This can be explained by the fact that in this case, the number of bytes that the incremental approach use is doubled, as these approaches have to send both the old and the new stream values. The number of communication overhead however is less than double, since in most cases, this additional information can still be fit into the same 4kb message.

6   Conclusions

Summarizing this work, we introduced the first solution to efficiently update probabilistic SUM queries on probabilistic streams. To allow efficient updates, we developed an incremental algorithms allowing to handle large number of stream updates efficiently. These incremental algorithms used an adapted variant of polynomial division, allowing to divide in the case where polynomials of the numerator had already been pruned. Our experiments clearly show that the incremental approach outperforms our basic approach, if the number of streams that change their value in each round is sufficiently small. Furthermore, the experiments showed that in a WSN environment, and in-network variant of our algorithm can even more reduce the communication cost.

References

1. Agrawal, P., Benjelloun, O., Sarma, A.D., Hayworth, C., Nabar, S., Sugihara, T., Widom, J.: Trio: A system for data, uncertainty, and lineage. In: Proc. VLDB (2006)
2. Li, J., Saha, B., Deshpande, A.: A unified approach to ranking in probabilistic databases. In: Proc. VLDB, vol. 2(1), pp. 502–513 (2009)
3. Soliman, M.A., Ilyas, I.F., Chang, K.C.-C.: Top-k query processing in uncertain databases. In: Proc. ICDE, pp. 896–905 (2007)
4. Cheng, R., Kalashnikov, D.V., Prabhakar, S.: Evaluating probabilistic queries over imprecise data. In: Proc. SIGMOD, SIGMOD 2003, pp. 551–562 (2003)
5. Murthy, R., Ikeda, R., Widom, J.: Making aggregation work in uncertain and probabilistic databases. TKDE 23(8), 1261–1273 (2011)
6. Widom, J.: Trio: A system for integrated management of data, accuracy, and lineage. In: CIDR, pp. 262–276 (2005)
7. Tao, Y., Xiao, X., Cheng, R.: Range search on multidimensional uncertain data. ACM Trans. Database Syst. 32(3) (August 2007)
8. Cheng, R., Kalashnikov, D.V., Prabhakar, S.: Querying imprecise data in moving object environments. IEEE Trans. Knowl. Data Eng. 16(9), 1112–1127 (2004)
9. Iijima, Y., Ishikawa, Y.: Finding probabilistic nearest neighbors for query objects with imprecise locations. In: Proc. MDM, pp. 52–61 (2009)
10. Cormode, G., Li, F., Yi, K.: Semantics of ranking queries for probabilistic data and expected results. In: Proc. ICDE, pp. 305–316 (2009)
11. Cheng, R., Kalashnikov, D.V., Prabhakar, S.: Evaluating probabilistic queries over imprecise data. In: Proc. SIGMOD, pp. 551–562 (2003)
12. Jampani, R., Xu, F., Wu, M., Perez, L.L., Jermaine, C.M., Haas, P.J.: Mcdb: a monte carlo approach to managing uncertain data. In: Proc. SIGMOD, pp. 687–700 (2008)

13. Hubig, N., Züfle, A., Emrich, T., Renz, N.M.M., Kriegel, H.-P.: Continuous probabilistic sum queries in wireless sensor networks with ranges. In: Proc. SSDBM, pp. 96–105 (2012)
14. Cranor, C., Johnson, T., Spataschek, O.: Gigascope: a stream database for network applications. In: SIGMOD, pp. 647–651 (2003)
15. Balazinska, M., Balakrishnan, H., Stonebraker, M.: Load management and high availability in the medusa distributed stream processing system. In: SIGMOD, pp. 929–930 (2004)
16. Tran, T.T.L., McGregor, A., Diao, Y., Peng, L., Liu, A.: Conditioning and aggregating uncertain data streams: Going beyond expectations. In: PVLDB, pp. 1302–1313 (2010)
17. Muthukrishnan, S.: Data streams: algorithms and applications. Now Publishers (2005)
18. Jayram, T.S., McGregor, A., Muthukrishnan, S., Vee, E.: Estimating statistical aggregates on probabilistic data streams. ACM Trans. Database Syst. 30, 26:1–26:3 (2008)
19. Jayram, T.S., Kale, S., Vee, E.: Efficient aggregation algorithms for probabilistic data, in SODA, pp. 346–355. Society for Industrial and Applied Mathematics (2007)
20. Tobler, W.: A computer movie simulating urban growth in the detroit region. Economic Geography 46(2), 234–240

Efficient Processing of Probabilistic Group Nearest Neighbor Query on Uncertain Data

Jiajia Li, Botao Wang, Guoren Wang, and Xin Bi

College of Information Science & Engineering, Northeastern University, P.R. China
{jiajia4487,edijasonbi}@gmail.com,
{wangbotao,wangguoren}@ise.neu.edu.cn

Abstract. Uncertain data are inherent in various applications, and group nearest neighbor (GNN) query is widely used in many fields. Existing work for answering probabilistic GNN (PGNN) query on uncertain data are inefficient for the irregular shapes of uncertain regions. In this paper, we propose two pruning algorithms for efficiently processing PGNN query which are not sensitive to the shapes of uncertain regions. The *spatial pruning* algorithm utilizes the *centroid* point to efficiently filter out objects in consideration of their spatial locations; the *probabilistic pruning* algorithm derives more tighter bounds by partitioning uncertain objects. Furthermore, we propose a space partitioning structure in order to facilitate the partitioning process. Extensive experiments using both real and synthetic data show that our algorithms are not sensitive to the shapes of uncertain regions, and outperform the existing work by about 2-3 times under various settings.

1 Introduction

As one of the variants of nearest neighbor(NN) query, a group nearest neighbor(GNN) query returns the object which minimizes the sum of distances to multiple query points. This type of query has many applications in clustering [1], outlier detection [2] and facilities management [3], and has been studied extensively on certain data [3–5].

However, uncertain data are inherent in various applications due to limitations of measuring equipment, delayed data updates, or privacy protection. In this paper, we focus on the problem of probabilistic GNN(PGNN) query which is to find all the uncertain objects whose probabilities of being the GNN results exceed a user-specified threshold. For example, several researchers in a city may choose one of the military units to inspect, and they hope the sum of their distances to the unit is minimal. For privacy and military secrecy reasons, only the blocks or the regions where the units locate are available. In this case, each unit is represented by a set of instances instead of a precise point, and each instance is assigned with an appearance probability. Figure 1 shows an example of uncertain objects and query points. Unit A has two possible locations a_1 and a_2, and their appearance probabilities are 0.7 and 0.3 respectively. PGNN returns the units whose probabilities of being GNN results exceed a specified threshold.

S.S. Bhowmick et al. (Eds.): DASFAA 2014, Part I, LNCS 8421, pp. 436–450, 2014.

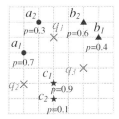

Fig. 1. Example of uncertain objects and query points

Since the location of an object is uncertain, the sum of distances between the uncertain object and a query set is not a fixed value but are multiple values with different probabilities. Therefore, the previous approaches for answering GNN query on certain data cannot be directly applied to PGNN problem. Additionally, as stated in [4], GNN query is more complex compared to traditional NN query because multiple query points are specified, which requires more distance computations and poses new challenges in designing efficient pruning algorithms. In summary, neither the traditional approaches for GNN query on certain data nor the approaches for NN query on uncertain data could apply to the processing of PGNN query directly. A naive approach is to enumerate all the possible worlds and compute their probabilities. Apparently, this approach which requires exponential time is impractical with respect to computation as well as I/O cost. New solutions are required to improve the efficiency for PGNN query.

To the best of our knowledge, the only algorithm which can handle PGNN query on uncertain data is proposed by Chen et al. [6]. However, it has the following limitations: (1) Since each uncertain object is approximated by a sphere, the algorithm is not efficient if objects have arbitrary shapes of their uncertain regions; (2) The spatial pruning algorithm bounds the query set with one MBR which weakens the pruning power, especially in the case that the query points are distributed sparsely; (3) The probabilistic pruning algorithm pre-computes the $(1-\beta)$-hypersphere using β, which might be not adapted to the user's preferences (user-specified threshold). In order to overcome these limitations, we propose two pruning algorithms for PGNN query which are efficient for arbitrary shapes of uncertain regions. The contributions of this paper are summarized as follows:

1. We propose a spatial pruning algorithm CBM (*Centroid Based Method*), that replaces the query set by the *centroid* point to reduce computation costs.
2. We propose a probabilistic pruning algorithm PBBP (*Partition Based Bounds Pruning*), which derives tighter bounds by partitioning uncertain objects.
3. We propose a space partition structure BA-Quadtree (*Bounded Aggregate Quadtree*) to facilitate the partitioning of objects for algorithm PBBP.
4. Extensive experiments are conducted to study the performance of the proposed algorithms. The results show that our algorithms are not sensitive to the shapes of uncertain regions, and outperform the existing work by about 2-3 times under various settings.

The rest of this paper is organized as follows. Section 2 briefly reviews the related work on GNN query on certain and uncertain data respectively. Section 3 formally defines the PGNN query on uncertain database. Section 4 presents the techniques of index structures, spatial pruning, probabilistic pruning and refinement for efficiently processing PGNN query. Extensive experiments are conducted in Section 5 to evaluate the efficiency of the algorithms. Finally, Section 6 concludes the paper and sketches some future works.

2   Related Work

GNN query on certain data has attracted a lot of research interests in the past few years [3–5, 7]. As uncertain data are inherent in many applications, new techniques for probabilistic GNN query on uncertain data should be developed. As far as we know, the algorithm proposed in [6] is the only algorithm for probabilistic GNN query so far. In this section, the existing work on GNN query over both certain and uncertain data are briefly reviewed.

In certain databases, GNN query was first proposed by Papadias et al. [3]. Three methods, namely MQM (multiple query method), SPM (single point method) and MBM (minimum bounding method), were proposed to answer GNN query. Li et al. [4] improved the above methods by taking the distribution of query points into account. Hashem et al. [5] proposed an efficient algorithm for GNN query considering privacy preserving. Zhu et al. [7] solved the GNN query in road networks based on network voronoi diagram.

In uncertain databases, the only algorithm for probabilistic GNN query was proposed by Chen et al. [6], where uncertain objects are represented by continuous probability density function (pdf). The spatial pruning algorithm bounds all the query points as an MBR to reduce the computation of distances. The probabilistic pruning algorithm pre-computes several ($1\text{-}\beta$)-hyperspheres for each object in a preprocessing step. The idea of this algorithm is that the object can be pruned with a probability of at least ($1\text{-}\beta$) if its corresponding sphere is pruned.

The algorithms in [3–5, 7] can be used to efficiently handle GNN query assuming that data are precise, but cannot be directly applied to answer probabilistic GNN query. Although the algorithm in [6] can handle uncertain objects, it has some limitations as described in Section 1. While our pruning algorithms for PGNN query are efficient for arbitrary shapes of uncertain regions.

3   Problem Definitions

In this section, the uncertainty model for an uncertain object is first introduced, and then the probabilistic group nearest neighbor (PGNN) query in uncertain databases is formally defined. The notations and their definitions used throughout the paper are summarized in Table 1.

An uncertain object is usually represented in two ways, either *continuously* which uses a probability density function (pdf) [6, 8, 9] or *discretely* which uses

Table 1. Notations

Notation	Definition
$U, |U|$	the uncertain objects set and its size
U_i	an object or i^{th} uncertain object in U
u_i^j	j^{th} instance of object U_i
$Q, |Q|$	the query points set and its size
q_i	the ith query point in Q
P_i', P_i^j	a partition or j^{th} partition of U_i
p_i^j	the appearance probability of u_i^j or P_i^j
λ	the user-specified probability threshold

a set of discrete values associated with assigned probabilities [10–16]. Note that a continuous *pdf* can be converted to a discrete one by sampling methods.

In this paper, we follow the discrete model, where each uncertain object U_i is represented by a set of d-dimensional points (instances) $u_i^1, ..., u_i^s$, and each instance u_i^j is assigned with an appearance probability denoted by p_i^j. Assuming that the probability distribution of each object is independent and the probability of each instance is independent of other instances as well, the equation $\sum_{j=1}^s p_i^j = 1$ holds. Since the probability distribution of an uncertain object is arbitrarily complex, the cost of distance computation is very expensive. In order to provide efficient computation in pruning phases, we use minimum bounding boxes (MBRs) to cover the uncertainty regions in the same way as [14, 17].

Definition 1. (Group Nearest Neighbor, GNN) *Given two d-dimensional objects sets P and Q, a GNN query of Q returns the object in P with the smallest sum of distances to all objects in Q. Formally, we use p to represent the result of a GNN query of Q, and it satisfies:* $\forall p' \in (P - p)), dist(p, Q) < dist(p', Q)$.

$dist(p, Q)$ denotes the sum of *euclidean distance* between a data object p and Q, and is defined as $dist(p, Q) = \sum_{i=1}^n dist(p, q_i)$.

Definition 2. (Probabilistic GNN, PGNN) *Given an uncertain objects set U, a precise query objects set Q, and a user-specified probabilistic threshold $\lambda \in (0, 1]$, a PGNN query of Q retrieves a set of data object $U_i \in U$, such that they are expected to be the GNN of Q with a probability no less than λ , that is,*

$$PGNN_Q = \{U_i | P(GNN_Q(U_i) \geq \lambda)\} = \{U_i | \sum_{\omega \in \Omega} p(\omega) \cdot \delta(GNN_Q^\omega(U_i)) \geq \lambda\} \quad (1)$$

$p(\omega)$ is the appearance probability of the possible world ω, and $\delta(GNN_Q^\omega(U_i))$ is an indicator function that is 1 if U_i is the group nearest neighbor of Q in possible world ω and 0 otherwise.

Example. Consider the example of Figure 1 which shows three uncertain objects A, B and C and the query set containing three query points q_1, q_2 and q_3. Table 2 illustrates the distances between instances and query points, and Table 3 demonstrates the process of calculating $PGNN_Q$. For example, in the possible world $\{a_1, b_1, c_1\}$, c_1 has the shortest sum distance 7.2, thus, object C is the GNN result in this possible world. The results of the other possible worlds can be obtained in a similar way. In the end, by adding up the probabilities 0.028 and 0.042, the probability of objects A being the result is obtained. Similarly, the probability of B and C are 0 and 0.93 respectively. Because the threshold λ=0.7, C is the only result.

Table 2. The distances between instances and query points in Figure 1

	a_1	a_2	b_1	b_2	c_1	c_2
q_1	$\sqrt{5}$	$\sqrt{2}$	3	$\sqrt{5}$	3	4
q_2	2	$\sqrt{17}$	$\sqrt{34}$	$4\sqrt{2}$	2	$\sqrt{5}$
q_3	$\sqrt{17}$	$3\sqrt{2}$	$\sqrt{5}$	3	$\sqrt{5}$	$2\sqrt{2}$
sum$\approx$	8.4	9.8	11.1	10.9	7.2	9.0

Table 3. Illustration of calculating $PGNN_Q$ of Figure 1 with $\lambda = 0.7$

Possible world ω	Probability $p(\omega)$	result
$\{a_1, b_1, \mathbf{c_1}\}$	$0.7 \times 0.4 \times 0.9 = 0.252$	C
$\{\mathbf{a_1}, b_1, c_2\}$	$0.7 \times 0.4 \times 0.1 = 0.028$	A
$\{a_1, b_2, \mathbf{c_1}\}$	$0.7 \times 0.6 \times 0.9 = 0.378$	C
$\{\mathbf{a_1}, b_2, c_2\}$	$0.7 \times 0.6 \times 0.1 = 0.042$	A
$\{a_2, b_1, \mathbf{c_1}\}$	$0.3 \times 0.4 \times 0.9 = 0.108$	C
$\{a_2, b_1, \mathbf{c_2}\}$	$0.3 \times 0.4 \times 0.1 = 0.012$	C
$\{a_2, b_2, \mathbf{c_1}\}$	$0.3 \times 0.6 \times 0.9 = 0.162$	C
$\{a_2, b_2, \mathbf{c_2}\}$	$0.3 \times 0.6 \times 0.1 = 0.018$	C

4   Probabilistic GNN Processing

In this section, we introduce the details of our approach (namely ISPR) which consists of four phases: ***I***ndex construction, ***S***patial pruning, ***P***robabilistic pruning and ***R***efinement.

4.1   Index Construction

There are two kinds of index structures used in our work to store uncertain objects and facilitate the execution of the pruning algorithms. The *global index* is used to manage the whole uncertain objects which mainly facilitates the spatial pruning algorithm, while the *local index* is used to organize the instances for each uncertain object which mainly facilitates the probabilistic pruning algorithm.

Global Index: R*-Tree. Without loss of generality, we use one of the most popular indexes R*-tree [18]. Each intermediate node of R*-tree is represented as an MBR, which bounds all the data in the subtree. The leaf node containing the MBRs of an uncertain object points to the local index of the object.

Local Index: Bounded Aggregate Quadtree. In the probabilistic pruning phase of our approach, those uncertain objects, who cannot be pruned in the spatial pruning phase, will be further partitioned to achieve even higher pruning ability. In order to facilitate the partitioning and pruning process, we design a novel space partitioning structure called BA-Quadtree (*Bounded Aggregate Quadtree*) based on the traditional quadtree for hierarchically organizing the instances of each uncertain object.

Our BA-Quadtree extends the traditional quadtree [19] in the following two aspects: (1) each non-leaf entry augments an aggregate (sum) probability of all instances in its corresponding subtree, and (2) when dividing the space into 2^d subregions, the subregions are bounded by the instances in its subtree to shrink the areas, which could make the bounds obtained more tighter.

Example. As an example, a BA-Quadtree is shown in Figure 2. U_1 contains 10 instances and each has the same appearance probability (0.1). The *height* of its BA-Quadtree is 4 and each entry is a two-tuples $<e.R, e.p>$, where $e.R$ represents the coordinates of the entry (partition), and $e.p$ is the sum probability of instances in its corresponding subtree. For example, the partition P_3 has three instances totally in its subtree, so the sum probability is 0.3. Obviously, the $e.p$ of root is 1 and for the leaf entry, it is the appearance probability of the instance.

It can be seen that BA-Quadtree divides the uncertain object U_i into several disjoint partitions $\{P_i^1, ..., P_i^s\}$ and any instance $u_i^j \in U_i$ is contained in at most one partition. When traversing the corresponding BA-Quadtree in a breadth-first manner, different granularities of subregions can be obtained, and each level of the index provides disjoint and complete partitions such that all the instances are included in these partitions. Since different regions are partitioned into different shrunk subregions by BA-Quadtree, it makes the probabilistic pruning algorithm applicable to arbitrary shapes of uncertain regions.

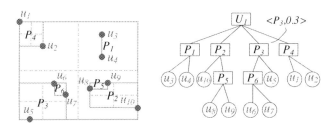

Fig. 2. Example of an uncertain object and its corresponding BA-Quadtree

4.2   Spatial Pruning

The objective of the spatial pruning phase is to prune the objects as many as possible only utilizing their spatial locations without taking the probability distributions into account. In this subsection, we propose a novel spatial pruning method, namely CBM(*Centroid Based Method*), which replaces a query set by the *centoid* point to reduce the computations of distances.

The idea of CBM is somewhat similar to that of SPM (*Single Point Method*) [3]. However, compared to SPM, which is designed for GNN problem on certain data, CBM is more complex due to the uncertainty, and the pruning rule should be redefined. In particular, CBM consists of three steps: (1) Figure out the *centroid q* of the query set Q by the same way as [3]. The *centroid* of Q is a point in space with a small value of $dist(q, Q)$ (ideally, q is the point with the minimum $dist(q, Q)$). (2) Search the candidate objects around q until that the remaining objects are definitely not the final results. The following theorem provides the details of the pruning rule. (3) Post-process the candidate objects which cannot be pruned to obtain the actual candidate objects set.

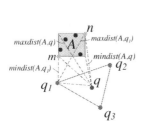

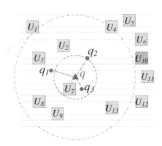

Fig. 3. Replace Q by q **Fig. 4.** Example of CBM

Theorem 1. *Given a set of uncertain objects U, a set of n certain query objects $Q = \{q_1, ..., q_n\}$ with q as the centroid point, object U_i can be pruned safely if there exists another object U_j in U that satisfies*

$$maxdist(U_j, q) < mindist(U_i, q) - 2dist(q, Q)/n.$$

Proof. Assume w.l.o.g. that there is an uncertain object A and three query points q_1, q_2 and q_3, as shown in Figure 3. Considering triangle $\triangle mq_1q$, due to the triangle inequality, $|mq_1| > |mq| - |qq_1|$ can be obtained, where $|mq_1|$ is equal to $mindist(A, q_1)$. Since point m belongs to the uncertain region of A, $|mq| \geq mindist(A, q)$. So, the inequation $mindist(A, q_1) > mindist(A, q) - |qq_1|$ holds. Similarly, the two inequations "$mindist(A, q_2) > mindist(A, q) - |qq_2|$" and "$mindist(A, q_3) > mindist(A, q) - |qq_3|$" also can be concluded. Sum up the three inequations, i.e.

$$MinDist(A, Q) > 3mindist(A, q) - dist(q, Q).$$

In the same way,

$$MaxDist(A, Q) < 3maxdist(A, q) + dist(q, Q)$$

can also be concluded. Therefore, if there are two uncertain objects U_i and U_j satisfy: $n \cdot maxdist(U_j, q) + dist(q, Q) < n \cdot mindist(U_i, q) - dist(q, Q)$, that is,

$$maxdist(U_j, q) < mindist(U_i, q) - 2dist(q, Q)/n,$$

the inequation $MaxDist(U_j, Q) < MinDist(U_i, Q)$ can be deduced. Therefore, U_i cannot be the result object. Hence, the theorem is proved.

In the second step, CBM prunes objects based on Theorem 1, and the uncertain objects that cannot be pruned are inserted into the candidate objects set S_{cnd}. Though it guarantees that all the discarded objects are definitely not the final results, *false positives* still exist in S_{cnd} (i.e. those objects that are not query results but are in S_{cnd}).

The third step of CBM further filters out the *false positives* to obtain the actual candidate objects set S_{acnd}. In particular: (1) the smallest maximum sum distance in S_{cnd} to all query points is first computed, denoted as $final_{max}$; (2) the objects whose minimum sum distances are larger than $final_{max}$ are pruned.

Algorithm 1. CBM spatial pruning algorithm

Input: Uncertain objects set $U = \{U_1, ..., U_m\}$; Query objects set $Q = \{q_1, ..., q_n\}$;
Output: the spatial candidate objects set S_{acnd};
 1: compute the approximate centroid point q;
 2: search the $MaxNN$ to q, U_{best} (use BF [20]);
 3: double $best_{max}=maxdist(U_{best}, q)+2dist(q, Q)/n$;
 4: initialize the priority queues PQ by the root of R*-tree and the sorting key of the
    entries in PQ are their $mindist$ to q;
 5: **while** $PQ \neq \emptyset$ **do**
 6:    pop an entry e from PQ;
 7:    **if** e is an intermediate node **then**
 8:        insert the entries Ns in e into PQ;
 9:    **else if** e is a leaf node **then**
10:        **if** $mindist(e, q) > best_{max}$ **then**
11:            break;
12:        **else if** **then**
13:            insert e into set S_{cnd};
14:        **end if**
15:    **end if**
16: **end while**
17: post-process S_{cnd} and obtain S_{acnd};
18: **return**;

Example. Algorithm 1 provides the pruning process discussed above and Figure 4 shows an example of CBM. CBM first computes the centroid point q and searches the $MaxNN$ of q (smallest maximum distance to q), denoted as $maxdist(U_{best}, q)$ (lines 2-3). In Figure 4, U_7 is the $MaxNN$ of q, and the result of $maxdist(U_7, q)+ 2dist(q, Q)/n$ is about 9.94. Then, it traverses R*-tree in the *best-first* manner, where the key of priority queue is the $mindist$ to the centroid point q (line 4). When the first object U_i satisfies $mindist(U_i, q) > maxdist(U_{best}, q)+2dist(q, Q)/n$ is found, the remaining objects can be pruned (lines 10-11). In the example, U_{10} is the first object satisfying the above inequation, thus, U_{10} and the remaining objects$\{U_5, U_6, U_{11}, U_{12}\}$ that have not been encountered are all discarded. Finally, CBM post-processes the objects in S_{cnd} and the actual candidate objects set is obtained (line 17).

4.3   Probabilistic Pruning

This phase is performed on the objects in S_{acnd}, and takes the probability distributions into account. In this subsection, an efficient probabilistic pruning algorithm PBBP (*Partition Based Bounds Pruning*) is proposed. By partitioning the uncertain regions of objects, PBBP derives more tighter upper and lower bounds of probabilities (denoted as UB and LB) of them being GNN results.

It can be found that each candidate object satisfies that its minimum sum distance to a query set (*mindist*) is certainly smaller than the maximum sum distances (*maxdist*) of other objects in S_{acnd} (otherwise, it is discarded by CBM, e.g. U_6 in Figure 5). In consideration of the partitioning, there must exists a partition (but not all the partitions) in U_i that its *mindist* is larger than the *maxdist* of an object in S_{acnd} such that this partition can be pruned. The main idea of PBBP is to partition uncertain regions into several subregions, and prune the objects based on the more tighter bounds.

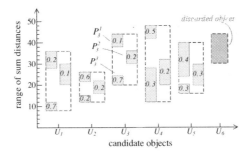

Fig. 5. The partitions of candidate objects and their ranges of sum distances

Example. Figure 5 shows an example of candidate objects obtained by CBM. The horizontal axis indicates candidate objects, and the vertical axis represents the corresponding range of sum distances to Q. Currently, each object is divided into 3 partitions (the partition strategy is derived by the local index, BA-Quadtree, cf. Section 4.1), and their aggregate appearance probabilities are illustrated inside the region. Because the *mindist*s of partitions P_3^1 and P_3^2 are both larger than the *maxdist* of U_2, they can be pruned. Therefore, the probability of U_3 being the GNN result is not larger than 1-0.1-0.2=0.7. In the following, the equations for deriving bounds of probabilities are introduced in details.

For convenience, we introduce an indicator function $\delta(P_i', P_j')$ which demonstrates the *mindist* and *maxdist* relationship between two objects or partitions:

$$\delta(P_i', P_j') = \begin{cases} 1, & if\ MaxDist(P_j', Q) < MinDist(P_i', Q) \\ 0, & else \end{cases}$$

A partition P_i^j of U_i can be pruned if "at least one of the other candidate objects U_x has a partition P_x^y satisfying $\delta(P_i^j, P_x^y) = 1$". Thus, the probability of

P_i^j being pruned can be computed by: $p_{pruned}(P_i^j) = 1 - \prod_{x=1}^{Onum-1}(\sum_{y=1}^{Pnum} p_x^y \cdot (1 - \delta(P_i^j, P_x^y)))$, where $Onum$ is the number of objects in S_{acnd} and $Pnum$ is the number of partitions. Furthermore, by combining all the partitions, the UB of U_i can be derived by Equation 2, where s is the number of partitions of U_i:

$$UB(U_i) = 1 - \sum_{j=1}^{s}(p_i^j \cdot p_{pruned}(P_i^j)). \tag{2}$$

To efficiently derive UB of U_i and reduce the search space, PBBP accesses the partitions of U_i in descending order of their aggregate probabilities, while accesses the other objects in ascending order of the probabilities that they cannot prune the partition of U_i. Once the current value of $UB(U_i)$ is smaller than the probability threshold λ, the remaining instances don't need to be considered any more and U_i can be pruned safely.

Similarly, P_i^j can be verified to be the GNN result if and only if, for every object, there exists at least one partition that can be pruned by P_i^j. Therefore, the maximum probability of P_i^j being the result is equal to the probability that those partitions exist simultaneously. That is, the lower bound of P_i^j being the result is: $LB(P_i^j) = \prod_{x=1}^{Onum} \sum_{y=1}^{Pnum}(p_x^y \cdot \delta(P_x^y, P_i^j))$. Note that, once there exists an object which has no such partition, the probability of P_i^j being the result is zero. In the similar way, the LB of U_i is derived as follows:

$$LB(U_i) = \sum_{j=1}^{s}(p_i^j \cdot LB(P_i^j)). \tag{3}$$

By traversing BA-Quadtrees in a breadth-first manner with different $depth$, different granularities of subregions can be obtained. Obviously, a larger value of $depth$ means more smaller regions of the partitions and more tighter bounds. But more probability computations are required to derived the bounds. Our experiments show that choosing $depth = Height/2$ provides a good heuristic. Algorithm 2 illustrates the details of PBBP. In particular, if the UB of U_i is less than λ, U_i can be pruned (line 5); if the LB of U_i is larger than λ, U_i is certainly one of the PGNN results (line 9); otherwise, the object should be further verified by the refinement phase and is inserted into S_{rfn} (line 11).

4.4   Refinement

Those objects in S_{rfn} which cannot be pruned by CBM and PBBP should be further verified in the refinement phase. That is, their exact qualification probabilities require to be computed. Actually, when all the objects are partitioned into instances, the UB derived by Equation 2 or the LB derived by Equation 3 is just the exact probability. If the value is larger than the threshold λ, it will be returned; otherwise it will be discarded. By sorting the instances of the object in descending order according to their probabilities, the algorithm can avoid enumerating all possible worlds. It can be seen that the number of uncertain objects in S_{rfn} directly determines the cost of refinement phase.

Algorithm 2. PBBP probabilistic pruning algorithm

Input: Candidate objects set S_{acnd}; Query objects set Q; BA-Quadtree; λ;
Output: The refinement set S_{rfn}; The result objects set S_{rs};
 1: traverse BA-Quadtree in a breadth-first manner and obtain several partitions;
 2: **for** each object $U_i \in S_{acnd}$ **do**
 3:     compute $UB(U_i)$ using Equation 2;
 4:     **if** $UB(U_i) < \lambda$ **then**
 5:         break;
 6:     **else if then**
 7:         compute $LB(U_i)$ using Equation 3;
 8:         **if** $LB(U_i) > \lambda$ **then**
 9:             insertU_i into S_{rs};
10:         **else if then**
11:             insertU_i into S_{rfn};
12:         **end if**
13:     **end if**
14: **end for**
15: **return**;

---

5   Experimental Evaluation

5.1   Experiment Setup

In this section, we compare our algorithm (called ISPR) with the state-of-the-art PGNN query algorithm (called LC) [6] using both real and synthetic data sets. In particular, we use two real data sets, CAR and TCB, which contain MBRs of 2,249,727 streets (polylines) of California, and 556,696 census blocks (polygons) of Iowa, Kansa, Missouri and Nebraska[1], respectively. The data space is normalized to $[0, 1000]$ on each dimension, and these MBRs are treated as *uncertainty regions*. We generate synthetic uncertain objects in a $[0, 1000]^d$ space as follows[14]: First the positions of uncertain objects are uniformly selected, and then a rectangle (MBR) is generated around the position with a fixed total sum of side lengths (referred to as *Dextent*). Finally, the instances within the MBR are distributed uniformly generated and their corresponding appearance probabilities are generated following Gaussian distribution. These generated synthetic data sets are denoted as SD. The extent is distributed uniformly on the dimensions, so there is a diversity of MBR shapes, some that are nearly cuboid, while others have a very large extent in few dimension only. For the query set, we first randomly select a point in the space, and generate a hyperrectangle centered at this point with a fixed total sum of side lengths (referred to as *Qextent*). And then n query points is randomly generated in the hyperrectangle.

For a fair comparison, we replace R-trees by R*-trees (with a page size of 4 Kbytes) wherever they are used in LC, and adapt LC to the discrete case. The involved parameters and their default values (bold) are summarized in Table 4. For each setting, we issue 100 queries and average the measures. All the

---

[1] Downloadable at http://www.chorochronos.org

algorithms are implemented in C++ with STL library support and compiled with GNU GCC. The hardware platform is one IBM X3500 sever with 2 Quad Core 1333MHz CPUs and 16G bytes memory under linux (Red Hat 4.1.2-42).

Table 4. Specifications of Parameters

parameter	values
# uncertain objects	2000, **4000**, 6000, 8000
# instances per object	50, **100**, 200, 400
extent of MBR, $Dextent$	20, **40**, 60, 80
# query points, n	5, **10**, 20, 40
extent of Q, $Qextent$	200, **400**, 600, 800
dimensionality	2, **3**, 4, 5
depth	0, 1, 2, **3**, 4
Probability threshold λ	0.1, 0.2, **0.4**, 0.8

5.2 Evaluation Results

The Effect of Uncertain Region. Figure 6 illustrates the effect of the extent of uncertain region on the algorithms under synthetic data sets. In particular, the performances of both algorithms degrade as $Dextent$ increases. This is because a larger value of $Dextent$ indicates much more overlapping objects, which needs more distance and probabilistic computations. Figure 7 shows the effect of different extensions in dimensions on the algorithm performance. The *Random* one randomly set extension in each dimension with a fixed maximum $Dextent$. The *Equal* one set each dimension with the same extension, while the *Dominant* one set the extension of one dimension 5 times higher than the others. The results show that LC is more sensitive to the distribution of extension. This is because that, the sphere is a more tighter approximation for an uncertain object when the MBRs are nearly cuboid. In other experiments, we use *Random* as default.

The Effect of the Depth of Partitioning, *Depth*. Figure 8 illustrates the query time of the probabilistic pruning and refinement phases of ISPR under synthetic data sets with different values of *depth*. It can be seen that *depth* offers a tradeoff between the computation costs for the two phases. In particular, a smaller value of *depth* means that the regions of the partitions are larger and the probabilistic pruning power is more weaker. Thus, the more objects require to be further verified in the refinement phase, which results in more expensive cost. On the contrary, a larger value of *depth* strengths the pruning power but needs more computational cost due to the increasing number of partitions in the probabilistic pruning phase. The total query time is fewest when *depth* is set to 3, which is about half the height of our BA-Quadtree. And how to determine a better value of *depth* for different objects and queries is our future work.

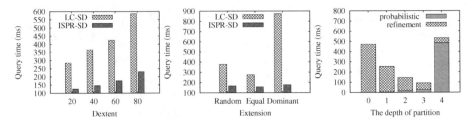

Fig. 6. Query time vs. *Dextent*

Fig. 7. Query time vs. extensions

Fig. 8. Query time vs. *depth*

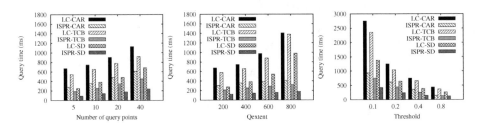

Fig. 9. Query time vs. size of query set

Fig. 10. Query time vs. *Qextent*

Fig. 11. Query time vs. λ

The Effect of Query Set, Size n and Extent $Qextent$. Figure 9 investigates the performances of LC and ISPR with different number of query points. Since both algorithms need to compute the *mindsit* and *maxdsit* from objects or partitions to n query points, it is not surprised to see that the query time increases smoothly as the query size gets larger. Figure 10 studies the effect of the extent size of query set on both algorithms. The results show that the query time of LC increases as $Qextent$ becoming larger while ISPR is not sensitive. A larger $Qextent$ means that the distribution of query points is more decentralized. Because LC bounds all the query points as an MBR to prune objects, more candidate objects would be left after the spatial pruning phase, which results in higher probabilistic pruning cost. While ISPR replaces the query points by the centroid point, thus, $Qextent$ has little effect on the query time.

The Effect of Probability Threshold, λ. In Figure 11, we increase the probability threshold λ from 0.1 to 0.8. The query times of both algorithms decrease as λ increases. This is mainly because that with a larger value of λ, a majority of objects are pruned in the probabilistic pruning phase and not required to be processed by expensive refinement, which reduces the total cost. Note that, the performance of LC yields extremely worse with a smaller threshold value while ISPR degrades more smoothly. This is because that the $(1-\beta)$-hypersphere pre-computed by LC is nearly the whole uncertain region with a small threshold which weakens the pruning power, while ISPR utilizes both upper and lower

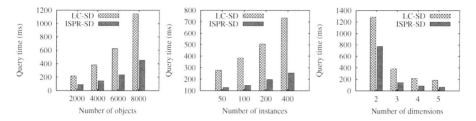

Fig. 12. Query time vs. # objects

Fig. 13. Query time vs. # instances

Fig. 14. Query time vs. dimensionality

bounds of probability to prune objects and most objects can be verified to be the results by the comparison of their lower bounds and λ.

Scalability. Figure 12-14 illustrate the scalability of the algorithms against the number of uncertain objects (m), the average number of instances of each object (s) and dimensionality (d) respectively under the synthetic data. Figure 12 and Figure 13 show that the query times of both algorithms increase regarding to the growth of the number of objects and instances. This is reasonable because, the larger of m and s, the larger number of candidate objects returns and the more complex are the probabilistic pruning and refinement phases. Nevertheless, the scalability of ISPR is better than LC. As expected, Figure 14 shows that the query times of both algorithms degrade rapidly against dimensionality. This is because higher dimensionality leads to less overlapping objects and fewer candidate objects need to be considered.

6   Conclusions and Future Work

In this paper, we focused on the problem of probabilistic GNN on uncertain data. We proposed two pruning algorithms which are not sensitive to the shapes of uncertain regions. Furthermore, in order to facilitate the pruning algorithms, we propose a space partition structure called BA-Quadtree based on traditional quadtree. Our experiments show that our proposed algorithms have a better performance and scalability than the existing solution. In particular, our algorithms outperform the existing solution by about 2-3 times under various settings.

In the future, we will improve our probabilistic pruning technique by dynamically setting the parameter *depth* according to the distributions of queries and the instances within different objects.

Acknowledgement. This research was partially supported by the National Natural Science Foundation of China under Grant Nos. 61173030, 61073063, 61025007, 61332006, 61272182 and 61100024; the Public Science and Technology Research Funds Projects of Ocean Grant No. 201105033; the National Basic Research Program of China Grant No. 2011CB302200-G.

References

1. Jain, A.K., Murty, M.N., Flynn, P.J.: Data clustering: a review. ACM Computing Surveys (CSUR) 31(3), 264–323 (1999)
2. Aggarwal, C.C., Yu, P.S.: Outlier detection for high dimensional data. ACM Sigmod Record 30, 37–46 (2001)
3. Papadias, D., Shen, Q., Tao, Y., Mouratidis, K.: Group nearest neighbor queries. In: ICDE, p. 301. IEEE Computer Society (2004)
4. Li, H., Lu, H., Huang, B., Huang, Z.: Two ellipse-based pruning methods for group nearest neighbor queries. In: Proceedings of the 13th Annual ACM International Workshop on Geographic Information Systems, pp. 192–199. ACM (2005)
5. Hashem, T., Kulik, L., Zhang, R.: Privacy preserving group nearest neighbor queries. In: EDBT, pp. 489–500. ACM (2010)
6. Lian, X., Chen, L.: Probabilistic group nearest neighbor queries in uncertain databases. TKDE 20(6), 809–824 (2008)
7. Zhu, L., Jing, Y., Sun, W., Mao, D., Liu, P.: Voronoi-based aggregate nearest neighbor query processing in road networks. In: GIS, pp. 518–521. ACM (2010)
8. Beskales, G., Soliman, M., Ilyas, I.: Efficient search for the top-k probable nearest neighbors in uncertain databases. VLDB 1(1), 326–339 (2008)
9. Cheng, R., Chen, J., Mokbel, M., Chow, C.: Probabilistic verifiers: Evaluating constrained nearest-neighbor queries over uncertain data. In: ICDE, pp. 973–982. IEEE (2008)
10. Hua, M., Pei, J., Zhang, W., Lin, X.: Ranking queries on uncertain data: A probabilistic threshold approach. In: SIGMOD, pp. 673–686. ACM (2008)
11. Kriegel, H.-P., Kunath, P., Renz, M.: Probabilistic nearest-neighbor query on uncertain objects. In: Kotagiri, R., Radha Krishna, P., Mohania, M., Nantajeewarawat, E. (eds.) DASFAA 2007. LNCS, vol. 4443, pp. 337–348. Springer, Heidelberg (2007)
12. Pei, J., Jiang, B., Lin, X., Yuan, Y.: Probabilistic skylines on uncertain data. In: VLDB, pp. 15–26. VLDB Endowment (2007)
13. Soliman, M., Ilyas, I., Chen-Chuan Chang, K.: Top-k query processing in uncertain databases. In: ICDE, pp. 896–905. IEEE (2007)
14. Bernecker, T., Emrich, T., Kriegel, H., Renz, M., Zankl, S., Züfle, A.: Efficient probabilistic reverse nearest neighbor query processing on uncertain data. VLDB 4(10), 669–680 (2011)
15. Li, J., Wang, B., Wang, G.: Efficient probabilistic reverse k-nearest neighbors query processing on uncertain data. In: Meng, W., Feng, L., Bressan, S., Winiwarter, W., Song, W. (eds.) DASFAA 2013, Part I. LNCS, vol. 7825, pp. 456–471. Springer, Heidelberg (2013)
16. Yuan, Y., Wang, G., Chen, L., Wang, H.: Efficient keyword search on uncertain graph data. IEEE Transactions on Knowledge and Data Engineering 25(12), 2767–2779 (2013)
17. Cheema, M., Lin, X., Wang, W., Zhang, W., Pei, J.: Probabilistic reverse nearest neighbor queries on uncertain data. TKDE 22(4), 550–564 (2010)
18. Beckmann, N., Kriegel, H.P., Schneider, R., Seeger, B.: The R*-tree: an efficient and robust access method for points and rectangles, vol. 19. ACM (1990)
19. Finkel, R.A., Bentley, J.L.: Quad trees a data structure for retrieval on composite keys. Acta Informatica 4(1), 1–9 (1974)
20. Hjaltason, G.R., Samet, H.: Distance browsing in spatial databases. TODS 24(2), 265–318 (1999)

Popularity Tendency Analysis
of Ranking-Oriented Collaborative Filtering
from the Perspective of Loss Function

Xudong Mao[1], Qing Li[2], Haoran Xie[1], and Yanghui Rao[1]

[1] Department of Computer Science, City University of Hong Kong,
Kowloon, Hong Kong
{xdmao2-c,hrxie2-c,yhrao2-c}@my.cityu.edu.hk
[2] Multimedia Software Engineering Research Centre,
City University of Hong Kong, Kowloon, Hong Kong
itqli@cityu.edu.hk

Abstract. Collaborative filtering (CF) has been the most popular approach for recommender systems in recent years. In order to analyze the property of a ranking-oriented CF algorithm directly and be able to improve its performance, this paper investigates the ranking-oriented CF from the perspective of loss function. To gain the insight into the popular bias problem, we also study the tendency of a CF algorithm in recommending the most popular items, and show that such popularity tendency can be adjusted through setting different parameters in our models. After analyzing two state-of-the-art algorithms, we propose in this paper two models using the generalized logistic loss function and the hinge loss function, respectively. The experimental results show that the proposed methods outperform the state-of-the-art algorithms on two real data sets.

Keywords: Collaborative filtering, matrix factorization, loss function.

1 Introduction

Collaborative filtering (CF) is a technique of making predictions about users interests by analyzing the previous preferences from many users. CF has been the most popular approach for recommender systems due to its effectiveness. In general, techniques of CF fall into two categories: memory-based approach [11,14] and model-based approach [2,4,7]. As memory-based approach suffers from the shortcomings of low accuracy and expensive computation, model-based approach has been gaining popularity recently. Model-based approach includes matrix factorization (MF) [7], Bayesian networks [2] and latent semantic analysis [4]. Most of early work aims at improving the accuracy of global rating prediction. However, in many practical applications, it is more common that business companies recommend only several items with largest scores to users. These applications are called ranking-oriented recommendation, which pursues the ranking accuracy rather than the global accuracy. Under this context, the rating prediction

S.S. Bhowmick et al. (Eds.): DASFAA 2014, Part I, LNCS 8421, pp. 451–465, 2014.

for ranking-oriented applications is unnecessary strict. Ranking-oriented CF is one of the most popular approaches for ranking-oriented recommendation. Many studies have been done to improve the performance of ranking-oriented CF, where a number of efficient models have been proposed from different perspectives of intuitions such as extending the existing rating-oriented CF to ranking-oriented CF [5], applying the study of learning to rank [8,16,17] and optimizing some evaluation metrics directly [12,15,19].

However, on the other hand, we found the non-personalized and popularity-based algorithm can work relatively well for ranking-oriented recommendation for some datasets, which is also stated in [3,18]. We demonstrate that some previous models work well because their algorithms have the property of boosting the popular items. We call this property of boosting the popular items *popularity tendency*. This paper proposes to use rank similarity between a given algorithm and PopRec algorithm [15] to measure the popularity tendency of the given algorithm, where Kendall's τ coefficient [6] is used to calculate the rank similarity. According to this line of reasoning, we found that some sophisticated models works better than PopRec because they incorporate personalized recommendation (personalization) while keeping popularity tendency (non-personalization). Thus the trade-off between personalization and popularity tendency is crucial to the performance.

This paper proposes to control the trade-off between personalization and popularity tendency by adjusting the loss function. Analyzing algorithms from the perspective of loss function is helpful because it can analyze the property of an algorithm directly regardless of its complicated modeling and compare the differences of various algorithms. With the objective of controlling the trade-off between personalization and popularity tendency, we propose two new models with the generalized logistic loss function and the hinge loss function respectively. By changing the parameters in our proposed models, the trade-off between personalization and popularity tendency can been adjusted such that the optimal trade-off can been found. The main contributions of our work are:

1. We introduce Kendall's τ coefficient to measure the popularity tendency of a specific algorithm.

2. We propose to control the popularity tendency of algorithms by adjusting loss function. By analyzing algorithms from the perspective of loss function, we are able to analyze the property of the algorithms directly.

3. We propose two models for ranking-oriented CF with the generalized logistic loss function and the hinge loss function respectively. The experimental results show that our proposed methods outperform the current state-of-the-art algorithms.

The rest of the paper is organized as follows. Section 2 discusses the related work. In Section 3, we present the problem definition and the analysis of two previous models. Our proposed two models are presented in Section 4. The experimental results are given in Section 5. Finally, we conclude our work in Section 6.

2   Related Work

Just like the rating-oriented CF, the approaches for ranking-oriented CF also fall into two categories, memory-based approach [9] and model-based approach [5,8,10,12,15,19]. The hallmark of memory-based approach is that it needs to compute the similarity between users or items. EigenRank [9] is a typical example of memory-based approach. It first measures the user similarity using rank correlations and then generates the ranking by fusing the preferences of a user's neighbors. Unlike the memory-based approach, model-based approach tries to learn a specific model using the training data and then it can predict the preferences of new data without training data. Many models in this category have been proposed including extension of rating-oriented matrix factorization model [5], ordinal model [8], probabilistic latent preference analysis (pLPA) model [10] and models with optimizing some evaluation metrics approximately such as normalized Discounted Cumulative Gain (nDCG) [19], Area Under the Curve (AUC) [12] and Mean Reciprocal Rank (MRR) [15]. In particular, [5] tries to solve the ranking problem by viewing it as a regression problem. It converts the ratings into binary preferences and defines the confidence variables to weight the error function. This method has a high time complexity because it needs to manipulate the whole user-item matrix rather than the sparse user-item matrix.

Recently, the technique of learning to rank has been widely applied in ranking-oriented CF, where the approaches of pointwise, pairwise and listwise all have been studied in ranking-oriented CF literature. Maximum-Margin MF , OrdRec [8] and CLiMF [15] belong to pointwise approach. Maximum-Margin MF and OrdRec both try to map the ratings into several ordering bins. Specifically, Maximum-Margin MF utilized the hinge loss function to model the ranking problem as a semi-definite optimization problem. Similarly, OrdRec utilized logistic regression to model the ranking problem as a maximum likelihood problem. However, both models need to determine the thresholds of bins manually. CLiMF is proposed to maximize a smooth version of MRR approximately. However, we found that the final approximated objective function of CLiMF is actually not what it is meant to be, and we will show that CLiMF has a strong popularity tendency. On the other hand, pairwise approach tries to order each pair of items. pLPA [10] is a probabilistic pairwise model that adopts the Bradley-Terry distribution to model the user preference of two items and then maximizes the likelihood of the observed data. Whereas, BPR [12] uses the logistic function to convert the user's preference difference of each pair to a probabilistic value and then maximize the log-likelihood of the observed data. To avoid the high computational cost of pairwise approach, a listwise method for ranking-oriented CF is proposed in [16]. BPR [12] is the most similar work to ours. One of our proposed models is a generalization of BPR. After the generalization, the new model can be more flexible in various data and perform better. The hinge loss function has been applied in the pointwise approach for ranking-oriented CF [17]. But for the pairwise approach, there is no study on it. The second model we propose is a pairwise model based on the hinge loss function.

3   Problem and Analysis

In this section, we first introduce the problem definition and notation. Then two state-of-the-art models for ranking-oriented CF are analyzed from the perspective of loss function.

3.1   Problem Formalization

The task of ranking-oriented recommendation can be defined as follows: *Given explicit or implicit feedbacks from users, the objective of the ranking-oriented recommendation is to provide each user with an optimal rank of items.* Ranking-oriented collaborative filtering (CF) adopts the methods of collaborative filtering to solve the problem of ranking-oriented recommendation. The MF-based models [8,12,15] are widely used in ranking-oriented CF. The formalization of MF-based model can be summarized as:

Let $P \in \mathbb{R}^{M \times N}$ be a user-item preference matrix, where M and N are the number of users and items, respectively. The entry p_{ui} denotes the preference degree of user u to item i. If p_{ui} is larger than p_{uj}, it means that user u prefers item i to item j. Each user u is associated with a user latent factor vector $U_u \in \mathbb{R}^d$ and each item i is associated with an item latent factor vector $V_i \in \mathbb{R}^d$, where d is the dimensionality of the latent factor. The preference value p_{ui} is predicted by the inner product of U_u and V_i, i.e., $p_{ui} = U_u^T V_i$. By adding the user and item bias terms, p_{ui} is predicted by $p_{ui} = U_u^T V_i + b_u + b_i$. For each user u, items can be ranked according to the preference values p_{ui}, $i \in \{1, 2, \cdots, N\}$.

In this section, we will analyze two state-of-the-art models from the perspective of loss function, which belong to pairwise and pointwise approach, respectively. Pointwise approach focuses on the observed positive feedbacks and tries to classify the positive feedbacks into observed preference levels [8] or to boost the preference values of the positive feedbacks [15]. Whereas, pairwise approach focuses on the positive/negative feedback pairs and tries to model the preference difference $x_{uij} = p_{ui} - p_{uj}$ correctly [12].

3.2   Analysis of Existing Models

As stated before, analyzing an algorithm from the perspective of loss function can get the property of this algorithm despite of the complicated modeling procedure. We will show the property of two state-of-the-art models, BPR [12] and CLiMF [15], in this section.

Bayesian Personalized Ranking (BPR). BPR [12] is one of the pairwise approaches. Pairwise approach aims at modeling positive/negative feedback pair (u, i, j) correctly, where (u, i, j) denotes that user u prefers item i to item j. BPR uses the logistic function to convert preference difference $x_{uij} = p_{ui} - p_{uj}$ to a probability value, and then maximizes the log-likelihood of all the observed pairs. However, the number of positive/negative feedback pairs is huge and it is hard to

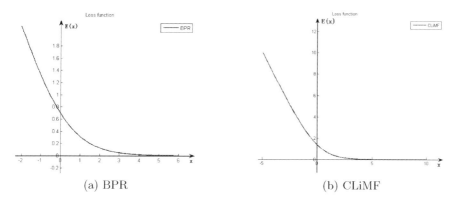

(a) BPR   (b) CLiMF

Fig. 1. The loss function of BPR and CLiMF

maximize log-likelihood of all pairs directly. So BPR adopts the stochastic gradient ascent where only one pair is sampled in each iteration. When applying the stochastic gradient ascent, the objective function in each iteration is given by

$$L(U_u, V_i, V_j) = ln(\frac{1}{1 + e^{-x_{uij}}}), x_{uij} = U_u^T(V_i - V_j) + b_i - b_j \tag{1}$$

By putting a minus sign of the objective function, we can derive the loss function of BPR as

$$E(x_{uij}) = ln(1 + e^{-x_{uij}}) \tag{2}$$

Actually the above loss function of BPR is identical to the one of logistic regression classifier [1], as plotted in Figure 1. We see from Figure 1 that for $x_{uij} < 0$, BPR applies a penalty that is approximately linear to x_{uij}. For $x_{uij} \geq 0$, BPR still applies a penalty that is an exponential decaying penalty though the preference order between item i and item j is correctly assigned.

Collaborative Less-is-More Filtering (CLiMF). CLiMF [15] tries to model the data by directly maximizing the Mean Reciprocal Rank (MRR). However, the final objective function (formula 8 in the original paper [15]) is obtained after two main approximations including smoothing the reciprocal rank and taking a lower bound of smooth reciprocal rank. When applying the stochastic gradient ascent, the objective function in each iteration[1] is given by

$$L(U_u, V_i) = ln(g(p_{ui})) + \sum_{k=1}^{N} ln(1 - Y_{uk}g(p_{uk} - p_{ui})) \tag{3}$$

where $g(p_{ui})$ is the logistic function and $Y_{uk} = 1$ denotes the positive feedback. By putting a minus sign of the objective function, we can derive the loss function of CLiMF as

---

[1] In [15], the authors use stochastic gradient ascent to optimize a batch of positive feedbacks of user u. We analyze it with optimizing a single positive feedback for simplicity. Our analysis is applicable to the batch learning. Our experimental results of CLiMF are obtained using batch learning.

$$E(p_{ui}) = -(ln(g(p_{ui})) + \sum_{k=1}^{N} ln(1 - Y_{uk}g(p_{uk} - p_{ui}))) \qquad (4)$$

By a simple rearrangement, Eq.(4) can be written in the form of

$$E(p_{ui}) = -ln(g(p_{ui}) \times \prod_{\{k,Y_{uk}=1\}} (1 - g(p_{uk} - p_{ui}))) \qquad (5)$$

Note that p_{uk} is not a random variable in Eq.(5). As the logistic function is monotonically increasing and $1 - g(p_{uk} - p_{ui})$ is always greater than zero, $g(p_{ui})$ and $\prod_{\{k,Y_{uk}=1\}}(1-g(p_{uk}-p_{ui}))$ are both monotonically increasing functions with respect to p_{ui}. In addition, the natural logarithm is also monotonically increasing function. According to the properties of monotonically increasing function, we can obtain that $E(p_{ui})$ is a monotonically decreasing function with respect to p_{ui}. So minimizing $E(p_{ui})$ is equivalent to increasing the value of p_{ui}. We plot the loss function of CLiMF in Figure 1 by setting several specific values of p_{uk}. From Figure 1, we can find that CLiMF tries to boost the preference values of p_{ui}.

From this analysis, we can conclude that in each iteration, CLiMF just increases the preference value p_{ui} to minimize $E(p_{ui})$. On the other hand, popular items will be sampled more times than unpopular items. Intuitively, the preference values of popular items will be larger than the unpopular items. This will lead to a strong popularity tendency effect. In order to verify our analysis, we calculate the Kendall's τ coefficient [6] between the rank of CLiMF and PopRec, and the experimental results show that the rank of CLiMF and PopRec are indeed very close.

4   Proposed Model

4.1   Popularity Tendency Analysis

We propose to use Kendall's τ coefficient [6] between a given algorithm and PopRec algorithm [15] to measure the popularity tendency of the given algorithm. Kendall's τ coefficient is a measure of the similarity between the orderings of the data:

$$\tau = \frac{N^+ - N^-}{\frac{1}{2}n(n-1)} \qquad (6)$$

where N^+ and N^- are the numbers of concordant pairs and discordant pairs. Pairs (x_i, y_i) and (x_j, y_j) are said to be concordant if the ranks for both elements agree: $\{x_i > x_j$ and $y_i > y_j\}$ or $\{x_i < x_j$ and $y_i < y_j\}$. Otherwise (x_i, y_i) and (x_j, y_j) are said to be discordant. τ ranges from -1 to 1. τ equals 1 if the two ranking are the same, -1 if one ranking is the reverse of the other one. We compute Kendall's τ coefficient for each user and then take the average over all users. We found that different loss functions can lead to different popularity tendency and result in different performances. We take loss function for pairwise

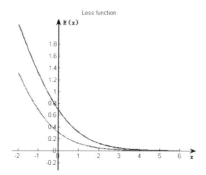

Fig. 2. Comparison of two loss functions

approach for example because our proposed models belong to pairwise approach. As shown in Figure 2, for $x_{uij} < 0$, both the red curve and blue curve have a approximately linear penalty to x_{uij}. But for $x_{uij} \geq 0$, the red curve is "pushed" to be closer to x-axis than the blue one. The red one increases the value of x_{uij} less than the blue one when the preference order between item i and item j is correctly assigned. Based on the experimental results, which will be shown in Section 5, we found the blue one has less popularity tendency than the red one.

4.2   Ranking-Oriented CF with the Generalized Loss Function

As stated before, the shape of the loss function can affect the performance. For the purpose of adjusting the shape of loss function curve, we propose the model with the generalized logistic loss function [13], given as:

$$E(x_{uij}) = ln(1 + qe^{-ax_{uij}+t}) \tag{7}$$

Note that we omit several parameters in the generalized logistic function as some parameters have little influence on the curve shape, and some others for simplicity of deviation. The loss functions with different parameters q, a and t are shown in Figure 3. We will show the relation between the parameters and popularity tendency in Section 5.

To optimize all the pairs (u, i, j) of Eq.(7), the number of pairs (u, i, j) would be $O(MN^2)$, which is too large to optimize the overall loss function. We apply the stochastic gradient descent with sampling to solve the problem same as in previous work [8,12,15]. With the regularization terms to avoid overfitting, we obtain the objective function for each pair (u, i, j) as follows:

$$L(U_u, V_i, V_j, b_i, b_j) = ln(1 + qe^{-ax_{uij}+t}) + \Omega,$$
$$\Omega = \frac{\lambda}{2}(\|U_u\|^2 + \|V_i\|^2 + \|V_j\|^2 + b_i^2 + b_j^2), \tag{8}$$
$$x_{uij} = U_u^T(V_i - V_j) + b_i - b_j$$

where λ is the regularization coefficient and $\| \cdot \|$ denotes the L2-norm.

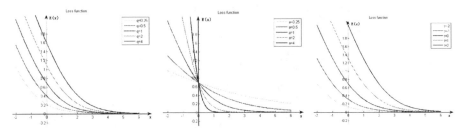

(a) Influence of parameter q (b) Influence of parameter a (c) Influence of parameter t

Fig. 3. Influences of parameters

The gradients for each variable can be derived as

$$\frac{\partial L}{\partial U_u} = \frac{aqe^{-ax_{uij}+t}}{1+qe^{-ax_{uij}+t}}(V_j - V_i) + \lambda U_u \tag{9}$$

$$\frac{\partial L}{\partial V_i} = -\frac{aqe^{-ax_{uij}+t}}{1+qe^{-ax_{uij}+t}}U_u + \lambda V_i \tag{10}$$

$$\frac{\partial L}{\partial V_j} = \frac{aqe^{-ax_{uij}+t}}{1+qe^{-ax_{uij}+t}}U_u + \lambda V_j \tag{11}$$

$$\frac{\partial L}{\partial b_i} = -\frac{aqe^{-ax_{uij}+t}}{1+qe^{-ax_{uij}+t}} + \lambda b_i \tag{12}$$

$$\frac{\partial L}{\partial b_j} = \frac{aqe^{-ax_{uij}+t}}{1+qe^{-ax_{uij}+t}} + \lambda b_j \tag{13}$$

The update rule can be derived as

$$\Theta_{new} \leftarrow \Theta_{old} - \gamma \frac{\partial L}{\partial \Theta} \tag{14}$$

where Θ can be U_u, V_i, V_j, b_i or b_j, and γ is the learning rate. The optimization algorithm for this model is listed in Algorithm 1.

4.3   Ranking-Oriented CF with the Hinge Loss Function

The second model we propose is to apply the hinge loss function for pairwise approach, where we incorporate a parameter into the hinge loss function to control the position of the inflexion point such that it can control the trade-off between personalization and popularity tendency. Our proposed hinge loss function for ranking-oriented CF is given by:

$$E(x_{uij}) = max(0, \frac{\alpha - x_{uij}}{\alpha}) \tag{15}$$

Algorithm 1. Optimization For Our First Model

---

1 Initialize U, V, B randomly;
2 **repeat**
3   **for** $i = 1 : |D^*|$ **do**
4   %$|D^*|$ is the number of positive feedbacks Sample (u, i, j);
5   Update U_u, V_i, V_j, b_i and b_j according to Eq.(14);
6   **end**
7 **until** *convergence*;

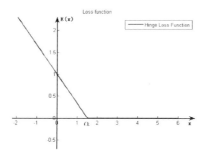

Fig. 4. The hinge loss function

As shown in Figure 4, for $x < \alpha$, $E(x_{uij})$ is differentiable and the gradients for each variable can be derived as:

$$\frac{\partial L}{\partial U_u} = \frac{(V_j - V_i)}{\alpha} + \lambda U_u \tag{16}$$

$$\frac{\partial L}{\partial V_i} = -\frac{U_u}{\alpha} + \lambda V_i \tag{17}$$

$$\frac{\partial L}{\partial V_j} = \frac{U_u}{\alpha} + \lambda V_j \tag{18}$$

$$\frac{\partial L}{\partial b_i} = -\frac{1}{\alpha} + \lambda b_i \tag{19}$$

$$\frac{\partial L}{\partial b_j} = \frac{1}{\alpha} + \lambda b_j \tag{20}$$

and for $x \geq \alpha$, we just skip the update of x_{uij}.
    The update rule can be derived as

$$\Theta_{new} \leftarrow \begin{cases} \Theta_{old} - \gamma \frac{\partial L}{\partial \Theta} & x_{uij} < \alpha \\ \Theta_{old} & x_{uij} \geq \alpha \end{cases} \tag{21}$$

where Θ can be U_u, V_i, V_j, b_i or b_j, and γ is the learning rate. The optimization algorithm for the second model is listed in Algorithm 2.

Algorithm 2. Optimization For Our Second Model

---
1 Initialize U, V, B randomly;
2 **repeat**
3   **for** $i = 1 : |D^*|$ **do**
4   %$|D^*|$ is the number of positive feedbacks Sample (u, i, j);
5   **if** $x_{uij} < \alpha$ **then**
6   Update U_u, V_i, V_j, b_i and b_j according to Eq.(21);
7   **end**
8   **end**
9 **until** *convergence*;

---

5 Experiments

5.1 Datasets

We evaluate the proposed models on two publicly available datasets. The first one is the *Yahoo! Music ratings for User Selected and Randomly Selected songs version 1.0*[2] dataset. This dataset contains $365,704$ ratings from $15,400$ users on 1000 songs. The rating values are on a scale from 1 to 5. Note that the test set of this dataset is collected from 5400 survey participants and the test songs for each participant are randomly selected. The second dataset is the *MovieLens 100K*[3] dataset. This dataset contains $100,100$ ratings from 943 users on 1682 movies. The rating values also are on a scale from 1 to 5.

5.2 Baseline Methods

We compare the proposed models with three methods including Popular Recommender (PopRec) [15], Bayesian Personalized Ranking (BPR) [12] and Collaborative Less-is-More Filtering (CLiMF) [15]. PopRec is a non-personalized algorithm that orders the items by the popularity. BPR is a MF-based pairwise approach that aims at modeling positive/negative feedback pairs correctly. CLiMF is a MF-based pointwise approach that tries to maximize MRR approximately. We have implemented PopRec and BPR, and the code of CLiMF is obtained at the author's personal website. As the model with generalized sigmoid loss function contains many hyper-parameters, it is expensive in computation to search the optimal hyper-parameters. So we optimize a, t and q in Eq.(8) separately, and treat them as three models denoted as *Gen(a)*, *Gen(t)* and *Gen(q)*. The proposed model with hinge loss function is denoted as *Hinge*.

5.3 Evaluation Metrics

We adopt three evaluation metrics to compare the performance of different methods. The first metric is Mean Reciprocal Rank (MRR) that is the average of the multiplicative inverse of the rank of the first relevant item:

---

[2] available at http://webscope.sandbox.yahoo.com/index.php
[3] available at http://www.grouplens.org/node/73

$$MRR = \frac{1}{|U|} \sum_{i=1}^{|U|} \frac{1}{rank_i} \tag{22}$$

where $rank_i$ is the rank of the first relevant item.

The second metric is Area Under Curve (AUC) which is equivalent to the proportion of correctly ordered items in the rank result, namely,

$$AUC = 1 - \frac{N^*}{N^+ \times N^-} \tag{23}$$

where N^+ and N^- are the number of relevant and irrelevant items respectively and N^* denotes the number of disordered pairs. We calculate AUC for each user and then take the average over all users.

The third metric is Precision@K (P@K). It equals to the proportion of relevant items in top-K items, that is,

$$P@K = \frac{N_k^+}{K} \tag{24}$$

where N_k^+ is the number of relevant items in top-K items. In this study, we set K to 5. P@5 is calculated for each user and the final value is the average over all users.

5.4   Experiment Protocol

For both datasets, we treat ratings whose values are at least 4 as relevant items or positive feedbacks. After this filtering, two datasets remain $129,748$ and $55,375$ ratings respectively. Both datasets are split into training, validation and test sets. For validation and test sets of each dataset, each user has similar quantities of ratings so that it can avoid the bias towards some users. At last, the details of the datasets partition are listed in Table 1. The validation set is used for parameter tuning including learning rate γ, regularization rate λ, dimensionality of latent factors as well as parameters in generalized loss function and hinge loss function. We adopt grid search for hyper-parameter optimization to determine the optimal ones. The hyper-parameters with best performance on the validation set are used to evaluate the performance on the test set.

Table 1. Details of two datasets

Dataset	#Users	#Items	Training set	Validation set	Test set
Yahoo!Music	15400	1000	125077	1529	3142
Movielens	943	1682	49906	1524	3945

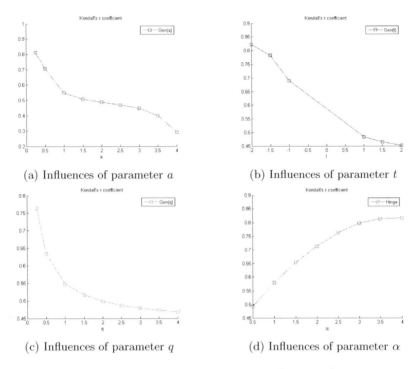

(a) Influences of parameter a (b) Influences of parameter t

(c) Influences of parameter q (d) Influences of parameter α

Fig. 5. Influences of parameters on popularity tendency

Table 2. Kendall's τ coefficient of each method

Dataset	BPR	CLiMF	Gen(a)	Gen(t)	Gen(q)	Hinge
Yahoo!Music	0.379016	0.844905	0.328537	0.259012	0.290082	0.446594
Movielens	0.568992	0.910833	0.698189	0.589724	0.682228	0.618641

5.5   Results and Discussion

Popularity Tendency. As stated before, popularity tendency is the tendency of an algorithm to recommend the popular items. We use Kendall's τ coefficient [6] between the ranking results of a specific method and PopRec to measure the popularity tendency of a specific algorithm. Table 2 shows the Kendall's τ coefficient values of each method, where the values are calculated using the hyper-parameters with best performance on the validation set. Based on the results, we have found the following interesting observations: (1) The τ value of CLiMF is much grater than others and this verifies that CLiMF has a strong popularity tendency. (2) As the test set of Movielens is randomly selected from the real dataset, it should contain many popular items. Thus the τ values with the optimal hyper-parameters setting are relatively large. In contrast, as the test set of Yahoo! Music is collected by an online survey, the τ values for Yahoo! Music are relatively small. (3) For Movielens, the τ values of our proposed models are slightly greater than BPR. This verifies that the optimal τ value for Movielens needs to be relatively large. Furthermore, we evaluate the influence of parameters of our proposed models. The results are shown in Figure 5. For each sub-figure in Figure 5, only one parameter changes and other parameters are fixed. We can observe that with a increasing, t increasing, q increasing or α decreasing, the popularity tendency will decrease.

Table 3. Evaluation results of each method

(a) Results on the Yahoo! Music dataset

	PopRec	BPR	CLiMF	Gen(a)	Gen(t)	Gen(q)	Hinge
MRR	0.029322	0.045216	0.029339	0.045988	0.049001	**0.049520**	0.044443
AUC	0.639391	0.707782	0.632281	0.706607	0.704747	**0.708467**	0.699434
P@5	0.007302	0.010911	0.007302	0.011078	0.012170	**0.012337**	0.010407

(b) Results on the Movielens dataset

	PopRec	BPR	CLiMF	Gen(a)	Gen(t)	Gen(q)	Hinge
MRR	0.187232	0.261335	0.187476	0.276889	0.271115	**0.277810**	0.273176
AUC	0.872349	0.924293	0.871982	0.925321	0.927878	0.926214	**0.928234**
P@5	0.062313	0.096574	0.062313	0.103426	0.101713	0.102570	**0.104069**

Performance Comparison. Table 3 shows the evaluation results on the test set of each method, where the hyper-parameters are determined by choosing the ones with best performance on the validation set. From the results, we have found the following interesting observations: (1) For both datasets, Gen(q) has the best performance for MRR. It improves 6.3% and 9.5% for MRR compared with BPR on Movielens and Yahoo! Music, respectively. (2) The performance of the proposed model with hinge loss function is comparable to BPR and even better on Movielens. (3) The evaluation results of CLiMF and PopRec are very similar. Actually, the Kendall's τ coefficient of the two methods is close to 1. (4) The performance on Movielens is much better than Yahoo! Music dataset because the test set of Movielens contains many popular items while the test set of Yahoo! Music is collected by an online survey and the items in the test set are selected randomly.

6   Conclusions

In this paper, we analyze the models of ranking-oriented CF from the perspective of loss function. Based on the analysis, we propose two models with generalized logistic loss function and hinge loss function, respectively. In the future, we will investigate the loss functions of listwise approach. We also plan to make better use of ratings as now we only treat ratings as binary feedbacks.

References

1. Bishop, C.M.: Pattern Recognition and Machine Learning (Information Science and Statistics). Springer-Verlag New York, Inc., Secaucus (2006)
2. Breese, J.S., Heckerman, D., Kadie, C.: Empirical analysis of predictive algorithms for collaborative filtering. In: Proceedings of the Fourteenth Conference on Uncertainty in Artificial Intelligence, UAI 1998, pp. 43–52. Morgan Kaufmann Publishers Inc., San Francisco (1998)
3. Cremonesi, P., Koren, Y., Turrin, R.: Performance of recommender algorithms on top-n recommendation tasks. In: Proceedings of the Fourth ACM Conference on Recommender Systems, RecSys 2010, pp. 39–46. ACM, New York (2010)
4. Hofmann, T.: Latent semantic models for collaborative filtering. ACM Trans. Inf. Syst. 22(1), 89–115 (2004)
5. Hu, Y., Koren, Y., Volinsky, C.: Collaborative filtering for implicit feedback datasets. In: ICDM 2008, pp. 263–272 (2008)
6. Kendall, M.G.: A new measure of rank correlation. Biometrika 30(1/2), 81–93 (1938)
7. Koren, Y., Bell, R., Volinsky, C.: Matrix factorization techniques for recommender systems. Computer 42(8), 30–37 (2009)
8. Koren, Y., Sill, J.: Ordrec: an ordinal model for predicting personalized item rating distributions. In: Proceedings of the Fifth ACM Conference on Recommender Systems, RecSys 2011, pp. 117–124. ACM, New York (2011)
9. Liu, N.N., Yang, Q.: Eigenrank: a ranking-oriented approach to collaborative filtering. In: Proceedings of the 31st Annual International ACM SIGIR Conference on Research and Development in Information Retrieval, SIGIR 2008, pp. 83–90. ACM, New York (2008)
10. Liu, N.N., Zhao, M., Yang, Q.: Probabilistic latent preference analysis for collaborative filtering. In: Proceedings of the 18th ACM Conference on Information and Knowledge Management, CIKM 2009, pp. 759–766. ACM, New York (2009)
11. Ma, H., King, I., Lyu, M.R.: Effective missing data prediction for collaborative filtering. In: Proceedings of the 30th Annual International ACM SIGIR Conference on Research and Development in Information Retrieval, SIGIR 2007, pp. 39–46. ACM, New York (2007)
12. Rendle, S., Freudenthaler, C., Gantner, Z., Schmidt-Thieme, L.: Bpr: Bayesian personalized ranking from implicit feedback. In: Proceedings of the Twenty-Fifth Conference on Uncertainty in Artificial Intelligence, UAI 2009, pp. 452–461. AUAI Press, Arlington (2009)
13. Richards, F.J.: A flexible growth function for empirical use. Journal of Experimental Botany 10(2), 290–301 (1959)
14. Sarwar, B., Karypis, G., Konstan, J., Riedl, J.: Item-based collaborative filtering recommendation algorithms. In: Proceedings of the 10th International Conference?on World Wide Web, WWW 2001, pp. 285–295. ACM, New York (2001)
15. Shi, Y., Karatzoglou, A., Baltrunas, L., Larson, M., Oliver, N., Hanjalic, A.: Climf: learning to maximize reciprocal rank with collaborative less-is-more filtering. In: Proceedings of the Sixth ACM Conference on Recommender Systems, RecSys 2012, pp. 139–146. ACM, New York (2012)

16. Shi, Y., Larson, M., Hanjalic, A.: List-wise learning to rank with matrix factorization for collaborative filtering. In: Proceedings of the Fourth ACM Conference on Recommender Systems, RecSys 2010, pp. 269–272. ACM, New York (2010)
17. Srebro, N., Rennie, J.D.M., Jaakola, T.S.: Maximum-Margin Matrix Factorization. Advances in Neural Information Processing Systems 17, 1329–1336 (2005)
18. Steck, H.: Item popularity and recommendation accuracy. In: Proceedings of the Fifth ACM Conference on Recommender Systems, RecSys 2011, pp. 125–132. ACM, New York (2011)
19. Weimer, M., Karatzoglou, A., Le, Q.V., Smola, A.: Cofirank, maximum margin matrix factorization for collaborative ranking. Advances in Neural Information Processing Systems 20 (2007)

Rating Propagation in Web Services Reputation Systems: A Fast Shapley Value Approach

An Liu[1], Qing Li[2], Xiaofang Zhou[1,3], Lu Li[4], Guanfeng Liu[1], and Yunjun Gao[5]

[1] Soochow University, Suzhou, China
[2] City University of Hong Kong, Hong Kong, China
[3] University of Queensland, Brisbane, Australia
[4] University of Science and Technology of China, Hefei, China
[5] Zhejiang University, Hangzhou, China
anliu@suda.edu.cn

Abstract. A new challenge in Web services reputation systems is to update the reputation of component services. As an emerging solution, rating propagation has received much attention recently. Current rating propagation algorithms either fail to fairly distribute the overall rating to component services or can realize a fair rating distribution at the cost of exponential time complexity. In this paper, we propose a fast Shapley value approach to propagate the overall rating of a composite service to its component services. Our approach ensures the fairness of rating propagation by using the advantage of the Shapley value, but significantly decreases its computational complexity from exponential to quadratic. Its fairness and efficiency are validated by experiments.

Keywords: service composition, Shapley value, reputation propagation.

1 Introduction

Reputation systems enable trust establishment among unknown entities in an open environment. Every entity in a reputation system has associated with a reputation typically derived from the rating that it has received during its previous interactions with others. Reputations are usually numerical values so that direct comparisons can be made between entities. The primary functionality of reputation systems is to collect, maintain, and disseminate reputations of each entity correctly and effectively even against attack from malicious entities. Over the past decade, reputation systems have been employed in a variety of applications. Some examples include e-commerce (e.g., eBay and Amazon), peer-to-peer (P2P) networks [5, 19, 25], content creation [1], web search [11], to name a few.

Recently, a considerable amount of research work has been done to build reputation systems for Web services. While most efforts aim at improving system accuracy by integrating multiple aspects of reputation sources, such as user ratings, personal experience, and personal QoS preference, and resilience to malicious user attack by assigning proper weights to user ratings, very little attention has been given to *rating propagation* which is a new challenge in the formation

S.S. Bhowmick et al. (Eds.): DASFAA 2014, Part I, LNCS 8421, pp. 466–480, 2014.

of Web services reputation systems. More specifically, a user is required to give a rating to a Web service after each interaction with the service so that its reputation can be updated, but when the service is composite (i.e., it consists of several component services), a rating propagation model is expected to distribute the rating of the composite service to its component services for the update of their reputations. However, it is nontrivial to design an efficient rating propagation model due to the following reasons:

- *Contribution.* Component services may have different contributions to the overall rating of a composite service. In particular, some component services may outperform what they have promised and some fail to deliver their advertised QoS, though the composite service has a good performance overall which results in a high rating. A good rating propagation algorithm should ensure that a component service is not awarded (or penalized) for the good (poor) performances of the other component services.
- *Delegation.* The user of a composite service cannot perceive the existence of component services, so the composite service needs to be delegated by the user to give ratings to its component services. Generally, there is a large variety of users, including conservative (tend to give a low rating), progressive (tend to give a high rating), and neutral [6], and different types of users often give different ratings to the same service. As a delegate, the composite service should also inherit the rating preference of the user.
- *Multi-level composition.* A composite service may be multi-level in the sense that some component services of a composite service are also composite and consist of other component services. This feature makes rating propagation more complex, as the above two issues should be properly dealt with no matter what level of composition exists.

To address the above challenges, we designed a Shapley value based approach called SVA for Web services rating propagation in [8]. SVA can achieve a *fair* rating propagation which is guaranteed by the fact that the Shapley value satisfies three axioms: *efficiency, symmetry,* and *balanced contribution* [15]. However, the calculation of the Shapley value typically involves an exponential time complexity, which has restricted its use mostly as a reference for theoretical interest, and limited its appeal as an implementable rating propagation mechanism for very large Web services composition. To overcome this weakness, we propose in this paper a Fast Shapley Value Approach (FSVA) for Web services rating propagation. Compared with the exponential time complexity of SVA, FSVA only has quadratic time complexity, which significantly improves the feasibility of Web services rating propagation based on the Shapley value.

The remainder of the paper is organized as follows. Section 2 describes a Web services reputation model which provides necessary background knowledge for the discussion of Web services rating propagation. Section 3 briefly introduces the original Shapley value approach, while Section 4 describes in detail how the fast Shapley value approach is designed. The fairness and performance of the fast Shapley value approach is studied by experiments in Section 5. Section 6 discusses related work, and finally, Section 7 concludes the paper.

2   Web Services Reputation Model

2.1   Preliminaries

A Web service is a software application whose functionality can be viewed as a set of operations where each operation consumes input values and generates output values. For the sake of focus and clarity, we assume every service has only one operation. The operation of every service has a set of numerical QoS parameters, for example, response time and invocation fee. Each QoS parameter has two types of values: *promised* (Q_p) and *delivered* (Q_d). The former is those that are advertised by a service before a user invokes it, while the latter represents the actual values of different quality parameters after the user invokes the service.

The structure of a composite service can be modeled by a *service graph* where its component services are connected by four types of composition patterns: *sequential, parallel, structured conditional*, and *loop* [24]. A service graph has only one start node and only one end node. The executed part of a service graph is defined to be a *service execution graph* (SEG) which includes the start node, the end node, all branches of every parallel pattern, the executed branch of every structured conditional pattern, and the unfoldment of every loop pattern (the unfoldment is feasible as the loop time of every loop pattern is determined after execution) [21]. Clearly, a SEG only contains two types of composition patterns: sequential and parallel.

Given a composite service c and the QoS of its component services, the computation of c's QoS is an important problem. The computational characteristics of Web services QoS can be divided into three types: *additive, multiplicative*, and *concave*. Given c and its component services $\{s_1, \cdots, s_n\}$, additive QoS means $Q(c) = \sum_{i=1}^{n} Q(s_i)$, multiplicative QoS means $Q(c) = \prod_{i=1}^{n} Q(s_i)$, and concave QoS means $Q(c) = \max(Q(s_1), \cdots, Q(s_n))$ or $Q(c) = \min(Q(s_1), \cdots, Q(s_n))$. When the value of a QoS parameter is deterministic and represented as a real number, some aggregation formula can be used to compute c's QoS [22, 24]. When the value of a QoS parameter is non-deterministic and modeled as probability distributions, extensive probabilistic calculation is needed [23].

2.2   Reputation Model

The reputation of a Web service is the aggregation of the ratings provided by different users, so the metric or representation of rating is a core component of reputation systems [2, 7]. In this study, the rating that a user gives to a Web service is defined as a vector $V = [v_1, \cdots, v_n]$ where n is the number of quality parameters of the service and $v_i (1 \leq i \leq n)$ is a real number between -1 (the worst rating) and 1 (the best rating). Note that continuous representation is chosen here for its simplicity since most formula in reputation models result in real number results [2], but other kinds of representation such as *binary* or *discrete* can also be adopted without altering our rating propagation algorithms. It is also notable that the rating could be a single value (i.e., an overall judgement of the service over multiple quality parameters), however, the user given the rating is also required to explicitly describe her quality preference, for example, in the form of Reputation Significance Vector [9] or Quality Aspects Importance [16], to increase the accuracy of reputation assessment. Moreover, each rating is assigned with a timestamp so that temporal factor can be taken into account in reputation computation. More details of this will be shown later.

The reputation of a service s, as seen by a user u, consists of two parts: R_I, an indirect evaluation based on other users' experience of s, and R_D, a direct evaluation based on u's direct experience of s, given by the following formula:

$$R(s, u) = \omega_I R_I(s, U) + \omega_D R_D(s, u) \tag{1}$$

where:

- U is a set of users selected by u;
- ω_I and ω_D are associated weights such that $0 \leq \omega_I(\omega_D) \leq 1$ and $\omega_I + \omega_D = 1$.

To compute R_I, u needs to select a set of users who have interacted with s in the past and are willing to share their personal evaluations of s. More specifically, R_I is calculated by the following formula:

$$R_I(s, U)[k] = \sum_{u_i \in U} \omega_{u_i} C(u_i) Sim(u, u_i)[k] R_D(s, u_i)[k] \tag{2}$$

where:

- $C(u_i)$ is the credibility of u_i such that $0 \leq C(u_i) \leq 1$;
- $Sim(u, u_i)$ is the similarity vector between u and u_i, where the kth element $Sim(u, u_i)[k]$ is the similarity between u and u_i in terms of the kth quality parameter such that $0 \leq Sim(u, u_i)[k] \leq 1$.
- ω_{u_i} is the associated weight such that $0 \leq \omega_{u_i} \leq 1$ and $\sum_{u_i \in U} \omega_{u_i} = 1$.

In the above formula, $R_D(s, u_i)[k]$, which is u_i's direct evaluation of s in terms of the kth QoS parameter, is weighted by the product of u_i's credibility and the similarity between u and u_i in terms of the kth QoS parameter. Specifically, ratings given by highly credible users have larger weights than that given by users with low credibilities. For malicious users, the weights of their ratings could even be set to 0. Moreover, even honest users may have different rating preference [6], for example, conservative users tend to give low ratings while progressive users tend to give high ratings, so the similarity between u and u_i is also taken into account to decrease the discrepancy of their rating preference. The computation of $C(u_i)$ and $Sim(u, u_i)[k]$ is quite challenging but is not the focus of this paper, so we simply adopt the method proposed in [9] for $C(u_i)$ and the technique introduced in [18] for $Sim(u, u_i)[k]$.

Besides other users' experience, a user's direct experience is also important for reputation assessment as the user is likely not to consider a service with whom she had a bad experience in the past, even if the service receives good ratings from other users [9, 13, 16]. Considering the fact that u may invoke s multiple times and each time gives it a rating, $R_D(s, u)$ is defined as:

$$R_D(s, u) = \sum_{l=1}^{n} \frac{\phi(\Delta t_l)}{\sum_{m=1}^{n} \phi(\Delta t_m)} V_l(s, u) \tag{3}$$

where:

- n is the total number of past invocations;
- $V_l(s, u)$ is the rating that u gives to s in the lth invocation;
- ϕ is an aging function and Δt_l is the difference between the current time and the time of the lth invocation.

Note that, after the lth invocation, u compares the delivered quality Q_d of s with the promised quality Q_p of s to produce a rating $V_l(s, u)$ [9]. In this sense, $V_l(s, u)$ can be seen as a function of $Q_d(s)$ and $Q_p(s)$ as follows:

$$V_l(s, u) = f(Q_d(s), Q_p(s)) \tag{4}$$

Also note that the aging function ϕ is user-specific. For example, u can assign equal weights to all ratings if she considers them to have the same importance. However, if she perceives the ratings in recent invocations to be more important as they reflect up-to-date quality, she can give them larger weights. She can also set the weights of some old ratings to be 0 as they are outdated.

3   Shapley Value Based Web Services Rating Propagation

We first give the definition of *Web services rating propagation problem* as follows:

Problem 1. Given a user u, a composite service c, and a rating $v = f(Q_d(c), Q_p(c))$ provided by u after invoking c, find a proper rating v_i for every component service $s_i \in c$ so that $s_i's$ reputation can be updated accordingly.

For completeness and readability, we briefly introduce SVA proposed in our previous work [8]. Considering the rating v that u gives to c depends on $\Delta(c) = Q_d(c) - Q_p(c)$ which is the QoS fluctuation of c in that invocation, SVA computes $\Delta_i(c)$ which is the QoS fluctuation of c caused by s_i in that invocation and assigns $f(\Delta_i(c))$ to s_i as its rating. If s_i is composite, SVA will continue to rate its component services. The cornerstone of SVA is to solve the *QoS fluctuation distribution problem* defined as follows:

Problem 2. Given a composite service c and its QoS fluctuation $\Delta(c)$ in one invocation, calculate $\Delta_i(c)$ for every component service $s_i \in c$, that is, its contribution to the overall QoS fluctuation $\Delta(c)$.

The above problem is, however, non-trivial as the overall QoS fluctuation has a complicated relationship with the QoS fluctuation of component services. Instead of deriving specific relationship via analyzing composition structure and QoS computational characteristics, we notice that the nature of QoS fluctuation distribution is similar to the cost sharing problem which is important in cooperative game theory [14], so SVA adapts the Shapley value for the computation of $\Delta_i(c)$. More specifically, the QoS fluctuation of c caused by s_i is calculated by the following formula:

$$\Delta_i(c) = \sum_{T \subseteq c \setminus \{s_i\}} \frac{|T|!(n - |T| - 1)!}{n!} (\Delta(c \mid T \cup \{s_i\}) - \Delta(c \mid T)) \tag{5}$$

where n is the number of component services in c, T is a set of component services in c, and $\Delta(c \mid T)$ is the QoS fluctuation of c conditional on T, that is, the QoS fluctuation of c when only the QoS fluctuations of component services in T are taken into account.

In formula 5, it is clearly seen that the marginal contribution of s_i to the QoS fluctuation of c conditional on T is $\Delta(c \mid T \cup \{s_i\}) - \Delta(c \mid T)$. The weight that is used in front of $\Delta(c \mid T \cup \{s_i\}) - \Delta(c \mid T)$ is the probability that T is just the set of services in front of s_i in an ordering of $\{s_1, \cdots, s_n\}$. In this case, there are $|T|!$ ways of positioning the services in T at the start of the ordering, and $(n-|T|-1)!$ ways of positioning the remaining services except s_i at the end of the ordering. Moreover, the probability of such an ordering occurs is $\frac{|T|!(n-|T|-1)!}{n!}$ when all orderings are equally probable. Therefore, the Shapley value computed by the formula is the marginal contribution of s_i to the QoS fluctuation of c conditional on all component services, that is, the QoS fluctuation of c caused by s_i.

4 Fast Shapley Value Approach

Clearly, the calculation of the Shapley value through formula 5 involves an exponential time complexity, which is the primary weakness of SVA. As an important extension of our previous work, we propose in this section a fast calculation method with quadratic complexity. Note that our current focus is homogeneous QoS with the same computational characteristic, and we leave fast calculation of the Shapley value for heterogeneous QoS as one important direction of our subsequent research.

4.1 Additive QoS Fluctuation Distribution

We first consider the computation of $\Delta_i(c)$ for additive QoS. Clearly, $Q_p(c) = \sum_{s \in c} Q_p(s)$ and $\Delta(c \mid T) = \sum_{s \in T}(Q_d(s) - Q_p(s))$ hold in this case, so we have:

$$\Delta_i(c) = (Q_d(s_i) - Q_p(s_i)) \sum_{T \subseteq c \setminus \{s_i\}} \frac{|T|!(n - |T| - 1)!}{n!}$$

$$= Q_d(s_i) - Q_p(s_i)$$

Regarding the time complexity of additive QoS fluctuation distribution, we immediately have the following theorem based on the above deduction:

Theorem 1. *For additive QoS, $\Delta_i(c)$ can be calculated in a constant time.*

4.2 Multiplicative QoS Fluctuation Distribution

Now, we consider the computation of $\Delta_i(c)$ for multiplicative QoS. First note that the following two formulae hold where $H = c \setminus \{T \cup \{s_i\}\}$:

$$\Delta(c \mid T) = Q_p(s_i)(\prod_{s \in T} Q_d(s))(\prod_{s \in H} Q_p(s)) - \prod_{s \in c} Q_p(s)$$

$$\Delta(c \mid T \cup \{s_i\}) = Q_d(s_i)(\prod_{s \in T} Q_d(s))(\prod_{s \in H} Q_p(s)) - \prod_{s \in c} Q_p(s)$$

By subtracting $\Delta(c \mid T)$ from $\Delta(c \mid T \cup \{s_i\})$, we obtain the difference:

$$(Q_d(s_i) - Q_p(s_i))(\prod_{s \in T} Q_d(s))(\prod_{s \in H} Q_p(s))$$

by which formula 5 can be rewritten as follows:

$$\Delta_i(c) = (Q_d(s_i) - Q_p(s_i)) \sum_{T \subseteq c \setminus \{s_i\}} \frac{|T|!(n - |T| - 1)!}{n!}(\prod_{s \in T} Q_d(s))(\prod_{s \in H} Q_p(s))$$

$$(6)$$

According to the size of T (denoted as $|T|$), formula 6 can be further rewritten as follows:

$$\frac{\Delta_i(c)}{Q_d(s_i) - Q_p(s_i)} = \sum_{j=0}^{n-1} \sum_{\substack{T \subseteq c \setminus \{s_i\} \\ |T|=j}} \frac{|T|!(n - |T| - 1)!}{n!}(\prod_{s \in T} Q_d(s))(\prod_{s \in H} Q_p(s)) \quad (7)$$

To facilitate later discussion, we introduce here some new notations: c' and c'_k. In particular, $c' = c \setminus \{s_i\} = \{s_{i1}, \cdots, s_{in-1}\}$ where s_{ik} is the kth ($1 \leq k \leq n-1$) element in c'; $c'_k = \{s_{i1}, \cdots, s_{ik}\}$ and we clearly have $c' = c'_{n-1}$. In addition, we consider a new variable $\delta_i(c')$ which is defined as follows:

$$\delta_i(c') = \sum_{j=0}^{n-1} \sum_{\substack{T \subseteq c' \\ |T|=j}} (\prod_{s \in T} Q_d(s))(\prod_{s \in \overline{T}} Q_p(s)) \quad (8)$$

Based on the notation of c'_k, $\delta_i(c')$ can be rewritten as follows:

$$\delta_i(c') = \delta_i(c'_{n-1}) = \sum_{j=0}^{n-1} \delta_i^j(c'_{n-1}) \quad (9)$$

$$\delta_i^j(c'_{n-1}) = \sum_{\substack{T \subseteq c'_{n-1} \\ |T|=j}} (\prod_{s \in T} Q_d(s))(\prod_{s \in \overline{T}} Q_p(s)) \quad (10)$$

The above rewrite is important for realizing an efficient computation of $\Delta_i(c)$ due to a recursive relationship between $\delta_i^j(c'_k)$ and $\delta_i^j(c'_{k-1})$ when $k \geq 3$, which is given by the following equation:

$$\delta_i^j(c'_k) = \begin{cases} \delta_i^0(c'_{k-1})Q_p(s_k), & \text{if } j = 0 \\ \delta_i^{j-1}(c'_{k-1})Q_d(s_k) + \delta_i^j(c'_{k-1})Q_p(s_k), & \text{if } 1 \leq j \leq k - 1 \\ \delta_i^{k-1}(c'_{k-1})Q_d(s_k), & \text{if } j = k \end{cases} \quad (11)$$

The proof of equation 11 (through mathematical induction) is straightforward and omitted here due to limited space. The time complexity of multiplicative QoS fluctuation distribution is given by the following theorem:

Theorem 2. *For multiplicative QoS, $\Delta_i(c)$ can be calculated in quadratic time.*

Proof. Based on formula 11 and its basic case where $k = 2$, it is clear that $\delta_i(c')$ and $\delta_i^j(c'_{n-1})(0 \leq j \leq n-1)$ can be calculated in quadratic time. Then, from formulae 7, 8, 9, we can see that $\Delta_i(c)$ can be calculated in quadratic time.

4.3   Concave QoS Fluctuation Distribution

In this section, we present the calculation of $\Delta_i(c)$ for concave QoS. Note that the method that will be shown subsequently aims at *max* (i.e., one type of concave QoS), but it is also applicable to *min* (the other type of concave QoS) only with some minor adjustments. Given a subset $T \subseteq c' = c \setminus \{s_i\}$ and $\overline{T}$ such that $T \bigcup \overline{T} = c'$, the difference between $\Delta(c \mid T \cup \{s_i\})$ and $\Delta(c \mid T)$ is:

$$\Delta_T = \max_{\substack{s_j \in T \\ s_k \in \overline{T}}}(Q_d(s_i), Q_d(s_j), Q_p(s_k)) - \max_{\substack{s_j \in T \\ s_k \in \overline{T}}}(Q_p(s_i), Q_d(s_j), Q_p(s_k)) \qquad (12)$$

Let $M = \max_{\substack{s_j \in T \\ s_k \in \overline{T}}}(Q_d(s_j), Q_p(s_k))$, formula 12 can be simplified to:

$$\Delta_T = \max(Q_d(s_i), M) - \max(Q_p(s_i), M) \qquad (13)$$

The final result of Δ_T is in the form of $Q_*(s) - Q_*(s')$ where $Q_* \in \{Q_p, Q_d\}$, $s \in c$, and $s' \in c$. Consequently, $\Delta_i(c)$ can be seen as a weighted sum of the promised and delivered QoS of every component service in c. Hence, we have:

$$\Delta_i(c) = \sum_{s \in c} \omega_{Q_p(s)} Q_p(s) + \sum_{s \in c} \omega_{Q_d(s)} Q_d(s) = \sum_{s \in c} \omega_{Q_*(s)} Q_*(s) \qquad (14)$$

To calculate the weight of $Q_*(s)$, we reorder the elements in the sequence

$$\langle Q_p(s_1), Q_d(s_1), \cdots, Q_p(s_{i-1}), Q_d(s_{i-1}), Q_p(s_{i+1}), Q_d(s_{i+1}), \cdots, Q_p(s_n), Q_d(s_n)\rangle$$

to construct an ordered sequence $QS = \langle qs_1, \cdots, qs_{2n-2}\rangle$ such that $qs_1 \geq qs_2 \geq \cdots \geq qs_{2n-2}$. We also define two functions: $\phi : qs \rightarrow s$ which returns the service whose (promised or delivered) QoS equals to qs and $\varphi : qs \rightarrow Q_*$ which returns the exact QoS type (i.e., Q_p or Q_d). Clearly, we have $\varphi(qs)(\phi(qs)) = qs$.

We first consider the case in which all elements in QS are different, that is, $qs_1 > qs_2 > \cdots > qs_{2n-2}$. We have the following lemmas:

Lemma 1. *If $qs_k < \min(Q_p(s_i), Q_d(s_i))$, then $\omega_{Q_*(\phi(qs_k))} = 0$.*

Lemma 2. *If $qs_k > \max(Q_p(s_i), Q_d(s_i))$, then $\omega_{Q_*(\phi(qs_k))} = 0$.*

Lemma 3. *If $\exists s_m \in c$, $\min(Q_d(s_m), Q_p(s_m)) > qs_k$, then $\omega_{Q_*(\phi(qs_k))} = 0$.*

Lemma 1 and Lemma 2 are both clearly hold based on formula 13. For Lemma 3, if $\omega_{Q_*(\phi(qs_k))} \neq 0$, then neither $Q_p(s_m)$ nor $Q_d(s_m)$ is considered in M, which is clearly a contradiction.

Lemma 4. *A necessary condition of* $\omega_{Q_*(\phi(qs_k))} \neq 0$ *is,* $\forall j,\ 1 \leq j \leq k-1,$ $\phi(qs_j) \notin T$ *if* $\varphi(qs_j) = Q_d$ *and* $\phi(qs_j) \in T$ *if* $\varphi(qs_j) = Q_p.$

Proof. If $\varphi(qs_j) = Q_d$ and $\phi(qs_j) \in T$, then the calculation of M must consider $Q_d(\phi(qs_j)) = qs_j > qs_k$ which indicates $\omega_{Q_*(\phi(qs_k))} = 0$. Similarly, if $\varphi(qs_j) = Q_p$ and $\phi(qs_j) \notin T$, then the calculation of M must consider $Q_p(\phi(qs_j)) = qs_j > qs_k$ which indicates $\omega_{Q_*(\phi(qs_k))} = 0$.

Algorithm 1. Compute concave QoS fluctuation distribution

---

1 **begin**
2     $\Delta_i(c) \leftarrow 0$, $S_p \leftarrow \emptyset$, $S_d \leftarrow \emptyset$, needStop $\leftarrow$ false
3   construct an ordered sequence $QS = \{qs_1, \cdots, qs_{2n-2}\}$;
4   **for** $k \leftarrow 1$ **to** $2n - 2$ **do**
5   **if** needStop = true **then** break; **if** $\varphi(qs_k) = Q_p$ **then**
          $S_p \leftarrow S_p \bigcup \{\phi(qs_k)\}$; **if** $\varphi(qs_k) = Q_d$ **then** $S_d \leftarrow S_d \bigcup \{\phi(qs_k)\}$; **if**
          $S_d \bigcap S_p \neq \emptyset$ **then** needStop $\leftarrow$ true; **if** $qs_k > \max(Q_p(s_i), Q_d(s_i))$
          then do nothing here; **else if** $qs_k > \min(Q_p(s_i), Q_d(s_i))$ **then**
6   calculate $\omega_{Q_*(\phi(qs_k))}$ according to formula 15;
7   **if** $Q_d(s_i) > Q_p(s_i)$ **then** add $\omega_{Q_*(\phi(qs_k))}(Q_d(s_i) - qs_k)$ to $\Delta_i(c)$; **if**
              $Q_d(s_i) < Q_p(s_i)$ **then** add $\omega_{Q_*(\phi(qs_k))}(qs_k - Q_p(s_i))$ to $\Delta_i(c)$;
8   **else**
9   calculate $\omega_{Q_*(s_i)}$ according to formula 15;
10            $\Delta_i(c) = \Delta_i(c) + \omega_{Q_*(s_i)}(Q_d(s_i) - Q_p(s_i))$; break;
11   **end**
12   **if** $\varphi(qs_k) = Q_p$ **then** $S_p \leftarrow S_p \setminus \{\phi(qs_k)\}$; $S_d = S_d \bigcup \{\phi(qs_k)\}$; **if**
          $\varphi(qs_k) = Q_d$ **then** $S_d \leftarrow S_d \setminus \{\phi(qs_k)\}$; $S_p = S_p \bigcup \{\phi(qs_k)\}$;
13   **end**
14   **return** $\Delta_i(c)$;
15 **end**

We are now in the position to calculate $\omega_{Q_*(\phi(qs_k))}$ for every qs_k such that $\min(Q_p(s_i), Q_d(s_i)) < qs_k < \max(Q_p(s_i), Q_d(s_i))$. Let $S_d = \{\phi(qs_j) | \varphi(qs_j) = Q_d\}$, $S_p = \{\phi(qs_j) | \varphi(qs_j) = Q_p\}$, $\forall j,\ 1 \leq j \leq k-1$. We first note that, when $S_d \bigcap S_p \neq \emptyset$, $\omega_{Q_*(\phi(qs_k))}$ must be 0 according to Lemma 3, so we only need to consider the case where $S_d \bigcap S_p = \emptyset$. According to Lemma 4, the subset T which provides a non-zero contribution to $\omega_{Q_*(\phi(qs_k))}$ satisfies the following two conditions: 1) it must contain the elements in S_p; and 2) it cannot contain the elements in S_d. However, it should be noted that T must contain $\phi(qs_k)$ if $\varphi(qs_k) = Q_d$ and cannot contain $\phi(qs_k)$ if $\varphi(qs_k) = Q_p$. The above membership is crucial to the calculation of $\omega_{Q_*(\phi(qs_k))}$, so we add $\phi(qs_k)$ into S_d (or S_p) if $\varphi(qs_k) = Q_p$ (or $\varphi(qs_k) = Q_d$). Let $S_{dp} = c' \setminus \{S_d \bigcup S_p\}$. Clearly, multiple subsets of c' could be T as there is no requirement for the set membership between T and the elements in S_{dp}. To construct a possible T, we select h elements in S_{dp} and put them into T, which results in T containing $h + |S_p|$ elements. Note that the corresponding weight of T is $\frac{|T|!(n-|T|-1)!}{n!}$ as defined in formula 5, and h can be varied from 0 to $|S_{dp}|$, so we have:

$$|\omega_{Q_*(\phi(qs_k))}| = \sum_{h=0}^{|S_{dp}|} \binom{|S_{dp}|}{h} \frac{(h+|S_p|)!(n-h-|S_p|-1)!}{n!} \qquad (15)$$

So far, we have calculated $|\omega_{Q_*(s)}|$ for all component services in c except s_i. To calculate $\omega_{Q_*(s_i)}$, we first note that formula 13 can be rewritten as follows:

$$\Delta_T = \begin{cases} Q_d(s_i) - M, & \text{if } Q_d(s_i) > M > Q_p(s_i) \\ M - Q_p(s_i), & \text{if } Q_p(s_i) > M > Q_d(s_i) \\ Q_d(s_i) - Q_p(s_i), & \text{if } M < \min(Q_d(s_i), Q_p(s_i)) \\ 0, & \text{if } M > \max(Q_d(s_i), Q_p(s_i)) \end{cases} \qquad (16)$$

which indicates $\omega_{Q_*(s_i)}$ consists of two parts: one part comes from the first two cases and is equal to $-|\omega_{Q_*(\phi(qs_k))}|$ as M must be one element in QS say qs_k, and the other part comes from the third case and its exact value can be calculated through formula 15 as long as the sets S_d and S_p are available.

Based on the above discussion, we give the pseudocode of computing $\Delta_i(c)$ for concave QoS in Algorithm 1. The main idea is to iteratively access an ordered sequence QS to accumulate the contribution of every element in QS to $\Delta_i(c)$. The boolean variable $needStop$ is used to determine whether the iteration needs to be terminated in advance. In particular, $needStop$ is set to be $true$ if $S_d \cap S_p \neq \emptyset$, as in this case the weights of the elements that have not been accessed in QS must be 0 according to Lemma 3, which makes it unnecessary to access these elements. The iteration also needs to be terminated when accessing an element $qs_k < \min(Q_p(s_i), Q_d(s_i))$ since the weights of all elements smaller than qs_k must be 0 according to Lemma 1. Before exiting the iteration, however, we must consider the contribution of s_i to $\Delta_i(c)$ in the form of $Q_d(s_i) - Q_p(s_i)$. At the beginning of every iteration, S_p and S_d are updated for later weight calculation. Then, $\Delta_i(c)$ are updated by accumulating the contribution of qs_k or the contribution of s_i according to formula 16. Before starting next iteration, we need to move $\phi(qs_k)$ from S_p to S_d or from S_d to S_p as explained earlier. The following theorem shows the time complexity of concave QoS fluctuation distribution:

Theorem 3. *For concave QoS, $\Delta_i(c)$ can be calculated in quadratic time.*

Proof. In Line 3 of Algorithm 1, the ordered sequence QS can be constructed by a well-known sorting algorithm such as Heapsort in linearithmic time. The calculation of formula 15 can be done in linear time, and it is embedded in the main iteration, so the running time of Algorithm 1 is $O(n^2)$, which completes the proof.

4.4 Numerical Example

We use a numerical example to illustrate FSVA, the fast Shapley value approach. Let a composite service c have five component services $\{s_1, \cdots, s_5\}$. The QoS of every service is a vector $[q_1, q_2, q_3]$ in which q_1 is additive, q_2 is multiplicative, and q_3 is max (or concave). Table 1 shows the promised/delivered QoS of these services in one invocation, the QoS fluctuation of c, and the QoS fluctuation of c caused by every component service generated by SVA.

Now, we compute the QoS fluctuation of c caused by s_2 through FSVA. For the additive QoS, we have $\Delta_2(c) = Q_d(s_2) - Q_p(s_2) = 6 - 8 = -2$ which equals to the result computed by SVA. For the multiplicative QoS, we have $c' = \{s_{21}, s_{22}, s_{23}, s_{24}\}$ where $s_{21} = s_1$, $s_{22} = s_3$, $s_{23} = s_4$, and $s_{24} = s_5$, and $c'_2 = \{s_1, s_3\}$, $c'_3 = \{s_1, s_3, s_4\}$, and $c'_4 = \{s_1, s_3, s_4, s_5\}$. We first compute $\delta_2^j(c'_2)$ for $0 \leq j \leq 2$. According to formula 10, we have $\delta_2^0(c'_2) = Q_p(s_1)Q_p(s_3) = 3$, $\delta_2^1(c'_2) = Q_p(s_1)Q_d(s_3) + Q_d(s_1)Q_p(s_3) = 18$, and $\delta_2^2(c'_2) = Q_d(s_1)Q_d(s_3) = 15$. Then, according to formula 11, we can obtain the following results: when $k = 3$, $\delta_2^0(c'_3) = 12$, $\delta_2^1(c'_3) = 78$, $\delta_2^2(c'_3) = 96$, and $\delta_2^3(c'_3) = 30$; when $k = 4$, $\delta_2^0(c'_4) = 60$, $\delta_2^1(c'_4) = 402$, $\delta_2^2(c'_4) = 558$, $\delta_2^3(c'_4) = 246$, and $\delta_2^4(c'_4) = 30$. From formulae 7, 8, and 9, we have: $\Delta_2(c) = (Q_d(s_2) - Q_p(s_2))(\frac{1}{5}\delta_2^0(c'_4) + \frac{1}{20}\delta_2^1(c'_4) + \frac{1}{30}\delta_2^2(c'_4) + \frac{1}{20}\delta_2^3(c'_4) + \frac{1}{5}\delta_2^4(c'_4)) = 138$ which is just equal to the result generated by SVA.

For the concave QoS, we first construct an ordered sequence based on which only three elements $qs_1 = Q_p(s_4)$, $qs_2 = Q_p(s_5)$, and $qs_3 = Q_d(s_1)$ are in the range $[35, 60]$. We have $|\omega_{Q_p(s_4)}| = 0.5$ (as $|S_p| = 1$, $|S_d| = 0$, and $|S_{dp}| = 3$), $|\omega_{Q_p(s_5)}| = \frac{1}{6}$ (as $|S_p| = 1$, $|S_d| = 1$, and $|S_{dp}| = 2$), and $|\omega_{Q_d(s_1)}| = 0.25$ (as $|S_p| = 3$, $|S_d| = 0$, and $|S_{dp}| = 1$). Besides, the weight for $Q_d(s_2) - Q_p(s_2)$ is $\frac{1}{12}$ as $|S_p| = 2$, $|S_d| = 1$, and $|S_{dp}| = 1$. Hence, $\Delta_2(c) = 0.5(60 - 55) + \frac{1}{6}(60 - 50) + 0.25(60 - 45) + \frac{1}{12}(60 - 35) = 10$ which equals to the result generated by SVA.

Table 1. QoS fluctuation distribution in a composite service

service	Q_p	Q_d	$\Delta(c)$	$\Delta_i(c)$
s_1	$[5, 1, 15]$	$[5, 5, 45]$	N/A	$[0, 312, \frac{5}{6}]$
s_2	$[8, 2, 35]$	$[6, 4, 60]$	N/A	$[-2, 138, 10]$
s_3	$[2, 3, 30]$	$[5, 3, 25]$	N/A	$[3, 0, 0]$
s_4	$[4, 4, 55]$	$[3, 2, 10]$	N/A	$[-1, -138, -\frac{25}{6}]$
s_5	$[6, 5, 50]$	$[8, 1, 20]$	N/A	$[2, -312, -\frac{5}{3}]$
c	$[25, 120, 55]$	$[27, 120, 60]$	$[2, 0, 5]$	N/A

5　Experiments

In this section, we compare the performance of FSVA with SVA [8] and the algorithm proposed by Nepal et al. in [12]. All experiments all performed on a PC with 2.8GHz CPU, 4GB RAM, Windows 7, and JDK 7. We first study the fairness of these rating propagation algorithms by considering a composite service c which invokes three services s_1, s_2, and s_3 one after another. For simplicity, we only consider one QoS attribute: response time. In particular, the response time of s_1, s_2, and s_3 is assumed to follow Gamma distribution

Gamma(200,2), Gamma(20,10), and Gamma(40,10), respectively. That is, their expected response time is 400, 200, and 400, respectively. Besides, we assume that their advertised response time is 500, 200, and 300, respectively. Clearly, s_1 would get a good reputation, while s_3 would get a bad reputation. We first directly invoke s_1, s_2, and s_3 5,000 times, and then invoke c 5,000 times. After each direct invocation, every service gets a rating directly. After each invocation of c, an overall rating given to c is propagated to s_1, s_2, and s_3 by a rating propagation algorithm.

Fig. 1 shows the reputation evolution when different rating propagation algorithms are used. After the first 5,000 invocations, s_1, s_2, and s_3 all have a stable reputation. In the last 5,000 invocations, the reputation of s_1 becomes worse while the reputation of s_3 becomes better when using the method proposed by Nepal et al.. This trend indicates that s_1 is penalized for the poor performance of s_3, and s_3 is awarded for the good performance of s_1. In contrast, all services have stable reputations when SVA or FSVA is used for rating propagation, which shows that the overall rating is fairly distributed by SVA and FSVA.

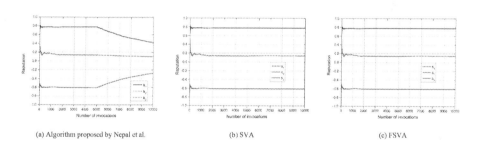

(a) Algorithm proposed by Nepal et al.   (b) SVA   (c) FSVA

Fig. 1. Reputation evolution under different rating propagation algorithms

Next, we compare the computation cost of FSVA and SVA since they both ensure the fairness of rating propagation. From Fig. 2(b), we can see that the computation cost of SVA increases rapidly as the number of component services grows, which just reflects the exponential time complexity of SVA. In particular, when there are 20 component services, SVA needs about 8 seconds to distribute the overall rating. Combining Fig. 2(a) and (b), it is clear that FSVA is much more efficient than SVA. Even for 150 component services, FSVA only needs less than 0.5 second to realize fair rating propagation for multiplicative QoS. Another interesting observation is the computation cost of FSVA for concave QoS is much lower than that for multiplicative QoS, though the theorems in Section 4 state that both cases have $O(n^2)$ time complexity. The reason is that, for concave QoS, the actual computation cost of formula 15 depends on not only the size of $|S_{dp}|$ but also the time exiting the main iteration, and these two factors are both dependent on the QoS of component services. In other words, the time complexity of FSVA for concave QoS is $O(n^2)$ in the worst case, and its time complexity in the average case is an interesting problem for future research.

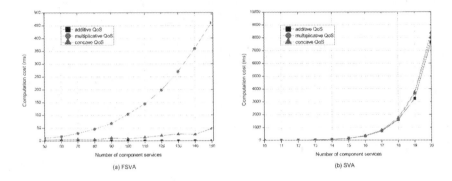

Fig. 2. Computation cost of FSVA and SVA

6   Related Work

In this section, we review some related works in the area of Web services reputation systems. The achievements of reputation systems in other research fields such as P2P network can be found in [2]. Web services reputation calculation is a fundamental issue and has received much attention recently. In [10], the authors propose an approach to compute reputation according to QoS history and user preferences over QoS attributes. In [4], the authors estimate reputation based on user ranking, QoS compliance, and QoS verity which is in fact a variance of QoS compliance. In [17], the authors also regard reputation as an aggregation of users' reports about QoS conformance, but they improve the accuracy of reputation computation by first clustering users into groups and then assigning to a user's report a weight defined by the size and stability of the group to which the user belongs. In [13], the authors present a method of adopting Bayesian network to compute reputation, which works well even if some users are malicious. These efforts provide useful methods to compute Web services reputation, but cannot be applied to component Web services which are transparent to users.

Another important research issue in Web services reputation systems is quality of rating. In [9], the authors state that the quality of rating depends on user credibility, and propose several techniques to assess user credibility accurately. In [3], the authors propose an award mechanism to encourage users to submit objective feedbacks. In [18], the authors adopt a Cumulative Sum Method to detect malicious ratings and use Pearson Correlation Coefficient to improve the accuracy of ratings. These efforts are beneficial to build an efficient Web services reputation system as the one introduced in this paper.

The rating propagation problem was first identified in [12] where the rating of a composite service is distributed based on the importance of component services defined by users and their QoS consistency. However, the authors do not provide formal fairness proof, and this approach cannot guarantee a complete fairness. In [19], the authors apply graph theory to calculate the importance of component services based on the structure of composition, but fairness proof remains absent. Our previous work [8] can realize a fair rating propagation at the cost of high computational complexity. In this paper, we significantly decrease the computational complexity from exponential to quadratic.

7   Conclusion

In Web services reputation systems, it is important to devise a rating propagation model by which the reputations of component services that are transparent to users can be updated. In this paper, we have employed the Shapley value to calculate the marginal contribution of every component service to the QoS fluctuation of a composite service, based on which the overall rating given to the composite service can be fairly propagated to its component services. We have distinguished three types of QoS with different computational characteristics and proposed for each an efficient QoS fluctuation distribution algorithm. We have shown that the Shapley value based QoS fluctuation distribution can be done in quadratic time, which is a significant improvement as the calculation of the Shapley value typically involves an exponential time complexity.

One limitation of the work reported in this paper is that we only consider homogeneous QoS with the same computational characteristic. However, heterogeneous QoS is also common in Web services composition. For example, response time is an additive QoS in sequential pattern, but a concave QoS in parallel pattern, so if a composite service contains both patterns, some extension need to be made for the proposed approach, which is one direction of our future work.

Acknowledgments. The work is supported in part by the National Natural Science Foundation of China under Grant Nos. 61003044, 61379033 and 61303019.

References

[1] Adler, B.T., de Alfaro, L.: A content-driven reputation system for the wikipedia. In: WWW 2007, pp. 261–270 (2007)

[2] Hoffman, K.J., Zage, D., Nita-Rotaru, C.: A survey of attack and defense techniques for reputation systems. ACM Comput. Surv. (CSUR) 42(1) (2009)

[3] Jurca, R., Faltings, B., Binder, W.: Reliable QoS monitoring based on client feedback. In: WWW 2007, pp. 1003–1012 (2007)

[4] Kalepu, S., Krishnaswamy, S., Loke, S.W.: Reputation = f(User Ranking, Compliance, Verity). In: ICWS 2004, pp. 200–207 (2004)

[5] Kamvar, S.D., Schlosser, M.T., Garcia-Molina, H.: The Eigentrust algorithm for reputation management in P2P networks. In: WWW 2003, pp. 640–651 (2003)

[6] Limam, N., Boutaba, R.: Assessing Software Service Quality and Trustworthiness at Selection Time. IEEE T. Software Eng. 36(4), 559–574 (2010)

[7] Liu, L., Shi, W.: Trust and Reputation Management. IEEE Internet Computing (INTERNET) 14(5), 10–13 (2010)

[8] Liu, A., Li, Q., Huang, L., Wen, S.: Shapley Value Based Impression Propagation for Reputation Management in Web Service Composition. In: ICWS 2012, pp. 58–65 (2012)

[9] Malik, Z., Bouguettaya, A.: RATEWeb: Reputation Assessment for Trust Establishment among Web services. VLDB J. (VLDB) 18(4), 885–911 (2009)

[10] Maximilien, E.M., Singh, M.P.: Conceptual Model of Web Service Reputation. SIGMOD Record (SIGMOD) 31(4), 36–41 (2002)

[11] McNally, K., O'Mahony, M.P., Coyle, M., Briggs, P., Smyth, B.: A Case Study of Collaboration and Reputation in Social Web Search. ACM T. Intelligent Systems and Technology (TIST) 3(1) (2011)

[12] Nepal, S., Malik, Z., Bouguettaya, A.: Reputation Propagation in Composite Services. In: ICWS 2009, pp. 295–302 (2009)

[13] Nguyen, H.T., Zhao, W., Yang, J.: A Trust and Reputation Model Based on Bayesian Network for Web Services. In: ICWS 2010, pp. 251–258 (2010)

[14] Nisan, N., Roughgarden, T., Tardos, E., Vazirani, V.V.: Algorithmic Game Theory. Cambridge University Press (2007)

[15] Shapley, L.S.: A Value for n-Person Games. Contributions to the Theory of Games II (Annals of Mathematics Studies 28), pp. 307–317. Princeton University Press (1953)

[16] da Silva, I., Zisman, A.: A framework for trusted services. In: Liu, C., Ludwig, H., Toumani, F., Yu, Q. (eds.) Service Oriented Computing. LNCS, vol. 7636, pp. 328–343. Springer, Heidelberg (2012)

[17] Vu, L.-H., Hauswirth, M., Aberer, K.: QoS-Based Service Selection and Ranking with Trust and Reputation Management. In: OTM Conferences 2005, pp. 466–483 (2005)

[18] Wang, S., Zheng, Z., Sun, Q., Zou, H., Yang, F.: Evaluating Feedback Ratings for Measuring Reputation of Web Services. In: SCC 2011, pp. 192–199 (2011)

[19] Wen, S., Li, Q., Yue, L., Liu, A., Tang, C., Zhong, F.: CRP: context-based reputation propagation in services composition. Service Oriented Computing and Applications 6(3), 231–248 (2012)

[20] Xiong, L., Liu, L.: PeerTrust: Supporting Reputation-Based Trust for Peer-to-Peer Electronic Communities. IEEE T. Knowl. Data Eng. (TKDE) 16(7), 843–857 (2004)

[21] Yu, T., Zhang, Y., Lin, K.-J.: Efficient algorithms for Web services selection with end-to-end QoS constraints. TWEB 1(1) (2007)

[22] Zeng, L., Benatallah, B., Ngu, A.H.H., Dumas, M., Kalagnanam, J., Chang, H.: QoS-Aware Middleware for Web Services Composition. IEEE T. Software Eng. (TSE) 30(5), 311–327 (2004)

[23] Zheng, H., Yang, J., Zhao, W., Bouguettaya, A.: QoS analysis for web service compositions based on probabilistic qoS. In: Kappel, G., Maamar, Z., Motahari-Nezhad, H.R. (eds.) Service Oriented Computing. LNCS, vol. 7084, pp. 47–61. Springer, Heidelberg (2011)

[24] Zheng, H., Zhao, W., Yang, J., Bouguettaya, A.: QoS Analysis for Web Service Compositions with Complex Structures. IEEE T. Services Computing (TSC) 6(3), 373–386 (2013)

[25] Zhou, R., Hwang, K.: PowerTrust: A Robust and Scalable Reputation System for Trusted Peer-to-Peer Computing. IEEE T. Parallel Distrib. Syst. (TPDS) 18(4), 460–473 (2007)

CLUSM: An Unsupervised Model for Microblog Sentiment Analysis Incorporating Link Information

Gaoyan Ou[1,2], Wei Chen[1,2,*], Binyang Li[4], Tengjiao Wang[1,2], Dongqing Yang[1,2], and Kam-Fai Wong[1,3]

[1] Key Laboratory of High Confidence Software Technologies, Ministry of Education, China
[2] School of Electronics Engineering and Computer Science, Peking University, Beijing, China
[3] The Chinese University of Hong Kong, Shatin, NT, Hong Kong, China
[4] University of International Relations, Beijing, China
pekingchenwei@pku.edu.cn

Abstract. Microblog has become a popular platform for people to share their ideas, information and opinions. In addition to textual content data, social relations and user behaviors in microblog provide us additional link information, which can be used to improve the performance of sentiment analysis. However, traditional sentiment analysis approaches either focus on the plain text, or make simple use of links without distinguishing different effects of different types of links. As a result, the performance of sentiment analysis on microblog can not achieve obvious improvement. In this paper, we are the first to divide the links between microblogs into three classes. We further propose an unsupervised model called Content and Link Unsupervised Sentiment Model (CLUSM). CLUSM focuses on microblog sentiment analysis by incorporating the above three types of links. Comprehensive experiments were conducted to investigate the performance of our method. Experimental results showed that our proposed model outperformed the state of the art.

Keywords: sentiment analysis, link information, unsupervised learning.

1 Introduction

Recently, microblog has become a popular platform for people to share their ideas, information and opinions with each other. The large volumes of microblog data provide an unprecedented opportunity to extract, analysis and monitor people's sentiment towards mass incidents, products and services. Over the past decade, the problem of sentiment analysis has been widely studied on reviews, blogs and news articles [1–8]. Traditional sentiment analysis approaches focus on the pure text analysis but neglect the nature of microblog. To demonstrate the

---

* Corresponding author.

S.S. Bhowmick et al. (Eds.): DASFAA 2014, Part I, LNCS 8421, pp. 481–494, 2014.

disadvantages of existing text-based methods, consider the following microblog
from Sina Weibo[1]:

> "*Nonsense. I am not optimistic.* //@B: Good thing. The CEO is also coming
> out. I am optimistic about Microsoft. //@A: Good News: Microsoft today an-
> nounced that it will acquire Nokia's devices & services business, license Nokia's
> patents and mapping services. " (by user D)

The microblog consists of two parts: the user-posted part and the repost part.
The user-posted part, which is underlined in the above example, is extremely short.
Based on our statistics on a large microblog data set, the average length of this part
is only 8-9 Chinese characters. Thus, it may lead to severe data sparsity problem
when only considering the user-posted part for sentiment analysis. The repost part
may contain opinions of different users (user B and user A in the example). It may
lead to meaningless result if we add the repost part to the user-posted part for
sentiment analysis.

Based on the above analysis, we argue that traditional text-based methods are
not suitable for microblog sentiment analysis. Social relations and user behaviors
in microblogging system provide us additional link information, which can be
used to improve the performance of sentiment analysis. As shown in Fig.1(a),
the microblogs are connected by the following three different types of links:
(1) "Same user" link: two microblogs are posted by the same user. (2) "Friend"
link: the users of two microblogs are friends. (3) "Behavior" link: two microblogs
with repost, reply or comment relatioship. The social relation information has
showed effectiveness to improve microblog sentiment classification [9, 10]. For
example, it is obvious that two microblogs m_i and m_j posted by the same user
under a given topic tend to contain the same sentiment polarity. Given a certain
topic, similar users will hold similar opinions. Thus two microblogs with "friend"
link tend to contain the same sentiment polarity.

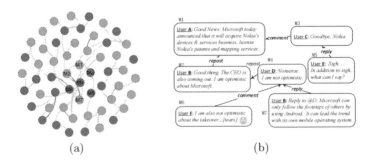

(a)   (b)

Fig. 1. An example of microblogs with links. (a) Green and red nodes are microblogs
with positive and negative sentiment. "Same user", "friend" and "behavior" links are
colored as red, green and gray, respectively. (b) Microblogs with "behavior" links. Only
the user-posted part is considered for each microblog.

---

[1] http://www.weibo.com

In addition to social relation links, there are more direct links between microblogs generated by user behaviors. Fig.1(b) shows the graph structure of microblogs with three kinds of social behavior links: repost, reply and comment. Existing link-based approaches [10, 9, 11, 12] are not appropriate for these user behavior links, since they all assume that microblogs with links tend to contain the same sentiment polarity. This assumption does not hold in the context of user behavior links. In Fig.1(b), m_2 and m_4 are connected by a "repost" link. It is obvious that their sentiment polarities are opposite. To exploit the user behavior link information for microblog sentiment classification, we propose a dictionary-based method to classify each user behavior link into agree/disagree/others, based on the content of the two linked microblogs.

In this work, we focus on the problem of microblog sentiment classification incorporating link information. We distinguish three types of links between microblogs: behavior, same user and friend. We propose a novel probabilistic model called Content and Link Unsupervised Sentiment Model (CLUSM). CLUSM integrates both textual content and link information into a unified framework. Comprehensive experiments on microblog data sets demonstrate that link information can obviously improve sentiment classification accuracy. The roles of different kinds of links are also studied.

The rest of this paper is organized as follows. Section 2 reviews the related work. In section 3, we first give some preliminary observations and then propose our model for microblog sentiment classification incorporating both content and link information. Section 4 describes how to perform inference and parameter estimation in our model. In Section 5 we introduce the data sets and show the experimental results. We conclude and describe future work in Section 6.

2 Related Work

In recent years, there are increasing number of interests on sentiment classification on microblog data [13–19]. [13] proposed an approach to automatically obtain labeled training examples by exploiting hashtags and smileys. [14] proposed an emoticon smoothed method to integrate both manually labeled data and noisy labeled data for Twitter sentiment classification. Most of these methods focus on the textual information but neglect the link information.

There are some work which incorporate link information to improve sentiment classification performance [20, 10, 12, 9, 11]. The first kind of method simply extracts extra features from links between microblogs. [10] extracted features from the user follower graph and then combined them with traditional textual features to represent the microblog. The second kind of method utilizes content and link information in a two-phase schema. [12] incorporated two kinds of behavior links and same user links. They applied a graph-based optimization algorithm based on the classifying result from SVM sentiment classifier. The third kind of method integrates content and link information in a unified framework. [9] considered two kinds of social relations by adding a regularization term to the least square cost function. [11] focused on user-level sentiment classification by incorporating the social networks between users.

One disadvantage of existing link-based approaches is that microblogs with links between each other should have similar sentiment polarity. However, as discussed in last section, the sentiment polarity of two linked microblogs do not need to be similar. They can be different and even opposite. Another disadvantage of existing methods is that they can not distinguish different kinds of links, while one of our aims is to study the roles that different kinds of links play in microblog sentiment classification.

3   The Proposed Framework

3.1   Preliminary Observation

The basic principles in our work are as follows. First, link information is important for microblog sentiment analysis. Second, different types of links play different roles in microblog sentiment analysis and need to be treated differently. To validate these principles, we demonstrate some statistics on real microblog data sets (Description of the data sets will be given in Section 5.1).

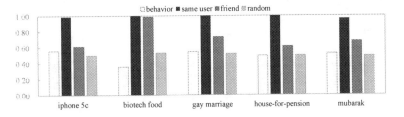

Fig. 2. Probability of two microblogs having the same sentiment polarity conditioned on different types of links. Random: two microblogs are selected randomly.

Fig.2 shows that it is almost sure that two microblogs posted by the same user sharing the same sentiment under a certain topic. The probability of two microblogs with "friend" link having the same sentiment polarity is also much higher than two randomly selected microblogs. This provides us direct motivation to utilize link information to improve the performance of microblog sentiment analysis. However, the "behavior" links show different characteristics. The probability of two microblogs with "behavior" link having the same sentiment polarity is not obviously higher than random. It is even lower than random in topic "biotech food" and "house-for-pension". Thus for the "behavior" links, we need to give insight into their contextual information (the contents of the linked microblogs) when exploiting them to improve microblog sentiment analysis. As a whole, these observations support our principles that link information is important and different types of links need to be treated differently for microblog sentiment analysis.

3.2   Framework Overview

Without loss of generality, suppose we are given a collection of microblogs for a certain topic T denoted as $X = \{X_1, ..., X_M\}$. For each microblog X_i, there is a corresponding sentiment variable $y_i \in \{pos, neg\}$ indicating the sentiment polarity of the microblog. Each microblog X_i is represented by a V-dimension sentiment feature vector, where V indicates the number of features. The links between microblogs are denoted as $\mathcal{E} = \mathcal{E}_b \cup \mathcal{E}_s \cup \mathcal{E}_f$. $\mathcal{E}_b$, $\mathcal{E}_s$ and $\mathcal{E}_f$ are "behavior" links, "same user" links and "friend" links, respectively. Given microblogs X and the links $\mathcal{E}$ for topic T, our goal is to determine whether each y_i is positive or negative with respect to T. To capture the fact that both content and link information are important, we integrate them to a unified objective. The objective is associated with the probability distribution of hidden variables Y given observed variables X:

$$P(Y|X) = \frac{1}{Z(X)} \prod_{i=1}^{M} \phi_i(y_i, X_i) \prod_{(i,j) \in \mathcal{E}} \phi_{i,j}(y_i, y_j) \tag{1}$$

where $Z(X)$ is the normalization function.

The content information is modeled through potential function $\phi_i(y_i, X_i)$. The link information is integrated in $\phi_{i,j}(y_i, y_j)$, which is the pairwise potential function between two neighboring sentiment variables y_i and y_j. Note that $\phi_{i,j}(y_i, y_j)$ is defined differently for different types of links.

3.3   Modeling Content Information

We define $\phi_i(y_i, X_i)$ as the conditional probability of the feature vector X_i given y_i, which is a multinomial distribution as follows:

$$\phi_i(y_i, X_i) = P(X_i|y_i) = \prod_{v=1}^{V} P(x_{iv}|y_i) = \prod_{v=1}^{V} \theta_{y_i v}^{x_{iv}} \tag{2}$$

where $\theta_{y_i v}$ is the model parameter and x_{iv} counts the number of occurrences of the v-th feature in microblog X_i.

We combine the features which have been proven effective for microblog sentiment analysis by previous work [17, 13], such as punctuation, emoticons and sentiment lexicons. Specifically, we use the following sentiment features to construct feature vector X_i: (1) Words appearing in the microblog content serve as word features, (2) Number of "!" characters and "?" in the microblog, (3) Emoticons in the microblog, such as [tears],[cry] and (4) Each term in a general sentiment lexicon serves as a sentiment lexicon feature.

3.4   Modeling Link Information

The link information is integrated in the potential function $\phi_{i,j}(y_i, y_j)$. We define $\phi_{i,j}(y_i, y_j)$ in the forms of energy function as follows:

$$\phi_{i,j}(y_i, y_j) \propto exp\{-\mu_{ij}E_{ij}(y_i, y_j)\} \tag{3}$$

Here, according to the link type between y_i and y_j, μ_{ij} can be one of the following three values: μ_b, μ_s and μ_f. μ_* is the weight parameter controlling the contribution of the corresponding type of link. $E_{ij}(y_i, y_j)$ is the energy function defined on the link between y_i and y_j.

Energy Function of "Behavior" Link. To capture the fact that the sentiment polarity of two microblogs with "behavior" link may be the same or opposite, we classify the "behavior" link into agree/disagree/others classes. If a link is classified as "agree", we expect that two adjacent sentiment variables y_i and y_j should be similar. If a link is classified as "disagree", we expect that two adjacent sentiment variables y_i and y_j should be different from each other. Based on the above considerations, we define the energy function for "behavior" link as:

$$E_{ij}^b(y_i, y_j) = \begin{cases} f(y_i - y_j) & \text{if } link_{ij} \in agree , \\ \lambda - f(y_i - y_j) & \text{if } link_{ij} \in disagree , \\ \delta \cdot f(y_i - y_j) & \text{others.} \end{cases} \quad (4)$$

where $\lambda > 0$, $\delta \in (0,1)$ is a hyper-parameter. $f(n) = 0$, if n $= 0$; otherwise $f(n) = \lambda$.

We use a simple dictionary-based method to perform agree/disagree classification of "behavior" links, although there exist alternative techniques [21, 22]. We first create two seed dictionaries that contain frequently used words, phrases and expressions to express agree and disagree, respectively. For example, "good idea" and "I agree with you" are in the seed agree dictionary, while "nonsense" and "I do not agree" are in seed disagree dictionary. Then the seed agree/disagree dictionary is expanded by replacing the words with their synonyms for each entry. Each entry in the expanded agree/disagree dictionary is then served as a pattern to match the microblog content. Suppose microblog m_i replies microblog m_j. If the content of m_i is matched by an entry in the agree dictionary, we classify the "behavior" link (i, j) as agree.

Energy Function of "Same User" Link. Intuitively, two microblogs m_i and m_j posted by the same user under a given topic tend to contain the same sentiment polarity. This is further validated by the statistics on real microblog data sets in Section 3.1. As shown in Fig.2, the probability of two microblogs with "same user" link having the same sentiment polarity is higher than 98% on all topics in our data sets. Thus it is reasonable to force the sentiment polarities of two microblogs to be the same if they are connected by a "same user" link. We define the energy function for "same user" as follows:

$$E_{ij}^s(y_i, y_j) = \begin{cases} 0 & \text{if } y_i = y_j , \\ \lambda & \text{if } y_i \neq y_j. \end{cases} \quad (5)$$

Energy Function of "Friend" Link. The "friend" links are also good clues to indicate sentiment similarity of the linked microblogs. However, they are not as strong as "same user" links. We define the energy function for "friend" link as follows:

$$E_{ij}^f(y_i, y_j) = \begin{cases} \frac{\lambda}{2} & \text{if } y_i = y_j, \\ \lambda & \text{if } y_i \neq y_j. \end{cases} \tag{6}$$

Based on the above potential function for content information and three types of energy function for link information, our object probability distribution in Eq.(1) can be rewritten as:

$$
\begin{aligned}
P(Y|X) = \prod_{i=1}^{M} \prod_{v=1}^{V} \theta_{y_i v}{}^{x_{iv}} \cdot exp \left\{ -\mu_b \sum_{(i,j) \in \mathcal{E}_b} E_{ij}^b(y_i, y_j) \right\} \\
\cdot exp \left\{ -\mu_s \sum_{(i,j) \in \mathcal{E}_s} E_{ij}^s(y_i, y_j) \right\} \cdot exp \left\{ -\mu_f \sum_{(i,j) \in \mathcal{E}_f} E_{ij}^f(y_i, y_j) \right\}
\end{aligned}
\tag{7}
$$

4   Inference and Parameter Estimation

In this section, we first present the inference algorithm based on Gibbs Sampling [23] to determine the sentiment polarity of microblogs. Then we describe how to learn the model parameters by utilizing the sampling result of the inference algorithm.

4.1   Model Inference

In order to perform Gibbs sampling, we need to compute the conditional probability $P(y_i = s | Y \setminus y_i, X)$, where $Y \setminus y_i$ indicates all the latent sentiment variables except for y_i. According to the Markov property, y_i is only dependent on its neighboring sentiment variables $Nei(y_i)$ and its corresponding obsevable variable X_i:

$$
\begin{aligned}
P(y_i = s | Y \setminus y_i, X) &= P(y_i = s | Nei(y_i), X_i) \\
&\propto P(X_i | y_i = s) \cdot \prod_{y_j \in Nei(y_i)} \phi_{i,j}(y_i = s, y_j)
\end{aligned}
\tag{8}
$$

Algorithm 1 summarizes the sampling process. For each sentiment variable y_i, $c(y_i, s)$ indicates the number of sampling times when y_i is sampled as value s. Given $c(y_i, s)$, the sentiment polarity of each microblog X_i can be determined by the following formula:

$$y_i^* = \arg \max_s c(y_i, s) \tag{9}$$

Algorithm 1. Microblog Sentiment Inference by Gibbs Sampling

Input: the link weight vector μ, hyper-parameter δ, λ and the number of sampling iterations K.

1: Initialize $c(y_i, s) \leftarrow 0$ for all $X_i \in X$ and $s \in \{pos, neg\}$.
2: Initialize y_i randomly for all $y_i \in Y$.
3: **for** each $k = 1, \cdots, K$ **do**
4:     **for** each $X_i = X_1, \cdots, X_m$ **do**
5:         Sampling a value s from $\{pos, neg\}$ for y_i according to Eq.(8).
6:         $c(y_i, s) \leftarrow c(y_i, s) + 1$
7:     **end for**
8: **end for**

Output: $c(y_i, s)$ for all $y_i \in Y$ and $s \in \{pos, neg\}$

4.2   Parameter Estimation

The parameters θ can be estimated by maximum likelihood method, in which the goal is to maximize the log likelihood function of the observed data X given θ:

$$\log P(X|\theta) = \log \sum_Y P(Y, X|\theta) \tag{10}$$

The above object function is hard to compute since the marginalization over the latent variables Y contains exponentially large number of terms. We use the expectation-maximization (EM) algorithm [24] to learn θ in an iterative way. The EM algorithm first computes the expected value of $logP(Y, X|\theta)$ with respect to $P(Y|X, \theta^{(k)})$ in the E-step, which is so-called Q-function. In the M-step, we obtain the revised parameter $\theta^{(k+1)}$ by maximizing the Q-function as:

$$Q(\theta, \theta^{(k)}) = \sum_{i=1}^{M} \sum_{y_i=s}^{S} \sum_{v=1}^{V} P(y_i|X, \theta^{(k)}) x_{iv} log\theta_{sv} \tag{11}$$

Following the idea of [25], we approximate $P(y_i|X, \theta^{(k)})$ by using the output $c(y_i, s)$ of Algorithm 1 as:

$$P(y_i = s|X, \theta^{(k)}) = \frac{c(y_i, s)}{\sum_{s=1}^{S} c(y_i, s)} \tag{12}$$

Note that (11) is associated with the constraints $\theta_{sv} > 0$. Lagrangian multiplier method is used to solve the constrained maximization problem, and the updating formula for θ is as follows:

$$\theta_{sv}^{(k+1)} = \frac{\sum_{i=1}^{M} c(y_i, s) x_{iv}}{\sum_{i=1}^{M} \sum_{s=1}^{S} c(y_i, s) x_{iv}} \tag{13}$$

The general sentiment lexicon can be used to initialize parameters θ_{sv} as follows. For sentiment feature v, if it is a positive term in the sentiment lexicon, we can set $\theta_{'pos',v} > \theta_{'neg',v}$. If v it is a negative term in the sentiment lexicon, we can set $\theta_{'neg',v} > \theta_{'pos',v}$.

5   Experiments

5.1   Data Sets

We collected our own data sets from Sina Weibo, which is a popular Chinese microblog service. We chose five popular and controversial topics: "iphone 5c", "biotech food", "gay marriage", "house-for-pension" and "mubarak". For each topic, we chose one representative microblog and then crawled all its reposting, replying and commenting microblogs.

Our data set consisted of the user ID of each microblog, thus the "same user" links between microblogs can directly be constructed. The "friend" links were constructed as follows. We implemented another crawler to extract the following relations of users. If user A follows user B and B also follows A, there exists a "friend" relationship between A and B. Further more, there is a "friend" link between each microblog posted by A and each microblog posted by B. The basic statistics of the data sets are shown in Table 1. The micrologs were manually labeled as *positive* or *negative* by two human annotators[2]. The annotation agreement is higher than 90%.

Table 1. Basic statistics of data sets

Topic	#Microblogs	#Users	#Links		
			Behavior	Same User	Friend
iphone 5c	1281	1159	1213	126	126
biotech food	477	428	351	8802	1922
gay marriage	451	433	388	65	91
house-for-pension	725	589	691	151	55
mubarak	634	505	626	168	52

5.2   Comparison with Different Methods

In the first experiment, we validate the effectiveness of the proposed model on exploiting link information to improve microblog sentiment classification. The compared methods are listed as follows:

- **HowNet**: We use the sentiment lexicon in the Chinese knowledge database HowNet[3] to construct a lexicon-based sentiment classifier.
- **MNB**: Multinomial Naive Bayes has been shown effective when classifying sentiment in microblogs [17]. In MNB, only content information is considered.
- **SANT**: The method proposed in [9]. They integrate "same user" and "friend" link information by adding a regularization term to the least square cost function. Both content and link information are considered.

---

[2] Since the chosen topics are controversial and most microblogs are connected by behavior links, the sentiment of the microblog is either positive or negative.

[3] http://www.keenage.com/

– **SVM+GraphOpt:** The method proposed in [12]. They first use a trained SVM to assign probabilistic sentiment scores for microblogs. Then a graph-based optimization is utilized to incorporate "behavior" and "same user" link information. Both content and link information are considered.

Note that *MNB*, *SANT* and *SVM+GraphOpt* are supervised methods which need annotated data for training. Although our model is generally unsupervised, we can integrate the labeled data by simply setting the corresponding sentiment variables as observed. In all models, we use the same set of sentiment features as described in Section 3.3. For each topic, we randomly select 30% of our data as training (observed) data. In our model, the hyper-parameters are empirically set as $\lambda = 1.0, \delta = 0.5$. We set $(\mu_b, \mu_s, \mu_f) = (0.5, 0.5, 0.5)$ for general experiment purposes. Table 2 shows the experimental result in terms of sentiment classification accuracy.

Table 2. Sentiment classification accuracy of different models

Method	iphone 5c	biotech food	gay marriage	house-for-pension	mubarak
HowNet	0.610	0.614	0.582	0.623	0.619
MNB	0.758	0.687	0.707	0.696	0.703
SANT	0.766	0.733	0.705	0.724	0.723
SVM+GraphOpt	0.768	0.772	0.706	0.731	0.724
CLUSM	**0.785**	**0.887**	**0.732**	**0.751**	**0.743**

From Table 2, we can see that the lexicon-based method *HowNet* achieves the worst performance. In all topics, the link-based methods (*SANT*, *SVM+GraphOpt* and *CLUSM*) achieve higher accuracy than purely content-based methods (*HowNet* and *MNB*). This suggests the effectiveness of link information for microblog sentiment analysis.

Our model outperforms *SANT* and *SVM+GraphOpt*. The good results of *CLUSM* benefit from the following two aspects: (1) We distinguish different types of links, while *SANT* and *SVM+GraphOpt* model different types of links in the same way. (2) *CLUSM* models the content and link information simultaneously, while link information is only used in the training phase in *SANT* and is only used to revise the output of the SVM classifier trained on content information in *SVM+GraphOpt*.

5.3 Effectiveness of Distinction of Link Types

The second experiment evaluated the effectiveness of distinction of link types. In CLUSM, we treat three types of links differently through the different definitions of energy function. To demonstrate the effectiveness of distinction of link types, we construct a modified version of CLUSM by setting three energy functions as shown in Eq.(5). The comparison results are shown in Fig.3. Note that in this experiment, no labeled data are used in both models.

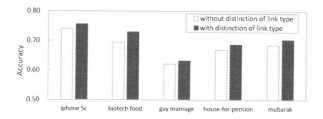

Fig. 3. Effectiveness of distinction of link type

As seen in Fig.3, our model with distinction of link types perform better than without distinction of link types. The reason is that different types of links capture different semantic meanings, thus should be treated differently.

5.4 Contribution of Different Types of Links

We also design the experiment to investigate the different roles of three types of links playing in microblog sentiment analysis. The experiments are conducted with separate link type. For example, if we are considering the "same user" link, μ_b and μ_f are set to be 0. Then we vary μ_s from 0 to 10. Since the results for different topics are similar, we only show the results for topic "iphone 5c" in Fig.4.

From Fig.4 we can see that the "same user" is most useful for microblog sentiment classification. Specifically, as μ_s increases, the sentiment classification accuracy increases and is consistently higher than only considering content (shown as gray line). When μ_s becomes large enough (larger than 0.6), the performance becomes stable. The "friend" link is also effective for improving sentiment classification performance. However, when μ_f is larger than 1, the performance decreases. The reason is that the opinions of friends can be different or even opposite. When μ_f is large, the model forces the sentiment polarity of two microblogs to be the same, which is not appropriate.

To further evaluate the effectiveness of the agree/disagree classification of the "behavior" links, we also show the result without agree/disagree classification. The results suggest that without agree/disagree classification, the "behavior" link may even hurt the performance. With agree/disagree classification, the "behavior" link can be helpful. The "behavior" link is most effective when μ_b is set with smaller value (between 0.3 and 0.5 in our data).

5.5 Integrating Labeled Data

Our model is generally unsupervised. To integrate the sentiment labels of some of the microblogs, we simply set the corresponding sentiment variables as observed variables. To investigate the effectiveness of different models of utilizing labeled data, we vary the size of labeled data from 0% to 50% . The average accuracy of five topics for each model are shown in Fig.5.

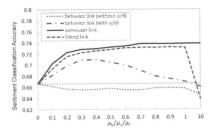

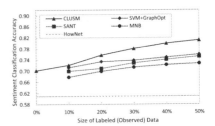

Fig. 4. Contribution comparison of link types for topic "iphone 5c"

Fig. 5. Effect of labeled (observed) data sizes

We can see that our model can utilize the labeled data more effective than other models. With no labeled data, the performance of our model is comparable to other supervised models with 10% training data. With the same size of training data, the link-based models outperforms content-based models. This again demonstrate that link information is important for microblog sentiment analysis.

6   Conclusion

We present a novel probabilistic model for microblog sentiment classification. Our model integrates content information and link information into a unified framework. Three types of links ("behavior", "same user" and "friend") are considered and modeled accordingly. Experiments conducted on real microblog data sets show that our model outperforms several baselines. The roles of different types of links are also studied. The results suggest that "same user" link can always improve performance. "Friend" link is also helpful. "Behavior" link can be helpful when appropriately utilized.

In the future, we plan to improve the agree/disagree classification algorithm and investigate its effect on the sentiment analysis performance. With the sentiment polarities of microblogs and the agree/disagree relations between them, it would also be possible to explore the sentiment influence between users.

Acknowledgments. This research is supported by the National High Technology Research and Development Program of China (Grant No. 2012AA011002), Natural Science Foundation of China (Grant No. 61300003), Research Foundation of China Information Technology Security Evaluation Center (No. CNITSEC-KY-2013-018) and Specialized Research Fund for the Doctoral Program of Higher Education (Grant No. 20130001120001). This research is partially supported by General Research Fund of Hong Kong (No. 417112) and Shenzhen Fundamental Research Program (JCYJ20130401172046450).

References

1. Turney, P.D.: Thumbs up or thumbs down?: semantic orientation applied to unsupervised classification of reviews. In: Proceedings of the 40th Annual Meeting on Association for Computational Linguistics, pp. 417–424. Association for Computational Linguistics (2002)
2. Pang, B., Lee, L., Vaithyanathan, S.: Thumbs up? sentiment classification using machine learning techniques. In: EMNLP, pp. 79–86 (2002)
3. Hu, M., Liu, B.: Mining and summarizing customer reviews. In: Proceedings of the tenth ACM SIGKDD International Conference on Knowledge Discovery and Data Mining, pp. 168–177. ACM (2004)
4. Godbole, N., Srinivasaiah, M., Skiena, S.: Large-scale sentiment analysis for news and blogs. In: ICWSM, vol. 7 (2007)
5. Devitt, A., Ahmad, K.: Sentiment polarity identification in financial news: A cohesion-based approach. In: ACL (2007)
6. Wilson, T., Wiebe, J., Hoffmann, P.: Recognizing contextual polarity in phrase-level sentiment analysis. In: Proceedings of the Conference on Human Language Technology and Empirical Methods in Natural Language Processing, pp. 347–354. Association for Computational Linguistics (2005)
7. Ou, G., Chen, W., Liu, P., Wang, T., Yang, D., Lei, K., Liu, Y.: Aspect-specific polarity-aware summarization of online reviews. In: Wang, J., Xiong, H., Ishikawa, Y., Xu, J., Zhou, J. (eds.) WAIM 2013. LNCS, vol. 7923, pp. 289–300. Springer, Heidelberg (2013)
8. Liu, B.: Sentiment analysis and opinion mining. Synthesis Lectures on Human Language Technologies 5(1), 1–167 (2012)
9. Hu, X., Tang, L., Tang, J., Liu, H.: Exploiting social relations for sentiment analysis in microblogging. In: WSDM, pp. 537–546 (2013)
10. Speriosu, M., Sudan, N., Upadhyay, S., Baldridge, J.: Twitter polarity classification with label propagation over lexical links and the follower graph. In: EMNLP, pp. 53–63 (2011)
11. Tan, C., Lee, L., Tang, J., Jiang, L., Zhou, M., Li, P.: User-level sentiment analysis incorporating social networks. In: KDD, pp. 1397–1405 (2011)
12. Jiang, L., Yu, M., Zhou, M., Liu, X., Zhao, T.: Target-dependent twitter sentiment classification. In: ACL, pp. 151–160 (2011)
13. Davidov, D., Tsur, O., Rappoport, A.: Enhanced sentiment learning using twitter hashtags and smileys. In: COLING (Posters), pp. 241–249 (2010)
14. Liu, K.L., Li, W.J., Guo, M.: Emoticon smoothed language models for twitter sentiment analysis. In: AAAI (2012)
15. Go, A., Bhayani, R., Huang, L.: Twitter sentiment classification using distant supervision, Stanford. CS224N Project Report, pp. 1–12 (2009)
16. Agarwal, A., Xie, B., Vovsha, I., Rambow, O., Passonneau, R.: Sentiment analysis of twitter data. In: Proceedings of the Workshop on Languages in Social Media, pp. 30–38. Association for Computational Linguistics (2011)
17. Bermingham, A., Smeaton, A.F.: Classifying sentiment in microblogs: is brevity an advantage? In: CIKM, pp. 1833–1836 (2010)
18. Pak, A., Paroubek, P.: Twitter as a corpus for sentiment analysis and opinion mining. In: LREC (2010)
19. Mukherjee, S., Bhattacharyya, P.: Sentiment analysis in twitter with lightweight discourse analysis. In: COLING, pp. 1847–1864 (2012)

20. Kim, J., Yoo, J.B., Lim, H., Qiu, H., Kozareva, Z., Galstyan, A.: Sentiment prediction using collaborative filtering. In: ICWSM (2013)
21. Lu, Y., Wang, H., Zhai, C., Roth, D.: Unsupervised discovery of opposing opinion networks from forum discussions. In: CIKM, pp. 1642–1646 (2012)
22. Murakami, A., Raymond, R.: Support or oppose? classifying positions in online debates from reply activities and opinion expressions. In: COLING (Posters), pp. 869–875 (2010)
23. Gelfand, A.E., Smith, A.F.: Sampling-based approaches to calculating marginal densities. Journal of the American Statistical Association 85(410), 398–409 (1990)
24. Dempster, A.P., Laird, N.M., Rubin, D.B.: Maximum likelihood from incomplete data via the em algorithm. Journal of the Royal Statistical Society. Series B (Methodological), 1–38 (1977)
25. Qi, G.J., Aggarwal, C.C., Huang, T.S.: On clustering heterogeneous social media objects with outlier links. In: WSDM, pp. 553–562 (2012)

Location Oriented Phrase Detection
in Microblogs

Saeid Hosseini, Sayan Unankard, Xiaofang Zhou, and Shazia Sadiq

School of Information Technology and Electrical Engineering,
The University of Queensland, Australia
{s.hosseini,s.unankard}@uq.edu.au, {zxf,shazia}@itee.uq.edu.au

Abstract. As a successful micro-blogging service, *Twitter* has demonstrated unprecedented popularity and international reach. Location extraction from micro-blogs (*tweets*) on this domain is an important challenge and can harness noisy but rich contents. Extracting location information can enable a variety of applications such as query-by-location, local advertising, crises awareness and also systems designed to provide information about events, points of interests (POIs) and landmarks. Considering the high throughput rate in *Twitter* space, we propose an approach to detect *location-oriented phrases* solely relying on tweet contents. The system finds associated phrases dedicated to each specific scalable geographical area. We have evaluated our approach based on real-world Twitter dataset from Australia. We conducted a comprehensive comparison between strong local terms (uni-word) and phrases (multi-words). Our experiments verify the system's capabilities using multiple trending baselines and demonstrate that our phrase based approach can better specify locality instead of words.

Keywords: location-oriented phrases, spatial data mining, micro-blogging.

1   Introduction

Location inference from $Twitter$[1] as a successful instance of *Micro-blogging* services is an important challenge. This can tackle *tweets'* noisy but fertile contents and prepare useful data for relevant individuals based on their geographical scope. Retrieving spatial information from *Twitter* is two-fold and includes user's location as well as geographic translation of location words or phrases found in the tweet contents, which may or may not be the same as the user's location (place from where the tweet was sent). The location information can be used to provide users with personalized services, such as local news, local advertisements, application sharing, crises awareness and systems designed to provide information about events, points of interest (POI), land-marks etc.

With a large number of *Twitter* users scattered in diverse geographical locations, the short messages can be analyzed to extract such region based information. User generated contents containing location names are more weighty

---

[1] http://twitter.com/

S.S. Bhowmick et al. (Eds.): DASFAA 2014, Part I, LNCS 8421, pp. 495–509, 2014.

and useful. These terms or phrases (multi-words) can be clear location names (e.g., Australia, Sydney, Gold Coast), a Landmark (e.g., a historical site, business name or a university) or unclear local terms or phrases that can pinpoint the target region. The region can be defined as an area suchlike a suburb, city, country or a geographical cell. We can provide examples of local terms here: *"howdy"* is a common word in *Texas* to greet and *"Gooday"* is the way of saying *"Hello"* in *Australia*. Also regarding local phrases, "here in brissie" highlights Brisbane with a high probable locality while "here on the Goldie" and "in the gc" signify the city of Gold coast in Australia.

Available resources for location extraction are limited in *Twitter* domain. A small minority of *tweets* (less than 6%) are attached with geographical coordinates (latitude,longitude) indicating that a small percentage of authors use location services. Also, the location field in user profile does not supply enough data to extract accurate information and the data retrieved from this field is not necessarily relevant to the location when a particular *tweet* is sent. Time-zone as another resource can be modified by users and is not very detailed because it reports locations as cities (e.g., Paris) and cannot be employed in applications dealing with high accuracy. Moreover, IP addresses are uncertain, as Virtual Private Network (*VPN*) may mask user's true location and an Internet Service Provider (*ISP*) may apply dynamic allocations. Therefore, *tweet* contents are the most comprehensive means that can be used in location extraction systems. However other resources can be considered as complements (e.g., location field can be used to bootstrap training data).

In order to extract location information, the first option to partition each *tweet* into valid segments is based on Named Entity Recognition (*NER*). However existing *NER* systems (e.g., `ANNIE, LingPipe and Stanford NER`) designed for ordinary traditional web documents fail or function poorly when they are tested on *tweet* contents. The key points of divergence are listed below:

- *Tweets* are short and do not comprise enough data uniquely.
- They are ungrammatical and also include misspellings.
- The messages include too many informal contents such as abbreviation and short-hands.
- A *tweet* contains limited information which cannot provide an appropriate descriptive context for its present entities.
- Capital letters are unreliable.
- There are too many terms regarded as Out Of Vocabularies (*OOV*) in *tweet* contents which are generated rapidly based on social trends and current events.
- Contents are lexically varied and scarce.
- Collecting an adequate volume of training data for named entity classification is a difficult task.

According to the problems listed above, our research challenges are two-fold:

1. We needed to present an effective solution to extract appropriate phrases from *tweets* in an unsupervised manner. We were inspired by [10] to tackle this issue.

2. We were also required to provide a method to identify local phrases associated to a region and establish their precise locations. A region in various levels of granularity (i.e., country, city, suburb and suchlike) can be determined as a geographical cell by top left and bottom right spatial points.

To define and address the research problem on this paper, our work builds upon notable previous works ([10],[3],[5]). Cheng et al. in [5] aim to devise a solution to detect local terms. They assume that there is an adequate number of terms dedicated to regions which gain a high frequency in their geographical focal points and low repetition rate in tweets which are generated in neighbouring locations while the frequency shrinks as moving away from the center. On the other hand there is a subset of terms which are mentioned frequently in their centers and appear less in other places. The model proposed by them excels prior language models however Chang et al. in [3] complete the concept of locally strong terms via considering multiple centers for each word. They estimate the probability of local terms in multiple focal points using Gaussian Mixture Model. Moreover they apply Total Variation and Symmetric Kullback-Leibler divergence to calculate the rate of non-localness via similarity to the stop words. We recommend the concept of local phrases similar to the previously well defined term based attitudes ([5],[3]).

From the other perspective, our work is inspired by the system so called as TwiNER[10]. Our research is similar to them as we calculate stickiness threshold to explore the most probable segments in tweet contents. TwiNER is similarly an unsupervised *NER* system which uses the statistics acquired from *Microsoft Web N-Gram* service[2] and *Wikipedia*[3] to exploit valid phrases. Originally, it detects Named entities without categorizing the types (e.g. location, names) but our task is concentrated on recognition of location segments which in another domain. We have utilized the idea of Symmetric Conditional Probability already used by Li et al. in [10]) to measure stickiness attribute between terms to optimize creation of plausible phrases.

The proposed system in this paper, investigates modelling of the location instances using phrases found in tweet contents. The process includes three main stages. For a given predefined region, the system *firstly* divides regional geo-tagged *tweets* into proper segments without requiring any prior labelling efforts. We also use *Microsoft Web N-Gram* service to compute stickiness probabilities between words comprising each of the segments. In the *second* stage, the system calculates the geographical covariance of *tweets* containing each of the phrases restricted to the current data set which attaches a *Focus Rate* attribute to any of the segments which are already detected. Anyhow, we have developed the *third* major partition of the system which evaluates how a location oriented phrase is distributed all around the world. It applies the result retrieved from *Twitter Search API* and computes the probability for a phrase to be connected to the region. This process is so called as *Distribution Rate Calculation*. Our contributions can be summarized as follows:

---

[2] http://web-ngram.research.microsoft.com/info/
[3] http://en.wikipedia.org/

1. We propose an unsupervised procedure to discover location multi-word entities in *tweet* contents. In spite of excessive lexical noise in context our method takes advantage of a web based service (*Microsoft Web N-Gram* service) to calculate stickiness probability of possible phrases using effective algorithms and afterwards admits local phrases which are associated to the predefined region.
2. Given a phrase, the proposed method estimates both geographical focus point and also distribution rate for the phrase. It also provides the locality strength evaluation for the phrase to be regarded as local phrase.
3. The system augments the recall rate in estimation of the user's location compared to mere term based approaches. Furthermore, based on any predefined geographical cell determined by two points and having various granularity, it can provide a set of local phrases which includes points of interest, popular places and also common phrases which are exclusive to the region.

The rest of the paper is organized as follows. Section 2 describes background and explains the research problem. Section 3 considers related work. The algorithms of our approach are presented in Section 4. The experiment and evaluation are shown in Section 5. The discussion is provided in Section 6. Finally, the conclusion is given in Section 7.

2   Research Problem

In this section we describe the concept behind the location oriented phrases. Firstly we define these phrases as they form the bases of our system. We then present the formulation of the problem addressed in this paper.

2.1   Location Oriented Phrase

In this research, our interest is to extract location oriented phrases which can be used as an indicator to highlight a region in *tweet* contents. We believe that location phrases explored in a user's tweets can conduct a relation between contents and a single or multiple regions. While a mere instance of a phrase can designate multiple related regions sharing the same phrase, another local phrase can disambiguate the first one and increase its accuracy in establishing an association to a specific region. For example assuming the region as a city, the phrase "surfers club" may be found in multiple cities in various countries but at least it can be dedicated to a series of cities. Anyhow, after identifying another instance like "movie world" on the same user's tweets we can conclude that both of the tweets belong to the same city (Gold Coast). Considering a spatial cell defined by surrounding geographical points, relevant location oriented phrases can be classified into two categories:

1. Genuine spatial phrases: This type of multi-words denote clear place names such as buildings, roads, bridges, shopping centres and in general every imagined instances for a point of interest which can signify the region distinctly

(e.g., *Story Bridge, Queen Street Mall* and *University of Queensland* which are all located in *Brisbane, Australia*)

2. Strong Local Phrases: There are special local phrases which can signify a particular locality. Such multi-words are suggested to be used commonly in a region or a set of regions.

Genuine spatial phrases are obvious location names while strong local phrases can be ambiguous. For example, *"in the GC"* and *"on the Goldie"* indicate to Gold Coast, however the terms *"GC"* and *"Goldie"* are not locally important.

2.2 Problem Statement

The problem that we address in this paper is to find an optimised process to identify location oriented phrases (i.e., Genuine spatial phrases and Strong Local Phrases(2.1)) in a set of tweets belonged to a spatial cell. The explored set of segments will then be associated to the predefined geographical area and will be used to link future tweets to the region. On the other hand, given a set of geo-tagged messages (M) and the geographical cell (C) defined by top-left and bottom-right corners, our task is to detect and distinguish a set of location oriented phrases linked to the cell C.

3 Related Work

Studying the geographical information estimation from user generated content like micro-blog service is becoming a compelling research issue. While our work focused on using micro-blog data such as *Twitter*, a similar task using multimedia data such as photos or videos has been explored (e.g., [8],[9],[14]). Also, the geographic location estimation problem has been studied on different platforms include web pages [1] and blogs [6].

   In spatial domain, Hecht et al. in [7] train a Multinomial Naïve Bayes model [12] to find users' location as accurate as the state in a content driven approach. Their algorithm called *Calgari*, instead of relying on mere counting (i.e., Term Frequency), assigns a score for each term mentioned in training dataset which is calculated based on the maximum conditional probability of the term in countries and states. Cheng et al. in [4] proposed an approach to use a term-based spatial classifier to estimate top k user's location as details as the city. Researches focused on the probabilistic distribution of the word over cities based purely on the content of the user's tweets. However, primary classifier involves two main problems in estimation of user's location. Firstly, majority of the terms are distributed similarly among all the cities and it implies that they cannot be used to effectively distinguish the location of the user. Secondly, small cities have a variety of words mentioned in tweet contents and the method cannot infer user's location properly.

   In order to solve the problem, Cheng et al. in [5] has employed the idea developed by Backstrom et al. in [2]. Backstrom et al. have investigated to figure out whether there is a proper subset of words which have a more focused geographical scope to be considered as local terms (e.g., *Aussie* for the people of Australia). Subsequently, they have devised a probabilistic model to exploit queries that have focus point on the map and a dispersion factor which indicates how the locality impact of the word decreases during receding from the center. On the other hand, they have devised a method to detect terms with spatial strength which are used more in their focal location that the number shrinks in documents generated by users further away from the center.

Cheng et al. in [5] proposed a classifier to identify terms with locality strength within tweet contents. Their probabilistic framework predicts top k locations of the user with an accuracy of the city-level and locates 51% of the users within 100 miles of their true locality merely based on the contents. Their approach competes *gazetteer* based frameworks because *gazetteer* may lack a few spatial vocabularies and also the *tweets* do not always contain clear location names (e.g., *Sydney*). Researchers aim to find local words with high local focus and rapid decrease as the twitter authors location move away from the central point. However, it requires choosing local words manually to train the classifier and rather than calculating the probability of the word usage, it specifies the focal points. The model proposed in [5] has competed previous language models, as it successfully detects strong geospatial terms (e.g., Casino) which are not found in *Gazetteers*. For example, Serdyukov et al. in [14] predict the location of *Flickr* photos based on the tags using probabilistic language models and Bayesian theory. However as they use the *Geo-names gazetteer* to detect spatial tags, they miss the terms which are not listed in *gazetteers* but have strong locality.

Considering the noisy nature of the *tweets*, providing a comprehensive manually selected dataset seems tedious. Furthermore the words might have more than one center. To address these issues, Chang et al. in [3] have used an unsupervised approach based on *Gaussian Mixture Model (GMM)*.

A few attempts have been made to accomplish the task of making a domain specific for *NER* for *tweets* in both unsupervised approach (e.g., [10]) and supervised approach (e.g., [11],[13]). Li et al. in [10] proposed unsupervised *NER* system for targeted *Twitter* stream as described in Section 1.

Current similar works on term-level approaches are mainly focused on developing probabilistic language models while treating the terms independently in equations. Adding extra features like population, local theme and co-occurrence can improve these models. Population can distinguish approaches utilized for small and big regions (e.g., cities). Authors Local-theme can improve performance in a way that if we recognise a user's general locality, it will require smaller subset of words from corpus in computations during inferring detailed locations. For example, if an initial locality of a person is exploited as *France* then the word *Paris* in her tweet would mean the capital city, rather than the *Paris, Texas* in *USA*. Regarding extraction of words with geo-scope strength, co-occurring terms and relationship between them in tweet contents can remove sparseness more efficiently, as it may find some strong dependent words which can be the clue when the words with string locality are absent.

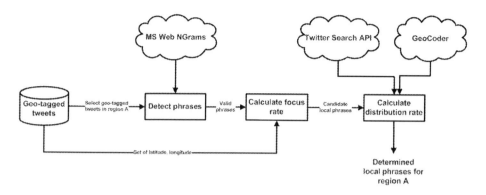

Fig. 1. Architecture of our system

4   Proposed Approach

To address the challenge mentioned in Section 1, we have devised a system which involves three main stages as can be seen in Fig. 1. Our approach is able to detect associated *Twitter* based multi-words/phrases for a geographical cell which is specified by top left and bottom right surrounding points. From other aspect, given a phrase, the system will determine the strength of the phrase usage in selected region. Based on a given region, *geo-tagged tweets* (i.e., *tweets* with location information) are the input of the system and each *tweet* will provide a high precision in granularity. The tweets are pre-processed by removing special characters (e.g., ":", "(", "-", ")", "#" and "@") to increase accuracy and reduce processing time. we also eliminate *re-tweets* as they are generated by people interested in a *tweet* and are not necessarily local.

The following sections provide information about the stages including the novelties applied.

4.1   Phrases Detection

The problem that we address in this step is how to extract segments from a given *tweet* message. A *tweet* segment can be either a single word (i.e., unigram) or a phrase (i.e., multi-gram). However, our interest is to recognise phrases in tweet contents. The initial statistical formula to detect phrases is to calculate how often the containing words appear together. As there is not any existing service provided by *Twitter* to calculate such required probabilities based on the whole *Twitter* database, we opted for *Microsoft Web N-Gram* which is an ongoing up to date cloud based platform that supplies access to the full corpus of indexed english documents by *Bing* in USA. Considering the dynamic nature of Bing, the service provides proper values for typical multi-words. Stickiness function is defined by using the generalized Symmetric Conditional Probability (*SCP*) for *n-grams* by Li et al. in [10]. The *SCP* is studied as a practical best practice to detect multi-words among other approaches such as Point Mutual

Information (PMI). We utilize the SCP as a successful multi-word stickiness evaluation algorithm for phrases up to *5-Grams* using retrieved probabilities from *Microsoft Web N-Gram*. Unlike [10], we excluded the statistical information from *Wikipedia* because it does not necessarily include all the terms mentioned in Twitter. Also, *Microsoft Web N-Gram* service as a bigger data set already comprises considerable number of words from internet domain so it may cover *Wikipedia* as well. For this reasons we decided to partition each tweet into a set of words using a particular separator (e.g. space) and study the rate of stickiness between them to find initial available segments.

SCP is defined to measure the cohesiveness of a segment by considering all possible binary segmentations, as shown in the following equation, where segment $s = \{w_1w_2...w_n\}$ and function $Pr()$ denotes the prior probability derived from *Microsoft Web N-Gram* service. We segment the message using $n - gram$. We observed that in our data, $n < 5$ is a proper bound as the maximum length of a segment s. The SCP function is now defined as follows:

$$SCP(s) = log\frac{Pr(s)^2}{\frac{1}{n-1}\sum_{i=1}^{n-1} Pr(w_1...w_i)Pr(w_{i+1}...w_n)} \tag{1}$$

As a matter of fact, Archiving results retrieved via *Microsoft Web N-Gram* service calls improves the segmentation process and reduces computation time noticeably. In the beginning, the segmentation process might be slow but as the number of processed *tweets* augments, the number of service calls is decreased as they are already captured in prior requests. Finally, we rank all possible segments by SCP values. Then we select all segments that have SCP value greater than a threshold (γ) as valid phrases.

4.2   Focus Rate Calculation

Not all segments detected in previous stage can be assigned as location oriented phrases. We need to formulate a bi lateral relation between a phrase and a geographical entity which can be as accurate as a spatial point (Latitude, Longitude). For example, the phrase *"here in brissie"* or *"Queen's mall"* are not obvious location names or piece of an address however they are repeatedly mentioned in *tweets* which are associated to "Brisbane, Australia". In this stage, we aim to remove segments that are commonly used everywhere by calculating their focus rate using our geo-tagged corpus. In order to calculate the focus rate for each phrase, we adopt the measure of linear relation between random variables namely Covariance. Covariance provides a measure of the strength of the correlation between two or more sets of random variates[4]. The covariance for two random variates X and Y (i.e., set of Latitude and Longitude for a given phrase p), each with sample size N is defined as follows.

$$Cov(p) = \frac{1}{N-1}\sum_{i=1}^{N}(x_i - \bar{x})(y_i - \bar{y}) \tag{2}$$

---

[4] http://www.encyclopediaofmath.org/index.php/Covariance

where $\bar{x}$ and $\bar{y}$ are the respective means.

We calculate the covariance considering the top left and bottom right spatial points which surround a given region then we use the value as a threshold (λ). A phrase is appointed as a candidate local phrase if its absolute covariance value is less than predefined threshold (λ).

4.3   Distribution Rate Calculation

In this stage, we need to identify a set of phrases per region that show strong locality. We assumed that phrases have some locations of interest where users tend to post messages extensively. In other words, each phrase has a number of centers where users post it more dense than users in other regions. When candidate local phrases are detected, the next step is to analyze their spatial distribution. Spatial locality is described as a set of messages mentioning some phrases, which are popular in a specific geographical area. For each of given phrases, we collected 500 *tweets* using *Twitter Search API*. We formulate the probability of a phrase p which belongs to a specific region r as Eq. 3.

$$P(p|r) = \frac{occur_{p,r}}{M_p} \times \frac{1}{|r|} \qquad (3)$$

$$occur_{p,r} = \beta|geotag\_occur| + (1 - \beta)|locfield\_occur|, \quad (0 < \beta < 1) \quad (4)$$

where the probability of the phrase p belonging to a region r is obtained by dividing the number of messages which contain phrase p in a region r ($occur_{p,r}$) by the total number of messages (M_p) which contain phrase p in all regions. The $\frac{1}{|r|}$ is the penalty factor to penalize a phrase widely used in many regions. The region can be found from geo-tagged information or location field in user's profile. *GeoCoder* is used to convert both geographical coordinates and the location field contents into the same hierarchical spatial format to accumulate the number regarding each region conveniently. The *geotag_occur* is the number of geo-tagged messages which contain phrase p in a region r while excluding geo-tagged messages, *locfield_occur* is the number of phrase p instances in the region r which are mentioned in author's location fields. We provide a higher accuracy coefficient ($\beta = 0.7$) to geo-tagged information which is chosen via trade-off. The phrase which has probability score greater than zero is selected to compute the distribution rate.

Finally, we rank local phrases according to their focus rate and the probability of occurrence in a given region. Local phrases should gain low focus rate but forceful likelihood in the region. Therefore, The distribution rate can be computed as Eq. 5.

$$Dist(p|r) = Cov(p) \times \frac{1}{P(p|r)} \qquad (5)$$

The lower value of distribution rate for a phrase assures its higher chance of being classified as a location oriented phrase.

5   Experiments and Evaluation

In this section, we initially evaluate the performance of segmentation module using *F1-Score*. We compare phrase detection versus two methods of Iterative and Rule based Greedy. Regarding recognition of location oriented phrases, we then assess our system's accuracy against two other. The Greedy baseline which reflects the local context and the term based NL Baseline.

5.1   Data Collection

Location information on tweets are scattered. So we initially listed certain clues (e.g. From, at, ...) regarding location context in *tweet* contents to gather more geo-tagged *tweets*, *tweets* attached with Latitude/Longitude. By using *Twitter streaming API* and considering the cues, from March 15 to July 1, 2013 we collected over 8 million *tweets* from *Spritzer Twitter Feed* which was sampled in different times of the day periodically. We used regular expressions formatted based on letters to exclude non English *tweets* and finally acquired 6,032,817 *tweets* in English language from which 1,577,240 were enclosed with geographical coordinates.

We aimed to implement and test our system in Australia but we found merely 13,519 geo-tagged *tweets*. As, the number was too small, we devised a procedure to enrich the dataset to include more tweets belonged to this country. We chose 4,519 distinctive users from Australian geo-tagged *tweets* observed in the primary collection and retrieved a maximum 1,000 past *tweets* per user via *Twitter API 1.1*[5]. Eventually we reached 659,009 geo-tagged *tweets* which were all of inland Australia. In order to extract location oriented phrases, we developed our system in two cities (Brisbane and Gold Coast). Based on the geographical cell, 58,000 geo-tagged *tweets* were recognised in Brisbane. Nonetheless we filtered them using cue words and retrieved 16,131 *tweets*. The goal was to see whether the system can give us the maximum results via processing fewer tweets. In a different approach we selected all geo-tagged *tweets* (i.e., 25,990 messages) belonged to Gold Coast geo-cell. In fact, we eventually realized that selection of all the tweets in a city ensures the better quality in detection of location oriented phrases. The better quality means that final discovered local phrases will be mostly generated by ordinary users. Therefore local phrases recognised in Gold Coast were more biased to the human natural intellect.
- *Data (MS SQL Server 2012 DB) and codes (In C#) are available upon request.*

5.2   Baseline Approaches

In this section, we introduce the baselines on evaluating segments/phrases detection and location oriented phrases detection.

---

[5] https://dev.twitter.com/docs/api/1.1

Phrases Detection Baselines: In order to evaluate the phrase segmentation without taking the locality into account, we compared the performance of our method with two baseline approaches.

1) Baseline Iterative: This approach creates any possible multiword out of the terms comprising each of the tweets. This method is an initial factor to generate maximum number of segments (2 to 5-grams). Apparently this approach owns the highest recall and the lowest accuracy as it does not rely on any knowledge base.

2) Rule Based Greedy: This greedy algorithm is suggested by Microsoft Web N-gram service [6] to form segments. It recursively constructs a set of words from the smallest possible constituent parts. The stickiness rate between words is evaluated to identify proper phrases. In each step of iteration, phrase likelihood is computed based on the value returned by *Microsoft Web N-Gram* service. Considering the tweet t containing the set of words $\{w1, w2, w3, w4\}$, if the method confirms $\{w1\ w2\}$ as a phrase then it will evaluate whether $\{w1\ w2\ w3\}$ can be a phrase or not. Given phrase $\{w1\ w2\}$, $Pr(\{w1w2\})$ denotes the prior probability derived from *Microsoft Web N-Gram* service for these two words to be mentioned together as a valid segment. The phrase is then detected if it satisfies four factors as follows.

First, the occurrence of $w1$ in a message makes the appearance of $w2$ in the same context (Eq. 6). Second, the prior probability difference is less than the level of significance refer to Eq. 7. Third, the probability that two words are joined together has a high significant using Eq. 8. Fourth, the word that has the prior probability greater than a certain number is detected as stop word (Eq. 9).

$$Log(Pr(w1)) + Log(Pr(w2)) < Log(Pr(w1w2)) \tag{6}$$

$$Log(Pr(w1)) - Log(Pr(w2)) < significance\_level \tag{7}$$

$$|Log(Pr(w1)) + Log(Pr(w2)) - Log(Pr(w1w2))| > 1 \tag{8}$$

$$Log(Pr(w1)) > stopword\_level \tag{9}$$

The baselines above were compared with SCP method used in [10] to select an optimum threshold regarding the stickiness rate to generate segments properly.

Location Oriented Phrase Detection Baselines: In order to evaluate our system, we compared it with *NL* and *Greedy* baselines as follows:

1) NL Baseline: While Cheng et al. in [5] considers one focal point for each of the local terms, [3] indicates that words might have multiple geographical centers. The *NL (Non-localness)* baseline used in [3] is an approach in exploring local words by NL factor in an unsupervised approach. The method filters non-local words out of the given corpus. Local words tend to have the farthest distance in spatial word usage pattern to the stop words (Same list used in [3]). So, we

---

[6] http://research.microsoft.com/en-us/collaboration/focus/
cs/web-ngram.aspx

Table 1. Cities used in Non-Localness baseline experiment

City	# of tweets
Brisbane	58509
Cairns	4200
Gold Coast	25990
Melbourne	169004
Sydney	268044
Townsville	4351

Table 2. Segmentation results compared against the baselines

Method	Precision	Recall	F1-Score
Baseline Iterative	0.056	0.846	0.104
Baseline Rule Based Greedy	0.065	0.844	0.121
SCP value > -5	0.104	0.026	0.041
SCP value > -10	0.060	0.181	0.090
SCP value > -15	0.076	0.529	0.133
SCP value > -20	0.077	0.758	**0.140**

opted for the NL measure computed in six cities (Table 1). For each word w, NL value is calculated as follows:

$$NL(w) = \sum_{s \in S} similarity(w, s) \frac{freq(s)}{\sum_{s' \in S} freq(s')} \tag{10}$$

where S is list of stop words, $freq(s)$ is frequency of stop word s in the corpus. The similarity of two terms, t_1 and t_2, is measured by Total Variation as Eq. 11.

$$similarity(t_1, t_2) = \sum_{r \in R} |P(r|t_1) - P(r|t_2)| \tag{11}$$

Where R is the list of regions in experiment and function $P(r|t)$ as the probability of the term t to appear in region r. We rank words using NL values in ascending order and ultimately Top-k words will be selected as the final list of the local terms.

2) Greedy Baseline: Phrase detection and focus rate calculation form the Greedy baseline. Phrase detection is the initial Rule based Greedy. Focus rate calculation also utilizes Eq. 2 to compute the correlation between latitude and longitude of phrases. Those having an absolute covariance less than (γ) will be set as local. γ is the covariance of the geo-cell's corners.

5.3   Evaluation

With regard to evaluation, we firstly assess segmentation process because it can influence the final results. We then estimate the strength of our system in exploring of the location oriented phrases.

Table 3. Location oriented phrase detection results compared against the baselines

Region	Method	Combine Accuracy
Brisbane	NL Baseline	7.54%
	Greedy Baseline	17.43%
	Our approach	22.04%
Gold Coast	NL Baseline	14.10%
	Greedy Baseline	27.75%
	Our approach	29.82%

Phrases Detection Evaluation: In order to evaluate segmentation procedure, we manually marked 745 *tweets* created by users around Brisbane (Australia) and retrieved 1,136 phrases as ground truth. We then assessed *F1-score*.

$$Prec = \frac{CDS}{TDS} \quad , \quad Recall = \frac{CDS}{TCS} \quad , \quad F1 = \frac{2 \times Prec \times Recall}{Prec + Recall} \qquad (12)$$

Where CDS is the number of corrected phrases detected from our approach, TDS is the total number of phrases detected from our approach and TCS is the total number of corrected phrases from the ground truth. Table 2 shows the segmentation results using various models. Nevertheless, the SCP method can effectively detect phrases better than others.

Location Oriented Phrase Detection Evaluation: Our prime system detected 419 location oriented phrases for Gold Coast and 186 for Brisbane (using 5.1). In order to evaluate our approach, we needed another dataset. Thus, we collected past *tweets* belonged to 200 users in Gold Coast (135,235 tweets) and other 200 users in Brisbane (134,584 tweets). We then searched in new dataset for local terms and phrases extracted by NL baseline and our own method. Subsequently, two annotators (i.e., local people) manually checked the correct terms and phrases. In the end, we selected those which were confirmed as true instances of local terms and phrases with minimum ambiguity by both of annotators. The accuracy of location oriented phrase detection is shown in Table 4.

However, we cannot directly compare our phrase-based model with term-based approach (i.e., NL Baseline). Based on initial experiments we figured out that the local terms are found more frequently than phrases. However they have a minimum accuracy in determining the regions exclusively. For example the number of times the term "Paradise" was recognised in tweets from Gold Coast is twice of "Surfers Paradise". However, it involves a high rate of ambiguity as it has multiple focal points. So, we measured the accuracy of the system using the following formula.

$$Combine\_Accuracy = \frac{C_p}{D_p} \times \frac{L_u}{S_u} \qquad (13)$$

where C_p is the corrected phrases detected by the approach, D_p is the total number of phrases detected by the approach, L_u is the number of users posted location oriented phrases in past tweets and S_u is the number of test users in the region. The results is shown in Table 3.

Table 4. The accuracies of location oriented phrase detection compared against the Greedy baseline

Region	Method	Accuracy
Brisbane	Greedy Baseline	48.47%
	Our approach	68.08%
Gold Coast	Greedy Baseline	28.34%
	Our approach	60.90%

6   Discussion

Regarding the experiment for segment detection, we can see that the threshold for *SCP* is very important. For example, phrases like *"the redcliffe tigers"* SCP: -12.88597, *"of Mt Cootha"* SCP: -10.18859, *"in Brisvegas"* SCP: -12.17709 and *"in Brissie"* SCP: -14.2984 gain a high probability in search results to be recognised as a local phrase for Brisbane, however, they will all be rejected if we consider the *SCP* threshold > -10. Anyhow, the SCP threshold needs to be selected based on a solid balance. For example, if it is selected as a very small number (i.e. SCP value > -40) then the number of insignificant phrases to process will increase and it will affect the whole performance of the system.

In order to detect location oriented phrases in a given region, as we can see from Table 4 our proposed approach can effectively detect location-oriented phrases against the baseline. The results of Gold coast dataset has less accuracy compared to Brisbane dataset both in the Greedy baseline as well as our approach because Gold coast dataset have more noise *tweets* than the Brisbane dataset.

The experiment on *NL* baseline is performed to show that phrases can better specify strong locality instead of using words. For example, the words like *"movie"* and *"world"* are not assigned as local words in Gold Coast region due to the values regarding probability distribution. On the other hand, the phrase *"movie world"* is likely to be a local phrase for Gold Coast region. Moreover, we found local phrases are more geographically focused. For example the word "Paradise" is detected as local term for Gold Coast city in Australia. However, the word is general and cannot be effectively used to distinguish Gold Coast independently. While in our phrase based approach, *"Surfers Paradise"* is detected which can pinpoint Gold Coast more accurately. Despite the fact that the current paper is purely based on detection of the local phrases in Twitter messages, as a future work we suggest that a new more comprehensive system may better operate if it takes advantage of both location oriented terms and segments. The reason is that in our approach, some of the words like "broadbeach" or even "movieworld" are missed as they are mentioned as single terms.

7   Conclusions

In this paper, we proposed an approach to detect location oriented phrases from geo-tagged tweets. Firstly, the method calculates stickiness parameters between words to form an initial set of proper segments. The values are computed using probabilities retrieved from an external web based knowledge-base. We then apply two steps of filtering to select location oriented phrases using both local context (corpus) and the global context (Twitter search API).

Based on our findings, Local phrases excel local terms in exploring spatial information in message contents, because they carry less ambiguity and are more scalable. Also, they augment both recall and precision as they gain less number of focal points.

With regard to the future work, we believe that a new bi-component system must be devised which will consider Location oriented terms and phrases jointly. Recognised Local phrases can support local terms to obtain less ambiguity. Together they have the potential to increase overall recall by supporting each other in the process of association to geographical cells. However, increasing the accuracy of the segmentation module and optimizing subsequent sections (i.e., Those involved in recognition of local phrases) remain as initial future tasks.

References

1. Amitay, E., Har'El, N., Sivan, R., Soffer, A.: Web-a-where: geotagging web content. In: SIGIR, pp. 273–280 (2004)
2. Backstrom, L., Kleinberg, J.M., Kumar, R., Novak, J.: Spatial variation in search engine queries. In: WWW, pp. 357–366 (2008)
3. Chang, H.-W., Lee, D., Eltaher, M., Lee, J.: @phillies tweeting from philly? predicting twitter user locations with spatial word usage. In: ASONAM, pp. 111–118 (2012)
4. Cheng, Z., Caverlee, J., Lee, K.: You are where you tweet: a content-based approach to geo-locating twitter users. In: CIKM, pp. 759–768 (2010)
5. Cheng, Z., Caverlee, J., Lee, K.: A content-driven framework for geolocating microblog users. ACM TIST 2, 1–2 (2013)
6. Fink, C., Piatko, C.D., Mayfield, J., Finin, T., Martineau, J.: Geolocating blogs from their textual content. In: AAAI Spring Symposium: Social Semantic Web: Where Web 2.0 Meets Web 3.0, pp. 25–26 (2009)
7. Hecht, B., Hong, L., Suh, B., Chi, E.H.: Tweets from justin bieber's heart: the dynamics of the location field in user profiles. In: CHI, pp. 237–246 (2011)
8. Kelm, P., Schmiedeke, S., Sikora, T.: Multi-modal, multi-resource methods for placing flickr videos on the map. In: ICMR, pp. 52:1–52:8 (2011)
9. Larson, M., Soleymani, M., Serdyukov, P., Rudinac, S., Wartena, C., Murdock, V., Friedland, G., Ordelman, R., Jones, G.J.F.: Automatic tagging and geotagging in video collections and communities. In: ICMR, pp. 51:1–51:8 (2011)
10. Li, C., Weng, J., He, Q., Yao, Y., Datta, A., Sun, A., Lee, B.-S.: Twiner: named entity recognition in targeted twitter stream. In: SIGIR, pp. 721–730 (2012)
11. Liu, X., Zhang, S., Wei, F., Zhou, M.: Recognizing named entities in tweets. In: Proceedings of the 49th Annual Meeting of the Association for Computational Linguistics: Human Language Technologies, HLT 2011, vol. 1, pp. 359–367. Association for Computational Linguistics, Stroudsburg (2011)
12. McCallum, A., Nigam, K.: A comparison of event models for naive bayes text classification. In: AAAI 1998 Workshop on Learning for Text Categorization, pp. 41–48. AAAI Press (1998)
13. Ritter, A., Clark, S., Mausam, Etzioni, O.: Named entity recognition in tweets: An experimental study. In: EMNLP, pp. 1524–1534. ACL (2011)
14. Serdyukov, P., Murdock, V., van Zwol, R.: Placing flickr photos on a map. In: SIGIR, pp. 484–491 (2009)

Author Index